Ecological Engineering for Wastewater Treatment

SECOND EDITION

Edited by

Carl Etnier
Department of Engineering
Agricultural University of Norway
Ås, Norway

Björn Guterstam
Stensund Ecological Center
Stensund Folk College
Trosa, Sweden

LEWIS PUBLISHERS
Boca Raton New York London Tokyo

Artwork used on the front cover was taken from the 1991 International Conference Poster, painted by Benjt-Erik Bengtsson.

Acquiring Editor: Neil Levine
Project Editor: Erica Orloff
Marketing Manager: Greg Daurelle
Direct Marketing Manager: Arline Massey
Cover design: Dawn Boyd
Manufacturing: Sheri Schwartz

Library of Congress Cataloging-in-Publication Data

Ecological engineering for wastewater treatment / [edited] by Carl Etnier and Bjorn Guterstam. -- [2nd ed.]
p. cm.
Includes bibliographical references and index.
ISBN 0-87371-990-5 (alk. paper)
1. Sewage--Purification--Biological treatment. 2. Constructed wetlands. 3. Ecological engineering. I. Etnier, Carl. II. Guterstam, Björn.
TD755.E43 1996
628.4′3--dc20 96–18015
CIP

Lewis Publishers is an imprint of CRC Press

International Standard Book Number 0-87371-990-5
Library of Congress Card Number 96–18015
Printed in the United States of America 1 2 3 4 5 6 7 8 9 0
Printed on acid-free paper

To the necessary alliance
between ecologists and engineers

FOREWORD TO THE SECOND EDITION

The first edition of *Ecological Engineering for Wastewater Treatment* was the report of a conference on the possibilities and practices within the eponymous new field. Now, three years later, the new science of ecological engineering is winning increasing acceptance worldwide. In established industrial economies like Sweden and the United States, initial skepticism and regulatory hurdles are giving way to burgeoning investments by companies and municipalities, increased research activity, and substantial interest from the public at large. Countries traditionally called Third World or less developed countries, like Chile and Uruguay, are investigating cheap and effective ways to build up a hitherto nearly nonexistent wastewater treatment infrastructure with ecological engineering. The former Second World countries of Eastern Europe and the former Soviet Union are turning to ecological engineering to solve their acute pollution problems from a time of great industrialization without concern for the environment. The field now even boasts its own journal, *Ecological Engineering*, published by Elsevier, and a professional society, the International Ecological Engineering Society, with headquarters in Switzerland.

In a longer-range perspective, the international community supports the concept of better resource management through Agenda 21, a useful handbook from the UN environmental conference in Rio de Janeiro in 1992. Today local politicians have a well-formulated reason to listen to their ecologically inspired and constructive engineers when their communities construct a sustainable society. Almost never before has there been such a challenge for engineers and biologists working together to contribute to the course of development.

Ecological engineering for wastewater treatment means that natural design is used to treat society's waste products in water in a way beneficial for both. The term must not be misconstrued as a claim that nature can perform alchemy and turn cadmium into gold *at the end of the pipe*. On the other hand, our wastes may be worth gold if managed wisely, for example, if nitrogen and phosphorus in wastewater are used to fertilize our crop plants. For this to be sustainable, wastewater needs to be *cleaner* than it is today. Changes made *at the beginning of the pipe,* such as removing toxic chemicals from the waste stream and conserving water to make a more concentrated waste, constitute a sort of "ecology in society," a type of resource management that could even be described as appropriate engineering. The book therefore covers a range of contributions from wastewater source separation at the household level to the use of huge, natural wetlands.

Ecological Engineering for Wastewater Treatment has provided the first collection of its breadth on the uses of and possibilities for the science of that name in industrial society. This edition has been updated and improved in several ways. Many of the articles have been updated, all have been re-edited, new chapters have been added, the size of the type has been increased for greater readability, and an index has been added. In this edition from Lewis Publishers, the information is now available for a wider circle of users than the first edition, published by Bokskogen, a small Swedish environmental publisher, in close cooperation with the editors.

This second edition is, then, an updated description of the state of the art, to be used in order to teach and inspire the management of wastewater as a resource, one of the keys to sustainable development. What can be more urgent than to improve the management of water and plant nutrients while at the same time cleaning up the natural environment? Changing a linear flow from resource to waste into a continuous circulation of materials, where today's urine becomes next spring's wheat and the bread of next autumn, is a compelling task, yet it has been forgotten by modern industrial society until recently. Our book is a tool for development in that direction, based on a reliable concept, tested for millions of years in nature.

Carl Etnier and Björn Guterstam
March 1994
Trosa, Sweden

P.S. Since the above was written, numerous projects have delayed the completion of the final editing by two years. I wish to thank the authors and the editors at Lewis Publishers for their patience in waiting for this edition to come out; I trust it will be seen as worth the wait.

A follow-up conference to the Stensund conference was held in Switzerland, in September 1995. The proceedings are due to be published in 1996, by TransTech. The third conference in the series is planned for June 1998, at the Agricultural University of Norway. These easiest way to keep posted on this conference and similar events is to join the International Ecological Engineering Society (Bahnhofstrasse 2, CH-6110 Wolhusen, Switzerland; heebjohannes@ping.ch).

Carl Etnier
March 1996
Ås, Norway

FOREWORD

Before urbanization and industrialization, wastewater was a bucket a day per family poured out somewhere, often where it was a nuisance for other citizens and sometimes to where it belonged in nature. The flow of wastewater has since swelled to become polluted underground rivers, contained in pipes, which course their way to the nearest outlet where they can be discharged into nature. The change from buckets to rivers encapsulates the resource management of our civilization. It is a symbol of a tragic and disastrous relationship between industrialized peoples and nature, a relationship in which vast resources are employed in trivial pursuits. Wastewater also is the sign of our inability to recognize a resource that we already have — it contains the plant nutrients that we pay dearly for when we manufacture fertilizer, plus an untapped reservoir of heat energy that could be exploited in cold climates. In many Western lands, we are just now developing knowledge of how to use these resources, instead of flushing them away, as we learn to reconsider the nature of the problem.

Wastewater has been treated as a technical problem to be solved with technical means. The engineering strategy was initiated to provide sanitation and later also developed to mitigate the environmental problems of pollution of waters and coastlines. But so long as wastewater is treated as something which just needs to be rendered less noxious or moved out of the way somewhere, we shall have it with us as a problem. It is perhaps only when we consider strategies for capturing the resources that it contains that we can truly solve, instead of merely hiding or moving, the hygienic and environmental problems wastewater poses.

As every engineer dreams of constructing good technical tools for humankind, the ecologist dreams of connecting the flows of energy and materials in society with those of nature for the benefit of both. It is here the two will meet in the joint project called ecological engineering.

The international conference on Ecological Engineering for Wastewater Treatment was held in Sweden at Stensund Folk College, March 24–28, 1991 to encourage this joint project not only between ecologists and engineers but also together with politicians, administrators, and the environmental movement. This conference gathered some of the first ecological engineers from three continents, North America, Asia, and Europe, with the aim of evaluating the potential of ecological engineering for wastewater treatment and promoting this technology in our respective societies. The cooperation between Stensund and the coorganizers, the U.S. non-profit organization "Center for the Restoration of Waters" @ Ocean Arks International in Falmouth, Massachusetts, made it possible to bring together 120 people to share during a highly creative and energetic week. The active and constructive participation of top officials from the U.S. and Swedish Environmental Protection Agencies showed that ecological engineering's ideas are needed now to solve practical problems.

The theoretical platform of the conference was formulated in five general questions:

- How big is the nutrient resource of wastewater and how can it be recycled?
- Is it possible in the long run to recycle nutrients without accumulating toxic substances?
- Are there climatic and other limitations for ecological wastewater technologies?

- What are the main obstacles when the ecological wastewater technologies are introduced into the market?
- What are the social impacts on attitudes, town planning, construction, and economics?

It is too much to conclude that these questions were answered. They were raised, discussed, and put on the table for the social debate on wastewater resource management strategies. The contributions to this conference are the state of the art of ecological engineering for wastewater treatment at the beginning of the 1990s. The principles work — we have come out of the wind tunnels and performed the first flight tests of many different models of engineered ecosystems. The wide spectrum of experiences reported here come from different climates and cultures and therefore they have approaches typical for those regional and local situations.

The characteristic in common for ecotechnologies for wastewater treatment is the use of natural processes and a great variety of organisms: bacteria, protozoans, zooplankton, algae, higher plants, fish, and crayfish. We meet ancient Chinese aquaculture and integrated farming practices which optimize their recycling with help from professors in ecology. We find a socio-ecological approach to using the wastewater resource in Calcutta to feed and give the poor population both work and pride. In Europe and North America, constructed wetlands are the rule, with a few ecosystems built indoors so they can function through the cold winters. In North America, in particular, ecological engineering competes with conventional technologies in a tough market.

The organization of this meeting at Stensund Folk College brought a wider and an activating aspect into the conference and to its participants. The Scandinavian tradition of social development through democratic and educational processes has over a century of history at the folk colleges. The "learning by doing" process was used to let the conference meet the society, a society with a great demand for ecological approaches to real problems. This was done for one day of the conference through the case studies, when participants left the school buildings behind to traipse through meadows and marshes or tour conventional treatment plants, looking for ways that the power of ecological engineering could be harnessed at seven sites in the area. Participants with many different approaches to ecological engineering co-operated (and argued) to develop plans that drew on all of their experience. The case study hosts — administrations, companies, communities, and organizations — also shared in the discussions, so that they might be able to adopt the plans the groups came up with.

Management and development of hardware for wastewater treatment mostly disregards the potential of local engagement and participation by the citizens and customers themselves. This is true for science as well. Ecological engineering, though, seems to be at a point where social and technological fields meet. The conference was therefore brought to a public meeting and festival during one evening with entertainment and science together on-stage at the town hall of Nyköping, a city on the Baltic coast 100 km south of Stockholm. Four hundred persons came for the fun and learning and, in addition to songs, jokes, and skits, were introduced to the Chinese dyke-pond aquaculture systems of the Pearl River Delta by George Chan of the Guangzhou Academy of Sciences. (This demonstrates the movement of know-how from so-called developing countries to industrialized nations, a flow which has played an important role in the development of ecotechnology.)

This book is organized according to the types of presentations at the conference: (1) Plenary lectures on such subjects as ecological engineering foundations, traditional roots, visions, and commercial prospects; (2) Workshops on indoor systems, outdoor systems, restoring natural water bodies of water, and political, social, and economic considerations; (3) Poster summaries; (4) Case study reports.

This conference on ecological engineering for wastewater treatment created an atmosphere of cooperation over borders of nations and disciplines, cooperation which is rare but urgently needed if we are going to restore what an exploiting civilization has destroyed for too long. We hope that the contacts made here during the conference and through the book will lead to much productive work and we look forward to another such meeting, somewhere, when new developments warrant our gathering together again.

Carl Etnier and Björn Guterstam
June 1991
Trosa, Sweden

ACKNOWLEDGMENTS

The International Conference on Ecological Engineering for Wastewater Treatment at Stensund Folk College and these proceedings were possible only through the tremendous engagement and cooperation among many people at the folk college, government agencies, private organizations, and businesses. The staff at Stensund Folk College, the Swedish Health and Sport Association (Svenska Frisksportförbundet), and our colleagues at the co-sponsoring organization, The Center for Restoration of Waters @ Ocean Arks International in Falmouth, Massachusetts, deserve the most credit for the success of the conference. It was a very ambitious task to create a meeting spot where, for one week, bioscience, engineering, politics, economics, myth, and play could mix harmoniously together. Thanks to this international troupe's high level of ideas, support, and plain hard work, we believe it was a success.

At Stensund, special thanks are due to Rigmor Breidemalm, headmaster, and Erik Samuelsson, chairman of the board, who proudly and bravely piloted the folk college tradition into new waters.

The conference was also only possible with the financial support of Apple Computer AB, the County of Södermanland, The German Marshal Fund of the United States, Mathema Konsult HB, New England Biolabs, Scandinavian Airlines Systems (SAS), Swedish Institute, Swedish International Development Authority (SIDA), Swedish Society for Nature Conservation (SNF), U.S. Environmental Protection Agency, Water Treatment Cooperation AB (WTC), and The World Wide Fund for Nature (WWF).

The splendid organization, the enthusiasm, and the hospitality of our hosts of the case studies symbolized how science and society must meet in the future. Their success was due to the work of the organizer Peter Ridderstolpe; the hosts Varis Bokalders, Håkan Sternberg, and Keith Kral of Nacka Environment and Health Authority at the Nacka ecovillage project; PeO Jägbeck and Erling Johansson, SMÅA, at the Farsta suburban ecovillage project; Per Carlsson and Håkan Jersenius at the Uttringe Hage project; Anders Eriksson and Hugo Thelin at Kabi Pharmacia-Biopharma, Strängnäs; Olof Gillberg, National Board of Public Buildings, at the Tullgarn Royal Castle site; Jan Älfvåg, Börje Adermark, and Hans Molander at the project Nya Nyköping; Per-Henry Markström and Tage Wiklund at the retrofitting study of Trosa sewage treatment plant; and the conference delegates who served as chairpersons and secretaries.

The workshop sessions were run with help from the coordination and wisdom of their chairpersons: Staffan Delin, Peter Ridderstolpe, John Teal, Nils Tiberg, and Bengt Warne.

The opportunity to bring ecological engineering to the people of Nyköping was impossible without the help of the staff at Culturum; the organizer and impressario for that evening, Torbjörn Karreskog; the governor of Södermanland, Ivar Nordberg; the artists Anders Linder, his "Ballågåbortgäng," and everybody else with the show *Moderna Tider*; Lasse Brandeby (alias Kurt Olsson); Birgitta Ekman and her children's chorus; and the lecturers Anders Wijkman, George Chan, and Ann-Mari Jansson. The chairman of the Nyköping City Council Göran Forssberg was the dinner host at the medieval Nyköping castle and founded the annual Nyköping Environmental Day of March 27.

Ingrid Scholander helped brainstorm the early planning of the conference. The students of Stensund Folk College generously lent out their rooms during Easter break to help house the delegates, and a number of them joined the conference staff. Maud Guterstam and Stefan Jonheimer helped run the conference office and Maud arranged travel for delegates. Study and Travel AB arranged the travel of the Asian delegates with patience and care. Bertil Hägerhäll of Coalition Clean Baltic and Gunnar Norén of the Swedish Society for Nature Conservation were the conference's department of East European connections. Gitte Jutvik imaginatively helped arrange the seaside promenade into the cultures of water and Scandinavia during the Celebration of Water on the first night, with help from Stefan Bjarnhagen and Göran Florell in the roles of Poseidon and Ran. Cliff Brännström and Galvours-WTC AB hosted the continuation of the Celebration with an animated party, and the folk chorus Rönninge Rhubarbs brightened the party with their songs from Sweden and South Africa.

There would be no book to edit without contributors. The fascinating and important work that their papers describe made editing the book a pleasure, and those whom we contacted helped bring this to press quickly by responding quickly and in detail to our comments and suggestions. The difficult process of trying to capture the essence of the conference between the covers of this book was made much easier by Claes Hedlund, who took most of the photos of the authors, prepared two of the book's figures, and helped in a dozen other ways; Bill Wilson, who worked way above and beyond the call of duty to prepare the Chinese manuscripts; Karen Schwalbe, who also worked long hours on those manuscripts and helped compile the glossary; Karlfriedrich Hofmann, Peter Ridderstolpe, and several others who contributed definitions to the glossary; Ralph Skagerling and Cortex Computer Center AB, who gave us the use of a scanner; Maureen Moir, whose typing and editing of the language of a number of the papers made our job much easier; Eva Jansson, who provided us with the drawing for the Nya Nyköping case study; and Nils Kautsky, our mussel consultant. Our publisher, Henry Kenamets, was indispensable — his painstaking work by our sides helped make the papers as readable and consistent as time pressures permitted.

Finally, we would like to thank our families, who supported us and put up with our long absences while the book was being assembled.

C.E. & B.G.

2nd edition: Thanks also to Albert Chan, who assisted with the editing and layout.

CONTRIBUTOR

Marek Agopsowicz
University of Agriculture and Technology of Olsztyn
10-957 Olsztyn
Poland

Udo Asmus
State Design Institute
"Eesti Projekt"
Lenini pst. 8
EE-0105 Tallinn
Estonia

Rob uit de Bosch
International Institute for the Urban Environment
Nickersteeg 5
NL-2611 EK Delft
The Netherlands

Marie-Claude Cantin
(deceased)
Aquaresearch Ltd.
P.O. Box 208
North Hatley, QC
Canada J0B 2C0

George L. Chan
9C Porchester Road
London W2 5DP
United Kingdom

Robert L. Crosby, Jr.
Biorealis Systems, Inc.
PO Box 77-2773
Eagle River, Alaska 99577

Tjeerd Deelstra
International Institute for the Urban Environment
Nickersteeg 5
NL-2611 EK Delft
The Netherlands

Karl F. Ehrlich
Aquaresearch Ltd.
P.O. Box 208
North Hatley, QC
Canada J0B 2C0

Carl Etnier
Agricultural University of Norway
Department of Engineering
Box 5065
N-1432 Ås
Norway

Carl Folke
Department of Systems Ecology Stockholm University
S-106 91 Stockholm, Sweden
The Beijer International Institute of Ecological Economics
The Royal Swedish Academy of Sciences
Box 50005
S-104 05, Stockholm
Sweden

Dhrubajyoti Ghosh
Calcutta Metropolitan Water and Sanitation Authority
370/1P
N.S.C. Bose Road
Calcutta-700047
India

Thomas Gumbricht
Royal Institute of Technology
Department of Land and Water Resources
S-100 44 Stockholm
Sweden

Björn Guterstam
Stensund Ecological Center
Stensund Folk College
S-619 00 Trosa
Sweden

Katrin Häberli
Eidgenössische Technische Hochschule
8 Brunngasse
CH-8001 Zürich
Switzerland

Johannes Heeb
Center for Applied Ecology Schattweid
CH-6114 Steinhuserberg
Switzerland

Jørgen Hinge
Landbrugets Rädgivningscenter
Udærsvej 15
DK-8200 Århus
Denmark

Karlfriedrich Hofmann
B.P.A., Biological Planning Associates
Dorfstrasse 4
D-72074 Tuebingen
Federal Republic of Germany

Alan Ismond, P.E.
A.T. Consultants
c/o Arctic Alaska Fisheries Corp.
Fishermen's Center, Fishermen's Terminal
P.O. Box 79021
Seattle, Washington 98119

Arvo Järvet
Institute of Geography
University of Tartu
46 Vanemuise
EE-2400 Tartu
Estonia

Petter D. Jenssen
Centre for Soil and Environmental Research
N-1432 Ås
Norway

Yan Jingsong
Research Professor
Nanjing Institute of Geography and Limnology
Academia Sinica
73 East Beijing Rd.
Nanjing
China

Gert Jungersen
Danish Technological Institute
Gregersenjev
P. O. Box 141
DK-2630 Taaastrup
Denmark

Nils Kautsky
Department of Systems Ecology Stockholm University
S-106 91 Stockholm
Sweden

William C. Kinsinger
The Gaia Institute & The Center for Ecological Studies
Cathedral of St. John the Divine
1047 Amsterdam Ave. @ 112th St.
New York, New York 10025

Tore Krogstad
Department of Soil Sciences
Agricultural University of Norway
P. O. Box 5028
N-1432 Ås
Norway

Valdo Kuusemets
Institute of Environment Protection
Estonian Agricultural University
5 Kreutzwaldi
EE-2400 Tartu
Estonia

Laima-Lilija Lazauskiene
Institute of Ecology
Lithuanian Academy of Sciences
Akademijos 2
232600 Vilnius
Lithuania

Karen H. Lerum
11620 E. Sahuaro Dr. #2087
Scottsdale, Arizona 85259

Trond Mæhlum
Centre for Soil and Environmental Research
N-1432 Ås
Norway

Ülo Mander
Institute of Geography
University of Tartu
46 Vanemuise
EE-2400 Tartu
Estonia

Julie A. Mankiewicz
The Gaia Institute & The Center for Ecological Studies
Cathedral of St. John the Divine
1047 Amsterdam Ave. @ 112th St.
New York, New York 10025

Paul S. Mankiewicz
The Gaia Institute & The Center for Ecological Studies
Cathedral of St. John the Divine
1047 Amsterdam Ave. @ 112th St.
New York, New York 10025

Olev Matt
State Research and Design Institute
Estonian Land Reclamation Project
Mustamäe tee 33
EE-200 006 Tallinn
Estonia

Tõnu Mauring
Institute of Geography
University of Tartu
46 Vanemuise
EE-2400 Tartu
Estonia

Maret Merisaar
Water Protection Laboratory
Tallinn Technical University
Järvevana tee 5
EE-0001 Tallinn
Estonia

William J. Mitsch
School of Natural Resources
The Ohio State University
2021 Coffey Road
Columbus, Ohio 43210-1085

Janusz Niemczynowicz
Department of Water Resources Engineering
University of Lund
Box 118
S-221 00 Lund
Sweden

Tiina Nõges
Limnological Station
Institute of Zoology and Botany
EE-2454 Rannu
Estonia

Urmas Nugin
Small Enterprise "Kobras"
Kastani 10
EE-202 400 Tartu
Estonia

Hanna Obarska-Pempkowiak
Technical University of Gdansk
Faculty of Hydrotechnics
Majakowskiego 11/13
80-952 Gdansk — Wrzeszcz
Poland

János Oláh
Fisheries Research Institute
P.O. Box 47
H-5541 Szarvas
Hungary

F. Pekár
Fisheries Research Institute
P.O. Box 47, H-5541 Szarvas
Hungary

Susan Peterson
Ecological Engineering Associates
13 Marconi Lane
Marion, Massachusetts 02738

Jesper Saxgren
Kolding Folk Highschool
Skovvangen 18
DK-6000 Kolding
Denmark

Ma Shijun, Ph. D.
Research Center for Eco-Environmental Sciences
Academia Sinica
7 Zhongguancun Rd.
Beijing
China

Li Sifa
Department of Aquaculture
Shanghai Fisheries University
334 Jun Gong Road
Shanghai
China

Jorge de Sousa Pires
Apple Computer AB
P. O. Box 31
S-164 93 Kista
Sweden

Hamish Stewart
Danish Folkecenter for Renewable Energy
Kammersgårdsvej 16
Sdr. Ydby
DK-7760 Hurup Thy
Denmark

Karin Sundblad
Dept. of Water and Environmental Studies
Linköping University
S-581 83 Linköping
Sweden

George Tchobanoglous
Department of Civil Engineering
University of California–Davis
Davis, California 95616

John M. Teal
Woods Hole Oceanographic Institution
Woods Hole, Massachusetts 02543

Peter Thomas
Sanitärsystemtechnik
Fritschestr. 27/28
D-10 585 Berlin
Germany

John Todd
The Center for Restoration of Waters @ Ocean Arks International
One Locust Street
Falmouth, Massachusetts 02540

Max Troell
Department of Systems Ecology
Stockholm University
S-106 91 Stockholm
Sweden

Andrée Turcotte
Aquaresearch Ltd.
P.O. Box 208
North Hatley, QC
Canada J0B 2C0

Arvo Tuvikene
Limnological Station
Institute of Zoology and Botany
EE-2454 Rannu
Estonia

Andris.V. Urtans
Laboratory of Hydrobiology
Institute of Biology
Latvian Academy of Sciences
Miera iela 3
Salaspils 229 021
Latvia

R. Vaba
Pärnu Sewage Treatment Plant
EE-3600 Pärnu
Estonia

Kirlys Vaizgantas
Institute of Geography
Lithuanian Academy of Sciences
Akademijos 2
LT-2600 Vilnius
Lithuania

Arild Vatn
Department of Economics and Social Sciences
The Norwegian Agricultural University
1432 Ås
Norway

Bengt Warne
Nature Housing-Warne Naturhus AB
Karlaplan 19 C
S-115 22 Stockholm
Sweden

Hans-B. Wittgren
Swedish Meteorological and Hydrological Institute
S-6501 76 Norrk6ping
Sweden

Joachim Zeisel
Sanitärsystemtechnik
Fritschestr. 27/28
D-10 585 Berlin
Germany

Brigitta Züst
Center for Applied Ecology Schattweid
CH-6114 Steinhuserberg
Switzerland

CONTENTS

Chapter 1

ECOLOGICAL ENGINEERING: THE ROOTS AND RATIONALE OF A NEW ECOLOGICAL PARADIGM

William J. Mitsch

ABSTRACT

Ecological engineering, or its synonym ecotechnology, is defined as the design of human society with its natural environment for the benefit of both. It has developed in the United States and the Western world since the early 1960s when it was first proposed by H.T. Odum. The concept also appeared in Chinese ecological literature over the past few decades, especially in articles by Ma Shijun and his colleagues. Case studies called ecological engineering include the design of several ponds and wetlands in the United States and the development of wetlands, farms, and riverine systems in China. There are similarities and differences of ecological engineering in the West and in China in ecosystem design, the management of ecosystem structure, recycling, diversity, and economic valuation for both approaches.

Ecological engineering is an additional aspect of applied ecology, but different from conventional approaches. It is also different than other related fields such as biotechnology and environmental engineering but encompasses much of what is discussed in ecological restoration, rehabilitation, and related fields. The bases of ecological engineering are self-designed ecosystems, a subsequent savings of non-renewable resources, a conservation ethic for ecosystems, and an integrative rather than reductionist approach to ecology. In the short term, ecotechnology could bring immediate attention to the importance of "designing and building ecosystems" for solving environmental problems, as a logical extension of the field of ecology. In the long term, ecotechnology will provide the basic science and applied approaches needed by environmental regulators and managers to control some types of pollution and to reconstruct the landscape in an ecologically sound way. The occurrence of this symposium and the introduction of a new journal *Ecological Engineering: The Journal of Ecotechnology,* are testimony to the fact that ecological engineering will become part of the ecological paradigm for the world in the near future.

INTRODUCTION

Environmental pollution (local, regional, and global) and pollution from poor land- and water-use practices often defy simple solutions. We have built resource-costly technological solutions to one type of pollution only to find that they create

0-87371-990-5/97/$0.00+$.50

TABLE 1
Classification and Examples of Ecotechnology According to Types of Applications

Application	Examples
1. Ecosystems are used to reduce or solve a pollution problem that otherwise would be harmful to other ecosystems.	Wastewater recycling in wetlands; sludge recycling.
2. Ecosystems are imitated or copied to reduce or solve a resource problem.	Mitigation wetland; integrated fishponds.
3. The recovery of an ecosystem is supported after significant disturbance.	Surface coal mine restoration; lake and river restoration; recovery of hazardous waste site.
4. Existing ecosystems are modified in an ecologically sound way to solve an environmental problem.	Biomanipulation of fish in reservoirs.

Adapted from Jørgensen and Mitsch, 1989.

another type of pollution. We now consider the ultimate degradation product, CO_2, to be a global pollutant and are faced with the virtual impossibility of controlling its emission in the next several decades. Many Third-World countries aspire to imitate current fossil-fuel based food production practices in developed countries, just as we are finding them ecologically threatening and energy intensive. In short, as resources, both renewable and non-renewable, are diminished, environmental problems are becoming more complex and less "solvable" by conventional technological means. H.T. Odum (1989a) has argued:

> As the resources of the world become increasingly limiting to the expansion of the human economy, the management of the planet must turn more and more to a cooperative role with the planetary life support system, sometimes called the stewardship of nature. The pattern of humanity and nature that prevails is symbiotic because two coupled systems have higher performance than two separate systems that are not mutually reinforcing.

Ecotechnology involves several approaches or applications to the designing of the landscape (Table 1). These applications range from constructing new ecosystems to solving environmental problems to ecologically sound harvesting of existing ecosystems.

Ecological engineering and ecotechnology combine basic and applied science for the restoration, design, and construction of aquatic and terrestrial ecosystems. The goals of ecological engineering and ecotechnology, which need to be carefully defined and implemented, are: (1) the restoration of ecosystems that have been substantially disturbed by human activities such as environmental pollution, climate change, or land disturbance; (2) the development of new sustainable ecosystems that have human and ecological value; and (3) the identification of the life-support value of ecosystems to ultimately lead to their conservation. Construction and restoration of ecosystems need to be done in an ecologically sound way.

HISTORY OF ECOLOGICAL ENGINEERING

EARLY DEFINITIONS

Ecological engineering was defined by H.T. Odum in the early 1960s as "those cases in which the energy supplied by man is small relative to the natural sources, but sufficient to produce large effects in the resulting patterns and processes" (Odum, 1962) and as "environmental manipulation by man using small amounts of supplementary energy to control systems in which the main energy drives are still coming from natural sources" (Odum et al., 1963). Odum elaborated on the breadth of ecological engineering by stating that "the management of nature is ecological engineering, an endeavor with singular aspects supplementary to those of traditional engineering. A partnership with nature is a better term" (Odum, 1971). Ecotechnology, which is sometimes used as a term synonymous with ecological engineering, was defined in Eastern Europe in the early 1980s as "the use of technological means for ecosystem management, based on deep ecological understanding, to minimize the costs of measures and their harm to the environment" (Uhlmann, 1983; Straskraba and Gnauck, 1985).

RECENT DEFINITIONS

Combining ecosystem function with human needs is the emphasis of ecological engineering (or its synonym "ecotechnology"), as presently defined by Mitsch (1988) and Mitsch and Jørgensen (1989a): "the design of human society with its natural environment for the benefit of both." It provides approaches for conserving our natural environment while at the same time adapting to and sometimes solving intractable environmental pollution problems. As stated in Mitsch and Jørgensen (1989a):

> [Ecological engineering] is engineering in the sense that it involves the design of this natural environment using quantitative approaches and basing our approaches on basic science. It is technology with the primary tool being self-designing ecosystems. The components are all of the biological species of the world.

This focus on, and utilization of, biological species, communities, and ecosystems is what distinguishes ecotechnology from the traditional technologies, which rely on devices and facilities to remove, transform, or contain pollutants but which do not consider direct manipulation of ecosystems.

CHINESE APPROACHES

Much of the approach to environmental management in China has remained an art, but in recent years there has been explicit use of the term ecological engineering to describe a formal "design with nature" philosophy (Ma, 1985). Ma (1988) suggests that ecological engineering was first proposed in China in 1978 and is now used throughout the whole country, with about 500 sites that practice agroecological engineering. At a recent Symposium on Agroecological Engineering in Beijing (Ma et al., 1988), Qi and Tian (1988) suggested that "the objective of ecological research [in China] is being transformed from systems analysis to system design and construction," stating that ecology has now a great knowledge base from observational and experimental ecology and is in the position to meet global environmental problems through ecosystem design, the main task of ecological engineering. Ma (1988) has since defined ecological engineering as: "... a specially designed system of production process in which the principles of the species symbiosis and the cycling and regeneration of substances in an ecological system are applied with adopting the system engineering technology and introducing new technologies and excellent traditional production measures to make a multi-step use of substance." Agroecological engineering is defined as an "application of ecological engineering in agriculture" (Ma, 1988). Yan and Yao (1989) describe integrated fish culture management as ecological engineering because of its attention to waste utilization and recycling.

ECOLOGICAL ENGINEERING OF WASTEWATER

Most recently, the term ecological engineering has been applied to the treatment of wastewater in ecologically based "green machines," with indoor greenhouse applications built both in Sweden and the United States in the late 1980s (Guterstam and Todd, 1990). Here the applications are described as "environmentally responsible technology [which] would provide little or no sludge, generate useful byproducts, use no hazardous chemicals in the process chain and remove synthetic chemicals from the wastewater" (Guterstam and Todd, 1990). Ma and Yan (1989) have described a Chinese ecological engineering system in the treatment of wastewater as "a series of techniques based on fundamental ecological principles established in a local region to adjust the structure and function in polluted bodies of water to maintain the ecobalance." They elaborate that ecological engineering can have the highest economic benefits in wastewater treatment because it does not depend on great operation and construction costs and because it involves the regeneration of abandoned resources (Ma and Yan, 1989).

ECOLOGICAL ENGINEERING IN WETLAND DESIGN

The design and construction of wetlands has been suggested as another application of ecological engineering (Mitsch, 1977, 1988; Mitsch et al. 1989; Fennessy and Mitsch, 1989a, b; Mitsch and Cronk, in press). In some cases, research of existing wetlands can aid in our understanding of patterns of chemical retention and transformations (Mitsch et al., 1989; Mitsch and Reeder, 1991). In other cases, the retention of wastewater to wetlands, both natural and constructed, has been instrumental in defining ecological engineering (Odum et al., 1977; Mitsch, 1977; Godfrey et al., 1985; Hammer, 1989). New applications of ecological engineering of wetlands are being found, including the control of other forms of water pollution such as coal mine drainage (Fennessy and Mitsch, 1989a, b) and nonpoint source pollution from agricultural areas (Mitsch and Cronk, in press).

CONTRAST WITH OTHER FIELDS

ECOLOGY

The field of ecology has a century of experience of the study of organisms and their environment. It is commonly viewed as a subset of the biological sciences and

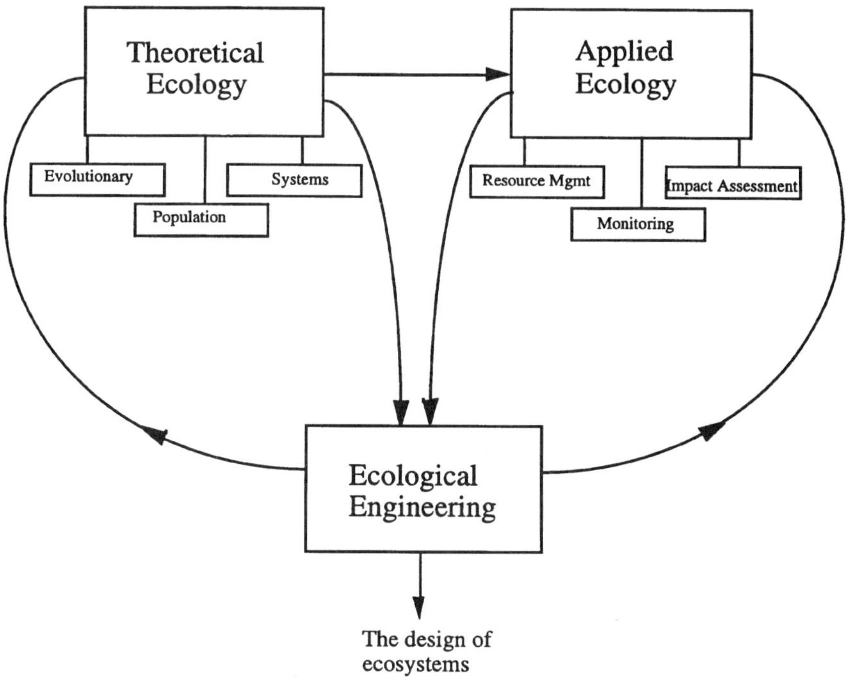

FIGURE 1. Relationships among theoretical ecology, applied ecology, and ecological engineering. Both theoretical and applied ecology provide fundamentals to ecological engineering but do not define it completely. Ecological engineering, in turn, feeds back ecological theory and management practices to theoretical and applied ecology.

is often divided into theoretical and applied ecology (Figure 1). Ecological engineering, as a third general area in ecology, is designing natural systems according to ecological principles learned over the past century (Figure 1). It is not merely the usual concept of applied ecology, often limited to monitoring and assessing environmental impacts or managing natural resources. Both basic and applied ecology provide fundamentals to ecological engineering but do not define it completely. Ecological engineering has its roots close to pure ecology, just as chemical engineering is close to chemistry and biochemical engineering is close to biochemistry. It logically remains a branch of ecology. Furthermore, ecological engineering can provide much to the establishment and testing of basic ecological theories; hence the feedback shown in Figure 1. Restoration ecology, a subset of ecological engineering if done properly, has been described as a field in which "the investigator is forced to study the entire system rather than components of the system in isolation from each other . . . One of the most compelling reasons for the failure of theoretical ecologists to spend more time on restoration ecology is the exposure of serious weaknesses in many of the widely accepted theories and concepts of ecology" (Cairns, 1988b). Bradshaw (1987) goes further in describing the restoration of a disturbed ecosystem as the "acid test of our understanding of that system." Hence, there is a high probability of advancing the understanding of ecological systems in ecological engineering because of the unique research approach that reconstructing ecosystems provides to scientists.

TABLE 2
Comparison of Ecotechnology and Biotechnology

Characteristic	Ecotechnology	Biotechnology
Basic Unit	Ecosystem	Cell
Basic principles	Ecology	Genetics; cell biology
Control	Forcing functions, organisms	Genetic structure
Design	Self-design with some human help	Human design
Biotic diversity	Protected	Changed
Costs	Reasonable	Enormous
Energy basis	Solar based	Fossil fuel based

From Mitsch and Jørgensen, 1989a

ENVIRONMENTAL ENGINEERING

Ecological engineering is not the same as environmental engineering, a respected field that has been well-established in universities and the workplace since the early 1960s and was called sanitary engineering before that. Environmental engineers are certainly involved in the application of scientific principles to solve pollution problems, but the concepts usually involve energy and resource intensive operations such as settling tanks, scrubbers, filters, and chemical precipitators.

BIOTECHNOLOGY

Ecological engineering and ecotechnology also should not be confused with biotechnology, which involves genetic manipulation to produce new strains and organisms to carry out specific functions. Some of the differences between ecotechnology and biotechnology relate to their basic principles, control, design, and ultimate possible costs to society (Table 2). Nevertheless, a comparison can be made between the development of ecotechnology and biotechnology. Ecotechnology is almost at the stage where biotechnology was 20 years ago. Molecular biology was just beginning then to define the basic science and techniques for the yet unborn field of biotechnology. Today, ecology is recognized as a fundamental science and is now developing the ecosystem-level tools to develop the field of ecotechnology. Despite the progress made in biotechnology (defined here as genetic manipulation of species), most researchers and environmental managers now agree that it will not solve the world's environmental problems and there may be some adverse environmental consequences from its development. Ecotechnology, which utilizes the existing array of species, communities, and ecosystems, may receive more attention as the limitations of biotechnology are experienced.

RESTORATION

Related fields such as ecosystem restoration are in their infancy, having been described in books only recently (Jordan et al., 1987; Cairns, 1988a; Buckley, 1989; Wali, in press) and in a few other publications (e.g., Bradshaw, 1987; Aber and Jordan, 1985). There is now an active Society for Ecological Restoration in the United States. Several of these fields are equivalent to or subsets of ecological engineering (Table 3) as they have the design of ecosystems as their theme. On the of biotic communities from estuarine, freshwater, and sewage sources. Formally, the question being asked was "whether the self-organization process [of species arrangements] occurs readily there with new conditions from wastewater influence and how much time is required" (Odum, 1989b). Odum (1989b) concludes that the design parameters such as inflows, loading rates, and other controls, along with the very important multi-seeding of as many species as possible, were important considerations in the ecological engineering of these estuarine ponds and that these parameters and seedings set the boundary conditions for a relatively rapid process of ecosystem development.

WETLANDS FOR RENOVATION OF DOMESTIC WASTEWATER

It is now common to see the application of ecological engineering concepts with the use of wetlands for treatment of domestic wastewater (cf. Godfrey et al., 1985; Mitsch and Gosselink, 1986; Hammer, 1989). One of the first experiments to investigate this idea was conducted in the early 1970s in Florida by H.T. Odum, K.C. Ewel, and colleagues (Odum et al., 1977; Ewel and Odum, 1984). Whole ecosystem experiments for the use of forested wetlands, particularly cypress domes dominated by *Taxodium ascendans* (also known as *T. distichum* var. *nutans*), for the treatment of high nutrient wastewater, were run for several years near Gainesville, Florida. Two cypress domes (0.5 to 1.0 ha in size) received treated wastewater from a trailer park at rates of about 2.5 cm/wk. A third dome received an equivalent groundwater inflow while a fourth served as a natural control. Measurements in shallow groundwater wells indicated that there was more than 90% removal of nutrients, organic matter, and minerals by the wetlands receiving wastewater (Dierberg and Brezonik, 1984) and nitrogen and phosphorus concentrations in the foliage and branches of the trees increased as wastewater was added and decreased again after the treatment stopped (Straub, 1984). This study, which also involved studies of wetland hydrology, soils, modelling, and wildlife, demonstrated that forested wetlands could be used in some cases to remove nutrients from wastewater with a minimum application of expensive and fossil energy consuming technology (Mitsch, 1977).

WETLANDS FOR RENOVATION OF COAL MINE DRAINAGE

Wetlands have been utilized for mine drainage pollution control from coal mines in the eastern United States (Brooks et al., 1985; Hammer, 1989; Wieder, 1989). Mine drainage forms when pyrite (FeS_2), a compound commonly associated with coal deposits, becomes exposed to the oxidizing forces of air and water during the mining process. Sulfide oxidation results, producing dissolved iron, sulfate, large amounts of acid and ultimately an orange precipitate, ferric hydroxide ($Fe(OH)_3$). Even after mining is stopped, mine drainage production may continue indefinitely. In one example of the application of ecological engineering principles to the treatment of mine drainage, a 0.22-ha constructed wetland and dominated by *Typha latifolia*, was evaluated for its ability to treat approximately 340 l/min of coal mine drainage from an underground seep in eastern Ohio (Fennessy and Mitsch, 1989a, b). Iron decreased by 50 to 60%, with slightly higher decreases during the growing season. In general, results from a number of studies have shown that improved treatment of mine drainage is correlated with longer retention times and lower iron

TABLE 3
Synonyms or Subsets of Ecological Engineering or Ecotechnology

1. Synthetic ecology	6. Ecosystem rehabilitation
2. Restoration ecology	7. Biomanipulation
3. Bioengineering	8. River restoration
4. Sustainable agroecology	9. Wetland restoration
5. Habitat reconstruction	10. Reclamation ecology

other hand, some of the approaches seem to lack one of the two major components of ecological engineering, namely: (1) the designed ecosystem should be self-sustaining (see below) and (2) there must be a theoretical ecology underpinning to the approach. Early work in Europe used the concept of bioengineering, or using plants as engineering materials (Schiechtl, 1980). More recently, much has been written on the recently described fields of restoration ecology (Aber and Jordan, 1985; Jordan et al., 1987; Buckley, 1989) and ecosystem rehabilitation (Cairns, 1988a; Wali, in press) for the restoration of disturbed landscape. Restoration ecology was defined by Cairns (1988b) as "the full or partial placement of structural or functional characteristics that have been extinguished or diminished and the substitution of alternative qualities or characteristics than the ones originally present with the proviso that they have more social, economic, or ecological value than existed in the disturbed or displaced state." The ecotechnology approach has also been applied to river and stream restoration (Gore, 1985) and to sustainable agriculture as agroecosystems (Lowrance et al., 1984).

CONCEPTS OF ECOLOGICAL ENGINEERING

SELF-DESIGN

When change occurs in natural systems, species shift as they substitute for each other and as food chains reorganize. As individual species change, ultimately new systems emerge that are better suited to the environments that are superimposed on them. The natural systems then begin to manipulate its physical and chemical environment to make it a little more palatable. Humans participate by providing choices of initial species, matching species with the environment. Nature does the rest. Odum (1989a) refers to this capability of self-design as *self-organization*, which "designs a mix of man-made and ecological components" in a "pattern that maximizes performance, because it reinforces the strongest of alternative pathways that are provided by the variety of species and human initiatives." Mitsch and Jørgensen (1989b) have described the concept as self-design. Multiple seeding of species into ecological engineered systems is one way to speed the selection process in this self-organization or self-design (Odum, 1989b).

SUSTAINABLE ECOSYSTEMS

Because most ecosystems are based directly or indirectly on the sun, they are self-sustaining systems. Once an ecosystem is designed and constructed, it should be able to sustain itself indefinitely through self-design, with only a modest amount of intervention by humans. Ecotechnology is based on the use of some non-renewable energy expenditure at the start but a dependence on solar energy after that. This means that the ecosystem, running on solar energy or the products of solar energy, does not need to depend on technological fossil energies as much as would a traditional technological solution to the same problem. If the system does not sustain itself, it does not mean that the ecosystem has failed us (its behavior is ultimately predictable). It means that ecological engineering has not yet developed to the point where we can facilitate the proper interface between nature and the environment. Modern agriculture and forestry also run on a combination of renewable (solar) and non-renewable (fossil fuel) energy but their basis is highly technological and fossil fuel based.

ECOSYSTEM CONSERVATION

Ecological engineering involves identifying those biological systems that are most adaptable to the human needs and human needs that are most adaptable to ecosystems. Ecological engineers have in their toolboxes all of the ecosystems, communities, populations, and organisms that the world has to offer. Therefore, a direct consequence of ecological engineering is that it would be counterproductive to eliminate or even disturb natural ecosystems unless absolutely necessary. This is analogous to the conservation ethic that is shared by many farmers, even though they may till the landscape. The ecotechnology approach will lead to a greater environmental conservation ethic than has been realized up to now. For example, when wetlands were recognized for their ecosystem values of flood control and water quality enhancement through ecological engineering experimentation and observation, wetland protection efforts gained a much wider degree of acceptance and even enthusiasm than they had before, despite their long understood values of habitat for fish and wildlife (Mitsch and Gosselink, 1986). In short, recognition of ecosystem values through ecological engineering provides greater justification to conserve natural ecosystems.

A SYSTEMS APPROACH

Odum (1989a) has stated that the practice of ecological engineering cannot be supported completely by reductive, analytic experimental testing and relating. Approaches such as modelling and cost-benefit analysis are more important as ecosystem design and prognosis cannot be predicted by summing parts to make a whole. We must learn to work with whole ecosystems rather than species by species. We must be able to synthesize a great number of disciplines to understand and deal with the design of ecosystems. Ecotechnology places much the same requirements on ecologist as systems modelling: having to consider the entire ecosystem, compartment by compartment, but ultimately to make a whole. Jordan et al. (1987) make an analogy that the best way to understand a system, whether a car or a watch, is to "attempt to reassemble it, to repair it, and to adjust it so that it works properly." Thus, ecotechnology is really a technique for doing fundamental ecological research and applying it in a systems approach.

TABLE 4
Contrast between Ecological Engineering in the West and China

	U.S. and West	China
Basic	**Principles of ecology**	
Design principles	Self-design with some human work	Heavy intervention by humans to maintain system
Emphasis on	Changing forcing functions to produce outcome	Changing ecosystem structure to produce outcome
People, #/ha	<1	>10
Subsidy type	Fussil fuel economy supporting ecologists	Human energy of resource managers
Subsidy size	Should be small	Sometimes surprisingly high, especially humans
Recycling	Acceptable	Absolute requirement
Commercial yield	Not usual	Always food production
Species diversity	Often monospecific	"Multi-layered" with many niches filled
Values considered	Aesthetics Conservation of natural resources Non-market values, e.g., water pollution control	Commercial products
Economic values	Willingness to pay, contributory, or replacement	Utilitarian
Experience	30 years (since early 1960s)	3,000 years

CASE STUDIES IN THE UNITED STATES AND CHINA

Ecological engineering in the United States and the West has stressed a partnership with nature and has been investigated primarily in experimental ecosystems rather than in full-scale applications (Mitsch and Jørgensen, 1989b). Some of the more significant experiments that have been conducted or are currently underway in ecological engineering relate to aquatic systems, particularly shallow ponds and wetlands. Ecological engineering in China, as pioneered by Ma (1985) and others, has been applied to a wide variety of natural resource and environmental problems, ranging from fisheries and agriculture, to wastewater control and coastline protection. The emphasis in the Chinese systems has been on applications rather than experimentation and on the production of food and fiber more than on environmental protection (Table 4). More details of the comparison are being published elsewhere (Mitsch, in press, a, b).

EXPERIMENTAL ESTUARINE PONDS

Odum (1985, 1989b) reported on one of the first studies, done in the 1960s, described as an experiment in ecological engineering. Estuarine ponds were built in Morehead City, North Carolina to investigate ecological changes as the ponds received secondarily treated municipal wastewater mixed with salt water. Three ponds received the sewage-salt water mixture and three ponds were controls, receiving only a tap water-salt water mixture. All ponds were "seeded" with a high diversity

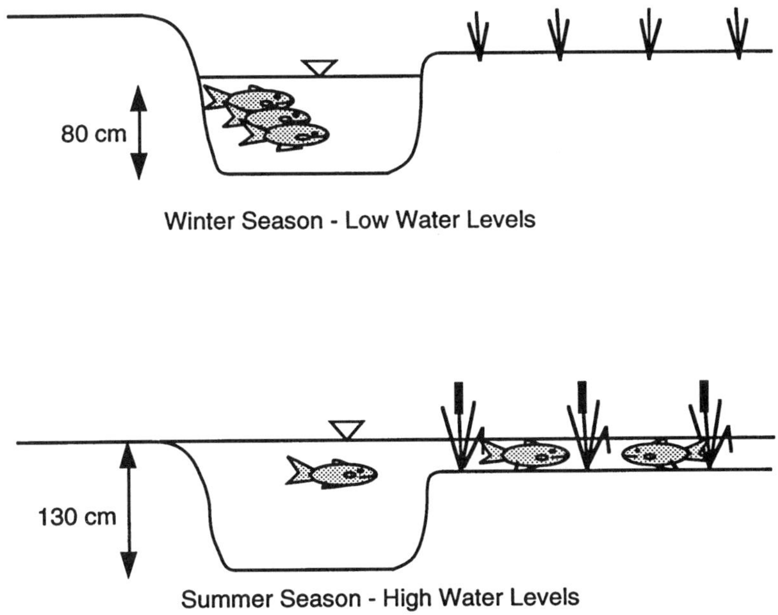

FIGURE 4. Seasonal patterns of water levels at Lake Go Fish-Wetland System in Yixing County, Jiangsu Province, China, showing winter and summer conditions. Water level manipulation maintains a dry wetland in January through March with about 80 cm of water in the channel. In April the water level in the channel is raised by pumping water from adjacent Lake Go, allowing some water to spill over into the wetland and fish to feed on the reeds. The water level is maintained at about 130 cm (50 cm in the wetland in May) and water is continually pumped through October until there is up to 1.0 m of water in the wetland. From October to December, the water level is decreased until the wetland is dry and fish retreat to the channels to be harvested.

of ecological engineering is accomplished primarily through water level manipulation that is synchronized with fish growing seasons and harvesting schedules and a series of deeper channels for fish overwintering extending through wetland areas that are much shallower (Figure 4). In April the water level in the channel is raised to about 130 cm (50 cm in the wetland). From October to December, the water level is decreased until the wetland is dry. Sometimes the channels are drained completely to harvest benthic fish. Fish yield is approximately 9 t/ha with 60% as grass carp and Wuchang fish, 30% as silver carp and bighead carp, and 10% as others, including crucian and common carp. Reeds not consumed by fish are used for fuel and thatching.

AGROECOLOGICAL ENGINEERING

Ma (1988) suggests that there are about 500 trial sites in agroecological engineering or ecoagriculture in China in more than 20 provinces. Examples of agroecological engineering are given by Wu et al. (1988) and Ma and Liu (1988) for villages in Hunan Province and Chai et al. (1988) for Jiangsu Province. Wu et al. (1988) discuss the production of the "three materials" of fertilizer, forage, and fuel at Wu Tang village (pop. 1505) through agroecological engineering. While 89.4%

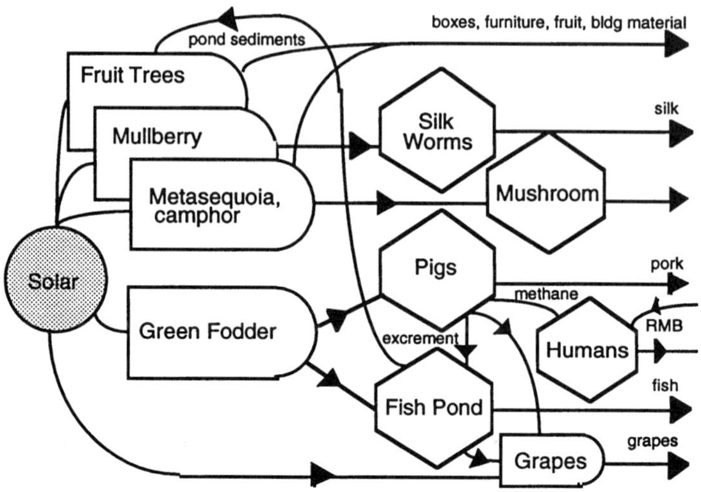

FIGURE 5. Energy flow diagram of agroecological engineering system in Tai Chang Village, Anhui Province, China. A multitude of plant and consumer systems are maintained with waste products from one providing support for another. Not shown but implied in system are the human feedback loops to manage and maintain each system. RMB is Chinese currency received for all products taken to market on right side of diagram.

of the energy of the system comes from solar-based organic production only 10.6% comes from industrial energy (no energy quality ratios used). Furthermore, 88% of the total organic production is consumed in the village, while only 12% is exported. This is described as a "self-supporting system" that does not depend greatly on outside industrial energy. Tai Chang Village, approximately 25 km west of Maanshan City on the south bank of the Yangtze River in Anhui Province, contains an agroecological engineering demonstration site. The village where the experimental site is located is approximately 700 ha in size with 390 ha under cultivation. Part of the village includes a demonstration site in agroecological engineering, including a pig farm, methane production and use, a fish pond, grape production, and tree plantations of oranges, *Metasequoia*, mulberry, and camphor (Figure 5). Silk worms consuming mulberry leaves and producing silk is another cottage industry in this agroecological engineering design. The emphasis is on recycling with pond detritus used as fertilizers for the plantations and pig waste fermented to produce methane for cooking.

SALT MARSH RESTORATION

Salt marshes, dominated principally by *Spartina anglica*, have been constructed along the east coast of China to stabilize the coastline, to accelerate accretion of sediments for reclamation of tidal land for agricultural or industrial development, to produce green manure, foodstuff and fuel, and to control stream siltation and pollution (Chung 1982, 1983, 1985, 1989). Large *Spartina* plantations were developed by first transplanting plants to rice paddies, a planting technique well developed in China. The reclaimed *Spartina* marshes along China's coastline have been described as providing many benefits to humans (Figure 6). The planting of *Spartina* in one location led to the accretion of 80 cm of sediments in 7 years (11.4 cm/yr) at one location and 66 to 68 cm more than at the control after 4 years (16.7 cm/yr)

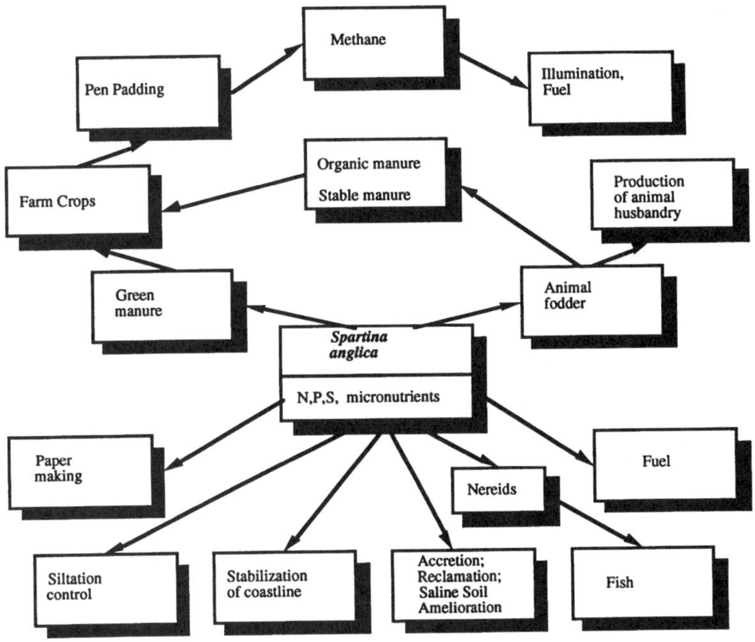

FIGURE 6. Benefits of reclaimed coastal wetlands in China, as summarized by Chung (1989).

at another site, leading to reclamation of the sea for crop production and coastal stabilization. The *Spartina* grass increased aeration and organic content and decreased the salt content of the soil in addition to dissipating wave energy and slowing currents. The newly created salt marshes were also used as a habitat for migratory birds, waterfowl, domestic fowl, nereids (worms), and crabs and as a pastureland for cattle and pigs. The grass is also harvested as a source of animal fodder, as an effective "green manure" for rice fields, as a source of fuel, and as a source of marsh gas (methane) for cooking and illumination.

RIVER POLLUTION CONTROL

A riverine water hyacinth (*Eichhornia crassipes*) ecosystem was investigated for its role as a water pollution control system in a series of ecological engineering experiments on a section of the Fumen River in an eastern suburb of Suzhou, Jiangsu Province (Ma and Yan, 1989). From May through December 1984, approximately 2.7 ha of the river were planted in water hyacinths. The benefits of this system includes the partial reclamation of polluted river water, particularly for nutrients, organic matter, and heavy metals, and the production of green fodder for a number of consumers (Figure 7). In the experimental hyacinth marsh, concentrations of COD, total nitrogen, total phosphorus, ammonia nitrogen, ortho-phosphate, organic nitrogen, and organic phosphorus decreased during the growing season. Microbial communities that develop on the root systems of the floating water hyacinths are responsible for a reduction of organic matter in the water. Approximately 2,500 t (fresh weight) of water hyacinths were harvested from the site to be used as fodder for fish, particularly for grass carp (*Ctenopharyngodon idella*) and Wuchang fish

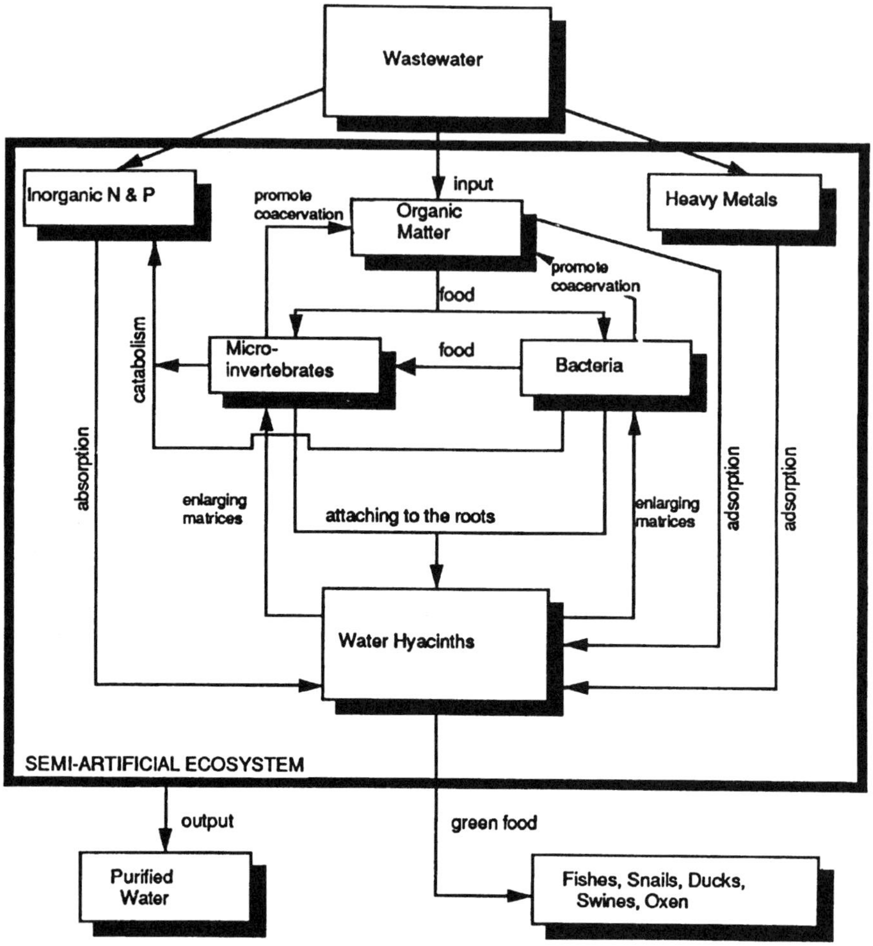

FIGURE 7. Model depicting processes and values of water hyacinth river restoration program. A riverine water hyacinth (*Eichhornia crassipes*) ecosystem was investigated for its role as a water pollution control system and as a subsequent source of fodder for ducks and fish. From May through December 1984, approximately 2.7 ha of the river were planted in water hyacinths (Ma and Yan 1989) and harvested at the end of the year.

(*Megalobrama amblycephala*) in culture ponds and for ducks, swine, and snails. Investigations of heavy metal content of fish and ducks that consumed the water hyacinths showed generally safe levels (Ma and Yan, 1989).

IMPORTANCE AND FUTURE OF ECOLOGICAL ENGINEERING

Case studies such as these suggest that an effort in ecology that focuses ecological research on designing and reconstructing ecosystems is now beginning. This ecotechnology needs to be directed to such environmental issues as: (1) preventing and mitigating the destruction of wetlands and other sensitive ecosystems; (2) adapting to global environmental changes such as global warming, increasing sea levels, and acidic deposition; (3) controlling point and nonpoint source water pollution from villages, municipalities, industry, agriculture, and mined lands; (4) protecting clean lakes and reservoirs and restoring those that are polluted; (5) adapting ecosystems to environmental improvements, such as river and lake clean-ups, that are a result of past environmental regulations; (6) cleaning up and stabilizing hazardous waste sites; and (7) reestablishing symbiotic patterns of recycling of wastes to produce valuable commodities. All of these and other environmental issues require that ecologists and engineers have an understanding of the kinds of ecosystem construction or rehabilitation best suited to a particular environment and, most importantly, the means of reconstructing the aquatic or terrestrial landscape to achieve desired goals.

The integration of ecology and engineering into an application that we have chosen to call ecological engineering or ecotechnology has not yet taken place to as great a degree as might be expected, given the great public concern for our major environmental problems and the clear need for innovative solutions. Ecologists still remain low on the learning curve compared to engineers when confronted after formal education with real-life issues of wetland construction, wastewater treatment design, river restoration, habitat reconstruction, or mine land rehabilitation. They often rise to the occasion quickly with homespun ecotechnology and novel approaches, but the integration is not there and techniques are relearned each time. Ecological theory may be mentioned but usually has not been integrated into a design framework necessary in the engineering profession. Furthermore, ecotechnology is, by definition, a combination of basic and applied research that requires interdisciplinary teams for its proper application. The development of such an interdiscipline and its methodologies and principles will require cooperative efforts from many related fields. The fruitful integration of these efforts will need a new administrative structure in universities and research laboratories and new ways of implementing technologies in our society. The need for the paradigm is clear; its implementation is still before us.

ACKNOWLEDGMENTS

I wish to particularly thank H.T. Odum for his initial seeding and continued encouragement of this work. Collaboration with Sven Erik Jørgensen was very helpful in focusing some of the ideas and case studies.

REFERENCES

Aber, J.D. and W.R. Jordan, III. 1985. "Restoration ecology: an environmental middle ground." *BioScience* 35:399.

Bradshaw, A.D. 1987. "Restoration: The acid test for ecology." In: Jordan, W.R., III, M.E. Gilpin, and J.D. Aber, eds. *Restoration Ecology: A Synthetic Approach to Ecological Research.* Cambridge Univ. Press, Cambridge. pp. 23-29.

Brooks, R.P., D.E. Samuel, and J.B. Hill, eds. 1985. *Wetlands and Water Management on Mined Lands. Proc. of a Conference, October 1985.* The Pennsylvania State University, University Park, Penn. 393 pp.

Buckley, G.P., ed. 1989. *Biological Habitat Reconstruction.* Belhaven Press, London. 363 pp.

Cairns, J., Jr., ed. 1988a. *Rehabilitation of Damaged Ecosystems, Volumes I and II*, CRC Press, Boca Raton, Florida.

Cairns, J., Jr. 1988b. "Restoration Ecology: the New Frontier." In: J. Cairns, Jr., ed. *Rehabilitation of Damaged Ecosystems, Volumes I and II,* CRC Press, Boca Raton, Florida. pp. 1-11.

Chai Tukui, Shi Wenchao, Lu Tienong, and Ye Maoxin. 1988. "Benefit analyses of agroecological engineering in Dongxu Village of Jiangsu Province." In: Ma Shijun, Jiang Ailiang, Xu Rumei, and Li Dianmo, eds. *Proceedings of the International Symposium on Agro-Ecological Engineering, August, 1988*, Ecological Society of China, Beijing.

Chung, Chung-Hsin. 1982. "Low Marshes, China." In: Lewis, R.R., ed. *Creation and Restoration of Coastal Plant Communities.* CRC Press, Boca Raton, Florida. pp. 131-145.

Chung, Chung-Hsin. 1983. "Geographical distribution of *Spartina anglica* Hubbard in China." *Bull. Mar. Sci.* 33:753-758.

Chung, Chung-Hsin. 1985. "The effects of introduced *Spartina* grass on coastal morphology in China." *Z. Geomorphol. N. F. Suppl.* Bd. 57:169-174.

Chung, Chung-Hsin. 1989. "Ecological engineering of coastlines with salt marsh plantations." In: Mitsch, W.J. and S.E. Jørgensen, eds. *Ecological Engineering: An Introduction to Ecotechnology.* J. Wiley & Sons, Inc., New York. pp. 255-289.

Dierberg, F.E. and P.L. Brezonik. 1984. "The effect of wastewater on the surface water and groundwater quality of cypress domes." In: Ewel, K.C. and H.T. Odum, eds. *Cypress Swamps.* University Presses of Florida, Gainesville. pp. 83-101.

Ewel, K.C. and H.T. Odum, eds. 1984. *Cypress Swamps.* University Presses of Florida, Gainesville. 472 pp.

Fennessy, M.S. and W.J. Mitsch. 1989a. "Design and use of wetlands for renovation of drainage from coal mines." In: Mitsch, W.J. and S.E. Jørgensen, eds. *Ecological Engineering: An Introduction to Ecotechnology.* J. Wiley & Sons, Inc., New York. pp. 231-253.

Fennessy, M.S. and W.J. Mitsch. 1989b. "Treating coal mine drainage with an artificial wetland." *Res. J. Water Poll. Control Fed.* 61:1691-1701.

Godfrey, P.J., E.R. Kaynor, S. Pelczarski, and J. Benforado, eds. 1985. *Ecological Considerations in Wetlands Treatment of Municipal Wastewaters.* Van Nostrand Reinhold Company, New York. 474 pp.

Gore, J.A., ed. 1985. *The Restoration of Rivers and Streams.* Butterworth, Boston, Massachusetts.

Guterstam, B. and J. Todd. 1990. "Ecological engineering for wastewater treatment and its application in New England and Sweden." *Ambio* 19:173-175.

Hammer, D.A., ed. 1989. *Constructed wetlands for wastewater treatment: Municipal, industrial and agricultural.* Lewis Publishers, Chelsea, Michigan. 831 pp.

Hey, D.L., M.A. Cardamone, J.H. Sather, and W.J. Mitsch. 1989. "Restoration of riverine wetlands: The Des Plaines River wetland demonstration project." In: Mitsch, W.J. and S.E. Jørgensen, eds. *Ecological Engineering: An Introduction to Ecotechnology.* J. Wiley & Sons, Inc., New York. pp 159-183.

Jordan, W.R., III, M.E. Gilpin, and J.D. Aber, eds. 1987. *Restoration Ecology: A Synthetic Approach to Ecological Research.* Cambridge Univ. Press, Cambridge. 342 pp.

Jørgensen, S.E. and W.J. Mitsch. 1989. "Classification and examples of ecological engineering." In: W.J. Mitsch and S.E. Jørgensen, eds. *Ecological Engineering: An Introduction to Ecological Engineering.* J. Wiley & Sons, New York. pp. 13-19.

Lowrance, R., B.R. Stinner, and G.J. House, eds. 1984. *Agricultural Ecosystems: Unifying Concepts.* J. Wiley & Sons, New York. 233 pp.

Ma Jinzhong and Liu Houpei. 1988. "Analysis of the functions of paddy-field ecosystem engineering in southern mountain and hilly areas of Hunan Province, China." In: Ma Shijun, Jiang Ailiang, Xu Rumei, and Li Dianmo, eds. *Proceedings of the International Symposium on Agro-Ecological Engineering, August, 1988*, Ecological Society of China, Beijing.

Ma Shijun. 1985. "Ecological engineering: application of ecosystem principles." *Environmental Conservation* 12:331-335.

Ma Shijun. 1988. "Development of agroecological engineering in China." In: Ma Shijun, Jiang Ailiang, Xu Rumei, and Li Dianmo, eds. *Proceedings of the International Symposium on Agro-Ecological Engineering, August, 1988*, Ecological Society of China, Beijing. pp. 1-13.

Ma Shijun, Jiang Ailiang, Xu Rumei, and Li Dianmo, eds. 1988. *Proceedings of the International Symposium on Agro-Ecological Engineering, August, 1988,* Ecological Society of China, Beijing.

Ma Shijun and Yan Jingsong. 1989. "Ecological engineering for treatment and utilization of wastewater." In: Mitsch, W.J. and S.E. Jørgensen, eds. *Ecological Engineering: An Introduction to Ecotechnology.* J. Wiley & Sons, Inc., New York. pp. 185-217.

Mitsch, W.J. 1977. "Energy conservation through interface ecosystems." In: *Proceedings — International Conference on Energy Use Managment*, Pergamon Press, Oxford. pp. 875-881.

Mitsch, W.J. 1988. "Ecological engineering and ecotechnology with wetlands: applications of systems approaches." In: A. Marani, ed. *Advances in Environmental Modelling*, Elsevier, Amsterdam. pp. 565-580.

Mitsch, W.J. 1995, a. "Ecological engineering: from Gainesville to Beijing: A comparison of approaches in the U.S. and China." In: C.A.S. Hall, ed. *Maximum Power*, University of Colorado Press, Boulder.

Mitsch, W.J., in press, b. "Ecological Engineering: Approaches to Sustainability and Biodiversity in U.S.A. and China." In: R. Costanza, ed. *The Ecological Economics of Sustainability*, Columbia University Press, New York.

Mitsch,W.J. and J.G. Gosselink. 1986. *Wetlands.* Van Nostrand Reinhold Company, New York. 539 pp.

Mitsch, W.J. and S.E. Jørgensen. 1989a. "Introduction to ecological engineering." In: Mitsch, W.J. and S.E. Jørgensen, eds. *Ecological Engineering: An Introduction to Ecotechnology.* J. Wiley & Sons, Inc., New York. pp. 3-12.

Mitsch, W.J. and S.E. Jørgensen, eds. 1989b. *Ecological Engineering: An Introduction to Ecotechnology.* J. Wiley & Sons, Inc., New York. 472 pp.

Mitsch, W.J., B. Reeder and D. Klarer. 1989. "The role of wetlands for the control of nutrients with a case study of Western Lake Erie." In W.J. Mitsch and S.E. Jørgensen, eds. *Ecological engineering: An introduction to ecotechnology.* J. Wiley & Sons, New York. pp. 129-158.

Mitsch, W.J., M.S. Fennessy, and J.K. Cronk. 1990. *Ecosystem Studies of the Des Plaines River Experimental Wetlands — 1989/90.* Final Report to Wetlands Research, Inc., Chicago, Illinois. 48 pp. + appen.

Mitsch, W.J. and B.C. Reeder. 1991. "Modelling nutrient retention of a freshwater coastal wetland: estimating the roles of primary productivity, sedimentation, resuspension and hydrology." In: *Ecological Modelling*, 54:151–187.

Mitsch, W.J. and J.K. Cronk. 1992. "Creation and restoration of wetlands: some design considerations for ecological engineering." In, R. Lal and B. A. Stewart (eds.) *Advances in Soil Science,* Vol. 7, *Soil Restoration.* New York: Springer-Verlag. pp. 217–259.

Odum, H.T. 1962. "Man in the ecosystem." Proc. Lockwood Conference on the Suburban Forest and Ecology. *Bull. Conn. Agric. Station* 652:57-75.

Odum, H.T. 1971. *Environment, Power, and Society.* J. Wiley & Sons, Inc., New York. 336 pp.

Odum, H.T., ed. 1985. *Self Organization of Ecosystems in Marine Ponds Receiving Treated Sewage.* North Carolina Sea Grant Office Publ. UNC-SG-85-04, North Carolina State University, Raleigh. 250 pp.

Odum, H.T. 1989a. "Ecological engineering and self-organization." In: Mitsch, W.J. and S.E. Jørgensen, eds. *Ecological Engineering: An Introduction to Ecotechnology.* J. Wiley & Sons, Inc., New York. pp. 79-101.

Odum, H.T. 1989b. "Experimental study of self-organization in estuarine ponds." In: Mitsch, W.J. and S.E. Jørgensen, eds. *Ecological Engineering: An Introduction to Ecotechnology.* J. Wiley & Sons, Inc., New York. pp. 291-340.

Odum, H.T., W.L. Siler, R.J. Beyers, and N. Armstrong. 1963. "Experiments with engineering of marine ecosystems." *Publ. Inst. Marine Sci. Univ. Texas* 9:374-403.

Odum, H.T., K.C. Ewel, W.J. Mitsch, and J.W. Ordway. 1977. "Recycling treated sewage through cypress wetlands." In F.M. D'Itri, ed., *Wastewater Renovation and Reuse.* Marcel Dekker Press, New York. pp. 35-67.

Qi Ye and Tian Han-qin. 1988. "Some views on ecosystem design." In: Ma Shijun, Jiang Ailiang, Xu Rumei, and Li Dianmo, eds. *Proceedings of the International Symposium on Agro-Ecological Engineering, August, 1988*, Ecological Society of China, Beijing.

Schiechtl, H. 1980. *Bioengineering for Land Reclamation and Conservation.* The University of Alberta Press. 404 pp.

Straskraba, M. and A.H. Gnauck. 1985. *Freshwater Ecosystems: Modelling and Simulation.* Elsevier, Amsterdam. 305 pp.

Straub, P.A. 1984. "Effects of wastewater and inorganic fertilizer on growth rates and nutrient concentrations in dominant tree species in cypress domes." In: Ewel, K.C. and H.T. Odum, eds. *Cypress Swamps.* University Presses of Florida, Gainesville. pp. 127-140.

Uhlmann, D. 1983. "Entwicklungstendenzen der Ökotechnologie." *Wiss. Z. Tech. Univ. Dresden* 32:109-116.

Wali, M.K. ed. 1992. *Environmental Rehabilitation: Preamble to Sustainable Development*, 2 Volumes, SPB Academic Publishing, The Hague, The Netherlands.

Wieder, R.K. 1989. "A survey of constructed wetlands for acid coal mine drainage treatment in the Eastern United States." *Wetlands* 9:299-315.

Wu Jin Fu, Yan Fu, Yang Shi Jie, Liang Hai Quan, and Yu Wu Jiao. 1988. "A primary study on coordinated ecological engineering of fertilizer and forage and fuel in Wu Tang Village of Changsha County." In: Ma Shijun, Jiang Ailiang, Xu Rumei, and Li Dianmo, eds. *Proceedings of the International Symposium on Agro-Ecological Engineering, August, 1988,* Ecological Society of China, Beijing.

Yan Jingsong and Yao Honglu. 1989. "Integrated fish culture management in China." In: Mitsch, W.J. and S.E. Jørgensen, eds. *Ecological Engineering: An Introduction to Ecotechnology.* J. Wiley & Sons, Inc., New York. pp. 375-408.

Chapter 2

THE FUNCTION OF ECOLOGICAL ENGINEERING IN ENVIRONMENTAL CONSERVATION WITH SOME CASE STUDIES FROM CHINA

Yan Jingsong and Ma Shijun

INTRODUCTION

For several decades attempts have been made to prevent pollution and protect the environment. Some engineering techniques already applied with more or less success are: (1) environmental engineering, such as dilution (as used in West Lake, Hangzhou, China) and dredging (as in Qinhuai River, Nanjing, China), and (2) artificial mechanical-chemical regeneration systems (as in many sewage treatment plants, including settling tanks, scrubbers, and sand filters). These measures are often too expensive, if possible at all, and many may cause extensive disturbances of other ecosystems. Their application is limited by some conditions (particularly economic), and they may not be freely used everywhere, especially in developing countries and regions. Therefore, alternative approaches are to be sought which can produce economic benefits and, at the same time, offer ecological, environmental, and social advantages. One such approach is ecological engineering.

DEFINITION OF ECOLOGICAL ENGINEERING

Ecological engineering may be defined as a specially designed and operated system for production processes combined with the management of the natural environment, for the benefit of both. The holistic principles of harmony, regeneration, and recycling of nutrients in an ecosystem (or co-ecosystem) are applied through the adoption of system engineering technology. Environmentally sound traditional production practices, introduced techniques, and new developments are applied, along with environmental protection measures, to create a multi-stepped and multi-layered use of nutrients, which should result in increased production and greater economy. The potential for useful products, including the waste from every production step, would be brought into full play and maximized, while environmental degradation is minimized. The objective in preventing and treating pollution while promoting a sustainable, stable, and harmonious development of production is to promote the effective and beneficial recycling of nutrients in nature. Economic benefits, ecological, environmental, and human social advantages are maximized while waste is minimized.

0-87371-990-5/97/$0.00+$.50

TABLE 1
Comparison of Ecological Engineering, Environmental Engineering, and Bioengineering

Characteristics	Ecological engineering	Environmental engineering	Bioengineering
Basic unit	Ecosystem — an organic whole	Local Environment — a part of the ecosystem	Cell — part of our organism
Basic principles	Ecology with economics	Civil engineering, biochemistry, and physics	Genetics and cell biology
Goal	Multi-objective	Mono-objective	Mono-objective
Control	Structure and function of ecosystem with forcing function	Forcing function	Genetic structure
Design	Self-design with some artificial regulation	Human design	Human design
Biotic diversity	Protected or increased		Change
Maintenance and development costs	Reasonable, often low	Expensive	Expensive
Energy basis	Solar based	Electric power or fossil fuel based	Electric power or fossil fuel based
Techniques and measures	Simple and convenient	Complicated and troublesome	Troublesome
Environmental advantage	Advantage to the whole environment	Advantage to the local environment but often disadvantage to the whole	Uncertain
Economic benefit	Getting some net profit	No direct net profit	Uncertain
Social advantage	To the advantage of the whole society	To the advantage of a local region but often to the disadvantage of the whole society	Uncertain

CHARACTERISTICS OF ECOLOGICAL ENGINEERING AS COMPARED WITH ENVIRONMENTAL ENGINEERING

It may be useful to know clearly the characteristics of ecological engineering and to compare it in detail with environmental engineering and bioengineering (Table 1). Some comparisons are briefly given here.

THE BASIC UNIT OF TREATMENT AND STUDY

The basic unit of treatment and study in ecological engineering is the ecosystem or co-ecosystem, which is recognized as an organic whole with integrity. The holistic nature of an ecosystem mainly appears in three aspects, as follows:

Internal Correlation

An ecosystem is an organic whole and a complex system. It is composed of many organisms and abiotic components, including some producers, consumers, and decomposers, and many kinds of inorganic substances. These organisms engage in the regeneration and cycling of material in the ecosystem, e.g., H_2O, O_2, CO_2, N, P, K, Ca, Mg, Na, Si, Fe, S, etc. There are some organic compounds, e.g., proteins, lipids, starches, cellulose, and humus, which connect living beings with abiotic components in an ecosystem while serving as boundaries between the biotic and abiotic components. These components, parts, or units of the ecosystem establish interdependence, interconnectedness, interinhibition, and causal multiplicities (iterative loops) among themselves to make an organic whole.

Without exception, the appearance, action, function, and change of each component, part, or unit, is always affected, more or less, by other components in the ecosystem. They are the synthetic effects of many components, and may be the result of forces having two or more different components, or of all the causes combined into the complex process itself. There are internal correlations, as, for example, in the case of the appearance, chemical morphogenesis, distribution, concentration, and change of a nutrient element in an ecosystem. They may be effected by some physical procedure, e.g., sedimentation, resuspension, or diffusion resulting from water (or air) flow, and/or by some chemical procedure, e.g., oxidation or reduction, combination, or complexing with other substances, and/or by some biological procedure, e.g., absorption, assimilation, dissemination, and digestion by some organism in the ecosystem. Another example is the appearance, density, distribution, growth, reproduction, production, and enrichment of plants in an ecosystem as they are affected by the chemical morphogenesis, concentrations, and ratios of the various chemical substances in the ecosystem. Various actions (the function of every component) only appear and develop in the ecosystem but cannot appear alone apart from the ecosystem. The whole function of an ecosystem is the integrated effect synthesized by all or most of components of the ecosystem, and this effect is usually larger than that of the simple sum of all elements which comprise the ecosystem. *Therefore, the object for study and treatment in ecological engineering is not only one part of the ecosystem but also the whole, and not only immediate interests but long-term benefits.*

External Correlation

Most ecosystems usually belong to some kind of open system, that is to say, they maintain their complex forms and functions through a continuous exchange of materials, energy, and information with other surrounding ecosystems. An ecosystem is an organic component of the co-ecosystem, i.e., the interacting ecosystems, in the region where the ecosystem is located and is also a part of nature as a whole. Nature, as well as the co-ecosystem, is also an organic whole and a system which is larger and more complex than the simple ecosystem. The ecosystem which we will subject to ecological engineering cannot and should not exist apart from co-ecosystems in nature. There are many interdependent interconnections, interinhibitions, and mutual causality multiplications (iterative causal loops) among many ecosystems in the co-ecosystem and nature as a whole. It is certain that an ecosystem depends on the co-ecosystem in the region where it is located for its existence and development. *Therefore, the object for study and treatment in ecological engineering is not only the ecosystem itself but also the co-ecosystem.*

Harmony between Structures and Functions

As a result of the long-term and evolutionary development of the ecosystem, any homeostatic ecosystem in nature has a relatively steady internal structural organization. The structures are the framework and channels which are necessary to complete its functions. The functions may be defined as the mechanism for the transformation of a substance from input to output as well as the action for making the change or transformation of the state of the ecosystem or component. Functions are the basis for maintaining the existence of structures and for transferring and transforming material, energy, and information for structural development. The harmony between structure and function is a prerequisite for achieving a relatively steady state and dynamic equilibrium in an ecosystem.

According to the viewpoints of ecology and philosophy, the fundamental causality of the development, including the dynamics and changes, of an ecosystem or an object is not external but internal. It lies in a contradiction within an ecosystem or within the object itself. This internal contradiction exists in every ecosystem or every single thing, in its motion and development. Contradiction within an ecosystem or inside an object is the fundamental cause of its development, while its interrelation and interactions with other external factors are secondary causes, i.e., external causes are the conditions of change and internal causes are the bases of change (Mao 1937).

External causes operate through internal causes. At a suitable temperature, an egg changes into a chicken, but no temperature can change a stone into a chicken, because each has a different basis. From this viewpoint, an ecosystem change is mainly determined by the structure and function of the ecosystem, the internal causes, and the forcing functions or external causes, e.g., input of pollutants, heat, or other factors that are conditions of change.

From this principle, the key measures in ecological engineering involve adjusting and controlling the ecosystem. The idea is to adjust and control the structure and function to encourage cycling, raise the production, and increase the self-purification capability of the ecosystem. It is only regarded as a supplementary and subsidiary measure to adjust and control the forcing functions to control or change conditions. In environmental engineering, this is often considered as the main source of control for protecting the environment of a local region, i.e., to adjust the forcing functions, such as pollution sources, volume of water flushed and energy input. In bioengineering, adjustment and control of the genetic structure are the principal control measures.

BASIC PRINCIPLES

The basic principle used in ecological engineering is ecology — especially systems ecology — while many kinds of technologies are adopted and synthesized, such as agriculture, applied chemistry, and economics. The guiding principle of ecological engineering is a holistic and systematic viewpoint, whereas the basic principle of environmental engineering is mainly civil engineering, along with physics, chemistry, and biochemistry.

GOAL

The goal of ecological engineering is multi-objective: to maintain homeostasis in an ecosystem while improving or maintaining environmental quality, thus achieving high yield, low consumption, good quality, high efficiency production and a full utilization of wastes. However, in environmental engineering, cleaning up or totally preventing pollution *problems* and improving the environmental quality in a local region is only a mono-objective. In bioengineering, the goal often is to create a new species to satisfy some human needs.

DESIGN

Just as an organism or population, an ecosystem has self-organization. It may be considered that the ecosystem can transform chaos to order without external control. The essence of ecological engineering is managing this self-organization. Traditional engineering, including environmental engineering, replaces nature with new structures, functions, and processes. Ecological engineering provides manmade, but ecologically engineered, designs that use ecosystem structures, functions, and processes. Self-organization designs a mix of manmade and ecological components. The role of successful design in ecological engineering is to create a self-managed system.

COST AND ENERGY BASIS FOR MAINTENANCE AND DEVELOPMENT

Multi-Step Use of Nutrients

When structures and functions harmonize well with each other within an ecosystem, organisms not only establish many mutual-condition relationships (as, for example, in a food chain or food web), they also become specialized in their living habits, such as two species (1) occupying different ecological niches in the ecosystem and (2) making gradational utilization of the nutrients offered by the natural world according to their own particular needs. In the metabolic processes of homeostatic ecosystems, waste and remnants from primary production by species (or agent) A are used as the raw and processed materials of secondary production by another species (or agent) B. The waste and remnants from the secondary production are used as the raw and processed materials of tertiary production by the species (or agent) C, and so on. Production wastes are utilized according to sets of such patterns until all are fully utilized, or until the waste and remnants from the final production cannot be utilized.

In this ecosystem, the potential of biomass production would be fulfilled by multi-layer and gradational utilization of the nutrients, including waste and remnants produced from all or most of the production process. Analyzed from an economic and environmental protection view, this multi-step use of nutrient, wastes, and remnants in a homeostatic ecosystem is not only efficient in utilization, but also results in less, or no, waste. It causes less pollution or is free of it.

Similarly, in ecological engineering, the structure of an ecosystem or some part of a co-ecosystem is recombined and regulated according to ecological principles, in order to make the whole engineering consist of several process links. These may be vertically structured (with simple linear flow) or a net system linked transversely by several vertical technological process chains. In ecological engineering, an effective and beneficial cycling would be made or promoted. The remaining parts or wastes from the first link on which the primary product has been output would be the raw materials for the second link and similarly at the next link. The number of production links will depend on the economic benefit from nutrient conversion between links and can be calculated on the basis of input/output ratio. As a result, the potential of biomass production would be brought into full play, while the waste or pollutants would be used as raw material for producing some utilizable material or commodity. The space and time considerations of ecosystem design during the ecological engineering process would be optimally utilized.

It is desirable, in ecological engineering for conversion of wastewater into utilizable resources, to apply the above principle of harmony between structure and function. Multi-step use of substances, with their full use of waste, would result in high economic benefits for (1) regulating the disharmonious structure and function in a polluted body of water and (2) establishing a semi-artificial ecosystem or co-ecosystem to combine sewage treatment and the full utilization of the wastewater. These principles and ideas should also be applied to agroindustry production systems if they are to make full use of limited natural resources, in order to speed up the development of the human economy while protecting the ecological environment.

Energy Basis

Since ecosystems are solar-based, either directly or indirectly, the concept of self-sustaining systems is fundamental to ecological engineering. Once an ecological engineering solution is designed and implemented, it should be able to sustain itself indefinitely with only a modest amount of intervention in terms of subsidiary energy. This means that an ecosystem, running on solar energy or the products of solar energy, does not need to rely on technological energy as much as high-tech solutions, so that the operational expenditure for energy consumption is usually low.

Equipment and Investment

Investing in ecological engineering is usually less costly because, unlike conventional engineering, its toolbox is the whole ecosystem, a community of organisms that nature can offer which costs nothing and has little need for artificial equipment. As a result of this characteristic, ecological engineering has high yields, low consumption, high efficiency, and good quality production, with less capital and operating investment, and larger economic profits.

Environmental engineering is one of the artificial engineering systems. All or most of its equipment is to be manufactured and brought in by man and, therefore, investment is large. It usually does not gain a net profit because its mono-objective does not allow direct production of some kind of commodity, only the control and treatment of pollution in a local environment. Energy consumption for operation would be expensive because its energy sources depend on electric power, water power, or other energies based on fossil fuel. This is the main reason that environmental engineering cannot be universally applied in the developing countries or even in some more developed countries where energy sources are poor or expensive. Its economics, for the reasons given above, are usually run at a loss and require allocating funds from government, business, or industry for construction and operational support. This may be one of the limiting factors to the spread and application of environmental engineering.

ENVIRONMENTAL ADVANTAGE

Since the natural environment consists of many ecosystems and is a dynamic system with self-organization and self-regulation of functions and effective and beneficial cycling of nutrients, not only can the damaged ecosystem be improved, it will be restored gradually to its normal material metabolism and regenerative state.

Although it is certain that some advantage to a local environment or region will occur through pollution treatment and control by environmental engineering, it may often indirectly give some disadvantage to the ecosystem or co-ecosystem, either through transference of pollutants produced by environmental engineering, such as flushed water and dredged sludge, or by some new pollution produced from another industry supplying the energy or equipment to the environmental engineering. It may result, more or less, in disturbance to the whole. Ecological engineering can not only minimize waste to prevent pollution and treat some pollution through the conversion of waste or pollutants into raw materials or utilizable resources, but also can promote the effective and beneficial cycling of nutrients within an ecosystem or regional co-ecosystem. If ecological engineering were to be widely applied in all or most regions, it should result in the advantage of environmental protection to the ecosystems of individual districts, as well as the world as a whole.

DEVELOPMENT AND POPULARIZATION OF ECOLOGICAL ENGINEERING, WITH CASE STUDIES FROM CHINA

ECOLOGICAL ENGINEERING IN CHINA

In China, ecological engineering was first independently proposed by Professor Ma, in the 1970s (Ma 1988). Its concept and main principle is accepted by the policy makers in the national government and many professionals. The principles have been put into play in the national strategy for environmental conservation, which contains overall plans for implementation and rational deployment. Comprehensive utilization and changing waste material into things of value turns bane into boon. Environmental conservation depends on the masses and application by everyone; protecting the environment is of benefit to the people. Under the guidance of this strategy, the dissemination of the knowledge and techniques of ecological engineering resulted in ecological engineering becoming popular quickly all over China. Before 1990, over 1600 sites of ecological engineering had been founded in 29 provinces or cities in China, including about 1200 agroecological engineering hamlets (or farms) and 273 ecological engineering villages belonging to 171 counties, and 19 agroecological engineering counties. Over 100 sites of ecological engineering for waste treatment and utilization were created. In practice the area encompassed by these sites is over 24,934 km^2 of farmland (1.69% of all the farmland in China), 1647 km^2 of forest and grassland, and 76 km^2 of inland water bodies, and has a population of 13,140,000.

CATEGORIES OF ECOLOGICAL ENGINEERING IN CHINA

According to the nature and main objective of the project, ecological engineering should fall roughly into one of five categories as follows: (1) Multi-layer and gradational utilization of waste in all or most of the process of production, such as the case study of the Dong Feng State Farm ecological engineering in Shanghai. (2) Pollution-free technology, the self-purifying and material-regenerating system in the factory, such as the case study of retrieving waste rare metal (silver) in a film studio in Wuxi. (3) The cycling and regeneration of nutrients such as the case study of a mulberry grove-fishpond model. (4) Regional multi-functional self-purification systems for sewage, such as the case study of sewage irrigation-treatment with utilization of sewage by the soil system, and water hyacinth. (5) The multi-functional agroindustrial combined production system, such as the case study of treatment and utilization of wastewater from an organophosphate pesticide and BHC (benzene hexachloride) manufacturing process, and some beneficial cycling in a village containing farming, forestry, animal husbandry, sideline production, and fish ponds. A synchronous development of economic benefit and ecological, environmental and social advantage have been gained in all ecological engineering sites (Dai et al. 1990, Ma 1985, Ma and Li 1987, Ma and Yan 1989, Yan and Yao 1989, Zhu et al. 1988).

SOME CASE STUDIES IN CHINA ON ECOLOGICAL ENGINEERING

Multi-Layer and Gradational Utilization of Waste in All or Most Processes of Production

The demonstration, study, and practice on Dong Feng State Farm Ecological Engineering in Shanghai — situated at middle part of Chong Ming Island in the mouth of the Yangtze River, about 80 km apart from the urban district — is a case study of this type. This farm possesses an area of 2664 ha of land, of which 1600 ha is under cultivation. The population of 15,000 inhabitants includes 12,000 employees and workers, among which 6000 persons are engaged in industrial work and 3000 in management, trading, marketing, and service for the above two production systems. The multi-functional agroindustrial combined production system (Figure 1) is as follows:

The first grade of this production system is farming. Most of the products, with all byproducts or waste from the first step of production, are used as the raw materials for the secondary grade of production, such as the annual production of 10,000 tons of grains processed into rice and flour, 8000 tons of green forage, and 12,500 tons of straw from rice and wheat fields which are used as cow feed in the dairy farm. Fifteen hundred tons of rape seed are sent to the oil extracting mill to produce vegetable oil, in addition to various kinds of vegetables which directly supply the needs of the market. Most of the products of the second step of production are sold directly, such as milk, vegetable oil, rice, flour, pork, eggs, and fish.

The waste or byproducts from secondary grade production are used as the raw materials for other production processes. For example, about 14,000 tons of manure are discharged annually by 750 cows in the dairy farm. Quite a lot of excrement and residues from the pig and poultry farms are used as the raw materials for a mesothermic biogas generation system. Therefore, a highly efficient and stable output of biogas is realized all year round. The average annual rate of gas output is above 0.7 m^3/m^3 of cells each day (the highest rate reaching 1.18 m^3/m^3/day). Sixteen tanks of biogas fermentation cells (altogether 1325 m^3) annually produce 572,240 m^3 of biogas, which supply about 1000 families with heat for their uninterrupted daily use all year round. Since 1985, when biogas generators were established and began operation, the production of biogas has provided more convenience in daily life for the inhabitants using biogas on the farm, just as most city dwellers use coal gas. This not only saved much time in preparing meals, it also gave a modernized daily life more leisure and saved about 700 tons of coal every year for other uses. Environmental progress has been made by reducing much of the air pollution, such as sulfur dioxide and nitrous oxide, produced by the combustion of coal and manures. This particular utilization avoided some organic pollution.

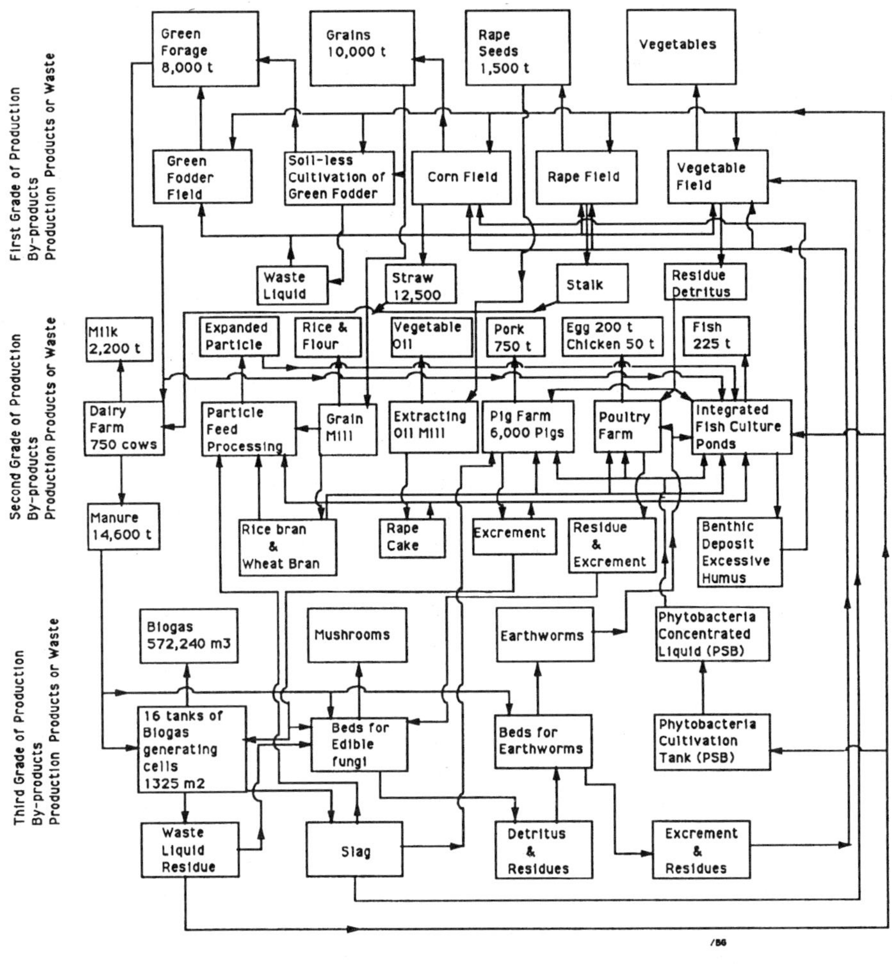

FIGURE 1. Model of a multi-step utilization of nutrients.

On the basis of testing and analysis, the solid slag from biogas fermentation, as a solid waste, is rich in nutrients, containing 17.58% raw proteins — more than that in its raw materials (8.10%) — and high in B vitamins. Its content of vitamin B_{12} exceeds that of fish meal by a factor of 11.0. The biogas slag has been used as feed for pigs to replace 20 to 25% of the fine feed. This slag is not only palatable for pigs, but also gains a higher feed efficiency by promoting the growth rate of pigs, up to 600 g/pig/day. In a production operation of 6000 pigs, this results in a total of 40,000 yuan in additional income for one year and saves 1300 tons of fine feeds for the pigs. The biogas slag can also be used as a part of the raw material in processing expanded particulate feed for fish. It promotes the fish production by 2 to 3% to feed with this kind of expanded particle feed, displacing the expensive commercial raw material used for food pellets.

Furthermore, the waste slag from biogas production is a high quality and quick-acting fertilizer, easily absorbed and utilized by vegetables. According to analysis, its composition consists of 30% organic matter, nitrogen, phosphorus, potassium, and humus, as well. When the slag was applied to cultivating superior quality vegetables, it not only reduced the use of chemical fertilizer (lowering costs and improving the soil quality) but also greatly increased yields, up to 60 tons of vegetables per hectare. Under these conditions, both the tomatoes and potatoes grew to 0.8 kg each and the largest cauliflowers grew to 3.5 kg each. They are worthy of the name "superior quality vegetables" because of their larger size and quality, without any distortion, pest control, or pollution by chemical fertilizer and pesticides.

The waste liquid residue from biogas production can be used as the culture solution for soilless cultivation, which should be neither influenced by the change of season and weather, nor limited by land. This would surely save the area of field needed for cultivating feed for cows and solve the problem of a seasonal lack of fresh green forage, especially in winter, when green fodder is in short supply and expensive.

The waste slurry from biogas production can also be used as fertilizer. In a study on the application of this slurry as a topdressing, it was used in the cultivation of edible fungi at 50% of the media. The annual yield of mushrooms in the experimental group exceeded that in the control group, which was irrigated with common water, by a factor of 2.4. The slurry can be also used as culture solution for cultivation of photosynthetic bacteria (PSB) for another kind of higher protein feed. The experiments were successful in feeding prawns, chickens, and hens by using the PSB as an additive, which also increased the disease resistance, survival rate, growth rate, and yield of these animals.

Rice bran, wheat bran, and rape seed cake are some byproducts or wastes of processing the grains and rape seeds. All of these byproducts are used as forage for pigs, poultry, and fish directly or through processing them into particle feeds mixed with other materials. The chicken manure is a waste from poultry farms. It contains, however, plenty of proteins and other nutrients. An experiment which raised hogs by use of different proportions of chicken manure to ordinary fodders showed that it did not lower either the growth rate or pork quality to add 15 to 30% chicken manure (fresh weight) to the forage and straw (in a ratio of one to one) with additional amendments of 4% rape cake, 4% plaster stone powder, and 0.5% calcium superphosphate. All are mixed together and decomposed until it becomes saprophytic.

The residue and detritus of edible fungus beds after harvesting the mushroom can be used as a medium for earthworm breeding or mixed with other kinds of waste, such as the manure of pigs, cows, and poultry. The earthworm is a good quality bait for fish and poultry feed, and it can also be processed into a kind of traditional Chinese medicine. The residue and detritus from earthworm beds is also a good fertilizer to improve soil.

This case study shows a type of system which simulates the principles of cycling and regeneration of nutrients in the co-ecosystem and in which the potential for production (including waste) is brought into full play and maximized while most of the waste (or pollutants) are converted to utilizable raw materials, resulting in less pollution. Their economic benefits and environmental advantages would be achieved synchronistically by way of combining several production links belonging to farming, sideline production, animal husbandry, fish culture, and industry, respectively.

Pollution-Free Technology: The Self-Purification and Material Regeneration System in the Factory or Industrial City

The case study of this type of ecological engineering is the study and practice of utilizing waste rare metal (silver) at Wuxi film studio. The water hyacinth (*Eichhornia crassipes*) and some other species of aquatic plants are used as a living medium to absorb and concentrate the silver from wastewater discharged from the film studio. The concentration of silver in this wastewater is only 20 to 280 ppb, which cannot be retrieved by general chemical techniques, but it will harm most species of aquatic organisms. This wastewater, which contains trace concentrations of silver, was cultured through water hyacinths for 46 to 72 h. The highest enrichment coefficient is 34,750. The highest accumulated amount of silver occurred in the root system of this floating hydrophyte where it reached 12,000 ppm in dry weight and 4.125% in its ash. The retrieval rate of silver attained 95 to 99%. Then the silver was recovered from the ash by general chemical methods. The result of this technique was not only the purification effect and the environmental advantage of the 90 to 98% removal of silver from the wastewater but also the reduction of BOD, COD, TN, TP, and TSS. The net result is the economic benefit of recovering the silver, and the residue after recovery could be used as the raw materials for making brick.

Another case study is the use of waste heat (Figure 2). Waste heat discharged from electric power plants and some factories is a common source of environmental pollution. Utilizing waste heat has occurred in many urban areas in China. If greenhouses which require different temperature gradients are built near factories that discharge waste heat, their waste heat can be used, based on the heat-diffusing coefficient, to cultivate certain aquatic plants and fishes for overwintering, accelerating breeding, and increasing production. For example, in Yangzhou, and Suzhou, China, the Nile tilapia (*Tilapia nilotica*), which cannot overwinter naturally in eastern China, is cultured in greenhouses built near electric power plants. The waste heat is used for overwintering, advancing maturity, breeding, and increasing production of this fish. In some industrial cities, such as in Shenyang, Beijing, and Maanshan, the waste heat from some factories is used as a heat source in winter in surrounding urban areas.

An important aspect of ecological engineering is that comprehensive pollution-free production in the factories can be attained by combining many kinds of specialized workshops or factories and unifying technological processes or production lines. This is based on simulating the principles of endless cycling of materials and constant regeneration of organisms in the ecosystem, in order to increase production and practice economy with a decrease in, or the absence of, pollution through the full utilization of waste materials and energy.

Cycling and Regeneration of Nutrients

Food chains involve the basic structure of an ecosystem. The regeneration and cycling of a nutrient in an ecosystem is completed through the whole metabolic process of primary production, secondary production, processing, and decomposition. The mulberry grove and fishpond system (Figure 3) is an effective multi-objective measure in agriculture in the Guangdong, Jiangsu, and Zhejiang provinces of China. The white mulberry tree (*Morus alba*) produces organic material (mulberry

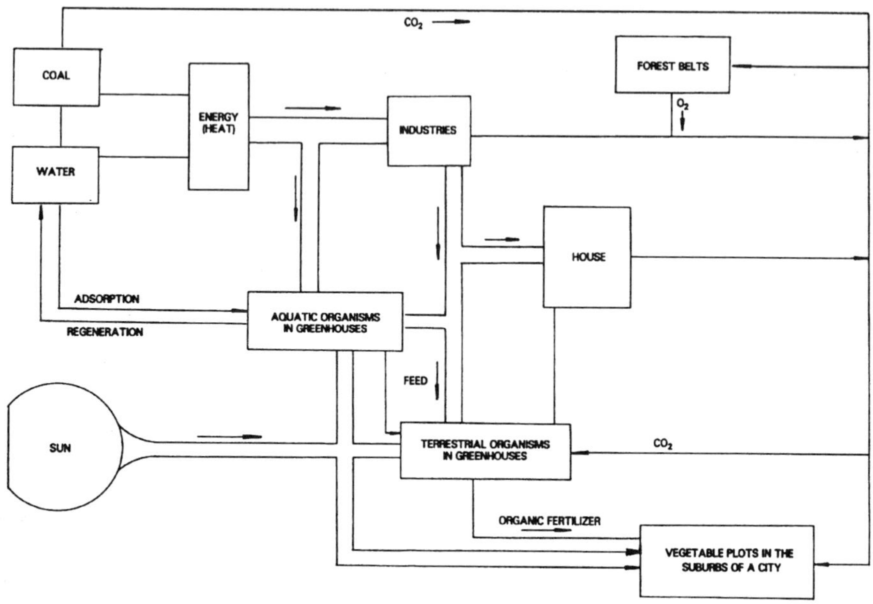

FIGURE 2. Model of the self-purifying system of the environment in industrial cities, our second type of application of ecological engineering principles (Ma 1985).

leaves, mulberry fruit, etc.) through photosynthesis. The mulberry leaves are used to feed silkworms (*Bombyx mori*), which in turn produce their silk and chrysalises; the mulberry fruit is one of the traditional Chinese medicines and is also used as a raw material in making wine, in conjunction with wine and distillers' grains. The withered and fallen parts of the mulberry and the distillers' grains from the brewery are applied to the fish pond, where they are converted into fish biomass through the process of another food chain.

In the integrated fish culture pond, polyculture, reasonable stocking, and rearing density are the key measures guaranteeing the regeneration and cycling of nutrients. Generally, five to eight (and sometimes more than ten) different species of fish are reared in the pond. These species of fish are different in feeding habits and inhabit many water layers. For example, the silver carp like to live in the upper layer of water and feed mainly on phytoplankton; the bighead carp and grass carp like to live in the middle and upper layers of water, with the former mainly feeding on zooplankton and the latter on grass and the withered and fallen parts of the mulberry trees. The black carp is carnivorous and feeds on zoobenthos and chrysalises of silkworms, and it inhabits the bottom of the water column. Common carp, pangolin, and tilapia are omnivorous and can eat the detritus of withered and fallen parts of mulberry, the excrements of the silkworms, and the distillers' grains from the brewery.

This kind of polyculture is not only a vertical utilization of various strata of water in the pond, but also promotes the regeneration and cycling of nutrients. These sideline wastes are products from mulberry groves and silkworm culture. Processing products are used to feed some species of fish in the pond, some excrements of

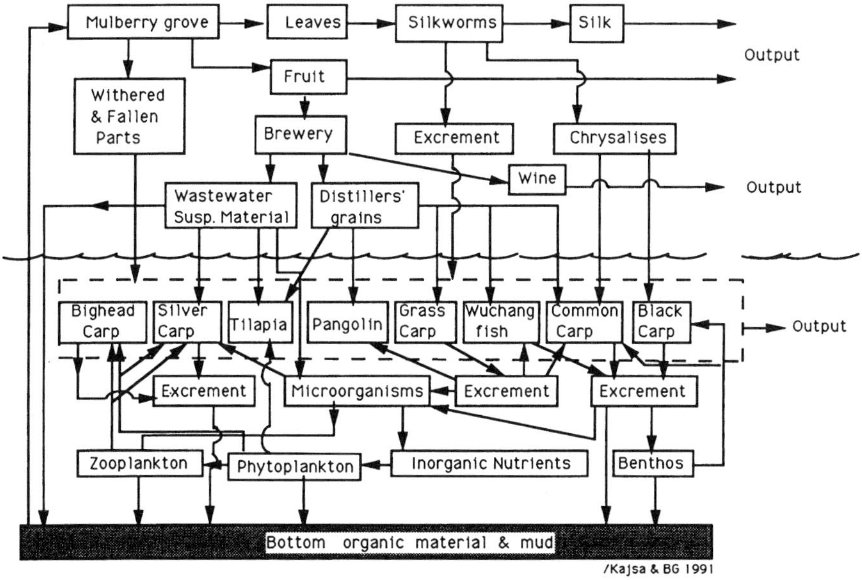

FIGURE 3. Model of cycling and regeneration in the mulberry grove-fishpond system.

black carp and grass carp are taken by omnivorous fish, and some benthic organisms such as aquatic oligochaetes and chironomid larvae are eaten by black carp, common carp, and crucian carp. Another part of the excrement of these fish and animals, as well as other unused organic matter, is decomposed by microorganisms and transformed into inorganic nutrients, and resulting in increased phytoplankton production, which is used to supply food for silver carp and many species of zooplankton. The zooplankton feed bighead carp and some omnivores. The remainder of organic matter and bottom mud is removed from the pond in winter and is returned to the mulberry grove, to be used as fertilizer.

Regional Multi-Functional Self-Purification Sewage Systems

One type of ecological engineering is irrigation treatment and utilization, as in the example of the case study of soil application for sewage treatment. The farmland, grassland, and forest ecosystems, all soil-based, can not only degrade and purify many kinds of pollutants in sewage, but also produce many useful products at the same time (Figure 4). There are numerous species of microorganisms that can decompose many kinds of organic materials in soil. Many species of plants grown on soil can absorb the nutrient salts decomposed from organic materials and contained in sewage. This results in an increased production of many useful farm products, forage grasses, or forest products, as well as sewage purification. The soil itself also seems to be a natural sieve, effectively filtering out most of the pollutants in sewage, including suspended solids, degradable organic pollutants, nutrient salts, and some pathogenic bacteria and viruses. The removal rate of pollutants may exceed 90% for BOD, TN, TSS, and coliform, and near 50 to 80% for TP.

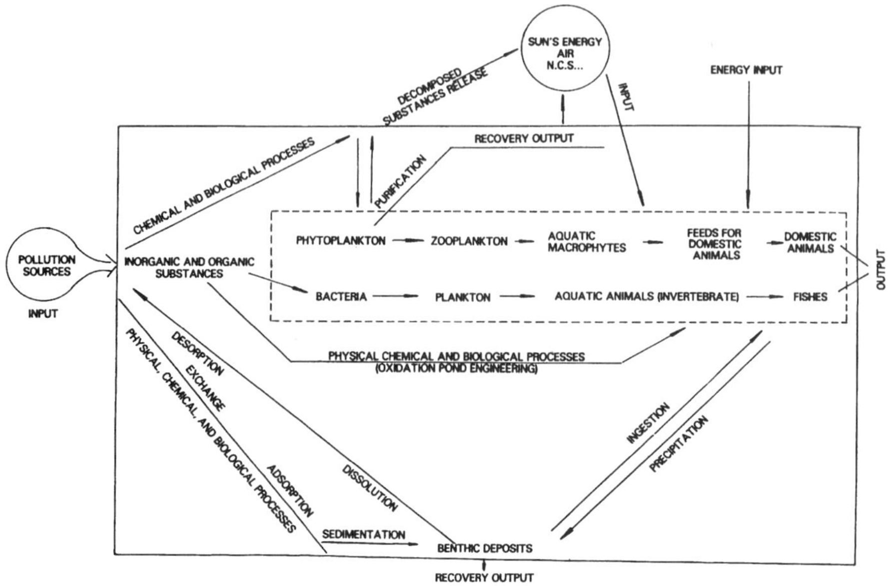

FIGURE 4. Schema of the self-purifying process of sewage, our fourth type of application of ecological engineering principles (Ma 1985).

In China, the area used for sewage irrigation was 42,000 ha in 1963, 105,333 ha in 1977, and developed to more than 1,400,000 ha in 1984. The use of sewage for irrigation is necessary due to the scarcity of water sources, especially in dry or semiarid regions, and supplies needed water for the development of industry and agriculture. It also increases farm production and reduces fertilizer expenses by utilizing nitrogen, phosphorus, potassium, and some trace elements from the sewage which are required by plants. The farm ecosystem in a sewage irrigation region plays an important role in decreasing pollution in the river system in some regions, because of the purification capacity described above.

Some heavy metals and man-made organic materials, such as pesticides, which are degraded with difficulty by the soil system, may become concentrated in the farm products grown in the regions of sewage irrigation. When the content of such heavy metals and man-made organic materials is greater than the standard for surface water, they can enter the food chain. In this case, it is necessary that pre-treatment of sewage should be undertaken to avoid harm to humans but still keep the benefits of utilization of wastewater. It is inadmissible for such sewage to be used to irrigate farms directly. Such pre-treatment may be through oxidation in waste stabilization (aerobic treatment) ponds or anaerobic ponds and sedimentation ponds. It is acceptable as well that some pre-treatment in discharge channels is used instead of sediment ponds or aerobic ponds, to bring self-purification of the channel into full play. It is also acceptable, as is happening in some regions in China, that sewage containing a large quantity of non-degradable pollutants should be discharged directly to timber forests or reed marshes for irrigation to purify the sewage, as absorption of these pollutants by non-food plants prevents them from entering the food chain, including humans.

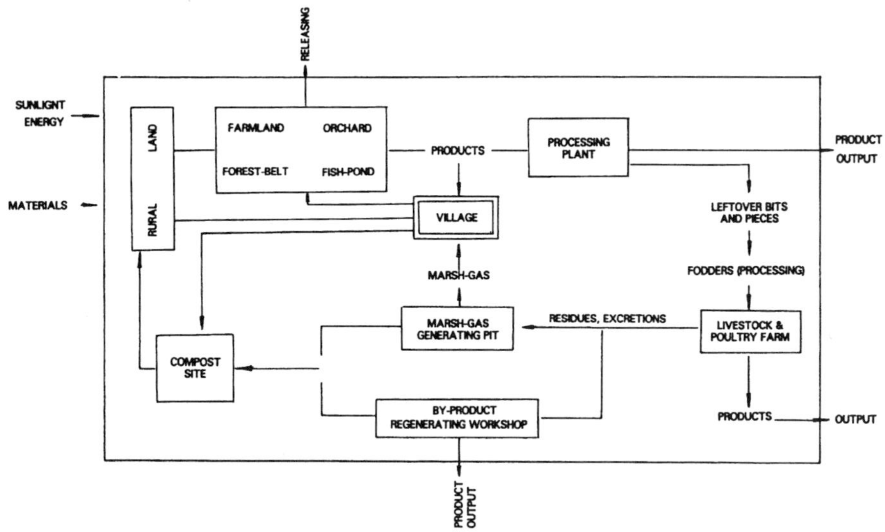

FIGURE 5. Model of the general structure of an agroindustrial combined production system (Ma 1985).

The Multi-Functional Agroindustrial Combined Production System

The principle of an endless cycling of materials and a continuous regeneration of organisms in ecosystems may be applied to the processing industry, with agricultural products as raw materials, so that agroindustrial products (including byproducts) compensate for raw materials. In this manner, the production system of a region may be kept relatively stable, with the reduction of wastes, prevention of pollution and improvement of rural ecological environments.

Figure 5 presents an ideal model of the general structure of an agroindustry combined with a production system, including farming, forestry, animal husbandry, sideline production, and fishponds, as well as the planning of the village. The quantity of products from farming, forestry, animal husbandry, sideline production, and fishponds as well as the scope of the processing industry, should be kept properly proportional to the human population and planned output values. This "organic whole" type of agroindustry combined with production systems may be regarded as an ideal model of rural development.

In fact, such technical systems have long been employed extensively in China, particularly in village industries. As a result, great benefits have accrued on both the economic and environmental sides, including a significant reduction of pollution.

REFERENCES

Dai, Q. Y., W. L. Dai, H. Lu, and X. Gao. 1990. "Study of the Purification and Accumulation of Silver in Wastewater by Aquatic Plants." *Acta Ecologica Sinica* 10(4): 343-349.

Jørgensen, S. E. and W. J. Mitsch. 1989. "Ecological Engineering Principles." In Mitsch, W.J. and Jørgensen, S.E. (eds.). *Ecological Engineering*. John Wiley & Sons, Inc. New York.

Ma, S. J. 1985. "Ecological Engineering. Application of Ecosystem Principle." *Environmental Conservation* 12(4): 331-335.

Ma, S.J. 1988. "Development of agroecological engineering in China." In: S.J. Ma, A. Jiang, R. Xu, and D. Li (eds.). *Proceedings of Int. Symp. on Agro-Ecological Engineering, August 1988.* Ecological Society of Beijing. pp. 1-13.

Ma, S. J., and S. H. Li, ed. 1987. *Agroecological Engineering in China*. Sci. Publ. House. Beijing.

Ma, S. J., and J.S. Yan. 1989. "Ecological Engineering for Treatment and Utilization of Wastewater." In Mitsch, W. and S. E. Jørgensen. *Ecological Engineering.* John Wiley & Sons, Inc. New York. pp. 185-217.

Mao, Zedong. 1937. "On Contradiction." *Selected Works of Mao Tse-Tung, Vol.1*:. pp. 274-312.

Yan, J. S., 1986. "The Main Principles and Types of Ecological Engineering for the Conversion of Wastewater into Resources." *Rural Ecologica Eco-environment.* 8:40-44.

Yan, J. S. 1987. "Studies on the ecological engineering for the conversion of wastewater into utilizable resources." *Bull. of Nanjing Inst. of Geography, Acad. Sinica.* 1:12-14.

Yan, J. S., and H. L. Yao, 1989. "Integrated Fish Culture Management in China." In W.J. Mitsch and S. E. Jørgensen. *Ecological Engineering.* John Wiley & Sons, Inc. New York. pp. 375-408.

Zhu, Z. Y., D. T. Li, and P. J. Yu. 1988. "A Demonstrative Study of Ecological Engineering in an Animal Farm." *Proceedings of the International Symposium on Agroecological Engineering.* Beijing, China.

Chapter 3

AQUACULTURE AND ITS ROLE IN ECOLOGICAL WASTEWATER TREATMENT

Li Sifa

ABSTRACT

The Chinese freshwater fish culture system is dominated by herbivorous filter-feeding fish (carp), in contrast to the systems in western countries, which are dominated by carnivorous fish (trout). The Chinese system uses polyculture techniques, managed as integrated fish farming. Grass carp is probably the most voracious herbivore, consuming relatively large macrophytes. Silver carp, bighead, and crucian carp utilize not only plankton, but also bacteria, thus improving water quality. When these fish are stocked together, a more effective polyculture is formed.

From a wastewater treatment point of view, a system dominated by herbivorous filter-feeding fish may be the answer where the nontoxic or detoxified wastewaters are first treated by macrophytes and algae. This offers a low-cost, highly efficient engineering approach.

INTRODUCTION

Aquaculture is the farming of aquatic organisms, including fish, molluscs, crustaceans, and aquatic plants. Farming implies some forms of intervention in the rearing process to enhance production, such as regular stocking, feeding, protection from predators, etc. (FAO 1990).

Aquaculture is widespread around the world. The world aquaculture production in 1988 was:

- Finfish: 7,114,597 metric tons
- Crustaceans: 612,200 tons
- Molluscs: 3,073,205 tons
- Seaweed: 3,632,281 tons (FAO 1990)

China is the leading country (6,658,686 t in 1988), followed by Japan (1,424,832 t in 1988). Of the yields from aquaculture, 49% are from marine water, 44% from freshwater, and 7% from brackish water (FAO 1990).

In addition to producing aquatic products to meet human demands, aquaculture takes up a very specific position among managed aquatic ecosystems. Chinese freshwater aquaculture systems have only recently been acknowledged as an approach to the ecologically engineered treatment of wastewater.

0-87371-990-5/97/$0.00+$.50

CHARACTERISTICS OF CHINESE FRESHWATER AQUACULTURE

From a wastewater treatment point of view, the following characteristics of Chinese freshwater aquaculture should be mentioned.

HERBIVOROUS FILTER-FEEDING FISH DOMINATED SYSTEMS

Asia has the widest range of commercially exploited herbivorous species. The grass carp (*Ctenopharyngodon idella*), probably the most voracious herbivorous fish in the world, is generally the dominant species in areas with abundant vegetation (Lin 1954). The herbivorous wuchang fish (*Megalobrama amblycephala*) and Chinese bream (*Parabramis pekinensis*) are also principal species in the Chinese fish polyculture system.

Indian carp, catla (*Catla catla*), rohu (*Labeo rohu*), and mrigal (*Cirrhina mrigala*) may feed to a certain extent on either fresh or decomposing higher plants (Huet 1972; Jhingran and Pullin 1985), but much less voraciously than the grass carp.

Among the fish employed for farming in Africa and in the majority of other countries in the tropics, various species of the genus *Tilapia* predominate. Due to a number of characteristic traits, the *Tilapia* are ideal species for fish culture in tropical regions. From the point of view of feeding, two groups of species have been identified:

1. Herbivorous and macrophytophagous forms (e.g., *T. rendalli, T. zillii, T. tholloni*);
2. Microphagous and omnivorous forms (e.g., *T. macrochir, T. mozambica, T. nilotica., T. nigra*. Huet 1966, 1972).

However, herbivorous tilapia are not cultured popularly. The tambaqui (*Colossoma bidens*), silver dollar fish (*Metynnis roosevelti*, *Mylossoma argenteum*), and pirapitinga (*Mylossoma bidens*) in the Amazon River basins in South America may also have potential as cultivated herbivores (Edwards 1980).

Herbivorous cultivated fish of European and North American origin which feed on macrophytes are practically non-existent (Huet 1972; McLarney 1984). According to Nikolsky (1963), there are no herbivorous fish in high latitudes because the growing period for vegetation is short, but at the lower latitudes there are many herbivores because of year-round vegetation growth.

Herbivorous macrophytic fish have a toothless mouth, but they do have strong and specialized pharyngeal teeth arranged in two rows, those on the lower row enhanced by doubled and flattened serrated cutting and rasping surfaces. The grass carp has no stomach and the gut is only two to three times the length of the fish, exceedingly short for an herbivore. The macrophytic tilapias have bicuspid and tricuspid teeth on the jaws and small sharp pharyngeal teeth which shred vegetation and rupture some of the cells (Trewavas, 1982).

Grass carp have a broad range of feeding preferences with regard to macrophytes. The most favored are soft plants such as: submerged aquatic macrophytes — *Vallisneria spiralis, Hydrilla verticillata, Potamogeton malainus, Potamogeton crispus, Potamogeton maackianus, Najas minor*; duckweeds — *Wolffia arrhiza, Lemna minor, Spirodela polyrrhiza*; grasses — such as perennial ryegrass (*Lolium perenne*),

TABLE 1
Daily Consumption of Macrophytes by Grass Carp

Fish size	Daily consumption (% body weight/day)	Food species	Author
1.00–1.25 kg	100–174	Aquatic macrophytes	Bhatia 1970
1 kg	100	Aquatic macrophytes	Anon. 1989
	100–125	Submerged aquatic macrophytes	Venkatesh & Shetty 1978
6 kg	26–28	*Hydrilla verticillata*	Shireman & Maceina 1980
1837 g	99	*Najas foveolata*	Mehta et al. 1976
930 g	131	*Najas foveolata*	Mehta et al. 1976
96 g	146	*Najas foveolata*	Mehta et al. 1976
1208 g	140	*Hydrilla verticillata*	Mehta et al. 1976
76 g	27	*Hydrilla verticillata*	Mehta et al. 1976
4.9–650 g	26.3–49.9	*Wolffia*, vegetable	Li et al. 1980

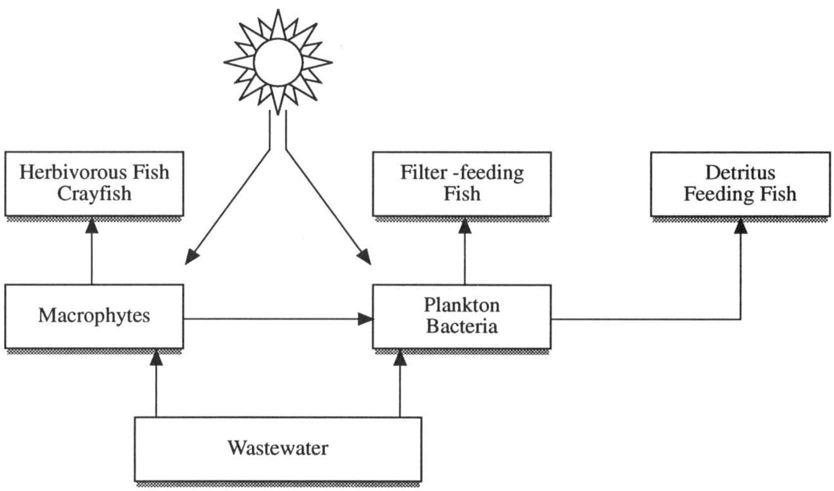

FIGURE 1. Wastewater–macrophytes–filter feeding fish system.

bunch grass (*Symphytum peregroinum*), sudan grass (*Sorghum sudanese*); vegetables; and the soft leaves of terrestrial plants.

Grass carp consume relatively large amounts of macrophytes (Table 1) due to the high water content of the ingested material and to the low digestion rate afforded by the short gut and a lack of cellulase. Body size and food type are important determinants of consumption rate.

Among planktivorous fish in Chinese polyculture systems, the principals are silver carp (*Hypophthalmichthys molitrix*), bighead (*Aristichthys nobilis*), and silver crucian carp (*Carassius auratus cuvieri*).

Silver carp has the shortest food chain, feeding on particles of 10 to 60 μm in size, including algae, bacteria floccule, and detritus, at the daily consumption rate of 10 to 20% of body weight. It is well known that plankton is the main food of silver carp, but the role of bacteria has only recently been studied. In fertile fish ponds, bacteria are abundant and form floccule by agglutination. Kyshezov (1977) reported that, in such waters, half of the pelagic bacteria floccules can be eaten directly by silver carp. Zhu (1989) studied the digestion and absorption rates of bacteria floccules by silver carp and Nile tilapia (*Oreochromis niloticus*) using Carbon-14 and found them to be 45.1% and 81.4%, respectively. In ponds with silver carp, the amount of bacteria is maintained at a level of 18 to 25 million cells/ml, indicating that cultivation of silver carp can improve water quality (He 1987). Bighead have similar feeding habits to silver carp but depend largely on zooplankton. This species is very efficient at using the food it ingests.

Because of the gill raker size and the presence of a translucent mucous coating on the gill rakers in these two species, silver carp and bighead can collect food particles as small as 10 μm in diameter. Large food particles such as zooplankton, large colonial algae, and large detrital particles have sufficient bulk to pass over the top of the gill raker coating directly into the gullet. Smaller food particles become embedded in the mucus and form aggregates that increase in size toward the distal end of the raker assembly and then pass to the gullet. Pharyngeal teeth grind plankton to allow more efficient digestion of usable protein.

Silver crucian carp are highly adaptable. In ponds, fish under the standard length of 26 mm feed mainly on zooplankton, such as rotifers (*Rolyarthra, Keratella, Triarthra, Diphanosoma, Brachionus, Polyarthra*), and crustaceans (*Daphnia, Moina, Bosmina, Alona*); when the length exceeds 26 mm, feeding shifts to detritus and phytoplankton (Ding 1988).

POLYCULTURE SYSTEMS

The traditional Chinese system of polyculture comprised seven to ten species with different feeding and habit niches and received a variety of organic inputs: macrophytes, manure (livestock and human), and agricultural byproducts (oil cakes, rice bran, silkworm pupae).

A mixed culture of Chinese carp — particularly grass carp, silver carp, bighead, bream, black carp, common carp, crucian carp, and mud carp — is an important principle and the technique is widespread. Its advantage is to utilize the space and food available in the pond as well as to maximize the beneficial interactions between compatible species with different feeding habits and ecologies.

The ecological interaction of a polyculture is expressed by a Chinese proverb: "one grass carp supports three silver carp" (Lin,1982). Due to the inefficient digestion of the grass carp, about half of the plant material, broken into fragments less than 3 mm in area, "passes out as feces which may support, directly or indirectly, an equal biomass of other species of fish, including plankton-eating cyprinids" (Hickling 1966). Thus, the grass carp has been referred to as a "living manuring machine" (Hickling 1971). On the other hand, if grass carp are not raised in a polyculture which includes filter-feeding fish, the water could become too fertile for grass carp to thrive in. Therefore, to optimize a macrophyte-fed aquaculture system, a polyculture of macrophagous and microphagous fish is necessary.

INTEGRATED FISH FARMING

Integrated fish farming is the comprehensive management of aquaculture, animal husbandry, and agriculture. These systems are guided by ecological theory and evaluated by economic and social efficiency. The water and land ecosystem is treated as a whole, with the aim of producing valuable protein from waste, converting ecological damage into benefit, and sustaining local recycling of resources. This strategy of ecological aquaculture can not only increase fish production and further improve ecological efficiency, but also improve social and economic efficiency. It is not only useful in the development of fish culture but will also help to mitigate environmental degradation, which is becoming serious.

USING WASTEWATER IN AQUACULTURE

The daily wastewater discharge in China is about 85 million tons. From the majority of cities in China, most of the wastewater is discharged directly into rivers, lakes, and seas without treatment. For example, in Shanghai, the largest industrial city in China, the daily discharge of wastewater is about 5 million tons, of which only 200,000 tons has been treated. About 4 million tons of untreated wastewater is discharged into the Wangpu River, a major river close to the city, and from there to the East China Sea. Since China is in its initial stage of industrialization, and its enormous population is steadily growing, the lack of wastewater treatment is a serious problem.

In general, there are three methods for the treatment of wastewater:

1. Natural self-purification of bodies of water. Since natural waters are limited and wastewater increases year by year, their ability to purify wastewater is limited and this kind of treatment should no longer be used, particularly in inland waters.
2. Artificial chemical or mechanical regeneration systems. Wastewater is held in a limited space and a series of physical and biochemical treatment processes are used for purification. This method regards wastes as non-recoverable, is managed at considerable cost, and needs large capital investment as well as energy, and is therefore difficult to conduct in developing countries.
3. Ecological engineering. This is based on ecological principles which convert wastewater into utilizable resources. This method should have a very bright future and the highest socioeconomic and ecological benefits.

Some schemes for ecologically engineered treatment have already been designed and employed. Examples include the Stensund aquaculture facility in Sweden, Solar Aquatics in New England (Guterstam and Todd 1990), and the conversion of wastewater into utilizable resources by water hyacinths (Zhang et al. 1986). Laboratory and pilot scale trials on wastewater-fed ponds have been successful in Thailand, Australia, the U.S., Israel, the Philippines, India, and Singapore (Lee et al. 1980). But to develop practical prototype-scale systems, a critical mass of expertise, facilities, and financial supports is required.

From a fisheries standpoint, Chinese fish culture is a well-known technique. From the wastewater treatment point of view, I believe that the polyculture system has great potential. If we can introduce its ecological principles into wastewater treatment, and improve its techniques properly, it will become a low-cost, highly efficient approach to wastewater treatment.

TABLE 2
Domestic Wastewater Quality of Some Large and Middle-Sized Cities in China[a]

	Beijing	Shanghai	Xian	Wuhan	Gungzhou	Saxi
SS	50–327	320.7	—	66–330	—	125
COD	30–88	—	—	52–64	25.4–36.3	117.1
BOD_5	90–180	360	—	320–338	18.2–23.8	63.7
NH_4-N	25–45	47.1	21.7–32.5	15–69.3	5.6–8.0	—
Cl	124–128	141.5	80–105	—	21–31	147–182.1
P	30–34.6	—	—	—	2.4–5.2	—
K	17.7–22	—	—	—	—	—
pH	7.4–7.7	7.3	7.3–7.9	7.1–7.6	6.7–7.5	7.0–7.8
Sludge proportion (%)	0.8–1.6	—	1.5–1.6	0.2	—	—

[a] Units are mg/l except for pH and sludge proportion.

Source: Jing and Zheng, 1989.

TABLE 3
Nutritient Content of Wastewater Averaged over 11 Cities in China (Beijing, Xian, etc.)

	TN	NH_4-N	P	K
mg/l	39.8–45.1	17.2–30.1	8.5–18.2	13.1–20.6

Source: Jin and Zheng 1989.

DIRECT UTILIZATION OF WASTEWATER

GENERAL CONSIDERATIONS

The wastewater can be classified as domestic wastewater, industrial wastewater, and mixed domestic and industrial wastewater. Domestic wastewater is composed of all the different wastewaters produced by family or household activities. It contains many nutrients (Tables 2 and 3). Industrial wastewaters vary in their nature and in general only those from slaughterhouses and some food processing contain no toxic substances.

With the development of industry, industrial wastewater inevitably increases. New detergents and washing machines have changed the contents of domestic wastewater. The expansion of urban populations also increases wastewater. The untreated mixed wastewaters are becoming unsuitable for raising fish.

A resource that could easily be overlooked is waste heat. Heat incident to the operation of electric power stations is dissipated through cooling systems, causing thermal pollution of nearby waterways. Employing this heat source to raise fish is one strategy for utilizing this waste.

Heat wastewater is used for: (1) Overwintering of tilapia, *Colossoma*, and other tropical fishes; (2) Raising brooders of Chinese carps for inducing reproduction earlier than the typical production schedule; (3) Raising marketable fish.

POND FISH CULTURE

Using wastewater to culture fish has a long history in China and other areas. Farmers store their cooking and washing wastes in small tanks, which then drain into nearby ponds. This is the earliest type of treatment of domestic wastewater.

Mulberry cultivation, silkworm breeding, and silk weaving go back about 2000 years in China. Systems integrating mulberry cultivation and dike fish ponds have been developed in southern and eastern China. Silkworm excrement and other wastes have been used to feed the fish. Wastewater from silk reeling processes (filature) is still in use as a fertilizer in fish ponds today.

In China, the use of wastewater for raising fish in ponds and lakes in the suburbs of some large cities began in the 1950s. Utilization can be grouped according to source as follows:

1. Utilization of domestic wastewater. The fish production can reach 3750 to 7500 kg/ha with no supplementary fertilization or feeding. When the concentration of N is 50 mg/l, with P at 10 mg/l, then 3.4 to 4 m^3 wastewater can produce 1 kg of fish.
2. Utilization of industrial wastewater. In Hangzhou, wastewater from a pig slaughtering factory (1 million head/yr) was used for pond fish culture with a total area of 100 ha. The nutrient concentration in the wastewater was $(NH_4)_2SO_4$, 193 g/t; $Ca_3(PO_4)_2$, 50 g/t; and K_2SO_4, 25.5 g/t. Annual discharge was 4.4 million m^3. The fish production was 7795 kg/ha.
3. Utilization of mixed domestic and industrial wastewater. At present, in most cities in China, domestic and industrial wastewater are mixed when discharged into rivers or lakes. Since small bodies of water like ponds have very limited purification capabilities and it is very difficult to monitor the harmful components in industrial wastewater, the quality of the fish produced is therefore not desirable. This kind of wastewater is therefore no longer used for fish pond culture.

LAKE FISH CULTURE

Some lakes in the vicinity of large cities, such as Donghu Lake in Wuhan, Xuanwuhu Lake in Nanjing, Xihu Lake in Hangzong, and Tianchi Lake in Gunming, have always been wastewater recipients. Because these lakes are near the city market, they also serve as large fish culture bases.

Case 1

Donghu Lake, 32.7 km^2, with a volume of 0.7×10^8 m^3 and an average depth of 2.1 m, had a wastewater input of 2.3×10^5 t/day (1981 to 1984). Since the 1970s, due to large amounts of discharged domestic wastewater, industrial wastewater, and surface runoff, high nutrient loading has occurred, causing increased nitrogen and phosphorus levels each year. From 1957 to 1978, the total nitrogen increased from 0.031 to 1.83 mg/l, and phosphorus increased from 0.008 to 0.78 mg/l. The annual input of nitrogen was 536.3 t and the input of phosphorus was 87.8 t, but the output of nitrogen and phosphorus was 213.1 and 20.1 t respectively, mainly by cultivated fish. This means that the annual accumulation of nitrogen and phosphorus was 323.2 and 67.7 t. Phytoplankton increased not only in biomass but also changed its community structure. Particularly, blue-green algae increased and an algal bloom appeared in summer and autumn. The water became imbalanced, and eutrophication became more and more serious (Chen, 1989).

Shortsighted aquaculture activity made this trend even more pronounced. In the early 1960s, macrophytes spread in most areas of the lake, playing a significant role in the purification of wastewater. Since the end of the 1960s, large additions of grass carp have destroyed the macrophytes. This destruction has inhibited the purification capacity of the lake, increasing the growth of phytoplankton. To increase fish production, species stocking has emphasized silver carp and bighead, and manure was added to promote their production. The total fish production increased from 364.35 t in 1973 to 801.55 t in 1978. Silver carp and bighead accounted for 83.73%.

Case 2

Jinjihu Lake is 717.9 ha, with an average depth of 1.5 m and a volume of 0.6×10^8 m^3. Since this lake accepted wastewaters from Suzhou, a city with 840,000 residents, the phytoplankton biomass increased from 11.7 million cells/l in 1977 to 49.8 million cells/l in 1980. The fish yield reached 2272.5 kg/ha in 1985, of which stocked fish accounted for 96.1 to 98.8%. Silver carp accounted for 41.2 to 57% and bighead accounted for 30.2 to 48.6%. The lake's fish yield might be the highest in China, compared to other lakes of similar size (Xia and Wang, 1988).

Case 3

Xihu Lake is 5.68 km^2, with a volume of 10.43×10^6 m^3 and a depth of 1.5 m. In the 1950s, macrophytes were abundant. In 1951, 400,000 grass carp were stocked, which destroyed the macrophytes. Although the stocking of grass carp was restricted, the macrophytes had no opportunity to recover. The continuous stocking of other herbivorous carp (such as wuchang fish), the increasing discharge of wastewater, and the dredging of the lake bottom further inhibited macrophyte growth.

INDIRECT UTILIZATION OF WASTEWATER

CULTIVATION OF GREEN FODDERS

Green fodders include aquatic and terrestrial plants. In China, so-called "three aquatic weeds" — water lettuce (*Pistia stratiotes*), water peanut (*Alternanthera philoxeroides*), and water hyacinth (*Eichhornia crassipes*) — are usually cultivated in rivulets, small bays, or swamps to avoid taking up land, and are fed to pigs and used as green manures. The production of green fodder is very high. For example, Zhang (1986) reported that, in a growing season (May to November), the water hyacinth yield was 90 to 100 t/ha (9 to 10 kg/m^2). Estimated absorbed nutrition from water by this production was 1580 kg nitrogen, 358 kg phosphorus, and 198 kg sulfur/ha, greatly increasing the efficiency of water purification.

Lin et al. (1990) reported the existence of a microecosystem in the roots of water hyacinths where pollutants in wastewater were first absorbed and fixed on the root surface, then degraded by the root microecosystem. The technology of oxidation ponds with water hyacinths would be a small investment and operating expense and have high purifying efficiency.

After degradation and reduction by macrophytic plants, some harmful contents, such as phenols, will not harm the fish. It has been indicated that duckweed, water lettuce, and water hyacinths have a very strong capacity to degrade and decrease phenols and other toxic materials (Ma et al., 1984).

When macrophytes are used for wastewater treatment, the plants produced must be treated properly to avoid becoming a source of new pollutants. Use of macrophytes for fish food is a better way to resolve this problem. According to the experiment by Zhang et al. (1986), the gross and net fish yield in ponds with water hyacinths as the only green fodder were 27.7 t/ha and 22.1 t/ha respectively. The food requirement was 11 kg water hyacinths and 1.3 kg fine food for 1 kg of grass carp and wuchang fish. After consumption of the water hyacinths by the herbivorous fish, which grow to 1 kg weight, the residue and excrement can promote an increase of 1 kg in other fish such as silver carp, bighead, common carp, crucian carp, and tilapia. It has been shown that even when the water hyacinth lives and grows in an aquatic environment slightly polluted by heavy metals (<1000 ppb), water hyacinth can be a safe and valuable green fodder (Ma and Yan 1989).

Fish do not like to eat the "three aquatic weeds" directly. But after proper processing, e.g., being minced into pieces or fermented, they are welcomed by fish. If 100 kg of rice bran and 0.5 kg yeast are sealed for fermentation, they can be brought out for use after 2 days at 26°C. Water peanuts contain the toxin saponin. It is processed by adding a little table salt at a concentration of 2 to 5% to eliminate the toxicity. If the plants are mashed with grass in a high-speed masher, these aquatic plants are desirable in fry culture, since fry can swallow the mesophyll cells in the paste, which are similar in size to zooplankton and phytoplankton. What is left serves as a nutrient source for the reproduction of plankton and other natural foodstocks.

CULTIVATION OF ALGAE

Liu and Zhang (1990) reported their experimental results of treatment by algae (Table 4). Generally, during the growing season (>18°C), for either macrophytes or algae, the efficiency of removal of phenol from water is high, for COD 70%, for TN, TP, and NH_4-N, 80%, and for BOD_5, 93% (Tong and Li 1986).

TABLE 4
Experimental Results of Treating Perfume Factory Wastewater with Algae (*Cladophora, Hydrodictyon*)[a]

	pH (mg/l)	COD (mg/l)	BOD_5 (mg/l)	DO
Before treatment	5	15,601.2	3,402	0
After treatment	7	5,414	1,127	6.5
Before treatment	4	6,026.5	3,186	0
After treatment	7	5,131	048	6.8
Before treatment	6	10,760.1	3,208	0
After treatment	7	5,273	1,080	7.2

[a] Temperature: 13–15°C, retention time: 36 h.

Source: Liu and Zhang 1990.

UTILIZATION OF BIOGAS POND WASTEWATER

Chen reported (1985) that manure water from biogas fermentation ponds can be used as fertilizer for raising phytoplankton. The water from biogas production could greatly change the composition of phytoplankton communities, and it could also regulate and sustain the naturally available food in different seasons.

IMPACT OF AQUACULTURE ON AQUATIC ECOSYSTEMS

IMPROPER STOCKING OF GRASS CARP MAY DESTROY MACROPHYTIC FLORA, THUS DECREASING THE PURIFICATION ABILITY OF AQUATIC ECOSYSTEMS

There are many lessons in China from those lakes exploited during the period from 1950 to the 1970s. They can be summarized as follows: (1) Biomass of macrophytes can be decreased seriously, even destroyed completely. (2) Structure of macrophytic communities changes towards undesirable species — desirable species (such as *Potamogeton maackianus*) disappear, while undesirable species (such as *Najas marina*) dominate. Based on the above lessons, Chinese scientists now pay more attention to the protection of macrophytes when utilizing lakes as fisheries.

IMPACT OF WASTEWATER FROM HIGHLY DEVELOPED AQUACULTURE AREAS

In high yield fishponds (gross fish yields of 15,450 to 30,350 kg/ha) in Jiangsu Province, from May to August, the average total nitrogen in pond water reached 1.24 mg/l and P_2O_5 reached 0.028 mg/l (Lei et al., 1984). This eutrophic pond water is discharged into nearby natural channels, and then into lakes. This promotes the eutrophication of natural bodies of water.

The rich effluent from the ponds, especially intensive ponds, is the primary culprit which destroys the ecosystems of rivers, lakes, estuaries, mangrove swamps, and bays. For example, in the fall of 1989, a red tide struck Bohai Bay, the center of the shrimp farming industry in northern China. Sewage from Tingjin, Beijing, and other large cities, plus the effluent from shrimp ponds, fueled the tide, which caused major losses at the shrimp ponds.

WASTES FROM CAGE FISH CULTURE

The food waste feeds from intensive cage culture is another source of organic loading. According to Beveridge (1987), when the food conversion ratio is 2.0, the P lost to the environment per ton of produced fish is 25.2 kg for trout, 22.6 kg for tilapia, and 55.7 kg for common carp.

CONCLUSIONS

1. Reasonable aquacultural utilization of nutrient rich and non-toxic wastewater is a low-cost, highly efficient method by which to utilize the wastes. It changes the wastes into useful resources.
2. The wastewater–macrophyte–fish pattern and wastewater–microphyte–fish pattern are two major kinds of ecologically engineered treatment systems which can efficiently transform various wastes into useful protein. From the

point of view of preventing eutrophication, the wastewater–macrophyte–fish pattern is a better way. Macrophytes are the best plants to degrade wastes and even toxic components. Grass carp in conjunction with silver carp and/or bighead is the best combination of fish to utilize the macrophytes.

3. The overstocking of grass carp and the large amount of wastewater discharged into lakes causes a serious destruction of macrophytes and stimulates the development of microphytes. This condition causes lakes to go from a grass-dominated pattern to phytoplankton pattern, thus decreasing its innate abilities for purification. Therefore, controlling the biomass of grass carp to sustain the macrophytes is a necessary management technique for maintaining the lakes' abilities for wastewater purification.
4. Pollution of rivers and lakes becomes more serious each year. This causes eutrophication and the accidental death of fish by oxygen depletion. This condition will affect the health and development of aquaculture.
5. Caution in the utilization of wastewater in aquaculture:
   a. It is important to prevent pollution by toxic industrial wastewater. Domestic wastewater may also have some toxic materials, but through dilution, oxidation, and naturalization, the concentrations of toxic materials can be controlled, thus producing fish fit for any national food hygiene standard.
   b. Before utilization of wastewater, it is necessary to investigate its source, discharge amount, and kinds and concentrations of toxic materials.
   c. Periodically monitor bodies water which comprise the local and regional fisheries and the concentrations of toxic materials in fish meals.

ACKNOWLEDGMENTS

The author would like to thank Dr. Björn Guterstam for his kind invitation to participate in the conference, Dr. Chao Liye, director of the Laboratory of Environmental Protection of Chinese Fisheries Academy, for providing references, and Tommy Fogelberg of Stensund Folk College and Bill Wilson of Environmental Planning and Design in Maui, Hawaii for their tireless work in typing and editing the manuscript.

REFERENCES

Anonymous 1989. *Prediction on the fisheries environment of China in 2000.* Fisheries Academy of China.

Beveridge, M. 1987. *Cage Culture.* Fishing News Books Ltd. pp. 149-156.

Bhatia, H.L. 1970. "Grass carps can control aquatic weeds." *Indian Farming* 20(2): 36-37.

Chapman, V.J., and B.J. Coffey. 1971. "Experiments with grass carp in controlling exotic macrophytes in New Zealand." *Hidrobiologia,* Bucharest, 12:313-323.

Chen, H. 1989. "Impact of aquaculture on the ecosystem of the Donghu Lake, Wuhan." *Acta Hydrobiologia Sinica* 13(4):359-368.

Chen, H. 1984. "An approach to the restoration of aquatic vegetation in the Xihu lake of Hangzhou, with reference to the water quality problem." *Acta Hydrobiologica Sinica* 8(2):237-244.

Chen, Y. 1985. "The effect of marsh gas-manure water on the change of the phytoplanktonic constituents and high yield in fish-ponds." *Journal of Fisheries of China* 9(1): 49-62.

Ding, Ruihua. 1988. "Growth and feeding habits of white crucian carp (Carassius cuvieri Tem & Schl.) in pond rearing." *Journal of Fisheries of China* 12(4):291-301.

Edwards, P. 1980. *Food potential of aquatic macrophytes.* ICLARM Studies and Reviews 5. 51pp. International Center for Living Aquatic Resources Management, Manila.

FAO 1990. Aquaculture Minutes No. 8.

Guterstam, B., and J. Todd. 1990. "Ecological engineering for wastewater treatment and its application in New England and Sweden." *Ambio* 19(3):173-175.

He, Z. 1987. "Further discussion on the food of silver carp (Hypophthalmichthys molitrix)." *Journal of Fisheries of China* 11(4):351-358.

Hickling, C.F. 1966. "On the feeding process in the white amur, Ctenopharyngodon idella." *J. Zool.* 148:408-419.

Hickling, C.F. 1971. *Fish Culture.* Faber and Faber, London.

Huet, M. 1966. "Methodes biologiques d'accroissement de la production pisciole Europe et Afrique." In *FAO World Symposium on Warm-Water Pond Fish Culture.* Rome. 39 pp.

Huet, M. 1972. *Textbook of fish culture breeding and cultivation of fish.* Fishing News Books Ltd. Surrey.

Jing, C., and L. Zheng. 1989. *Handbook of water quality technical work.* Energy Publisher, Beijing.

Jhingran, V.G., and R.S.V. Pullin. 1985. *A hatchery manual for the common, Chinese and Indian major carps.* Asian Development Bank, International Center for Living Aquatic Resources Managements, p. 191.

Lee, B.Y., K.W. Lee, M.G. McGarry, and M. Graham. 1980. "Overview of wastewater treatment and resource recovery." Report of a Workshop on High-Rate Algae Ponds, Singapore, 27-29 February, 1980. International Development Research Center, Canada. IDRC-154e. 47 pp.

Lei, Y., S. Yu, and J. Xu. 1984. "Studies on the water quality of the high-yield fish ponds in He Lie Commune, Wuxi Shi. Part 1. Chemistry of fishpond water and primary productivity." *Journal of Fisheries of China* 7(3):185-199.

Li, S., H. Yang, and W. Lu. 1980. "Preliminary research on diurnal feeding rhythm and the daily ration for silver carp, bighead carp and grass carp." *Journal of Fisheries of China* 4(3):275-283.

Li, S. 1987. "Energy structure and efficiency of a typical Chinese integrated fish farm." *Aquaculture* 65:105–118.

Lin, H.R. 1982. "Polycultural system of freshwater fish in China." *Can. J. Fish. Aquat. Sci.* 39:143-150.

Lin, S.Y. 1954. "Chinese systems of pond stocking." *Proc. Indo-Pacif. Fish Coun.* 5(2 & 3):113-125.

Liu, B., and J. Zhang. 1990. "An approach on algae in the treatment of wastewater." *Shanghai Environmental Sciences* 2:32-33.

Ma, S., and J. Yan. 1989. "Ecological engineering for treatment and utilization of wastewater." In W.J. Mitsch and S.E. Jørgensen (eds.), *Ecological Engineering: An Introduction to Ecotechnology.* John Wiley & Sons. Inc. pp. 185-217.

McLarney, W. 1984. *The Freshwater Aquaculture Book: A Handbook for Small Scale Fish Culture in North America.* Hartley & Marks, Publishers. pp. 23-24.

Mehta, I., R.K. Sharma, and A.P. Tuank. 1976. "The aquatic weed problem in the Chambal irrigated area and its control by grass carp." In Varshney, C.K. and L. Rzoska (eds.), *Aquatic weeds in Southeast Asia.* The Hague, Netherlands. pp. 307-314.

Nikolsky, G.V. 1963. *The Ecology of Fish.* Academic Press, London and New York. L. Birkett, translator.

Pomeroy, L.R. 1980. "Detritus and its role as a food source." In R.S.K. Barnes and K.H. Mann (eds.), *Fundamentals of Aquatic Ecosystems.* Blackwell Scientific Publications, Oxford. pp. 84-102.

Shireman, J.V. 1983. "The utilization of grass carp for hydrilla control in Lake Baldwin, Florida." J. Fish Biol. As cited in: J.V. Shireman and C.R. Smith. 1983. "Synopsis of biological data on the grass carp Ctenopharyngodon idella (Curiel and Valenciennes, 1844)." *FAO Fisheries Synopsis* No. 135, p. 86.

Trewavas, E. 1982. "Tilapias: taxonomy and speciation." In R.S.V. Pullin and R.H. Lowe-McConnell (eds.). *The Biology and Culture of Tilapias.* International Center for Living Aquatic Resources Management. Manila. pp. 3-13.

Venkatesh, B., and H.P.C. Shetty. 1978. "Studies on the growth rate of the grass carp fed on two aquatic weeds and a terrestrial grass." *Aquaculture* 13:45-53.

Xia Y, and X. Wang. 1988. "The fish population structure and potential production in Jinjihu lake." *Journal of Fisheries of China* 12(3):267-275.

Zhang Y., Q.Y. Day, X.Q. Li, and X.G. Zhang. 1986. "Studies on the water hyacinth used as fodders for cultured fish." *Jiangsu Ecology* 2:129-136.

Zhu, X. 1989. "Bacteria floccule as diet of filter-feeding fishes." *Journal of Fisheries of China* 13(4):339-345.

Chapter 4

ECOSYSTEMS APPROACH TO LOW-COST SANITATION IN INDIA: WHERE THE PEOPLE KNOW BETTER

Dhrubajyoti Ghosh

ABSTRACT

Waterlogged areas on the edges of a number of cities in India have been identified and taken over by the government for transformation into low-cost sanitation and resource-recovery ecosystems. The proposed low-cost sanitation and resource-recovery project is to convert unused or underutilized wetlands beyond the edge of a city into a cluster of ponds that will treat sewage and use it to grow fish, using primarily solar energy. This is followed by integrating suitable downstream utilization of the fish pond effluent with agriculture or forestry.

The initial set of projects have started in the state of West Bengal, where Calcutta has the largest traditional system of using city sewage: fisheries with an area of more than 3000 ha. The first series of ponds, P1, are batch-fed ponds and are lentic water bodies. They are used for irrigating adjoining vegetable fields. These are the first set of wastewater fisheries. The second set of ponds, P2, are mixed-flow ponds and are lotic water bodies. Wastewater reaches these ponds through a network of drainage channels. Effluent from these fisheries is either used in the adjoining paddy fields or drains into the outfall channels directly.

The ponds have demonstrated that non-formal initiatives can transform wetlands into efficient resource-recovery systems. Very little institutional support has been given by the government to any of the entrepreneurs or cooperative units running this resource-recovery system. Equally remarkable is the performance of the Mudially Fishermen's Cooperative Society (MFCS) in transforming a waterlogged area into a resource-recovery ecosystem and in demonstrating self-help sanitation model as a prototype in the southwestern fringe of the city of Calcutta.

INTRODUCTION

Waterlogged areas on the edges of cities are poised to gain newfound significance in India. For a number of municipalities such areas have been identified and taken over by the government for transformation into low-cost sanitation and resource-recovery ecosystems under the provisions of the Ganga Action Plan of the Ministry of Environment and Forests of the Government of India. This action is the beginning of an ecosystem approach for solving the problems of municipal sanitation in a developing country.

0-87371-990-5/97/$0.00+$.50

The proposed low-cost sanitation and resource-recovery project is to transform unused or underutilized wetlands beyond the edge of a city into a cluster of ponds that will treat the city's sewage and use it to grow fish, with the principal energy source being the sun. This is followed by integrating suitable downstream utilization of the fish pond effluent with agriculture or forestry. The initial set of projects have started in the state of West Bengal, where we have in Calcutta the largest traditional system of using city sewage: fisheries with an area of more than 3000 ha.

The adoption of an alternative technology option based on fisheries and agriculture in place of mechanical sewage treatment plants was first suggested by the author (Ghosh, 1985) at the onset of India's historic program of rescuing the River Ganga from the toxic assault of municipal wastewater. The formal project outline was subsequently submitted for three municipalities (Ghosh, 1987). These projects were subjected to stringent scrutiny and appraisal by specialists and are now ready for implementation. It is expected that a considerable amount of working experience will be gained through the implementation of these projects.

This, however, is not to underrate the existing experience in the Calcutta wetlands, which has shown that non-formal initiatives can transform wetlands into efficient resource-recovery systems. In all these years, very little institutional support has been given by the government to any of the entrepreneurs or cooperative units running this resource-recovery system. This is in direct opposition to conventional options, which are always awarded large amounts of capital. Therefore, if a vast metropolitan city's sewage is treated and reused by a system whose managers do not render fealty to any multinational bank assistance program and do not celebrate their vassalage as treatment plant workers of conventional designs, where is the reason to hesitate in imagining that municipalities can be self-sustaining in their task of sanitation? This is where we can look forward to finding a new era of self-help sanitation for municipalities in developing countries. Here is where the challenge begins.

There is no reason to pretend that the theoretical guidelines for translating the experience of traditional wetland practices into a reliable technology option are foolproof. In fact, wastewater treatment and reuse with wetlands is far more crude than by conventional methods. Yet, from the standpoint of ecological balance and economic viability, the former towers high above the latter in its rustic grace and is destined to shape the future grammar of sanitation technology options for tropical developing countries.

The ongoing and ensuing studies of the wetland option for wastewater treatment and reuse will significantly improve the technology for implementing such projects. However, it is necessary to develop a tentative guideline for the upcoming system and provide a closer view of the concept.

AN OUTSTANDING HERITAGE

The wetlands of east Calcutta are situated between the levee of the River Hooghly on the west and the Kulti Gong on the east. They are distributed nearly equally on the two sides of the dry weather flow channel reaching the Kulti Gong to the east. The wetland area lies approximately between latitudes 22° 25" to 22° 40" North and latitudes 88° 20" to 88° 35" East. The region is suitable for using solar radiation for improving wastewater quality. Each hectare of a shallow water body can remove about 237 kg of biological oxygen demand (BOD) per day. Winter sky clearance factor is adequate enough (about 90%) to sustain biochemical activities in water purification.

This area falls within the South Bengal ecotone, where the process of land formation is largely influenced by the Ganga River system that continues to create land both by upland and tidal deposition. This pattern of delta building itself has undergone significant changes due to natural and manmade causes.

Calcutta city grew on the Hooghly levee. While river banks are always raised first to allow human settlement and productive activities, the spill areas also gradually become higher to the tidal level from subsequent reclamation. Unfortunately, with the death of the River Bidyadhari, the tidal channel which used to deposit silt in this area, the process of natural deposition has completely stopped since the end of the last century. This incomplete process of delta building did not allow the low-lying areas behind the Hooghly levee, east of Calcutta, to rise higher.

Since the early 15th century, when the Ganga changed its main course from the Bhagirathi to the Padma, this eastward change of the course of the main flow of the River Ganga brought changes in the process of delta building in central and south Bengal. A number of distributaries and redistributaries were cut off from any upland flow, which signaled the end of those channels. However, human interference in the region further reduced the spill area and the channel beds filled up to quicken the process of decay. In this way, the Bidyadhari became defunct and the opportunity for silt deposition in its spill areas ceased.

Prior to wastewater fisheries, this region was a brackish water fishery area, gradually rendered derelict on account of the receding Bidyadhari spill channel. Interestingly, these fisheries were responsible for reducing the spill area and, compounded with the dwindling upland flow from the River Hooghly to the spill channel, caused the death of the tidal creek. During the 1930s, the Bidyadhari carried only city sewage and, in the process, was further choked, due to the high silt content of the city sewage.

In such a dire situation, one of the leading fish producers of that time was successful in growing fish in a large water area using city sewage in the same process we find today. The basic feature of this "mono pond" wastewater fishery was that it combined the processes of treating the sewage and recovering its nutrients through aquaculture within the same pond. Application of sewage was timed on the basis of retention time needed to improve the water quality appropriate for growing fish.

There has been an introduction of newer varieties of fish species. Fish producers in this area have shown keen interest in designing the optimum crop group. In fact, it is in these fisheries that we find one of the most efficient yield rates from a multiple species polyculture.

In spite of the early success, the lack of wastewater supply restricted the spread of such fisheries in this area. This had to wait until the construction of the Dry Weather Flow (DWF) channel to carry city sewage to the Kulti Gong as a new outfall receptacle with provision for fishery feeding. Since then, there has been a rapid spread of sewage-fed fisheries. However, the scenario is changing as the specter of Salt Lake City, a satellite city planned for in the vicinity, has loomed larger and larger.

Formally, one has to mark the year 1956 as the beginning of the loss of ecological stability in these vast wetlands. There had been instances of forcible attempts by violent farmers to take over the rights to land. Such actions were long overdue. For more than 50 years, owners of small parcels of land were systematically and heinously forced to give up their property rights in favor of a few large landowners. But the unfortunate fallout of such a forcible occupation of the big owners' land by the landless farmers lay elsewhere. Fisheries were mistaken for the fishery owners themselves. So, fisheries were drained and the cultivation of paddy was attempted. It required lengthy persuasion to convince the farmers that it was the fishery owners and not the fisheries that should face the might of distributive justice. Aside from the multiple benefits provided by the fishery, it was the most efficient ecosystem for the farmers who were natural cultivators of fish rather than paddy. At present, this realization is a predominant trend among the villagers and forcible takeovers have been largely on the wane. Organized fish farm labor movements have succeeded in obtaining better terms of labor from the employers and the fisheries have survived. Yet, with the enactment of the second amendment of the Land Acquisition Act, vast areas of land have become potentially ceiling surplus. Appropriate imagination and new policy planning are needed to recover the lost stability of these fragile wetlands.

The year 1980 signaled the beginning of an ecological appraisal of this ecosystem. Subjectively, this approach to studying the ecosystem took a completely separate standpoint of environmental significance, and, objectively, it adhered to a barefoot approach to understanding and actively involving the local villagers, village-based organizations, and similar local bodies to lead the challenge for its survival (Ghosh and Sen, 1987).

Bringing the environmental significance of this wetland ecosystem to the fore has resulted in a discernible change in the attitude of the governments, both at the central and state level. Conversion of these wetlands for urban development is an idea that is no longer automatically seen as the best option, although the possibility cannot be ruled out. And this new approach to the wetlands east of Calcutta has drawn global attention (e.g., Edwards 1985). Taken together, it can be said that ecological issues have well unfolded to step up the strength of conservation.

In the 1950s, the uncertainty of the future of wastewater fisheries concerned only a handful of entrepreneurs and a few thousand fish farmers. This situation has since undergone a metamorphic transformation. The tenacious interaction with the villagers and village-based organizations has armed them with a new consciousness and determination to protect their right to stick to their ancestral profession instead of accepting a resettlement program. It would not be so wrong to say that, from a period of "little hope" for these wetlands to survive, we have now come to one of "cautious optimism," looking forward to a scientific preservation of an outstanding ecosystem.

HOW DOES THE SYSTEM WORK?

To understand the operation of wastewater utilization in fisheries and agriculture in the wetlands of east Calcutta, it may be helpful to use a resource system framework (Figure 1). One hundred fifty million gallons (over 500,000 cubic meters) of wastewater

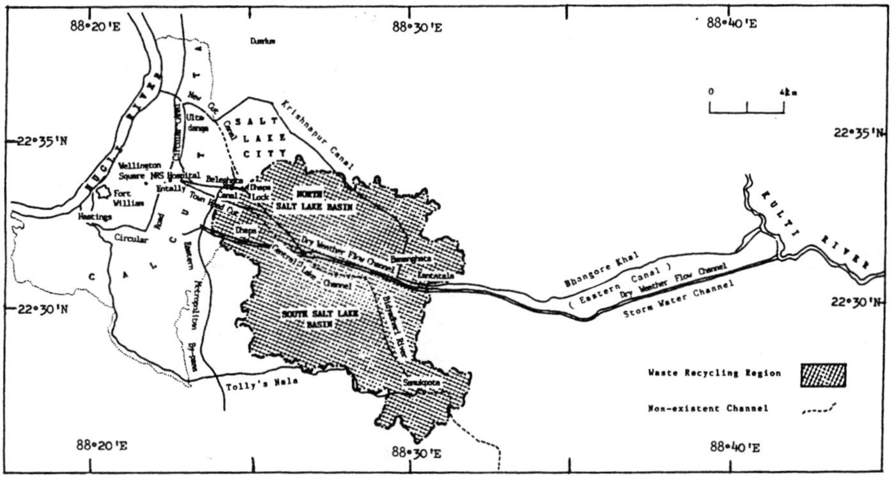

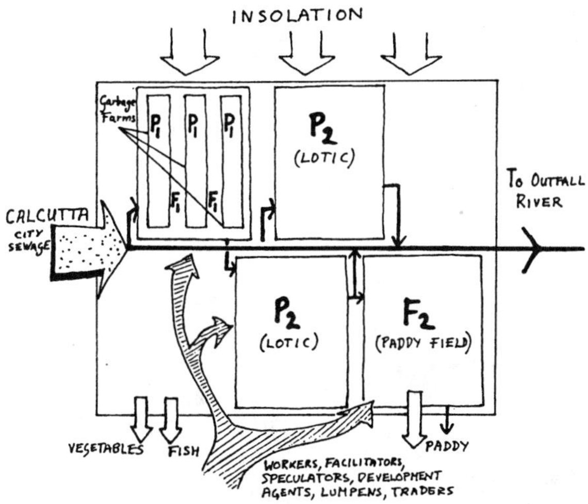

FIGURE 1. Resource system framework for the wetlands of Calcutta.

is pumped out daily through the outfall channels from Calcutta, eventually reaching the river Kulti Gong, 28 km east of the city. From here, some of it is taken into the wetlands.

The first series of ponds, P1, are batch-fed ponds and are lentic water bodies. They are used for irrigating adjoining vegetable fields. These are the first set of wastewater fisheries. The second set of ponds, P2, are mixed-flow ponds and are lotic water bodies. Wastewater reaches these ponds through a network of drainage channels. Effluent from these fisheries is either used in the adjoining paddy fields or flows directly to the outfall channels.

A wetland system transformed for improving the quality of sewage and trapping its nutrients seems to operate within a number of boundary conditions. Major boundary conditions are: (a) area available, (b) hydraulic regime, and (c) micro-climate.

All photos in this report taken by the Swedish photographer Heldur Netocny.

TABLE 1
Working Determinants

Boundary conditions	**Major determinants**
1. Area	Land cost, land use, future plans, awareness of wetland functions
2. Hydraulic regime	Land formation, groundwater, water retention
3. Micro-climate	Location (latitude, longitude), special conditions
Operational variables	**Major determinants**
1. Flow	Urban population, water supply, sewage network, depth of pond
2. Water quality	Detention time, limiting, number of ponds, diversion, diffusion, bed preparation
3. Fish recruitment and harvesting	Choice of species, size of recruitment and harvesting, stocking density, harvesting schedule
Institutional factors	**Major determinants**
1. Financial	Land use policy, project viability, hidden investment
2. Coordination	National and state plan, nodal initiative, people's awareness
3. Skill, workers' welfare, and terms of labor	Management policy, labor laws, trade unionism
4. Entrepreneurial risk	Land acquisition plans, chances of wetland reclamation, law and order
5. Training, information, and awareness	Policy support, scientific and technical program, NGO activities, missions

Within these boundary conditions, the biology works on the basis of skillful manipulation of three major operational variables: (a) water flow, (b) water quality, and (c) population balance.

Institutional factors of significance include (a) finance, (b) coordination, (c) skill, worker's welfare, and terms of labor, (d) entrepreneurial risk, and (e) training, information, and awareness.

Table 1 summarizes the major determinants found to influence wetland changes.

EXTENSION: A GOOD BEGINNING

In the wetlands of eastern Calcutta we have seen the world's largest cluster of wastewater fish ponds. Equally remarkable is the performance of the Mudially Fishermen's Cooperative Society (MFCS) in transforming a waterlogged area into a resource-recovery ecosystem and in demonstrating self-help sanitation model as a prototype in the southwestern fringe of the city of Calcutta.

About 100 members of the MFCS produce about 5.6 tons/ha of fish per year from about 50 ha of water area which once was a "wooded area." In the process, about 23 million l/day of municipal plus industrial sewage is treated in the wetlands. The entire ecosystem has been built without any external financial help. The money earned from the sale of fish has not only given a gross per capita income of about Rs.71.00 ($4.00)/day (which is more than three times the minimum daily agricultural wage in India) but has also allowed a development expenditure of Rs.50 Lakhs

TABLE 2
Wastewater Quality in the MFCS Pond System[a]

Parameters	Influent water	Effluent water
Flow $m^3 \cdot hr^{-1}$	993.3	947.0
Temp°C·Amb/Water	30/28	33/28
pH	7.95	7.50
Total solids	1152.01	788.00
Suspended solids	51.18	73.00
Dissolved solids	1099.90	715.00
Total voltile solids	340.41	210.00
BOD	77.58	15.00
COD	470.42	65.00
Total Alkalinity as $CaCo_3$	285.44	192.00
Total Nitrogen as N	114.07	31.00
Phosphate as P	0.20	0.04
Mercury as Hg, µg/l	4.52	Below detectable limit

[a] All parameters except pH, temperature, mercury and flow are expressed in $mg \cdot l^{-1}$

Source: NEERI, 1990.

($263,157.89) over the last 5 years. Municipal sanitation in any other similar setting need not be something requiring government expenditure of public money to operate the system. What was hitherto perceived as a liability can be an asset. We can very well talk about a viable resource-recovery ecosystem that provides food, sanitation, and jobs for the weaker section of society.

THE PONDS

At present, 23 million l/day of sewage are flowing through about 50 ha of water bodies (Tables 2 and 3). By the middle of this year, there will be 15 ponds of various sizes in the MFCS wetland area. Of these, nine smaller ponds will be used for improving the water quality before it enters the larger ponds, where the fish are cultivated and the water is treated further. All of these ponds have interconnecting culverts and the water eventually drains out into a canal leading to the River Hooghly. Unlike the barren landscape of the east Calcutta wetlands, where trees are scarce, a dense trail of organized planting marks the MFCS area as one of the most sought after open spaces in the region. From 1985 to 1989, about 96,000 saplings were planted on the pond dykes, of which 60% have survived.

GROWING FISH

In some of the ponds during the present year, fish have been found which are known to be unable to endure the toxic stress which carp and tilapia can. These species are chanda (*Chanda ronga* and *Chanda nama*), mourala (*Amblypharyngodon mola*), and punti (*Puntius sophore* and *Puntius conconius*), which are positive indicators

TABLE 3
Removal of BOD and Coliform in the MFCS Pond System

	Influent water	Effluent water	% of removal
BOD ($mg \cdot l^{-1}$)	77.00	15.00	8.052
Fecal coliforms (MPH/100 ml $\times 10^3$)	46,000.00	0.91	99.9980

Source: NEERI, 1990.

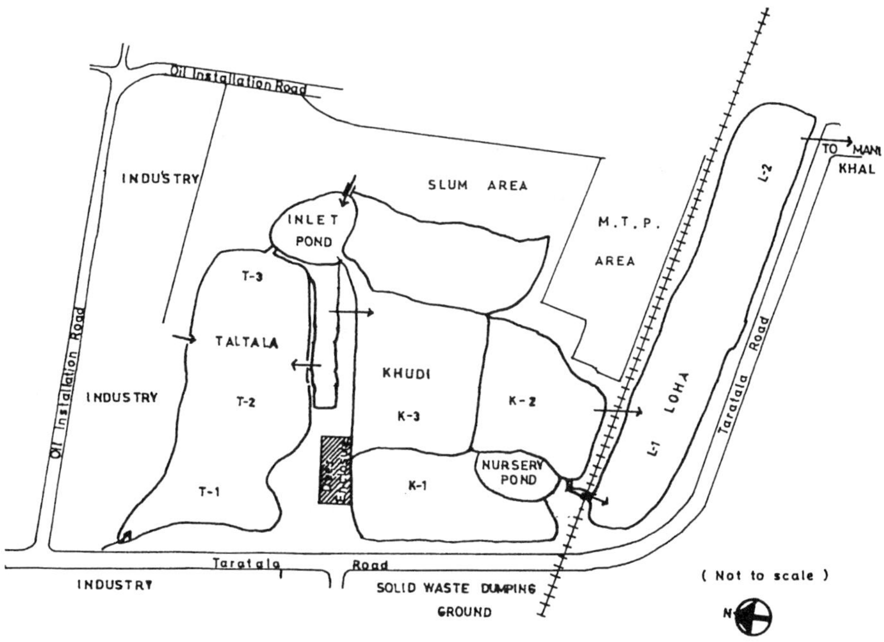

LOCATION MAP OF THE PONDS OF THE M.F.C.S.

FIGURE 2. Mudially Fishermen's Cooperative Society wastewater fish ponds.

of the level of improvement of the water quality that takes place in the pond system. Nevertheless, the dominant culture in all the ponds is that of carp and tilapia. It has been found that there is a direct relationship between the extent of purification of wastewater and the density and variety of plankton. Stocking density and the variety of fish cultured are also related to the quality of water. While air-breathing fish (*Ophiocephalidae* and *Channidae*) are found to survive in the ponds closer to the inlet, carp, prawn, and other more sensitive species grow in the subsequent ponds far from the inlet. It has also been found that the yield of fish is highest in the terminal pond, the Loha *jhil*, near the outlet (7.80 tons/ha/y) in comparison with 7.00 tons/ha/yr, 4.93 tons/ha/yr and 3.64 tons/ha/yr in the Khudi 1, the Taltala, and Khudi 2 and Khudi 3 *jhils*, respectively (Figure 2).

INCOME AND EXPENDITURES

Not many cooperatives, anywhere, would be able to show a standard of bookkeeping as efficient and candid as that of the MFCS. A striking feature of the income distribution is the near total absence of skewness. The highest earning member gets no more than 3.5 times the wage earned by the lowest earning member of the cooperative. If anything has contributed to the growth in earning, it is the hard labor given to the fishery by the members. The earnings from fish sales is phenomenal. Total earnings from sales has been increasing each year since the introduction of sewage into the ponds started. The return from sales of fish in 1986 was about Rs.35 Lakhs ($184,210.50) and it increased 150% in 1989. There is also a monthly variation in the earning pattern (Table 4).

ENVIRONMENTAL EDUCATION

The routine demonstration lectures by the farmer members of the MFCS are unforgetable. It is a pleasure to appreciate their level of understanding of the ecosystem and its problems. The plan is to develop a "Tutorial Ecosystem" for various levels of learners, from children to the decision makers. An information center has been built that houses seminar rooms, a laboratory, and a small museum. However, the most important aspect of the environmental education is the variety of life forms that coexist to form a unique assemblage of plants and animals, something which is fading very fast from congested cities like Calcutta.

By the end of this year a systematic program for environmental training is likely to be put into practice, training which is not otherwise available in an area so close to the center of the city of Calcutta.

The scope for carrying out research in a pond system for wastewater treatment and fish cultivation is endless. "Better" designs will definitely come on the basis of a gradually clearer understanding of the ecosystem. This, however, need not bar us from applying "good" designs right now. The example of the MFCS can be simulated elsewhere, with similar environmental conditions. It has enough information to motivate an interested planner to think for his own municipality and in terms of transforming a low-lying marsh on the edge of the city into an ecosystem of self-help sanitation.

CONCLUSION

In India, the decade of the 1980s has been most eventful for ecology as a shaper of sanitation technology options. The first half of the decade was spent in reassessing the wastewater fishponds to the east of Calcutta for their outstanding environmental significance. The second half of the decade was effective in countering the "universal" choice of mechanical sewage treatment. In a number of cases, this pond system option of treating wastewater and recycling nutrients has been adopted in place of mechanical sewage treatment plants in the country's most expensive river sanitation program. This pond system, which can be called the Resource Efficient Stabilization Tank (REST) system, is now being implemented in three municipalities with the prospect of being implemented in a few more shortly.

TABLE 4
Monthly Earning from Fish Sale (In Rs.)[a]

Year	July	Aug	Sept	Oct	Nov	Dec	Jan	Feb	Mar	April	May	June	Total
1980–81	42,346	49,517	41,312	52,528	57,509	61,601	54,146	47,301	41,537	51,201	103,621	195,842	798,461
1981–82	91,013	70,117	88,309	76,201	69,212	86,301	85,151	83,985	79,546	87,243	97,241	106,993	1,021,412
1982–83	98,306	67,576	69,307	112,503	107,202	78,507	76,594	91,322	87,507	134,509	137,564	157,028	1,217,525
1983–84	125,220	68,212	96,155	94,201	95,907	72,301	74,506	90,442	88,336	107,109	195,211	202,477	1,310,077
1984–85	86,422	60,501	123,222	98,451	42,221	69,340	86,937	92,301	75,405	97,485	132,982	223,208	1,188,833
1985–86	88,420	62,536	72,228	122,987	45,887	85,803	84,423	79,383	76,235	101,395	165,704	324,269	1,309,275
1986–87	372,176	428,828	266,926	265,506	205,651	180,070	184,898	165,021	193,459	205,589	352,592	609,855	3,429,926
1987–88	408,341	425,199	375,536	280,583	290,096	148,594	80,960	101,625	240,970	260,670	352,067	537,004	3,501,606
1988–89	253,949	310,524	284,934	304,986	223,241	206,067	205,676	162,765	165,064	289,872	489,957	883,682	3,902,625
1989–90	452,627	518,942	488,851	530,211	453,672	375,026	296,190	365,645	411,293	292,283	355,235	460,952	5,001,441

[a] $1.00 = Rs.19.00

Source: MFCS Records

Most of the assumptions in the design of the REST system projects are empirical. Upgrading the subjective basis of the design is a major task. While some of the information on this issue has already been collected, more will have to be collected to create a satisfactory basis for a design. Our vision, however, extends much beyond designing only a pond system. The actual task is understood as a unified regional

development plan to set "waste recycling districts" on the city's edge. Waste recycling districts will be a new kind of urban facility that will provide food, sanitation, and jobs for the largely impoverished rural folk of the fringe area villages. The wetlands to the east of Calcutta are the biggest storehouse of information to help us standardize the design of waste recycling districts.

Before we learn many more things, it will be appropriate to highlight the most significant lesson from the age-old management practices of the Calcutta wetlands. This, unmistakably, is the perception of a self-help sanitation model. Turning waste into wealth is a matter of simple reality in the ecosystem that has been developed. The task of giving a clearer shape to the model is already under way. Yet, an interim evaluation can bring about a number of new realizations which are capable of changing the prevailing assumptions in designing sanitation technology options. These realizations are that:

1. Wastewater aquaculture does not necessarily have to be a "batch-fed" process. It can also be a "mixed-flow" process and the process of water quality improvement can take place in the same reactor where the fish will also grow.
2. Fish survival is not a function of water quality (or, for that matter, of dissolved oxygen) alone, but very much a function of the choice of species cultivated.
3. WHO's recommendation of 1000 coliform bacteria per 100 ml of water needs very serious reviewing. It has failed to reflect the contribution of culture (eating fried fish instead of uncooked or undercooked fish) and also of the sterilization effect of incident sunshine and the rate of decomposition that vary from place to place.
4. Introducing aquaculture in stabilization tank systems can create enough wealth to reasonably invalidate the criticism of comparatively higher capital requirement for the acquisition of land compared to conventional mechanical treatment plants. The farmers have put up their best effort in bringing forth a newfound concept of "self-help sanitation." It will depend on us ecological engineers as to how best we can learn from them to give shape to the decisions in municipal sanitation planning in future.
5. In developing countries where population density is high and settlement pressure on urban edges is real, it is difficult to conserve peri-urban wetlands unless:
    a. Sufficient wealth is created from them.
    b. They provide stable job opportunities for the rural folk of the city's fringe.
    c. They are utilized for treating municipal sewage.

Such a transformation of wetlands is not in agreement with a particular trend of thinking in wetland management, which is sensitive to any kind of interference with the ecosystem. We, however, do not consider such a view as tenable if we have to conserve a wetland on the city's edge. For any wetland, left wild, will eventually be drained and subsequently reclaimed for urban expansion.

Transforming waterlogged marshes on the city's fringe for wastewater conservation is gaining a new social, economic, and technological relevance that we have not seen in many years. Never before was a peri-urban marsh allowed any more significance other than as a prospective site for urban takeover. Never before could a model for self-help sanitation be perceived. Never before was a job opportunity of any significant scale imagined for the rural folk in wastewater management technology. Never before had an open space facility coincided with wastewater utilization ecosystems. And never before had algae played such an important role in wastewater management for performing the dual task of nutrient recovery and food chain promotion for growing fish.

REFERENCES

Edwards, P. 1985. *Aquaculture: A component of low-cost sanitation technology.* World Bank Technical Paper #36.

Ghosh, D. 1985. *Cleaner rivers: The least cost approach.* State Planning Board, Govt. of West Bengal. Calcutta, India.

Ghosh, D. 1987. *Resource efficient stabilization tank system: The least cost technology option for municipal effluent sanitation.* Institute of Wetland Management & Ecological Design. Calcutta, India.

Ghosh, D. & Sen, S. 1987. "Ecological history of Calcutta's wetland conversion." In: *Environmental Conservation.* Elsevier Sequoia S.A., Lausanne, Switzerland, Vol. 14, No. 3, pp. 219-226.

NEERI, 1990. *Water quality studies of the Jeel in Calcutta port area.* National Environmental Engineering Research Institute, Nagpur, India, pp. 35 & 75.

WHO, 1989. *Health guidelines for the use of wastewater in agriculture and aquaculture.* Technical Report Series No. 778. World Health Organisation. Geneva.

Chapter 5

WASTEWATER-FED FISH CULTURE IN HUNGARY

J. Oláh and F. Pekár

ABSTRACT

An eight-year-long research project was conducted to purify and utilize the domestic sewage in fishponds from 1971 to 1978 in Hungary. Qualitative and quantitative analysis of the community structure and the establishment of the management parameters in order to develop and complete a technological package were the central targets of the project. Sewage loading rates from 50 to 150 m^3/ha/day sewage loading rate and fish stocking densities of 1300 to 12,000 fish/ha were tested. Optimal load and stocking densities were established at around 150 m^3/ha/day and 3500 to 4000 two-year-old fish/ha, respectively. The technology and management is as simple and natural as possible except for the mechanized pumping of sewage through tubing and rotary sprinklers. This fishpond ecosystem is able to produce 15 to 20 kg fish/ha/day and purify the sewage water to an acceptable level.

Since a detailed qualitative analysis on the seasonal dynamics of the amount of biomass of the fish food organisms gives only an approximation of their quantitative importance in the food web, a new research project was initiated and conducted from 1974 to 1987. Domestic sewage, pig, chicken, and sheep manure were used as organic carbon sources, and the production processes at the first and last parts of the energy flow were measured to quantify the efficiency of the energy transfer in this organic carbon-fed fishpond ecosystem. The quantity of the daily organic carbon load and the autochthonous organic carbon production by photosynthetic algae and bacteria were measured almost daily, while the fish production was estimated fortnightly. Hundreds of measurements showed that the conversion of the organic carbon introduced daily through bacterial production processes is the key production pathway in organic carbon-fed fishpond technology.

INTRODUCTION

In China and the Far East, the domestic sewage-fed aquaculture has a long tradition. The present day technologies are the results of empirical experiences accumulated and improved during centuries and based on the principle of polyculture. In Europe the first significant trials were started in Germany in the year 1929. However, the adapted monoculture technology with only common carp proved to be far less efficient. After a five-year-long research project was completed in Hungary, adapting the old Asian polyculture principles to the temperate European climate, organic-fed fisheries have gained importance. In Europe, the utilization of

0-87371-990-5/97/$0.00+$.50

sewage oxidation ponds for aquaculture with organic recycling benefits both resource conservation and wastewater disposal. In countries with developed economies and an endangered nature, the latter has high priority compared to the food production. Here, we present the ecosystem function, the efficiency, and the development of the management parameters of Hungarian sewage-fed fish culture technology.

PROJECT HISTORY

In Hungary, the first experimental trial to purify and utilize the domestic sewage in fishponds was carried out with the sewage of city Fonyód on a total water surface area of 211 ha in the years of 1971 and 1972 (Ponyi et al., 1973, 1974). The experimental design was established using the experiences of farmers and angling societies operating sewage-fed ponds in the cities of Gödöllö, Balatonföldvar, Szeged, Deszk, Fonyód. The research target was to describe and quantify the community structure from bacteria to fish.

Establishing the management parameters, in order to develop and complete a technological package, was the central target when a new research project was initiated — the construction of a sewage-fed fish farm as part of a five-year-long experiment from 1974 to 1978. The experimental farm was built nearby the old sewage fishpond system of the city of Fonyód and consisted of six large ponds (1.6 ha each) and six small ponds (0.18 ha each). Sewage was introduced daily through a pipe system supplied with rotary sprinklers.

In 1974 and 1975, a rather wide range of sewage loads from 5 to 250 m^3/ha/day were tested, with the stocking density held constant at 1300 to 1400 two-year-old fish/ha. In 1976 the sewage load was 50 m^3/ha/day in each pond and the stocking density was tested from 1400 to 5000 two-year-old fish/ha. In 1977 the loading rate was further improved, testing a more narrow load gradient from 75 to 175 m^3/ha/day. In the last year of the research project, the optimal stocking density was established further in a rather narrow density range of 2200 and 3000 two-year-old fish/ha when the optimal sewage load of the previous experiment was applied, 150 m^3/ha/day. After finishing the five-year-long research project, we monitored the commercial operation of the farm during the next five years, 1979 to 1983. Afterwards the Hungarian National Standards for Domestic Sewage Fishpond were written and approved by the government (NWB 1982). Unfortunately, the last five years from 1984 to 1988 were not very fruitful for distributing the technology all over the country due to inadequate money and administrative regulation. We have encountered more interest from abroad, in countries like Yugoslavia and France.

MANAGEMENT PARAMETERS

In Hungary, the sewage-fed fishpond technology is applica[illegible]eceive, process, utilize, and purify the domestic sewage from April until October. This is usually the peak period of sewage production. The summer peak is very pronounced at recreational or tourist areas like around Lake Balaton, where the technology itself was developed. The technological package is the final result of the five-year-long research

TABLE 1
Management Parameters and Culture Techniques

Pond bed	Earthen	
Pond area	1–10 ha	
Water depth	1 m	
Water supply	River	
Draining	Winter dry	
Sewage	Raw after sedimentation	
Introduction	Pumping through tubing and sprinklers	
Frequency	Daily during morning	
Sewage quantity	150 m³/ha/day	
Stocking density	3500–4000 fish/ha	
Stocking structure	Weight(g)	Density (fish/ha)
Hypophthalmichthys molitrix	200	2500
Cyprinus carpio	190	1500
Stocking time	April	
Growing period	120 days	
Harvesting time	October	
Fish production	12–20 kg/ha/day	

project plus the five years monitoring of the commercial operation of the same system (Table 1).

The technology and management is as simple and natural as possible except for the mechanized pumping of sewage through tubing and rotary sprinklers. To simplify the management practice, we have reduced the number of fish species in the polyculture as much as possible but still apply the basic principle of the complete grazing pressure both on the planktonic and benthic communities. With a properly high density, the silver carp and common carp are able to utilize completely both the planktonic and benthic fish food resources and are able to disturb and oxidize the sediment water interface and balance the plankton development as a result of the bioturbation of their feeding activity. A stocking density of 3500 to 4000 two-year-old fish/ha is high enough to process the daily introduction of sewage but low enough to reach a final fish weight of more than 1 kg. Higher stocking density increased the sewage purification efficiency but decreased the final fish weight below 1 kg, unacceptable on the European market for table fish. On commercial operations, farmers prefer to start stocking their sewage farm with two-year-old fish instead of one-year-old fingerlings, which are more liable to escape or die. If we start with fingerlings and want to reach two-year-old fish as a final product, the fingerlings must be stocked at the rate of 7000 to 8000 fish/ha.

Beside the stocking density, the most critical management parameter is the magnitude of the sewage load. The final figure of the optimal load in the two-year-long testing trials was established around 150 m³/ha/day.

Calculating from all of the pond loading experiments gives a rather clear average daily growing response of the total fish standing stock to the increasing load (Figure 1). Below the loading rate of 150 m³/ha/day, the fishponds were limited by organic carbon or energy and the fish growth rate decreased very significantly, from

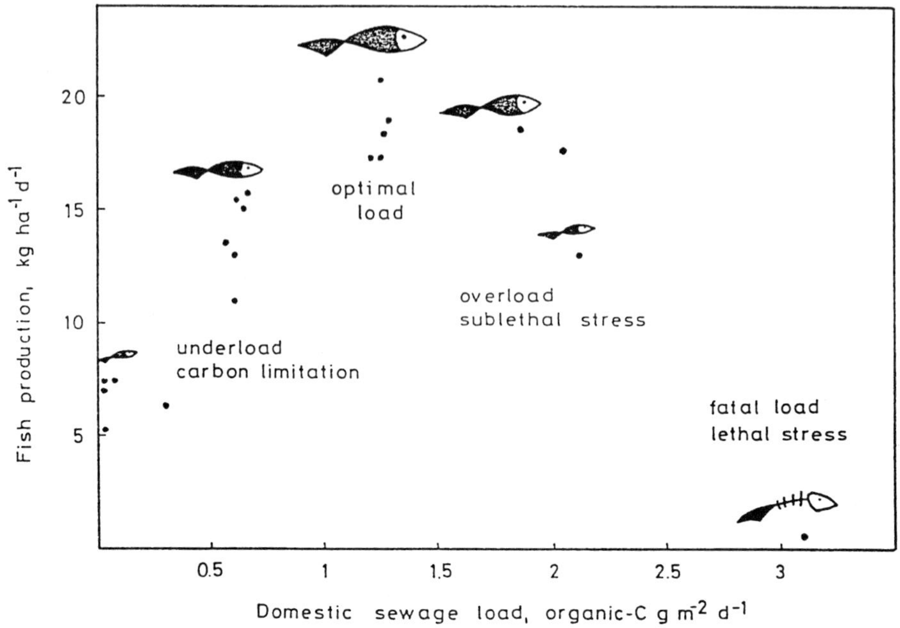

FIGURE 1. Relation between fish production and sewage loading rate.

20 to 5 kg/ha/day. The optimal load is around 150 m^3/ha/day. Higher loading rates create an overload situation which results again in a significant decrease of the average daily fish growth rate due to the sublethal stress of the unionized freely dissolved ammonia gas coupled with early morning low oxygen. The highest loading rate of 250 m^3/ha/day was accompanied by a mass fish kill in the middle of the growing season due to the lethal stress of the ammonia. The fatal load, however, may vary around this value depending primarily on the meteorological conditions in a given year or region.

Using this technological package, this fishpond ecosystem is able to produce 15 to 20 kg fish/ha/day and purify the sewage water to an acceptable level (Table 2). Both the COD (chemical oxygen demand, i.e., organic carbon) and the BOD (biochemical oxygen demand, i.e., easily oxidizable organic carbon) were reduced very significantly. Inorganic nitrogen and phosphorus were also very significantly reduced. The water may be released into any natural river or lake. The energy and nutrient content of the sewage were converted to fish tissue, which is edible food.

FISH FOOD COMPARTMENTS

The conversion of the daily introduced organic material into fish biomass is possible through the community's food web. Both the planktonic and benthic animal and plant communities were almost completely quantified during the first experimental trials on commercial ponds in 1971 and 1972 as well as during the

TABLE 2
River, Sewage, and Effluent Water Quality, mg/l

Parameter	River	Sewage	Effluent
Total solids	100–200	500–1000	150–200
COD_5	10–15	150–300	20–30
BOD_5	3–4	110–120	9–10
pH	7–8	7.3–7.4	8.1–8.3
O_2	7–8	0	8–10
NH_4-N	0.2–0.3	48–50	0.3–0.5
NO_3-N	0.8–1.0	0.4–0.5	0.2–0.3
Total-N	1–1.5	50–55	2–3
PO_4-P	0.02–0.16	8–8.5	0.4–0.5
Total-P	0.13–0.23	10–12	0.7–1.0
Detergent	0.04–0.08	4–4.5	0.15–0.18

five-year-long research project on the experimental ponds from 1974 to 1978. A team of Hungarian zoologists, algologists and bacteriologists was organized to quantify the community structure to species level. This time- and money-consuming effort was considered important in order to understand the functioning of this ecosystem. Almost two decades ago, when the original idea of this research project was born, this traditional holistic approach to describe the community structure at the level of species dominated, at least in Hungary. Here we synthesize the architecture of the pond ecosystem with emphasis on the natural food webs (Figure 2). We demonstrate the architecture of ponds with and without fish at different sewage loading rates. The figures on the curves are the load in the unit of g C/m²/day. All the ponds, regardless of whether they contain fish, have higher levels of each fish food organism at higher loading rates. The fish community impacts the food compartments dramatically: Their feeding and bioturbation smooth out what are otherwise significant fluctuations, peaks, succession of population, and high diversity. Ponds without fish become overgrown with macrophytes, macroalgae (*Cladophora*), algae, or zooplankton (*Daphnia*). These events are almost unpredictable. Without fish there is no balancing force.

In fishponds with a proper fish community, both the bacterioplankton and phytoplankton are highly productive. The only significant fluctuation is the tendency for a buildup of their population during the growing season and a significant biomass decrease at the end of the growing season, when the water cools before harvest. Among planktonic animals competing with silver carp for bacterioplankton and phytoplankton, rotifers dominate. Among the large zooplankton, copepods are present throughout the growing season and cladocerans, mainly *Moina,* exhibit a peak around midsummer. The grazing pressure of the common carp on the oligochaetes and chironomids living in the pond sediment is very pronounced compared to ponds without fish. There was also a detailed study of the small animals living in sediment, the meiobenthos, comprised mainly of nematodes and crustaceans. Their number may reach 200,000/m² but their biomass is low and their availability

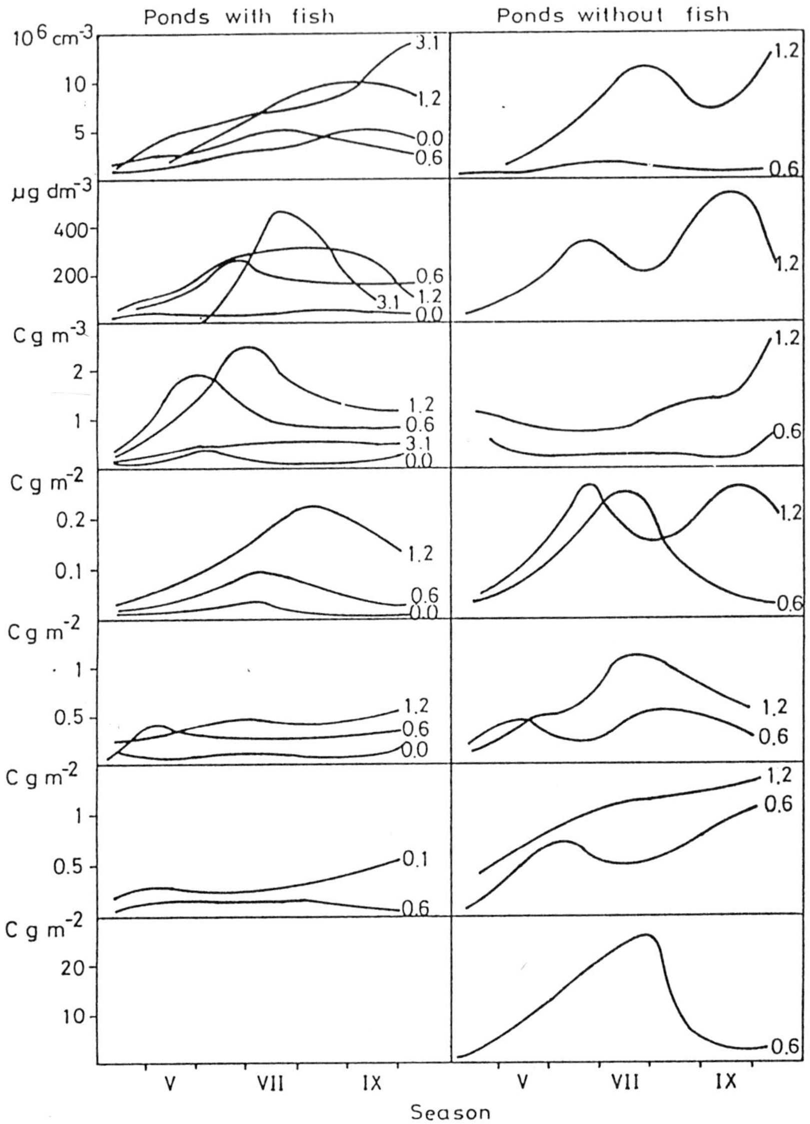

FIGURE 2. Trends of natural food compartments in sewage ponds with and without fish at different loading rates. Figures on curves are loading rates in g C/m²/day.

for common carp is not evident due to their small size. No difference was detected between ponds with and without fish in seasonal trends for meiobenthic biomass.

ENERGY TRANSFER

Even a detailed qualitative analysis of the seasonal dynamics of the number or biomass of the fish food organisms gives only an approximation of their quantitative

TABLE 3
Energy Transfer in Organic-Fed Fish Ponds[a]

Pathways	Domestic sewage	Pig manure	Chicken manure	Sheep manure
Organic carbon load	0.6–3.1	2.3–4.8	0.7–5.2	3.1
Primary production	3.9 ± 1.75 (n = 6)	4.68 ± 1.93 (n = 95)	5.22 ± 2.93 (n = 296)	2.87 ± 1.46 (n = 48)
Planktonic bacterial production	2.48 ± 0.15 (n = 6)	2.89 ± 1.9 (n = 34)	0.40 ± 0.25 (n = 83)	0.58 ± 0.41 (n = 77)
Benthic bacterial production	—	—	1.24 ± 0.48 (n = 5)	3.64 ± 2.34 (n = 19)
Fish production	0.11 ± 0.01 (n = 6)	0.18 ± 0.07 (n = 4)	0.18 ± 0.10 (n = 56)	0.16 ± 0.06 (n = 28)
Fish production × 100 / Primary production	3.37 ± 1.48 (n = 6)	4.15 ± 2.25 (n = 4)	6.42 ± 3.87 (n = 78)	5.71 ± 2.67 (n = 48)
Fish production × 100 / Primary + bacterial prod.	1.84 ± 0.46 (n = 6)	2.02 ± 0.49 (n = 4)	4.77 ± 2.66 (n = 52)	2.13 ± 1.26 (n = 47)

[a] All units are g C/m^2/day, except the last two rows, which are percentages.

importance in the production processes. To measure, however, biomass production per unit time for each species, a tremendous amount of work would be required by highly specialized workers. Another alternative is to convert the measured biomass values to biomass production by applying the production-per-unit-biomass coefficients for each species available in the literature. However, these coefficients fluctuate in a rather broad range, depending on available food concentration, competition, and many other factors. Instead of analyzing the detailed pathways of energy flow in the zooplankton and zoobenthos community, we have devoted the available money and manpower to measuring the production processes at the first and last stages of the energy flow. This way we were able to quantify the utilization efficiency of the planktonic and benthic animal food resources.

The quantity of the daily organic carbon load and the autochthonous organic carbon production through the algal primary production were both measured almost daily. The conversion of the daily introduced sewage through the heterotrophic bacterial production processes is the key production pathway in sewage-fed fishpond technology. Nevertheless, it had not been measured until the start of our research project. This quantification turned out to be very significant to understanding the production pattern in the almost entirely organic-fed aquaculture system. Here we present our results for domestic sewage as well as for pig, chicken, and sheep manure-fed fishponds (Table 3).

To quantify the bacterial production processes in an organic-fed fishpond we applied three different methods in parallel: dark ^{14}C-CO_2 fixation method of Romanenko (1964), ^{35}S-SO_4 uptake method of Monheimer (1974), and the

^3H-thymidine method of Moriarty (1986). Reliable results were reached in the water columns through both the ^{14}C-CO_2 and ^3H-thymidine methods, but for the sediment we relied just on the ^3H-thymidine method. In these ways, quantifying the production processes of bacterio- and phytoplankton and measuring the fish production, we were able to determine the energy transfer efficiencies in these systems. How much of the bacteria and primary produced organic carbon is converted to fish tissue carbon growth daily? We have answered this question based upon several hundreds of measurements.

The primary production was very high in all of the organically fed systems. The planktonic bacterial production was very high in the domestic sewage and pig manure-fed system and lower in chicken and sheep manure systems. The reason for this pronounced difference is explainable by the form and mode of introduction. The domestic sewage and pig manure was introduced in a very diluted liquid form, which was immediately suspended and remained mainly in the water column. The chicken and sheep manures were introduced in solid form. Although well-dispersed on the pond surface, most of the manure sank slowly or rapidly to the bottom, creating a very nutrient-rich condition for sediment bacteria. The measured benthic bacterial production was very high, especially in ponds receiving sheep manure. The average coefficient of conversion from primary production to useable biomass was almost 5% in chicken manured ponds and lowest (<2%) in domestic sewage-fed ponds (Table 3). The explanation is the much higher fish density in the former ponds, where we used fingerlings at a density of 12,000/ha instead of two-year-old fish of 4000/ha in the domestic sewage technology. So the efficiency is highly regulated by fish density through force feeding on the available fish food organisms.

SEWAGE PURIFICATION

Describing the management parameters, we have presented the routine values of the effluent water quality. All the ranges were well below the acceptable standards (Table 2). We have computed the relation between effluent water quality and the daily sewage load (Figure 3). The relations are rather tight. Chemical oxygen demand (COD) and biochemical oxygen demand (BOD_5) both were low in the effluent until the daily sewage load of around 150 m^3/ha. Above this value both parameters suddenly increased, indicating the decreasing purification capacity at the highest load. The total nitrogen, total phosphorus, and the detergent concentrations in the effluent water were still acceptable, even at the highest loads.

The removal efficiencies for water quality parameters were also computed along with the increasing daily nutrient loads. The relation was similar for all parameters (Figure 4). The removal percentages were small at low loading rates and increased with increasing loads. The removal percentage for total phosphorus reaches a very high value, almost 99% at the highest loadings. The bulk of the phosphorus removal is realized through the sediment's phosphorus absorbing capacity, which is so high only if the bioturbating activity of common carp is present. The removal through harvested fish biomass is around 8 to 10%. The macrophytes remove 6 to 8% (Figure 5). All these removal efficiencies compete very well with the conventional sewage treating methods of trickling filter and activated sludge.

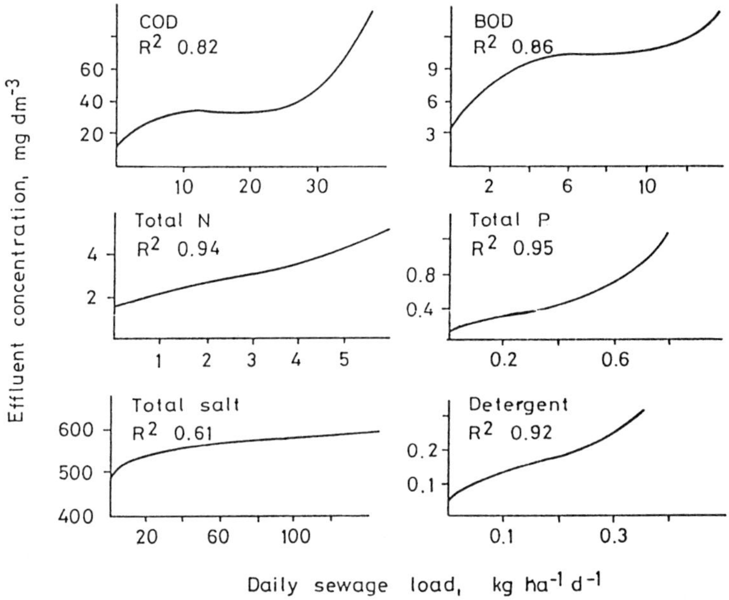

FIGURE 3. Relation between effluent water quality and daily sewage load.

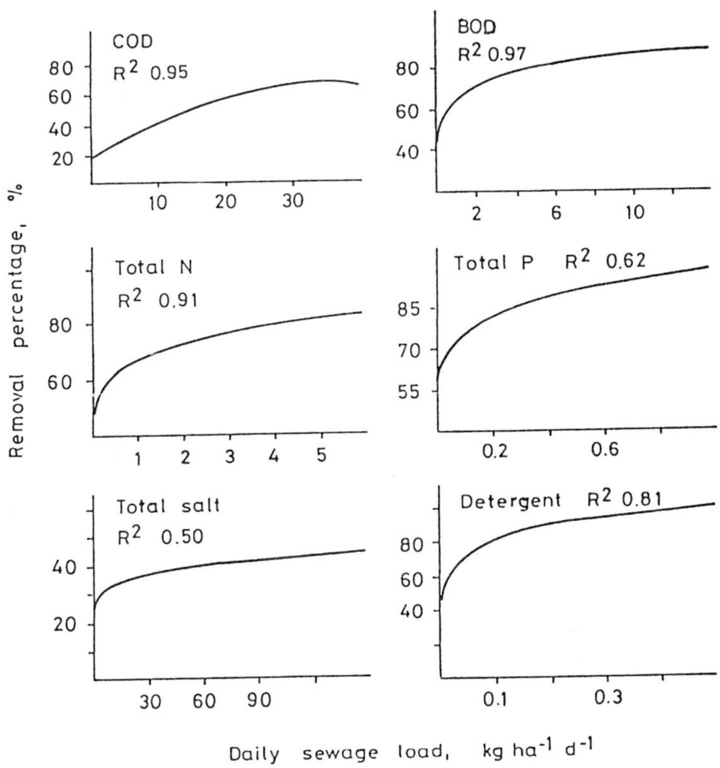

FIGURE 4. Relation between sewage purification capacity and daily sewage load.

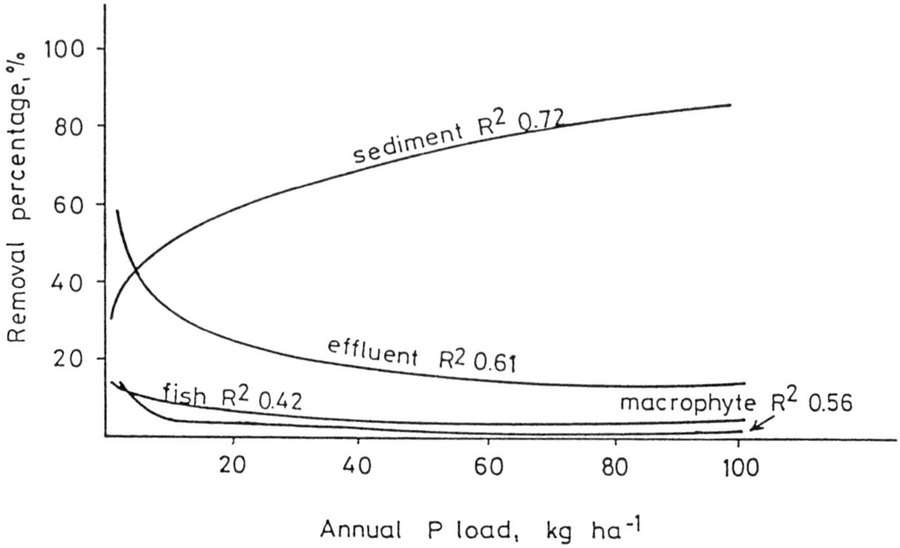

FIGURE 5. Phosphorus removal compartments (removal efficiency related to daily load).

REFERENCES

Monheimer, R.H. 1974. "Sulphate uptake as a measure of planktonic microbial production in freshwater ecosystems." *Can. J. Microbiol.* 20:825-831.

Moriarty, D.J.W. 1986. "Measurement of bacterial growth rates in aquatic systems using rates of nucleic acid synthesis." *Adv. Microb. Ecol.* 19:245-292.

N.W.B. 1982. *Sewage disposal in reservoirs stocked with fish and fishponds, technical guidelines.* OVH M1-10 419-82T. National Water Board. Budapest, Hungary. 27 pp. (In Hungarian).

Ponyi J., P. Bíró, J. Oláh, N.P.-Zánkai, G. Tamas, T. Csekei, Gy. Kiss, T. Morvai, and I. Bancsi. 1973. "Limnological investigations of a fish-pond supplied with sewage-water in the vicinity of Lake Balaton I." *Annal.Biol.Tihany* 40: 227-284.

Ponyi J., P. Bíró, N.P.-Zánkai, J. Oláh, G. Tamas, T. Csekei, Gy. Kiss and T. Morvai. 1974. "Limnological investigations of a fish-pond supplied with sewage-water in the vicinity of Lake Balaton II." *Annal. Biol. Tihany* 41: 235-288.

Romanenko, V.I. 1964. "Heterotrophic assimilation of CO_2 by bacterial flora of water." *Microbiol.* 33:610-614.

Chapter 6

LAND-BASED SYSTEMS, CONSTRUCTED WETLANDS, AND AQUATIC PLANT SYSTEMS IN THE UNITED STATES: AN OVERVIEW

George Tchobanoglous

ABSTRACT

The use of what might be termed "natural systems" for the treatment of wastewater has increased as reliable design information and data have been developed. The purpose of this paper is to present a brief overview of the types of natural systems that have been developed and used for this purpose. The topics to be discussed include: (1) a classification of natural systems, (2) a review of the important design requirements for aquatic systems, (3) a more detailed discussion of constructed wetlands and aquatic plant systems, and (4) a review of some of the types of questions that must be answered to improve the operation of treatment systems using plants to treat wastewater.

CLASSIFICATION OF NATURAL TREATMENT SYSTEMS

"Natural treatment systems" may be defined as treatment systems in which specific ecological environments found in nature are utilized for the treatment of wastewater. Natural treatment systems can be grouped into three major classifications: (1) land-based systems, (2) constructed wetlands, and (3) aquatic plant systems (see Table 1). As reported in Table 1, land-based systems may be classified as: (1) slow-rate irrigation systems, in which the applied wastewater is used for the irrigation of crops; (2) rapid infiltration systems, in which the wastewater is treated as it percolates through the soil horizon; and (3) overland flow systems in which the wastewater is allowed to flow over a planted, impervious surface in a thin layer. Because these systems have become so commonplace, they are not considered further in this paper. References 8, 10, and 11 may be consulted for further details on the land-based systems. The number of natural systems currently in operation in the U.S. is also reported in Table 1.

DESIGN CONSIDERATIONS FOR CONSTRUCTED WETLANDS AND AQUATIC PLANT SYSTEMS

The design of aquatic systems must be based on a consideration of the contaminants of concern, knowledge of the fate processes in the environment that affect the contaminants of concern, and the specific details involved in the design. Each of these topics is considered further in the following discussion.

0-87371-990-5/97/$0.00+$.50

TABLE 1
Classification and Number of Land-Based, Constructed Wetlands, and Aquatic Plant Systems

Type of system	Number
Land-based systems	>1,000
Slow rate systems	
Rapid infiltration systems	
Overland flow systems	
Constructed wetlands	
Free water surface (FWS)	56
Subsurface flow (SF)	98
Aquatic plant systems	
Systems employing floating plants	4–8
Systems employing submerged plants	?

Source: Adapted in part from U.S. EPA

TABLE 2
Contaminants of Concern in Wastewater Treatment

Contaminants	Reason for concern
Suspended solids	Sludge deposits and anaerobic conditions
Biodegradable organics	Depletion of natural oxygen resources and the development of septic conditions
Pathogens	Communicable diseases
Nutrients	Growth of undesirable aquatic life; eutrophication
Heavy metals	Metallic constituents added by usuage. Many metals are also classified as priority pollutants
Priority organic pollutants	Suspected carcinogenicity, mutagenicity, teratogenicity, or high acute toxicity. Many priority pollutants resist conventional treatment methods (known as refractory organics)
Dissolved inorganics	Inorganic constituents added by usage

CONTAMINANTS OF CONCERN

The principal contaminants of concern in wastewater are reported in Table 2. The first three contaminants reported in Table 2 (organic matter, suspended solids, and pathogenic organisms) are the traditional contaminants of concern in the treatment of wastewater. Today, the other contaminants listed in Table 2 are of equal or greater concern. The contaminants that must be removed will be site specific and must be evaluated in each case.

FATE PROCESSES IN THE ENVIRONMENT

To design effective and ecologically sound treatment processes, it is important to have a clear understanding of the mechanisms that are operative in the environment that serve to alter the contaminants of concern to make them benign, or to recycle them. The principal fate processes in the environment are reported in Table 3.

TABLE 3
Fate Processes in Aquatic Systems

Process	Comment
Bacterial conversion-aerobic/anaerobic	BOD_5, nitrification, denitrification, sulfate reduction, etc.
Algal synthesis	NH_4^+, PO_4^{-3}, etc.
Gas absorption/desorption	O_2, CO_2, CH_4, NH_3, H_2S
Sedimentation	SS, BOD_5 associated with SS; adsorbed contaminants
Filtration	SS, BOD_5 associated with SS, adsorbed contaminants
Natural decay	Bacterial, plant
Adsorption	Metals, trace organics, NH_4^+
Volatilization	VOCs, NH_3
Chemical conversion reactions	Hydrolysis, substitution, precipitation, photooxidation, ion exchange, oxidation/reduction, etc.

TABLE 4
Design Steps for Constructed Wetlands and Aquatic Plant Systems

Design step	Comment
1. Definition of treatment objectives	Contaminants to be removed and the degree of removal required must be defined
2. Site evaluation and selection	Topography, soil, flood hazard, existing land use, climate
3. Determination of pretreatment level	Degree of pretreatment depends on effluent requirements and the removal capability of the particular natural systems
4. Vegetation selection and management	Cattails, reeds, rushes, bulrush, sedges, cannas, water hyacinths, duckweed, etc.
5. Determination of design parameters	Hydraulic detention time, water depth, basin geometry, BOD loading rate, hydraulic loading rates (surface and cross-sectional)
6. Vector control measures	Biological and/or chemical control
7. Detailed design of system components	Preparation of detailed plans and specifications
8. Determinatin of monitoring requirements	Depends on local control agency and process control requirements

DESIGN STEPS

The important steps involved in the design of natural treatment systems are summarized in Table 3.[7] The importance of each of the steps cannot be overstressed. If the steps given in Table 4 are followed, the resulting design will usually be workable.

CONSTRUCTED WETLANDS

Wetlands constructed specifically for the purpose of treating wastewater are known as "constructed wetlands." The two principal categories of constructed wetlands are: (1) free water surface (FWS) wetlands and (2) subsurface flow (SF) wetlands. Each of these types of constructed wetlands is considered in the following discussion and in greater detail in References 1, 2, 3, 9, and 11.

FIGURE 1. Typical constructed free water surface wetland with bulrush, other sedges, and cattail used to treat the effluent from a series of oxidation ponds.

FREE WATER SURFACE WETLANDS

Free water surface wetlands consist of basins or channels with a suitable medium to support the growth of emergent vegetation, with open areas and with water flowing at relatively shallow depths. The key feature of a FWS wetland is the presence of a free water surface. A typical example of a FWS wetland is shown in Figure 1. Typical design data for FWS wetlands are presented in Table 5.

SUBSURFACE FLOW WETLANDS

In subsurface flow wetlands, the applied wastewater flows horizontally through basins or channels filled with rock or sand and planted with suitable plants (see Figure 2). In Europe, where soil is used, systems are known as root-zone method (RZM). In SF systems, the specific surface area and the porosity of the medium are important variables in the design of these systems. Typical design criteria for SF systems are presented in Table 5. As reported in Table 5, there is a significant variation in the design values used for both the FWS and SF systems. This wide variation in design values is due, in part, to the lack of scientifically based design criteria.

PERFORMANCE EXPECTATIONS FOR CONSTRUCTED WETLANDS

Constructed wetland systems are effective in the removal of biochemical oxygen demand (BOD), suspended solids (SS), and nitrogen. Pathogenic microorganisms, metals, and trace organics are also removed, but to a lesser extent. The basic removal processes operative for BOD, SS, and nitrogen include biological conversion, sedimentation, chemical precipitation, and adsorption (see Table 3).[4,7,9] Typical performance data for constructed wetlands are reported in Table 6. As shown in Table 6,

TABLE 5
Design Guidelines for Constructed Wetlands

Design parameter	Unit	Type of system FWS	Type of system SF
Hydraulic detention time	d	1–16	1–4
Water depth	ft	0.3–2.0	1.0–2.5
BOD_5 loading rate	lb/acre·d	20–40	10–60
Specific area	acre/(Mgal/d)	20–40	2–20
Cross-sectional hydraulic loading	gal/ft^2·d	NA	200–1000

Notes: FWS = free water surface; SF = subsurface flow. Excludes advanced wastewater treatment (AWT) and total retention systems.

Source: Adapted in part from US EPA.

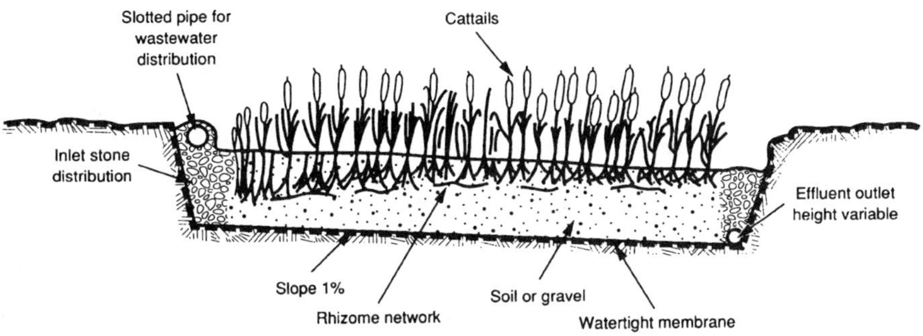

FIGURE 2. Cross section through typical subsurface flow wetland treatment system.[7]

TABLE 6
Typical Performance Data for Constructed Wetlands

Quality parameter	Unit	Type of system FWS	Type of system SF
BOD_5	mg/l	<20	<25
SS	mg/l	<20	<15
Total N	mg/l	<15	<12
Total P	mg/l	<6	<1–4

Notes: FWS = free water surface; SF = subsurface flow. Excludes AWT and total retention systems.

Source: Adapted, in part, from U.S. EPA.

TABLE 7
Capital Cost of Constructed Wetlands

Type	Cost, $/acre[a] Low	High	Mean	Median
SF	72,222	100,000	87,218	87,560
FWS	2,071	65,000	22,200	18,000

[a] Cost reference 1988–1989.

Source: U.S. Environmental Protection Agency.

the performance of constructed wetlands is quite good, with most parameter values being below 20 mg/l.

CONSTRUCTION COSTS FOR CONSTRUCTED WETLANDS

Representative construction cost data for constructed wetlands are presented in Table 7. As shown in Table 7, SF systems cost approximately four times as much as FWS wetlands. However, it should be noted that the surface area required for a SF system is approximately one quarter of that required for FWS wetland (see Table 5). At the present time, there is insufficient operating experience with these systems to develop meaningful operation and maintenance costs.

AQUATIC PLANT SYSTEMS

Aquatic plant systems are, typically, shallow ponds or channels with floating or submerged aquatic plants. At the present time, only a limited number of treatment systems employing aquatic plants are in use (see Table 1).

FLOATING AQUATIC PLANT SYSTEMS

The two plants that have been used most commonly in floating aquatic plant wastewater treatment systems are water hyacinths (*Eichhornia crassipes*) and duckweed (*Lemna* spp.). Typical data for four different types of floating aquatic plant treatment systems are presented in Table 8. As shown in Table 8, both unaerated and aerated water hyacinth systems have been used. A typical water hyacinth system employing supplementary aeration is shown in Figures 3 and 4. The loading rate for aerated systems (see Table 8) is significantly higher than for nonaerated systems.[5] A typical duckweed wastewater treatment system is shown in Figure 5.

SUBMERGED AQUATIC PLANT SYSTEMS

Submerged aquatic plant systems employ plants suspended in the water column or rooted in the bottom sediments. The photosynthetic parts of the plants are in the water column. The most common plant that has been studied is pennywort (*Hydrocotyle umbellata*). To date, no full-scale aquatic treatment systems employing submerged aquatic plants have been built.

TABLE 8
Typical Design Criteria for Floating Aquatic Plant Wastewater Treatment Systems[a]

	Type of water hyacinth treatment system			
Item	**Secondary aerobic (nonaerated)**	**Secondary aerobic (aerated)**	**Nutrient removal aerobic (nonaerated)**	**Duckweed treatment system**
Influent wastewater	Screened or settled	Screened or settled	Secondary	Facultative pond effluent
Influent BOD_5, mg/l	130–180	130–180	30	40
BOD_5 loading, lb/ac·d	40–80	150–300	10–40	20–30
Water depth, ft	1.5–3	3–4	2–3	4–6
Detention time·d	10–36	4–8	6–18	20–25
Hydraulic loading rate, Mgal/ac·d	0.035–0.075	0.125–0.300	0.040–0.160	0.060–0.090
Water temperature, °C	>10	>10	>10	>7
Harvest schedule	Annually to seasonally	Twice monthly to continuously	Twice monthly to continuously	Monthly for secondary treatment weekly for nutrient removal

[a] Adapted from References 7 and 9.

FIGURE 3. View of water hyacinth treatment system at San Diego, California, in which supplementary aeration is used to sustain higher organic loading rates.

FIGURE 4. Harvesting of water hyacinth using truck equipped with articulated pickup boom. Person in the background is redistributing the water hyacinths in the pond after harvesting.

FIGURE 5. Typical duckweed treatment system comprised of a pond with floating dividers and a duckweed harvester (Courtesy Lemna Corp.).

PERFORMANCE EXPECTATIONS FOR AQUATIC PLANT SYSTEMS

Aquatic plant systems are also effective in the removal of BOD, SS, and nitrogen. Typical performance data for aquatic plant systems are reported in Table 9. As shown in Table 9, the performance of water hyacinth systems is very good, with most parameter values being below 20 mg/l.

TABLE 9
Typical Performance Data for Floating Aquatic Plant Wastewater Treatment Systems

Quality parameter	Unit	Type of hyacinth system S(NA)	S(A)	N(NA)	DW
BOD_5	mg/l	<20	<15	<10	<30
SS	mg/l	<20	<15	<10	<30
Total N	mg/l	<15	<15	<5	<15
Total P	mg/l	<6	<1–2	<2–4	<6

Notes: S(NA) = secondary–nonaerated;
S(A) = secondary–aerated; N(NA) = nutrient–nonaerated;
DW = duckweed.

Source: Adapted in part from Reference 7 and U.S. EPA.

RESEARCH ISSUES

Although a number of constructed wetlands and floating plant systems have been built, much remains to be known before the design and construction of these systems can be considered routine. Important issues that must be defined include:

1. Improved understanding of how plants function.
2. New configurations and improved operating techniques.
3. Engineered applications of plants for the treatment of wastewater.
4. Improved management techniques for natural systems.

For example, it is well known that a number of aquatic plants are able to transport oxygen from the atmosphere to the root zone. What is unknown is how to optimize the rate of oxygen transport to the root zone selectively to meet specific treatment objectives. The beneficial effects of variable recycling rates in controlling the transport of contaminants to achieve effective treatment is just beginning to be appreciated more fully. To take advantage of the beneficial effects of flow recycling, new pond configurations (see Figure 6) are needed to minimize the cost of pumping.[6] As noted previously, the wide range of parameter values used for the design of constructed wetlands reflects the somewhat empirical approach that has been used for the design of these systems. Before engineered natural systems can be implemented routinely, rational design parameters based on defendable scientific principles must be developed. In the past, the management of natural treatment systems has been haphazard and not well defined. Improved management techniques must be developed that will

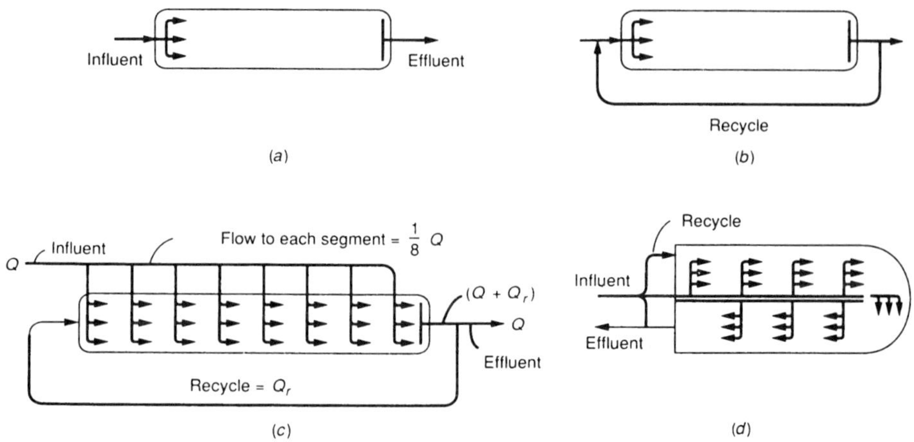

FIGURE 6. Alternative flow diagrams for constructed aquatic treatment systems: (a) Plug-flow, (b) Plug-flow with recycle, (c) Step-feed with recycle, and d) Step-feed with recycle in wraparound pond.[6,7]

allow for operational changes to be made in response to changing wastewater characteristics, climatic conditions, and effluent discharge requirements. The aforementioned issues are but a few examples of technical issues that need to be resolved if the use of aquatic systems for wastewater treatment is to become a viable alternative to conventional treatment.

SUMMARY

The beneficial use of natural systems for the treatment of wastewater is well established. However, the use of such systems is in its infancy. As the uncertainties now associated with the use of natural systems are resolved, these methods of treatment will assume their rightful place alongside more conventional technology. The challenge to the attendees of the First Conference on Ecological Engineering is to develop the data and information that will hasten the time when natural aquatic systems are commonly accepted for the treatment of wastewater.

REFERENCES

1. Hammer, D.A. (ed). *Constructed Wetlands for Wastewater Treatment: Municipal, Industrial and Agricultural.* Lewis Publishers. Chelsea, Michigan. 1989.
2. Reed, S.C., E.J. Middlebrooks, and R.W. Crites. *Natural Systems for Waste Management and Treatment.* McGraw-Hill. New York. 1988.
3. Reddy, K.R. and W.H. Smith (eds). *Aquatic Plants for Water Treatment and Resource Recovery.* Magnolia Publishing Inc. Orlando, Florida. 1987.
4. Stowell, R., R. Ludwig, J. Colt, and G. Tchobanoglous. "Concepts in Aquatic Treatment System Design." *J. Environmental Engineering Division, Proceedings ASCE,* Vol. 107, No. EE5, October 1981.

5. Tchobanoglous, G., F. Maitski, K. Thomson, and T.H. Chadwick. "Evolution and Performance of City of San Diego Pilot Scale Aquatic Wastewater Treatment System Using Water Hyacinths." *Journal WPCF,* Vol. 61, No.11/12, 1989.
6. Tchobanoglous, G. "Aquatic Plant Systems for Wastewater Treatment: Engineering Considerations," in: K.R. Reddy and W.H. Smith (eds), *Aquatic Plants for Water Treatment and Resource Recovery,* Magnolia Publishing Inc. Orlando, Florida. 1987.
7. Tchobanoglous, G. and F.L. Burton. *Wastewater Engineering: Treatment, Disposal, Reuse,* 3rd ed. McGraw-Hill. New York. 1991.
8. U.S. Environmental Protection Agency. *Process Design Manual for Land Treatment of Municipal Wastewater.* EPA 625/1-81-013. U.S. EPA. Cincinnati, Ohio. October 1981.
9. U.S. Environmental Protection Agency. *Design Manual for Constructed Wetlands and Floating Aquatic Plant Systems for Municipal Wastewater Treatment.* EPA 625/1-88-022. U.S. EPA. Cincinnati, Ohio. September 1988.
10. U.S. Environmental Protection Agency. *Process Design Manual for Land Treatment of Municipal Wastewater: Supplement on Rapid Infiltration and Overland Flow.* EPA 625/1-81-013a, U.S. EPA. Cincinnati, Ohio. October 1984.
11. Water Pollution Control Federation. *Manual of Practice, Natural Systems for Wastewater Treatment.* February 1990.

Chapter 7

ECOLOGICAL ENGINEERING FOR WASTEWATER TREATMENT: THEORETICAL FOUNDATIONS AND PRACTICAL REALITIES

Björn Guterstam

ABSTRACT

Ecological engineering is needed not only as a tool for solving a pollution problem but also for creating feedback loops of resources from wastewater to society. The foundations of this approach are the considerable amounts of the major plant nutrients nitrogen and phosphorus to be found in household and municipal wastewater, which can be reused in new biological production. Using a constructed food chain in an indoor, intensive aquaculture, a full-scale wastewater treatment facility has been in operation since late 1989 in a Nordic climate at 60° N. lat. in Sweden. Wastewater from 40 person equivalents (p.e.) has been treated from January 1990 to January 1993, i.e., an average flow of 7 m^3 daily.

From a calculation based on the 100 p.e. population of Stensund Folk College, which serves as a model society for studies of recycling technologies of wastewater in this study, the nutrient recycling was 8% of the phosphorus and 11% of the nitrogen in the aquaculture. Phosphorus in incoming water is 38% lower than the national average, presumably due to use of phosphate-free detergents. The effluent nearly meets Swedish bacterial and organic matter standards for swimming water. The concentration of seven monitored metals in the water was reduced 23 to 56% by anaerobic treatment. There was a 69% higher energy consumption in winter than in summer. The problems of scaling and productivity of the wastewater aquaculture are discussed, together with its operational costs.

THE SWEDISH WAY

Municipal wastewater treatment in Sweden was built to meet new pollution restrictions as they were formulated in the 1960s, i.e., to prevent eutrophication of lakes and coastal waters and to meet public health standards. Advanced wastewater treatment (AWT) uses a three-step treatment with mechanical, biological, and chemical treatment of the wastewater. Today 80% of urban residents are connected to this type of treatment. After a few years of treatment, recovery of the waters was observed. Algal blooms decreased in the inner coastal waters of the Baltic archipelagoes and swimming was possible even in the waters of the capital, Stockholm.

During the late 1980s, eutrophication in outer parts of Swedish coastal waters resulted in algal blooms mainly caused by nitrogen emissions from municipal

0-87371-990-5/97/$0.00+$.50

wastewater outlets, runoff of nutrients from agriculture, and atmospheric deposition (Rosenberg et al. 1990). New governmental efforts now stress reconstruction of municipal treatment plants, including a nitrogen elimination feature through denitrification (Prop. 1987/88:85). The first "megaproject" for advanced three-step wastewater treatment with nitrogen removal puts Sweden in a leading position at a price of about 25 billion SEK (UN 1987, Gandy 1982). The new denitrification project has a price of 50 to 100 SEK/kg/N, which will result in a total cost of about 4 billion SEK. The concept of traditional engineering deals with wastewater mainly as a problem in order to get rid of nitrogen, phosphorus, and organic matter. At the same time, agriculture tries to minimize the costs of fertilizers.

Up to 60% of the manuring potential of sludge was used in agriculture until 1989 (SCB 1987, p. 81). Since then, the Swedish Farmers Federation has stopped sludge spreading on members' farmland because of fear of toxic constituents such as metals and organochlorides (Denninger pers. comm.).

FROM TECHNOCRATIC OPPRESSION TO ECOLOGICAL LIBERATION

"The Swedish Way" of treating wastewater has its roots in a strongly centralized society, where the same technology is implemented in the whole country. Once decided on, the approach did not change or develop until the new denitrification decision. Unfortunately, modern ecological demands are not met in Sweden when taking aspects of toxic substances and recycling of nutrients into consideration.

Municipal wastewater is still not regarded ecologically — as a fertilizer resource for plant and even food production — although the idea was stressed as early as the 1950s by the ecologist and member of the Swedish Academy, Sten Selander (Selander 1987). He regarded the untreated flow of wastewater from the increasing urban population as a nutrient drain. On the global scale he predicted a future starvation if the limited mineral phosphorus was not recycled to agriculture. Scientists such as the microbiologist Staffan Delin in Sweden were stopped on political and administrational levels when they proposed the use of wastewater for production of new biomass instead of its nutrients being mineralized in modern treatment plants (Delin 1973). A first real signal from the government to enforce a resource management strategy for wastewater treatment was the decision in 1988 to give economic support to the Stensund Wastewater Aquaculture project, which was opened in late 1989. The governmental strategy to protect the coastal areas from eutrophication got a boost in that direction from the Stensund project.

THE WASTEWATER RESOURCE

The aim of our conference is to answer some general questions about the wastewater resource and to evaluate the conditions for introduction of recycling technologies for wastewater treatment, especially in northern, industrialized countries. This aspect was focused on during the 1970s, as wastewater was tested as a source for biomass production and for treatment in aquacultures and in artificial and constructed wetlands (Reed et al. 1988). Recommendations and conclusions from meetings stress the importance of recycling and management of natural resources like water, nutrients, and energy (D'Itri 1977, Bastian and Reed 1979). Recent scientific and traditionally engineered approaches in the industrial world unfortunately tend to regard natural systems more as a new kind of mechanical filter for wastewater — a good example is Cooper and Findlater (1990). Our conference on ecological engineering for wastewater treatment should be seen in the perspective of both 20-year-old recommendations from ecologists and the current political demands to find ways for a sustainable development using a resource management strategy based on ecological principles (Daly 1990).

Let me continue with Sweden as an example of the present situation when evaluating wastewater as a resource. The municipal wastewater is a mixture from households and industries diluted and cooled by leakages into the pipes of groundwater and water from other sources. This resource flow is about 300 l/capita/day, up to 15°C warm as a yearly average. The heat becomes more available as new pipes are built and cooling through leakages can be avoided. The theoretical, potential energy resource for heat pumping and methane production out of municipal wastewater is approximately 10 TWh, or 15% of the nuclear energy produced in Sweden. This is calculated on an urban population of 7 million people as 9.2 TWh heat from wastewater and 0.4 TWh potential energy from biogas production (SCB 1987, Anderson 1976).

Wastewater carries 7000 t phosphorus and 40,000 t nitrogen each year from the municipal treatment plants (Hasselgren 1988). The magnitude of the plant nutrient resource in the wastewater, when compared with the amounts of phosphorus (45,000 t) and nitrogen (240,000 t) used as fertilizers in the Swedish agriculture, is 16% and 17%, respectively (SOU 1983:10). These figures have been calculated in order to indicate the order of magnitude of resources in wastewater in an industrialized country with predominantly urban population (75% in Sweden). From a thermodynamic perspective it is expensive to buy new fertilizer in the form of ammonia synthesized from atmospheric nitrogen and hydrogen from oil by the Haber-Bosch method (at 400°C and 200 atm): The production of 1 kg of nitrogen costs 14 kWh, which makes nitrogen more expensive than fuel for tractors in the modern, mechanized agriculture (Anderson 1976).

As fresh water is seldom a limited resource in Sweden, this resource has not been taken into consideration. Distributing secondarily treated wastewater as nutrient-rich fresh water for irrigation purposes is still an unexplored market. Golf course and garden irrigation sometimes are restricted because of periodical shortage of water. This type of reuse may be better than discharging treated wastewater into recipient ecosystems during the summer when they are most sensitive to ecological disturbances.

TOXIC SUBSTANCES AND PATHOGENS

The problems of pathogens and of toxic chemicals in wastewater are increasing with increased amounts of wastes and use of chemical substances in our societies. Modern wastewater treatment plants are not constructed for detoxification of stable organic compounds or elimination of metals from wastewater. Pathogens are eliminated mostly by chlorination during the swimming season, but this is sometimes not sufficient even in Sweden for protecting swimmers from health risks.

The number of chemical substances used in Sweden is 20,000, including metals, pesticides, detergents, etc. (SOU 1984:77). In Sweden, though, an increasing awareness of chemical pollution has led to the introduction of environmentally sound soaps and detergents. The authorities also try to eliminate the most toxic chemicals from the market.

Understanding the limitations of recycling due to the problems with toxins is essential as our work continues with development of new technologies. Research in this field was done in the traditional Chinese aquaculture, where ecotoxicological aspects were investigated in "The Yar Lake Project" (Chang et al. 1981, Zhang et al. 1982). The fate and biodegradation of organochloride pesticides were followed in algae and fish ponds receiving industrial wastewater. The use of microbiological processes and the knowledge of the time and space they need to detoxify modern industrial wastewater shows the deep ecological know-how represented by Chinese scientists. Unfortunately, Western ecotoxicological experiences have not become transformed into new and better wastewater technologies.

FOUNDATIONS OF ECOLOGICAL ENGINEERING FOR WASTEWATER TREATMENT

New technologies taking ecological aspects into consideration, together with an increased social awareness of the environmental problem, are often described as ecological engineering. H.T. Odum, who coined the term nearly 30 years ago, has described ecological engineering as half science and half engineering: "techniques of designing and operating the economy with nature . . . Just as an engineer is asked to make a bridge that works and lasts, the ecological engineer should provide a pattern with nature that works and lasts"(pp. 80 to 81, Mitsch and Jørgensen 1989). Ecological engineering and ecotechnology are defined by Mitsch and Jørgensen (1989) as "the design of human society with its natural environment for the benefit of both." The concepts of ecology give engineering a new dimension in helping to find ways for our societies to develop sustainable strategies to achieve many of today's ambitious political goals. This constitutes a theoretical platform for us who are trying to create the first examples of those principles for wastewater. The roots and rationales of this paradigm are further presented by Mitsch (1996).

In order to give an overview of different approaches to wastewater treatment, they have been placed in a two-dimensional diagram where the abscissa shows the degree of treatment intensity and the ordinate shows the degree of recycling (Figure 1).

Conventional wastewater treatment as it is performed today, using advanced wastewater treatment, consists of intensive technologies with little or no recycling. In Sweden, sludge recycling is banned by the farmers (see above) and some treatment plants do extract heat out of the water by heat pumping. The recycling potential is indicated by the arrow. Soil filters and oxidation ponds are technologies to clean water and they do not demand much maintenance. Natural wetlands do recycle some of their nutrients in a way beneficial to the ecosystems. Planted soil filters, infiltration wetlands, and energy forests represent increasing degrees of recycling. They constitute examples of constructed lands and wetlands.

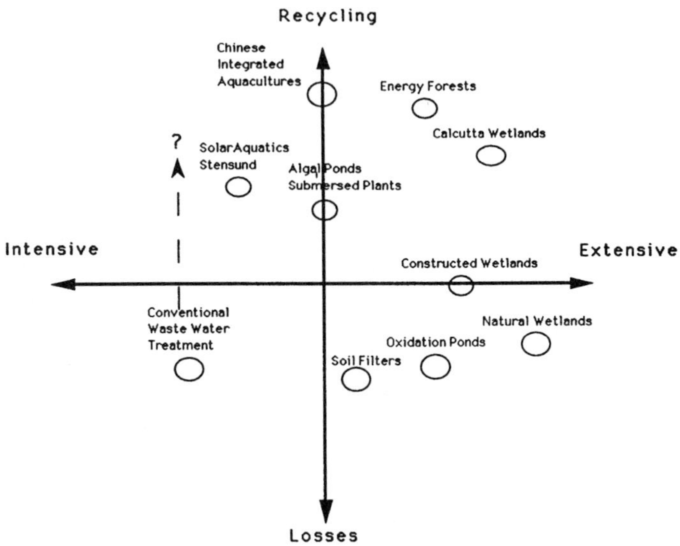

FIGURE 1. A classification of the different approaches to wastewater treatment, according to degree of treatment intensity and recycling. (From Guterstam and Ridderstolpe 1991.)

Algal ponds (Shelef et al. 1980) and channels with submersed plants (Gumbricht 1996) are partly constructed for recycling. Intensive indoor systems like Solar Aquatics and Stensund Wastewater Aquaculture (Guterstam and Todd 1990) represent extremes in maintenance as they are actively harvested and run during winter in cold climates. The most sophisticated recycling examples we know of are the Chinese integrated farms with aquaculture, agriculture, and energy production (Yan and Ma 1996, Li 1996). The Calcutta Wetlands represent a smart and extensive low-technology with the aim of producing food fish and vegetables for the local population (Ghosh 1996).

STENSUND WASTEWATER AQUACULTURE

The aquaculture approach for household wastewater goes back to the Chinese integrated farm. It was developed for warmer climates in the U.S. during the 1970s (Dinges 1976, Stewart and Serfling 1979). Stensund Wastewater Aquaculture is an intensive indoor technology built for demonstration, research, and development of recycling through natural processes using constructed food webs. The project aims at a sophisticated detoxification and recycling of the wastewater resources, including nutrients for aquacultural production. The greenhouse in which the system is housed permits operation throughout the year and has heat pumps installed for reuse of heat from air and water.

The aquaculture concept was tested in a small pilot plant in 1987 to 88 with encouraging results. Nutrient elimination efficiencies were, on average, 62% for phosphorus and 85% for nitrogen (Guterstam and Todd 1990). The results convinced the government to provide economic support to construct a full-scale plant in 1988 (see above).

FIGURE 2. Stensund Wastewater Aquaculture on the Baltic coast in Sweden. Photo: Max Plunger.

THE WASTEWATER TREATMENT AT STENSUND

The wastewater aquaculture was opened in October 1989 and is part of the wastewater treatment system for a population of about 100 persons living and working at the Stensund Folk College (Figure 2 and 3). The wastewater, which is comparable to municipal sewage, had a flow measured at the plant of 17 m^3/day during the first 3 years of operation, 1990 to 1992 (Guterstam 1993). Some wastewater was lost and groundwater gained through leaks in the pipes. The water comes from a local well. Stensund is representative of a small rural village on the Baltic coastline of Sweden, with thousands of similar sites in the region (see case studies, this volume). It constitutes a model society for studies of energy and material flows and the construction of sophisticated recycling technologies for waste and wastewater.

The wastewater is pretreated in a series of tanks, i.e., first in a mechanical treatment plant where floating fats and feces are separated as a raw sludge before entry into the aquaculture greenhouse. In order to meet Swedish phosphorus effluent standards, a chemical precipitation facility was also installed here and pretreated the water going to the aquaculture for 1½ years. Now the water is separated after the primary settling tank and only that water which bypasses the aquaculture is chemically treated. At present, 1993, the aquaculture treats 40 p.e. of secondarily treated wastewater (7 m^3/day). The times for the different types of pretreatment and the daily flows are shown in Figure 4.

During the first 3 years, the wastewater flowed by gravity through the following nine treatment steps, with a total volume of approximately 200 m^3 and a 29-day residence time (Figure 5):

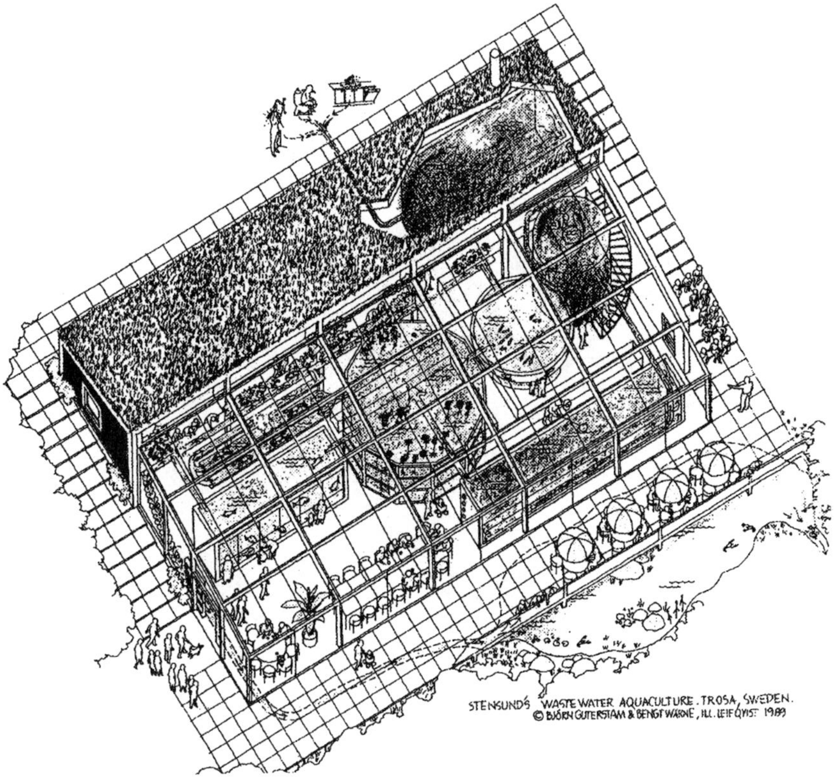

FIGURE 3. A cutaway diagram of Stensund Wastewater Aquaculture.

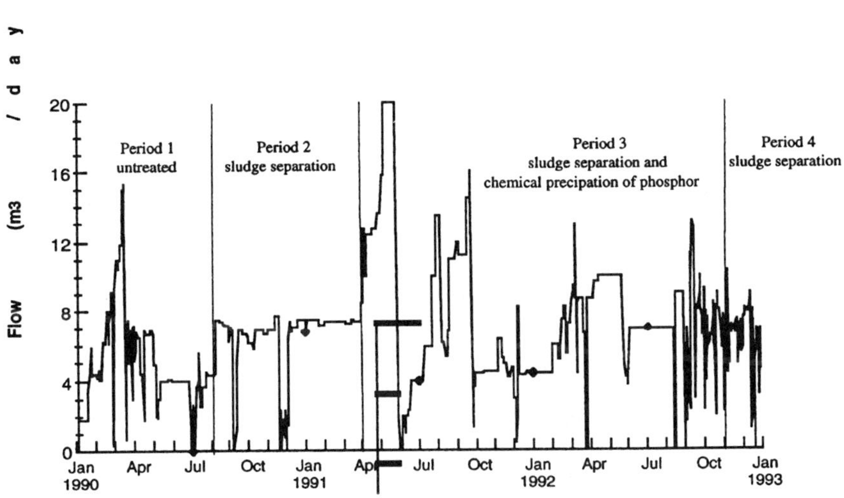

FIGURE 4. Variation in the daily flow into the aquaculture, 1990–1992.

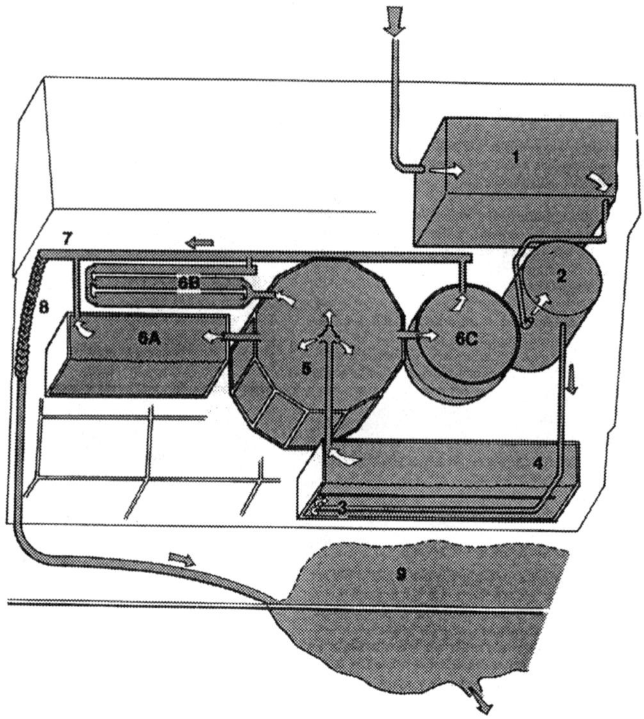

FIGURE 5. A schematic view of Stensund Wastewater Aquaculture.

1. The wastewater was collected in a storage tank of 28 m^3, which received it 15 h/day to suit the diurnal rhythm of higher organisms.
2. Anaerobic tank, 20 m^3, for bacterial degradation of organic compounds and precipitation of metals by bacterially released sulfides.
3. Aerobic (aerated) water, 9 m^3 in the bottom 1 m of a 3-m deep tank (total volume 27 m^3), a biofilter of gravel to continue microbiological mineralization and detoxification of the wastewater. (This part was extended in February 1993 with a separate biofilter and an aerator — Aquatrim — connected between the anaerobic tank and the algal tank, No. 4.)
4. Algal tank with cultivation of phytoplankton and bacteria behind the glazed part of the 3-m deep tank using the upper 2 m, or 18 m^3, as a first step of primary production in the constructed food chain.
5. Cultivation of zooplankton, grazers and detritus feeders in a 2.5-m deep, 40-m^3 tank for consumption of biomass from the preceding steps. Aquatic plants like water hyacinths (*Eichhornia crassipes*) and duckweed (*Lemna minor*) are cultivated on the surface.
6. Cultivation of fish in basins with tropical aquatic plants like *Eichhornia crassipes, Pistia stratiotes, Hydrocleis nymphoides, Salvinia auriculata, Azolla filiculoides,* and *Lemna minor.*

6A. Fish polyculture inspired by Chinese aquaculture principles, 2 m deep, 27 m^3.

6B. Two channels for hydroponics (see 7), 6 m long, 0.4 m wide and 0.3 m deep.
6C. "Fish Globe," 9 m^3, tank for fingerlings and goldfish (*Carassius auratus*).
7. Hydroponic cultivation in three 6-m long channels (see 6B) of vegetables and other useful plants like tomatoes (*Lycopersicon*) and willows (*Salix dasyclados*).
8. A Flowform "water staircase" aerates the water as it leaves the indoor part of the aquaculture.
9. Outdoors, a 40-m^3 crayfish pond for cultivation and hibernation of the noble crayfish (*Astacus astacus*). The water flows down a slope to an energy forest project (started in 1991) with willows (*Salix dasyclados*) and to a planned full-scale crayfish aquaculture before it enters the Baltic Sea.

PRACTICAL REALITIES AND RESULTS, 1990–92

The task for the wastewater ecological engineer is to fit the right dose of the "wastewater drink" to its consumers. The reality is a never-ending wastewater flow to be directed through the aquaculture or whatever system there is. The first 3 years of operation, 1990 to 1992, were a period of bringing the team of organisms together and learning to understand the conditions of this constructed ecosystem.

The abiotic realities have two areas of concern: the physics of the greenhouse and the chemistry of the wastewater. The climate of the greenhouse is presented by Warne (1996). The wastewater treatment concept stresses detoxification as a first step in order to avoid accumulation of, for example, metals in the food chain.

Metals (Al, Cd, Co, Cr, Cu, Fe, Hg, Mn, Ni, Pb, and Zn) were analyzed from different parts of the aquaculture one time after a year of operation (Hallin et al. 1990). They all showed a tendency to become accumulated on the bottom of the anaerobic tank (No. 2 in Figure 5) and decreased in concentrations through the aquaculture as far as they could be detected by factors of 4 to 30 times. The most prevalent metal was copper, which is used in Swedish tap water pipes.

The efficiency of the anaerobic metal precipitation was on average 40% for the metals reported in Table 1. The presence of cadmium is explained by its sources in food and smoking (SOU 1984:77), while mercury in household wastewater is more mysterious — broken thermometers may be a source. Plant uptake of copper was only noted in duckweed (*Lemna minor*); no accumulation of heavy metals was found in the harvested tomatoes.

The wastewater stream as a source for the indoor system at Stensund was originally planned to be treated as a whole, with only a mechanical filter for particles >2 mm. Reality taught us that in a small society like Stensund, with its short sewage pipes, much of fats and feces are not fragmented into small pieces. Therefore, they are now separated as a floating raw sludge "cake" before the aquaculture. The nutrient content trapped here was found to be 39% of the nitrogen and 28% of the phosphorus. Since 1991, part of the sludge comes from chemical precipitation (Figure 6). The other reality was that only 40% of the wastewater passed through the aquaculture the first 3 years, resulting in considerable amounts of nutrients lost, untreated by the aquaculture.

The total amount of phosphorus and nitrogen from the 100-person household indicates a lower nutrient output per capita with only 0.6 kg phosphorus compared

TABLE 1
Metal Reduction at Stensund Wastewater Aquaculture from Its Entrance in the Storage Tank, Passage through the Anaerobic Tank and after Passing through the Whole Facility[a]

Metal		Inflow values (Storage tank) mg/l	% reduction by anaerobic tank	% total reduction through Aquaculture
Iron	Fe	0.84	39	77
Aluminum	Al	0.65	40	89
Lead	Pb	0.018	56	72
Cadmium	Cd	0.0013	23	62
Copper	Cu	0.59	46	97
Mercury	Hg	0.005	40	—
Zinc	Zn	0.23	35	87

[a] Measured at a flow of 7.7 m^3/day, at a temperature range of 17.8–18.3°C from the bottom to the surface of the 4-m high anaerobic tank. Sulfide ion concentration = 12.3 ppm and pH = 7.2. November 11, 1990.

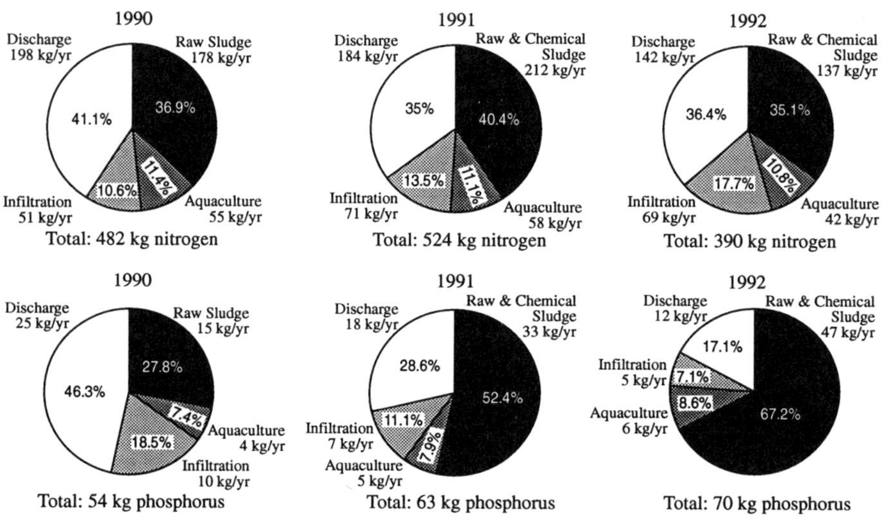

FIGURE 6. Phosphorus and nitrogen budgets for three years of the Stensund model society.

to 1.0 kg/capita, while the nitrogen output was a little higher with 4.7 kg/capita compared to the 4.0 kg nitrogen as yearly average values for untreated sewage in Sweden (SCB 1987). This coincides with an increased use of phosphate-free laundry detergents and dishwasing soap by the staff and students at the college.

The fraction recycled was about 10% of the nutrients from Stensund (8% of total phosphorus and 11% of total nitrogen; Figure 6). The recycled nutrients were found in a biomass fraction which was harvested annually as 1500 kg wet weight of water plants, an unmeasured amount of algal sludge and up to 50 kg tomatoes for consumption.

TABLE 2
Energy Consumption at Stensund Wastewater Aquaculture

Energy consumption (kWh/day)	Summer (S)	Winter (V)	Difference (S-V)	Δ% of S
Electricity	206	338	132	64
Heat[a]	−48	19	67	—

[a] The heat is accounted for as the net of export-import, where the negative values signify energy export and the positive signify an import of heat.

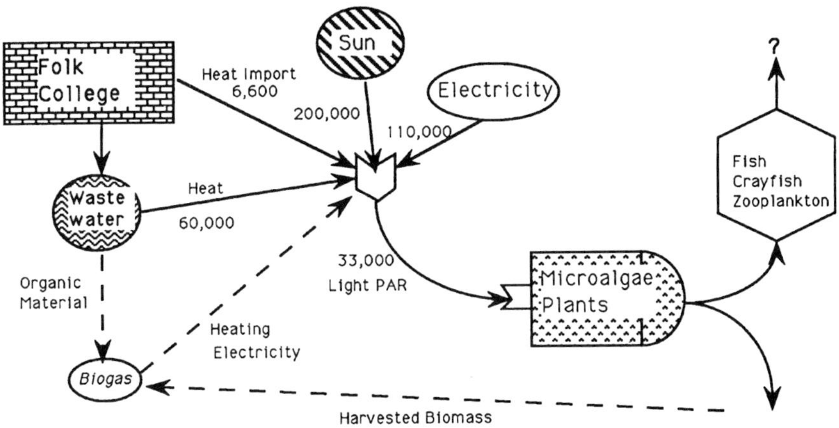

FIGURE 7. Energy flow model: Conceptual drawing of resource flows (kWh/y) in the Stensund Wastewater Aquaculture with the actual results for 1990.

ENERGETICS AND ECONOMY

The energy and monetary budgets of an ecologically engineered wastewater system are essential for evaluating the system's reliability and whether it can become widely adopted in northern countries. The Stensund facility is in operation continuously throughout the year using electricity for lamps in winter and for pumps, computers, laboratory equipment and heat pumps. The average daily electrical energy consumption was therefore 64% higher in winter than in summer (Table 2). The heat potential from solar radiation into the greenhouse of the aquaculture and from the warm wastewater is 250,000 kWh/y. This heat was recycled as 44,000 kWh with two heat pumps. This means that 18% of the available heat energy of the aquaculture was used for heating laboratory, office and tap water. Tap water was also exported to other Stensund buildings. During the winter the aquaculture was almost self-supporting with respect to heat; only a small net import of 1700 kWh/y was registered. The energy flow of the Stensund wastewater aquaculture is shown in Figure 7.

TABLE 3
Operational and Electricity Costs for Stensund Wastewater Aquaculture and Conventional Wastewater Treatment Facilities

Facility	Σ pe	Technology[a]	m^3/day	SEK/m^3	kWh/m^3
Aquaculture[b]	40	Ecological	7	34	39
Stensund	100	Conventional + Ecological	17	17	16
Helgarö	200	Conventional	40	21	13
Trosa	5300	Conventional	1700	3	0.3

[a] Conventional technology implies here a three-stage treatment process without additional nitrogen reduction like denitrification.
[b] Income for products of the Wastewater Aquaculture and exported heat are not included.

A more efficient use of the heat energy and the addition of recycled energy from composted raw sludge could turn our wastewater facility into a heat generator for the Stensund model society in the future.

A comparison of the operational costs to treat wastewater conventionally and ecologically was done using average data from the Stensund facility after three years (Table 3). The Stensund data were compared with a planned conventional wastewater treatment plant for 200 p.e. (Helgarö) and with an operating 5300 p.e. facility (Trosa). The cost for pure ecological treatment for 40 p.e. at Stensund was considerably higher than conventional costs both in terms of money and energy. The combined ecological and conventional treatment at Stensund, though, was in the same range as for the calculated 200 p.e. facility.

The high cost for ecological wastewater treatment at Stensund is explained by its small scale and by the fact that the potential for income from produced biomass, e.g., tomatoes and fish, was not fully exploited. Areas for plant cultivation and fish farming could be better used than in the Stensund demonstration facility. Pumps and greenhouse lamps could also be used more efficiently. The resources of heat for export and nutrients for biomass production are enough theoretically to economically balance the plant's operating costs. Producing a net income constitutes the future challenge of this and other ecologically engineered wastewater projects.

WATER QUALITY

The water quality parameters of bacteria, BOD, COD, and suspended solids have been the easiest to deal with in the aquaculture. Organic material reduction was > 95% (Table 4). Coliform bacteria (35°C), and *E. coli* (44°C) were reduced dramatically through the aquaculture, resulting in a water quality nearly acceptable for bathing (Figure 8).

Other physiochemical parameters of ecological importance are temperature, dissolved oxygen, pH, and ammonia. Low temperatures (<13°C) caused by rainwater leakage into sewer pipes in winter killed tropical fish (*Tilapia*) during the first winter. The leaks have now been eliminated by installation of new sewer pipes. Oxygen levels were sometimes recorded down to <1 ppm in algal, zooplankton and polyculture basins. The pH levels were stable and increased through the aquaculture

TABLE 4
Organic Matter Reduction in the Stensund Wastewater Aquaculture

Parameter (mg/l)	In	Out	% Reduction
BOD_7	330	7	98
COD	108 6	48	96
Suspended solids	372	4	99

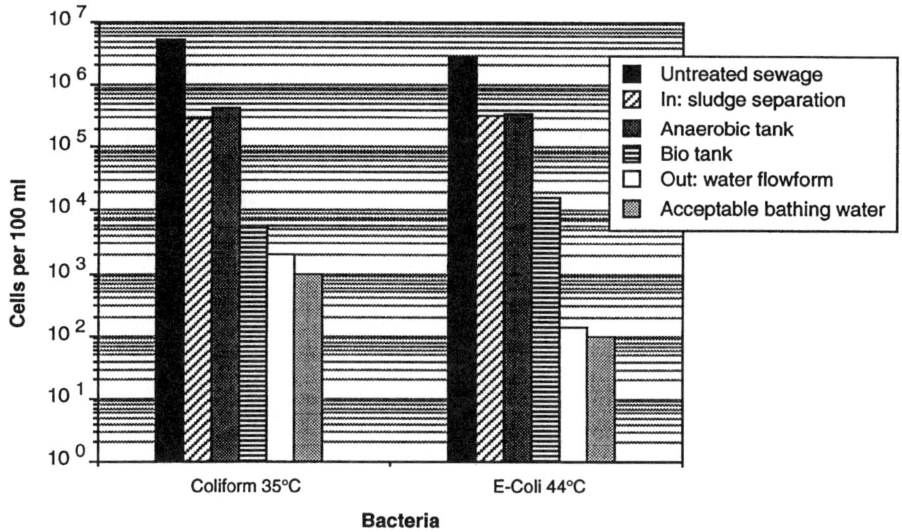

FIGURE 8. Bacteria in various parts of the aquaculture, as an average of measurements in 1990 and 1992.

from 6.8 to 7.5 in the incoming raw sewage to 7.8 to 8.2 in outgoing water at the Flowform water staircase. Levels of pH around 7.5 to 8.0 in combination with concentrations of 15 to 20 mg NH_4-N/l have probably not been toxic to zooplankton and fish populations since most of the ammonia is in the form of ammonium (McLarney 1984).

Alkalinity of the Stensund water is high because of limestone groundwater, around 300 mg/l as $CaCO_3$. This indicates that carbon dioxide is not limiting for the cultivation of phytoplankton.

ECOLOGICAL REALITIES

Populations of phytoplankton and zooplankton have not developed or reproduced in the aquaculture as was planned. We think that the main reasons for this are:

1. Phytoplankton populations (Figure 5, no. 4) have been limited by a toxic, partly unoxidized bottom water (Figure 5, no. 3) with a redox potential around

+100 mV and traces of hydrogen sulfide (0.02 mg/l). Light was limiting through shading by much detritus (recorded through daily Secchi disk readings). A complex competition situation between phytoplankton on the one hand and sessile algae, bacteria, and other microorganisms on the other hand is another possible reason for the absence of an algal bloom. A separation of the biofilter bottom organisms into a new tank, connected before the algal tank in early 1993, will hopefully solve this problem.

2. Zooplankton populations of daphnids and copepods have fluctuated and were most of the time replaced by smaller zooplankton such as rotifers and protozoans (Figure 5, no. 5). The instable conditions of the constructed food web with a poor phytoplankton diet for the daphnids partly explains this population change.

Fish populations have only been fed on the produced plants and animals from the aquaculture (Figure 5, no. 6A). Goldfish (*Carassius auratus*) and tilapia (*Tilapia marie*) have established reproducing populations, while the common carp (*Cyprinus carpio*) has not succeeded in reproducing. The lack of a real mud substrate and no real polyculture with grass carp and other unpopulated niches of the "fish basin" are signs of poor conditions for the carp. Introduced herbivorous tilapias in 1992 spawned and established a relatively strong population, which favors the development of a functioning fish polyculture.

Data from fish and crayfish aquaculturing were only qualitative during the reported period. During the first year eggs were hatched from salmon (*Salmo trutta*) and pikeperch (*Stizostedion lucioperca*) with no survival probably due to high temperatures (>20°C) resulting in fungal diseases. The noble crayfish, *Astacus astacus,* established a reproducing population in the crayfish pond (Figure 5, no. 9). The idea of growing young crayfish one season in the aquaculture for stocking out into natural waters was developed with support from the Swedish National Fisheries Board. It has yet to be implemented, due to lack of funding.

A general criticism of our aquaculture approach is the lack of species diversity within the different basins, i.e., a combination of phytoplankton, zooplankton, snails, and a diverse bottom fauna and fish is recommended for all three basins (Li pers. comm., Yan pers. comm.).

COMPUTER MODEL AS A TOOL

We have reported a combination of qualitative and quantitative results from the first year of operation of the Stensund wastewater aquaculture. In general, these results just indicate the principal qualities of an indoor facility in full operation. The reason for this is the absence of stable experimental conditions during a period of installations and corrections of pumps, filters, water flows, ventilation, populations of plants and fish produced and harvested, etc.

Computer simulation models are useful tools for ecotechnology, as they help us to understand and optimize all processes involved from waste treatment efficiencies to economics (Kimberly et al. 1991). The development of the aquaculture technology is now supported by computers logging flows, temperatures, water chemistry, and population dynamics. Theoretical values have been used to simulate the aquaculture

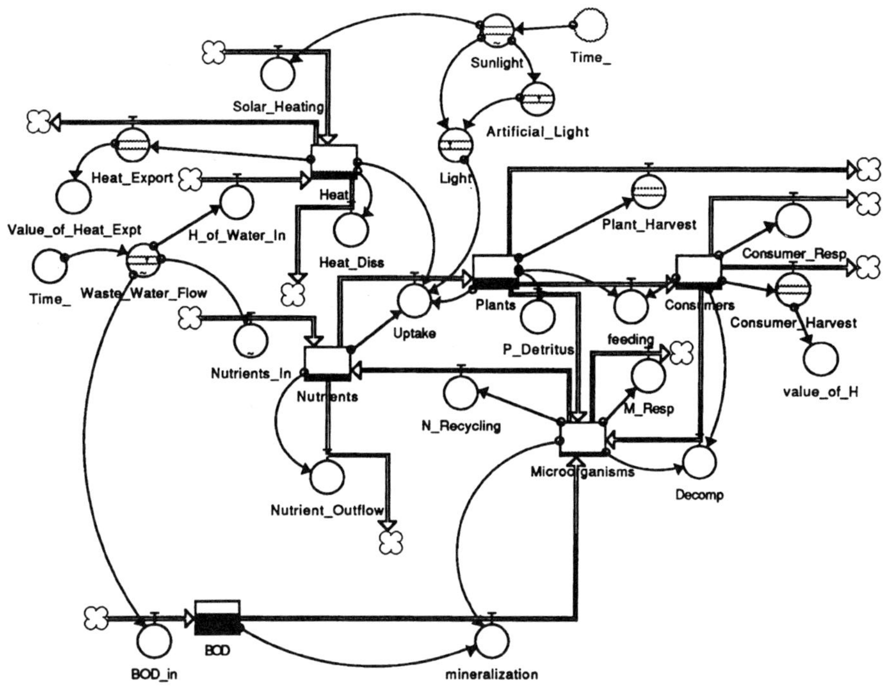

FIGURE 9. Computer model illustrating the energy and material flows of the Stensund Wastewater Aquaculture.

processes using the computer program STELLA™ (Structural Thinking, Experimental Learning Laboratory with Animation) for Macintosh computers. This gives us the possibility to generalize and to simulate the wastewater aquaculture. Experimental data will then be used for correcting a theoretical computer model of the aquaculture (Figure 9).

The computer model can also become a tool when planning new ecological wastewater treatment facilities (Nyman 1991). It is important to arrange the wastewater treatment according to the local conditions, considering wastewater qualities, climate, geography, possible species to cultivate, economy, know-how, and demands for energy, water, and nutrients. Therefore, it is important to give general aspects of biology and thermodynamics together with technical and construction considerations.

Today we have simulated the first theoretical values of an aquaculture using a systems ecology approach of energy and materials flows in a Stella model (Figure 10, Costanza pers. comm.). Thermodynamics helps us to scale the treatment facility as insolation sets a limit on the productivity and nutrient uptake by the green plants through photosynthesis. This process is the basis of the whole concept of recycling through ecological engineering for wastewater treatment. Algae and vascular plants such as submersed plants, floating plants, marsh plants, and terrestrial species are primary producers of new biomass resources out of the wastewater. The animals grown in the aquaculture, such as fish and crayfish, constitute pathways to improve the quality of the reused wastes and also indicate the economic values to be gained.

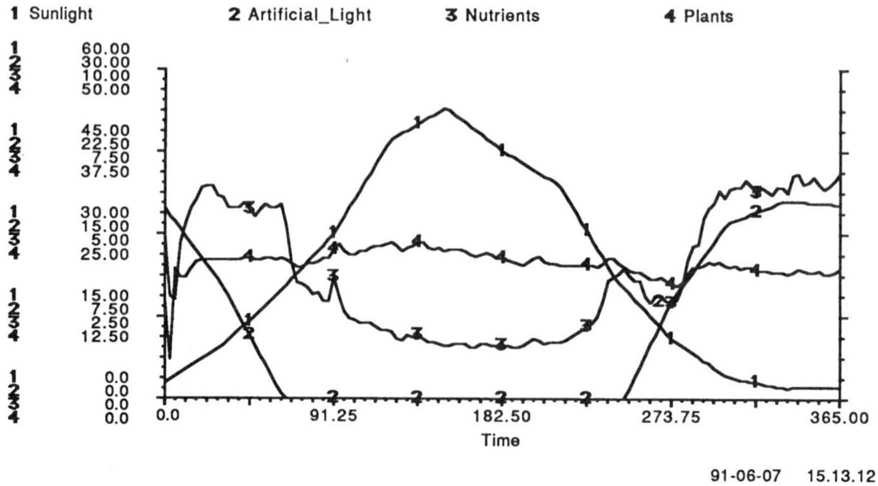

FIGURE 10. Simulations of four parameters of the Stensund Wastewater Aquaculture.

In the Stensund case, a high degree of recycling would incorporate outdoor systems to increase the biomass production, i.e., an energy forest and a crayfish aquaculture, a biogas reactor for wastewater and solid organics, and an optimal number of heat pumps in the greenhouse. Together these projects form the Stensund Ecological Center for development, demonstration of, and education on how to apply feedback loops from society's resource management to nature.

CONCLUSIONS

Ecological engineering for wastewater treatment has presented some of its first results in a modern industrial society in a cold climate using an intensive indoor system at Stensund. The first experiences show the principles for plant nutrients and energy to be recycled. Detoxification of metals and stable organic compounds is possible. Further emphasis must be put on optimizing the natural processes used, i.e., biomass production and harvest must be balanced with wastewater flows and microbiological processes of mineralization and detoxification to become better understood. The resource management approach of an ecologically engineered wastewater facility has positive feedbacks on the economy as well as the social welfare of the surrounding community.

Intensive indoor systems can serve as:

1. Wastewater treatment plants.
2. Greenhouses for production of new biomass, i.e., plants, fish, and crayfish in combination with outdoor systems for efficient nutrient reuse.
3. Heat distributors to the surrounding society.
4. Educational centers inspiring citizens to contribute to a sustainable water use.
5. An inspiring working environment for the wastewater engineering staff.

ACKNOWLEDGMENTS

This work is done in a team organized at Stensund Folk College. All colleagues who have taken part in this project in all aspects have made this report and the conference possible only through their engagement and trust in a sustainable future. Thanks to you all! Special acknowledgments go to lab assistants Lotta Andersson, Margareta Hagstad, Gudrun Karlsson and Monica Palin-Wilett; Stensund's ecological engineers Gunnar Lind and Ola Sundin; Peter Ridderstolpe, Firma Ekologisk Teknik, Östhammar, Sweden; Professor Jorge de Sousa Pires, Apple Computer AB; Rolf Bodeck, Svenska Merkanto AB; and Ekhagastiftelsen, Miljöpartiets Ekologiska Fond, and The Swedish National Board for Technical Development (STU/NUTEK), who made this work possible with financial support.

REFERENCES

Anderson, R. 1976. "Modern kretsloppsfilosofi: Små självförsörjande enheter — miljövänliga och energisparande." In: *NFR:s Årsbok 1976/77* (Yearbook of Swedish Natural Science Research Council), pp. 177-198.

Bastian, R. K. and S.C. Reed 1979. *Aquaculture systems for wastewater treatment: Proceedings and engineering assessment.* EPA 430/9-80-006.

Chang, Y.-y, S. Mei-juan, and T. Yu-un, 1981. "Transport and fate of BHC in Aquatic Environment," *Acta Scientiae Circumstantiae* 1(3): 242-249.

Cooper, P.F. and B.C. Findlater (Eds.). 1990. *Use of constructed wetlands in water pollution control.* Pergamon Press: Oxford.

Costanza, R. pers. comm. Center for Environmental and Estuarine Studies, University of Maryland, MD.

Daly, H.E. 1990. "Toward some operational principles of sustainable development," *Ecological Economics* 2(1):1-6.

Delin, S. 1973. "Naturvård eller naturmord," *Forskning och framsteg*, No. 2, 3-7.

Denninger, M., pers. comm., Swedish Farmers Federation (LRF), Stockholm.

Dinges, R. 1976. "A proposed integrated biological wastewater treatment system." In: Tourbier J. and R.W. Pierson. *Biological control of water pollution.* Philadelphia: University of Pennsylvania Press. pp. 225-239.

D'Itri, F.M. 1977. *Wastewater renovation and reuse: Proceedings of the international conference on the renovation and reuse of wastewater through aquatic and terrestrial systems, pollution engineering and technology.* New York and Basel: Marcel Dekker, Inc.

Gandy, E. 1982. "Kommunal avloppsvattenrening — utvärdering av en miljövårdssatsning," Report to the governmental investigation Maturresurs-och Miljökommittén 1982-07-01.

Ghosh, D. 1997. "Ecosystems approach to low-cost sanitation in India: Where the people know better." In: C. Etnier and B. Guterstam (eds.), *Ecological engineering for wastewater treatment.* Boca Raton, FL: CRC Press.

Gumbricht, T. 1997. "Submerged macrophytes for recovery of eutrophied waters." In: C. Etnier and B. Guterstam (eds.), *Ecological engineering for wastewater treatment.* Boca Raton, FL: CRC Press.

Guterstam, B. and J. Todd, 1990. "Ecological Engineering for Wastewater Treatment and its Application in New England and Sweden," *Ambio* 19(3):173-175.

Guterstam, B. and P. Ridderstolpe, 1991. "Avlopp — en förbisedd och misshandlad resurs," Stadsbyggnad 2, 1991, pp. 24-30.

Guterstam, B. 1993. *Stensunds vattenbruk — ekologisk teknik för avloppsvatten.* Slutrapport Innovatörsbidrag STU/NUTEK 1989-92, uppdaterad 2:a upplaga. 25 pp. In Swedish with English summary.

Hallin, S., A. Hedlund, and H. Olsson 1990. Wastewater treatment efficiencies in Stensunds Vattenbruk. Inst. f. Markvetenskap, avd. Växtnärningslära, Swedish University of Agricultural Sciences (mimeogr., in Swedish).

Hasselgren, K. 1988. "Utnyttjande av kommunala restprodukter i energiskogsodling." In: P. Hjort. *Samhällets ämnesomsättning.* Forskningsrådsnämnden Rapport 88:2, pp. 21-25.

Kimberly, A. B., M.S. Fennessy, and W. J. Mitsch. 1991. "Designing wetlands for controlling coal mine drainage: An ecological-economic modeling approach," *Ecological Economics* 3(1):1-24.

Li, S., pers. comm., Department of Aquaculture, Shanghai Fisheries University, Shanghai, China.

McLarney, W. 1984. *The freshwater aquaculture book: A handbook for small scale fish culture in North America.* Cloudburst Press, p. 134 and Table X-2-3.

Mitsch, W.J. 1997. "Ecological engineering: The roots and rationale of a new ecological paradigm." In: C. Etnier and B. Guterstam (eds.), *Ecological engineering for wastewater treatment.* Boca Raton, FL: CRC Press.

Mitsch, W.J. and S.E. Jørgensen, 1989. *Ecological engineering: An introduction to ecotechnology.* New York: Wiley-Interscience.

Nyman, R-M. 1991. "Examensprojekt eko" (U 88). Utvecklingsingenjörslinjen, Högskolan i Halmstad, in Swedish, mimeogr.

Prop. 1987/88:85. *Miljöpolitiken inför 90-talet.* Regeringens proposition 1987/88:85, pp. 162-63.

Reed, S.C., E.J. Middlebrooks and R.W. Crites 1988. *Natural systems for waste management and treatment.* McGraw-Hill.

Rosenberg, R., Elmgren R., Fleischer S., Jonsson P., Persson G. and H. Dahlin 1990. "Marine eutrophication case studies in Sweden," *Ambio* 3:102-108.

SCB 1987. *The natural environment in figures: Yearbook of environmental statistics 1986-87, official statistics of Sweden.* Stockholm. p. 80

Selander, S. 1987. *Det levande landskapet i Sverige*, Gothenburg, Sweden: Bokskogen.

Shelef, G., Y. Azov, R. Moraine and G. Oron,1980. "Algal mass production as an integral part of a wastewater treatment and reclamation system." In: G. Shelef and C.J. Soeder, *Algae biomass production and use.* Amsterdam: Elsevier/North-Holland Biomedical Press. pp. 163-189.

SOU 1983:10. *Användning av växtnäring.* Official governmental investigation. 310 pp.

SOU 1984:77. *Kemikaliekontroll — huvudbetänkande av kemikommissionen.* Official governmental investigation. p. 59.

Stewart, W. C. and Serfling S. A. 1979. "The Solar Aquacell System for primary, secondary or advanced treatment of wastewaters." In: R. K. Bastian and S.C. Reed, *Aquaculture systems for wastewater treatment: Proceedings and engineering assessment.* EPA 430/9-80-006, pp. 376-412.

UN 1987. *Environment statistics in Europe and North America: An experimental compendium.* UN Statistical Commission and Economic Commission for Europe, Conference of European Statisticians Statistical Standards and Studies — No 39. Tables I-23 and I-24.

Warne, B. 1997. "Stensund Wastewater Aquaculture: The building." In: C. Etnier and B. Guterstam (eds.), *Ecological engineering for wastewater treatment.* Boca Raton, FL: CRC Press.

Zhang, Yongyuan, Zhuang Dehui, Sun Meijuan, Tan Yuyun, Zhang Quanzheng, and Li Jianqiu 1982. 8 papers reporting from the "Yar Lake Project", *Acta Hydrobiologica Sinica* 7(4): 489-538. In Chinese with abstracts in English.

Yan, J.S. and Ma S.J. 1997. "The function of ecological engineering in environmental conservation with some case studies from China." In: C. Etnier and B. Guterstam (eds.), *Ecological engineering for wastewater treatment.* Boca Raton, FL: CRC Press.

Yan, J.S., Pers. comm., Nanjing Institute of Geography and Limnology, Academia Sinica, Nanjing, China.

Chapter 8

STENSUND WASTEWATER AQUACULTURE: THE BUILDING

Bengt Warne

INTRODUCTION

To design a building for ecological engineering such as the Stensund Wastewater Aquaculture at Trosa was quite a challenge (Guterstam 1991, Maechel 1991). From an architectural perspective, it must harmonize with the Baltic Sea and its beautiful archipelago near Stockholm. It also has to suit the little castle and the classical park at Stensund. It should create a link to the forest and coastal wildlife nearby. It is meant to be a vital part of the expanding Stensund Folk College, with its local, national, and international network. Technically it was even a tougher challenge. The building construction ought to take as much advantage as possible of naturally driven systems in order to establish a tropical to subtropical climate for the biological processing of used water. Stensund has wind, rain, and plenty of sunshine — 2000 h/yr — but the winters are harsh, cold, dark, and long.

THE STENSUND FACILITY

The ecological wastewater concept and performance is documented by Guterstam and Todd (1990) and the Stensund facility in particular by Guterstam (1991) and its potential in Nordic climate town planning by Lundahl (1989). Our goal was to create an informative adventure to anybody — an attractive building with a tropical garden, pouring water and open ponds, a lecture room with a café for meetings and workshops, and an inspiring environment for human ecological research and development.

I designed the building to be dug down in the southern slope to reduce heat transmission to the north. It is also a solar collector with a nearly doubled effect in winter when the sunshine is reflected by an ice- and snow-covered landscape at the Baltic Sea.

The construction from poison-free materials like wood and minerals can be used as thermal storage for heat from day to night and for cold from night to day. The foundation stores heat from week to week, thereby balancing climate extremes. The masses of soil and rock under the building give extra thermal storage with an even seasonal balance from summer to winter.

In order to give the thermal storage under the building a maximal volume, the ground insulation goes horizontally some meters all around the outer walls. The walls and the roofs are superinsulated and all glazing is triple.

The building was constructed utilizing some of the principles which were successfully used for homes in Europe and North America — Solar/Earth Buffering Double Envelope Housing (Larsson 1978, Booth D. 1980, Warne 1982).

0-87371-990-5/97/$0.00+$.50

EXPERIENCE

At Stensund, the concept of Solar/Earth Buffering Double Envelope Housing worked out well but the enormous increase of vapor from the warm, open microalgae basin surprised us (Figure 1). An intensive, constant flow of air bubbles through the water gives the greenhouse a radical addition of humidity. This became a task to deal with together with optimizing the biological processes at the Stensund project for development of an ecotechnology for wastewater treatment. The measurements of indoor and outdoor temperature, water temperature in the zooplankton basin in the center of the building, together with the relative humidity during the first year of operation are presented in Figures 2a and 2b (Björn Guterstam, pers. comm.). Humidity problems occur during cloudy and rainy days, especially in winter. Very low humidity from spring and on to autumn makes it possible to dry up condensation spots in the building.

During the first two winters the greenhouse was almost not ventilated at all in order to save energy. This worked very well as long as the air temperature could be held some degrees higher than the open ponds. When the indoor temperature fell without heating, a condensation machine was created with effects that might damage the building.

An uncontrolled tropical swampy indoor climate will sooner or later invite unwanted guests such as mold, fungi, and insects. If they arrive it can be tremendously costly to get rid of them. That has not yet occurred at the Stensund facility.

FUTURE WORK

It was found the thermal flow of the air in the greenhouse building did not necessarily have to pass around the laboratory with its double wall (Kupke pers. comm.). The walls of the ponds can give the same effect, which is shown in the drawings (Figure 1).

For the coming winter 1991 to 1992 we will make the following improvements in order to solve the humidity problem of the building:

1. The airspace between the ceiling of the laboratory part of the building and the outer roof will be closed. The northern wall will be extra-isolated at a weak point where it meets the roof. These reconstructions will essentially increase the insulation by 150 mm dead air space around the laboratory part of the building (Figure 1). The condensation in the walls will disappear and the airspace will be ventilated by relatively dry and fresh air, passing from tubes in the earth under the building into the aquaculture greenhouse, continuing via tubes under the laboratory cellar. The airflow within the greenhouse will go in one direction when driven by sunshine, and in the opposite direction at night and on cloudy days. These flows are shown in the drawings (Figure 1).

   The building itself constitutes the thermal storage. The relatively warm aquaculture basin/tank walls automatically creates an upward thermal upflow which is enhanced by the sinking cold air stream at the glass walls (when it is colder outdoors).

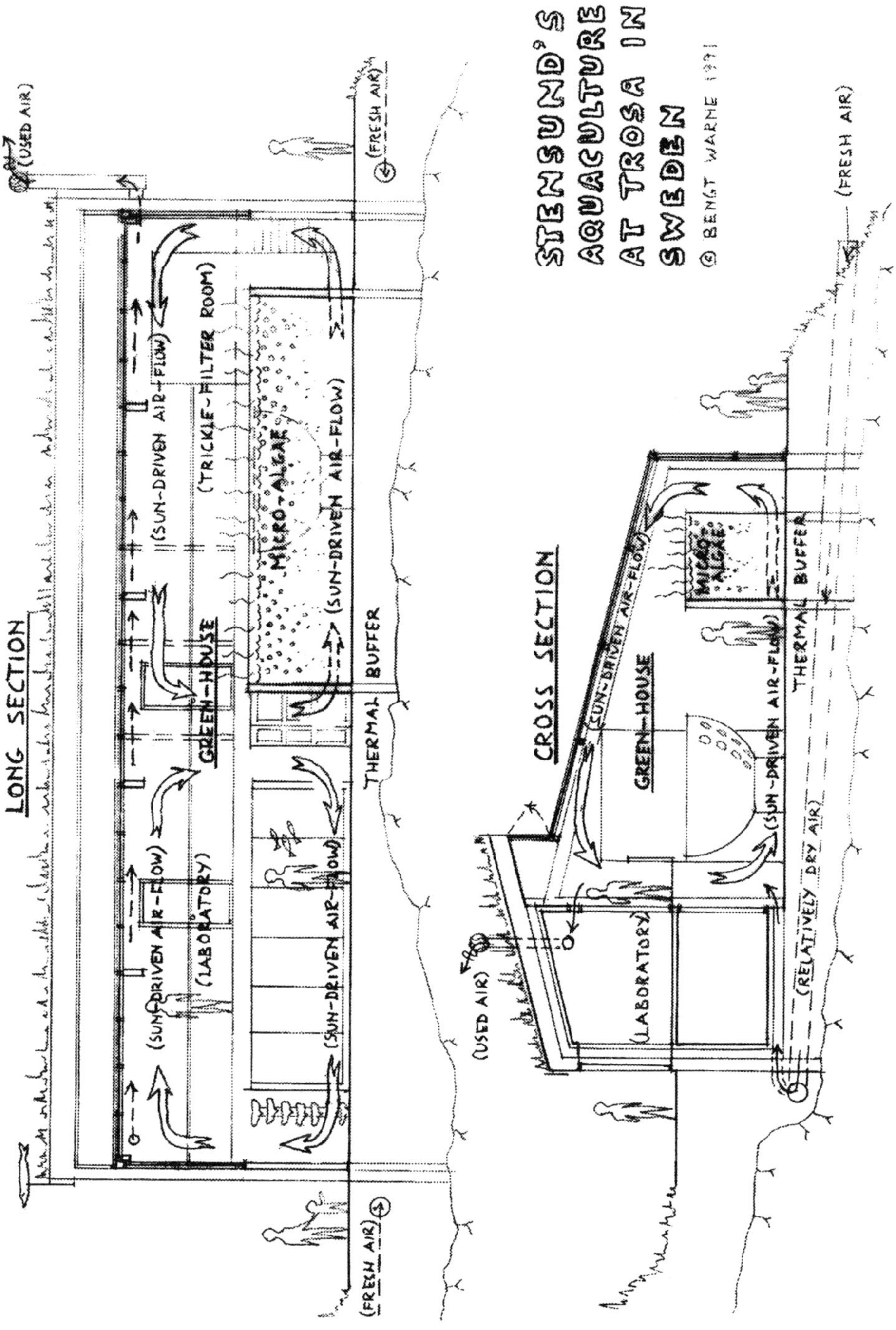
LONG SECTION
(USED AIR)
(FRESH AIR)
(SUN-DRIVEN AIR-FLOW)
(TRICKLE-FILTER ROOM)
MICRO-ALGAE
GREEN-HOUSE
(LABORATORY)
THERMAL BUFFER
STENSUND'S AQUACULTURE AT TROSA IN SWEDEN
© BENGT WARNE 1991
CROSS SECTION
(RELATIVELY DRY AIR)

FIGURE 1.

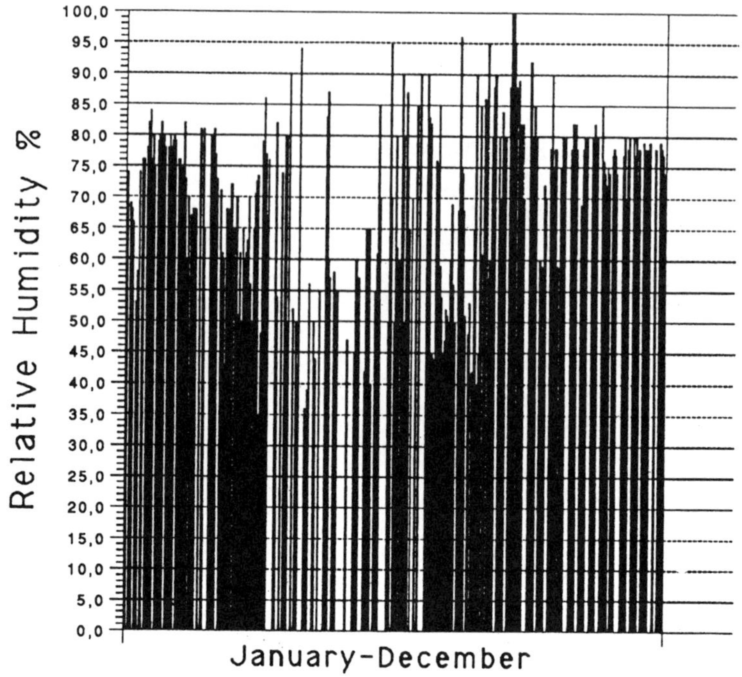

FIGURE 2A. Relative humidity inside Stensund Wastewater Aquaculture during 1990.

2. Vapor condenses on the coldest indoor surfaces. To take advantage of this phenomenon in an aesthetically pleasing way, a greenhouse heat pump for the treated wastewater was installed in series before the Flowform water staircase. In this way, the temperature of the water entering the Flowforms will have a temperature of 4°C instead of 20°C, which then creates a large and useful condensation surface.

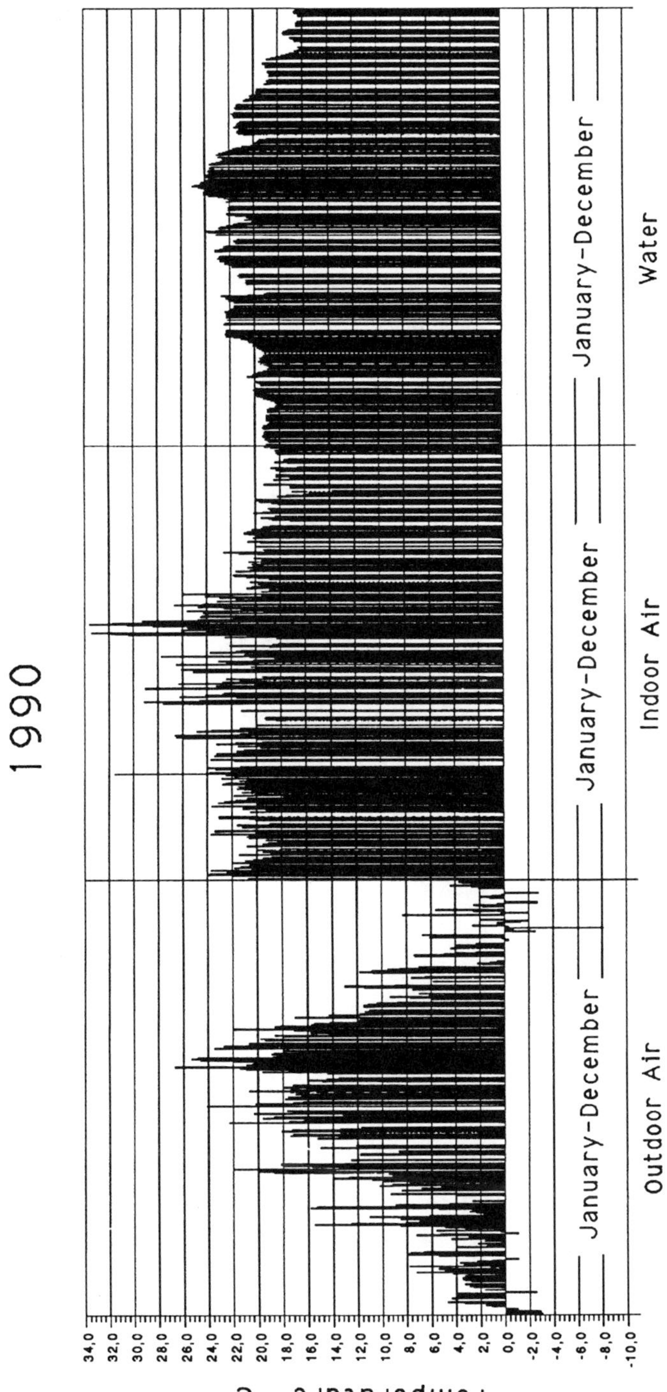

FIGURE 2B. Air temperatures outdoors and in the greenhouse and wastewater temperatures inside the greenhouse at Stensund Wastewater Aquaculture in 1990.

REFERENCES

Booth, D. 1980. *The Double Shell Solar House,* U.S. Department of Energy's Appropriate Technology Small Grants Program, Community Builders, Canterbury, NH.

Guterstam, B. 1997. "Ecological engineering for wastewater treatment: Theoretical foundations and practical realities." In *Ecological engineering for wastewater treatment.* Boca Raton, FL: CRC Press.

Guterstam, B. and J. Todd 1990. "Ecological engineering for wastewater treatment and its application in New England and Sweden," *Ambio* Vol. 19 No. 3, May 1990, pp. 173-175.

Kupke, C. 1991. Fachhochschule für Technik, Stuttgart, Germany, Pers. comm.

Larsson, H. 1978. *Feasibility Study of Thermal Storage in a Greenhouse Foundation*, Sweden Agency No. 78–4482, Smithsonian Science Information Exchange Inc., Washington, D.C.

Lundahl, G. 1989. "Vatten," *Arkitektur* No 10, pp. 18-19.

Maechel, I. 1991. "Vattenbruket som vände problemen till en resurs," Byggindustrin No 17,. pp. 26-31.

Pearson, D. 1989. *The Natural House Book*, New York: Simon & Schuster. pp. 62-63.

Warne, B. 1982. "The Symbio Home," Research Grant No 811532-9, Swedish Council for Building Research (BFR, in Swedish with English summary).

Chapter 9

ECOLOGICAL ENGINEERING, LIVING MACHINES, AND THE VISIONARY LANDSCAPE

John Todd

ABSTRACT

Ecological engineering has the potential to assist contemporary societies, bring them into balance with the natural world. Ecological engineering can provide the foundations for the design and development of living machines which can produce diverse foods, generate fuels, treat and recycle wastes, provide architectural heating and cooling, and aid in the restoration of soils, water, and air.

INTRODUCTION

I am addressing this Stensund meeting with a broad mandate, namely to show that ecological engineering, and the fruits of its potential, will shape the next century. As we visit the case study projects today please keep in mind that ecological concepts have the inherent potential to shape the future. Let me explain why I think that this can be so.

That modern societies are out of balance with the natural world has become a truism. It is my thesis that the contemporary technologies which support industrial societies are inadequate in terms of protecting the environment and dealing with resource scarcities and inequities between peoples and regions. What is needed is a fundamental technological revolution that will integrate advanced societies with the natural world to the mutual benefit of both. Pollution, atmospheric alteration, and the loss of soils and biotic diversity are artifacts of technological cultures estranged from the great natural systems of the planet. Modern cultures exploit the natural world and in so doing threaten their own long-term viability. It is essential that we create a truly symbiotic relationship with nature. This relationship will be predicated on new, highly evolved technologies I call living machines.

LIVING MACHINES

A living machine is a device made up of living organisms of all types and usually housed within a casing or structure of "gossamer" materials. Like a conventional machine, it is comprised of interrelated parts with separate functions used in the performance of some type of work. Living machines can be designed to produce fuels or food, to treat wastes, to purify air, or to regulate climates, or even all of these simultaneously. They are engineered with the same design principles used by nature to build and regulate its great ecologies in forests, lakes, prairies, or estuaries.

0-87371-990-5/97/$0.00+$.50

Their primary energy source is sunlight. As the planet, they have hydrological and mineral cycles.

They are, however, totally new contained environments. To create a living machine, organisms are reassembled in unique ways for specific purposes. Their parts or living components can come from almost any region of the planet and be recombined in utterly new ways. The reassembly or building of the machine is based upon knowledge of the natural history and the niches of the organisms that make up the parts, and upon a calculated determination of their individual fit into a unique constellation of organisms reassembled by the engineer for a given purpose.

I have, for example, in my office an ecological digester for treating wastes. It was designed to eliminate pulp and paper wastes without discharge of waste material into the external environment. Normally pulp and paper sludge is impossible to treat and is disposed of in landfills. The living machine in my office is small, housed within two interconnected, thin-walled translucent fiberglass cylinders filled with water. Sunlight penetrates the top and sides. The internal environment is almost entirely bathed in light during daylight hours. Inside the cylinders is a complex of rock minerals, diverse algae species from New England lakes and ponds, and bacteria, fungi, and protozoans from local environments. There are also "manufactured" (commercially cultured) bacteria, molluscs from Europe and North America, crayfish from North America, and fish from Europe, Africa, and South America. Higher plants from every continent except Antarctica are rafted on the surface or integrated within the system. The exotics in the system, like goldfish or carp, are either ubiquitous in this part of the world or incapable of surviving in the wild in a temperate climate. Examples of these include fishes from tropical Africa and South America and some of the floating water plants like water hyacinth, that cannot survive northern winters in the wild.

Within this ecological digester, microbial life attaches itself to the wastes, a heavy gooey sludge which is thereby preconditioned for animals in the system. The microbes are protected by the gases, especially the oxygen produced by algae and higher plants. Toxic substances are also processed by photosynthetic organisms which, in turn, are sustained by organics solubilized by the bacteria. This living machine requires animals to function. Animal filter feeders and detritus eaters pass the pulp mill sludge back and forth through their guts and intestines.

The original waste material is digested and transformed into a wide range of byproducts, some of which are useful. The whole system is a microcosmos comprised of a large number of species. Its ecological engineering is global in terms of the selection of organisms but very specific in design. In the case of the pulp mill waste digester, it was the quality and characteristics of the sludge itself that dictated the organisms and also the ratios of the various organisms to the others. Initially a balance was achieved by trial and error. In order to optimize performance, some of the original species had to be removed and replaced with others.

APPLICATIONS

Over the past 20 years with my associates I have designed and built a number of living machines to grow food, heat and cool buildings, treat sewage, sludge, septage, and boat wastes. It is possible to apply the same ecological engineering to the production of high-quality fuels, including hydrogen gas. Ecologically engineered machines can also produce byproducts that can be used in the manufacture of materials ranging from paper products to advanced composite construction materials. Linked living machines, which themselves form an engineered ecology of machines, can be designed to protect and even restore natural environments. They can also support human communities.

Living machines are fundamentally different from conventional machines. They represent, in essence, the intelligence of the forest or the lake, reapplied to human ends. Like the forest or the lake, their primary power is the sun, and like natural ecosystems they have self-design capabilities. They rely on biotic diversity for self-repair, protection, and overall system efficiency. Living machines need not be small nor isolated from larger natural systems. Scale need not be a factor as living machines, like nature, are made up from parts which fuse the "genius" of evolution. These include such independent qualities of life forms as repair, duplication, feeding, and waste excretion, dynamically balanced with the interdependent qualities of organisms like gas, mineral, and nutrient exchanges. The same design principles embodied in a cell extend to encompass the whole planetary biota. This allows living machines to vary greatly in size. I have designed a living machine to purify 300 million gal of drinking water daily for a major city, housed within a greenhouse-covered canal 27 miles in length.

The potential contributions of living machines to the 21st century is enormous. They are dependent on fossil fuels for the manufacture of long-lived materials, but not for combustion. They reintegrate wastes into larger systems and break down toxic materials or, in the case of metals, recycle them or lock them up in centuries-long cycles. Living machines have the potential to help feed large numbers of people, particularly in urban areas and this capability could address issues of equity between peoples and regions. Some of the poorer parts of the planet, especially the semi-arid subtropics, have enormous potential for living machines. The greatest reservoirs of "spare parts" are undoubtedly the tropics.

Living machines further have the potential to release natural systems from human bondage. By miniaturizing the production of essential human services, they would free wild nature to develop wilderness without human exploitation. This is significant in that the long-term survival of humanity may well be predicated on a dramatic increase in wilderness areas which are the great repositories of biological diversity.

BACKGROUND

Living technologies are a recent occurrence, scarcely two decades old. The intellectual pioneer of ecological engineering is Howard T. Odum, who coined the term in his seminal book, *Environment, Power and Society,* published in 1971. It was he who set the stage for the articulation and creation of living machines. In the same year, Bill McLarney and I built our first such device, housed within an 18-foot diameter, translucent, geodesic dome in a clearing in the woods on Cape Cod. It contained an aquatic ecosystem for the culture of foods. Tilapia, at the time a little-known fish, was the principal product of our primitive engineered ecosystem.

By 1980 ecological engineering had been applied broadly and put to practice by a small but growing group of scientists. In 1980 in *Tomorrow is Our Permanent Address: The Search for an Ecological Science of Design as Embodied in the Bioshelter* by Nancy Jack Todd and myself, we compared ecological engineering with conventional engineering, and argued that the latter, as currently practiced, was not sustainable over time. In 1984 in *Bioshelters, Ocean Arks, And City Farming: Ecology as the Basis of Design*, we attempted to articulate the basic precepts for sustainable ecologically engineered technologies. The most recent book on this subject, published in 1989, is *Ecological Engineering: An Introduction to Ecotechnology,* edited by William Mitsch and Erik Jørgensen, and dedicated to H.T. Odum. It is likely to establish ecological engineering as a legitimate academic discipline.

QUALITIES

Compared to conventional technologies, living machines have many unique qualities. It is their aggregate characteristics that most distinguish them. People accustomed to seeing mechanical moving parts, to experiencing the noise or exhaust of internal combustion engines or the silent geometry of electronic devices, often have difficulty imagining living machines. Complex life forms, housed within strange light-receptive structures are at once familiar and bizarre. They are both garden and machine. They are alive yet framed and contained in vessels built of novel materials, some of which are still in the developmental stages. Living machines bring people and nature together in a fundamentally radical and transformative way.

The difficulty many people may initially feel on first seeing living machines lies in the fact that although they serve human ends, they have unfamiliar attributes. A comparison between living machines and conventional technologies, including biotechnologies, which are not living machines, is presented in Tables 1 and 2.

TABLE 1
Energy Sources of Living Machines and Conventional Technologies

Energy	Living machines	Conventional technologies
Primary sources →	The sun	Fossil fuels, nuclear power
Secondary sources →	Radiant energy, internal biogenesis of gases	Combustion and electricity
Control →	Electricity, wind, and solar electric	Electrical, chemical and mechanical
Capture of external energy →	Intrinsic to design	Rare
Internal storage →	Heat, nutrients, and gases	Batteries
Efficiency →	Low biological transfer efficiency in subsystems, high overall aggregate efficiency	High in best technologies, low, when total infrastructure is calculated
Flexibility →	Inflexible with regards to sunlight Flexible, with adjunct energy sources	Inflexible
Pulses →	Tolerant and adapted	Usually intolerant, tolerant in specific instances

TABLE 2
Further Comparisons of Living Machines and Conventional Technologies

Living machines	Conventional technologies
Control	
Primarily internal throughout complex living circuits	Electrical, chemical, and mechanical controls applied to system
Threshold number of organisms for sustained control	External orchestration and internal regulation
All phylogenic levels from bacteria to vertebrates act as control organisms	
Disease is controlled internally through competition, predation, and antibiotic production	Through application of medicines
Feedstock both internal and external	Feedstocks external
Modest use of electrical and geseous control inputs orchestrated with environmental sensors and computer controls	Sophisticated control engineering
Pollution	
Pollution, if occurs, is an indication of incomplete design	Pollution intrinsically a byproduct, capture technologies need to be added
Positive environmental impact	Negative or neutral environmental impact
Management and repair	
Training in biology and chemistry essential	Specialists needed to maintain systems
Empathy with systems may be a critical factor	Empathy less essential
Costs	
Capital costs competitive with conventional systems	The standard
Fuel and energy costs reduced	High fuel and energy costs
Labor costs probably analogous — still to be determined	The standard
Lower pollution control costs	The standard
Operation costs lower because of reduced chemical and energy inputs	The standard
Potential reduction of social costs, in part because of potential transferability to lesser industrialized regions and countries	Social costs can be high
Design	
Parts are living populations	Hardware based
Structurally simple	Structurally complex
Complex living circuits	Circuit complexity often reduces
Passive, few moving parts	Multiple moving parts
Dependent entirely upon environmental energy and internal storage systems	Energy intensive
Long life spans . . . centuries	Short life spans . . . decades
Materials replacement	Total replacement

TABLE 2 (continued)
Further Comparisons of Living Machines and Conventional Technologies

Living machines	Conventional technologies
Internal recycling intrinsic	Recycling usually not present, pollution-control devices used
Ecology is scientific basis for design	Genetics is scientific basis for biotechnology, chemistry is basis for process engineering, physics for mechanical engineering
Materials	
Transparent climatic envelopes	Steel and concrete
Flexible lightweight containment materials	Reliance on motors
Electrical- and wind-powered air compressors/pumps	Structurally massive
Biotic design	
Photosynthetically-based ecosystems	Independent of sunlight
Linked sub-ecosystems	Unconnected to other life forms
Components living populations	Only biotechnologies use biotic design
Self-design	No self-design
Multiple seedings to establish internal structures	
Pulse driven	
Directed food chains: end points are products including fuels, food, waste purification, living materials, climate regulation	

Their adaptiveness and ability to protect the environment is due to their photosynthetic base which is powered by sunlight. Although secondary sources of energy can and often are used for control and light augmentation, the uniqueness and economic viability of living machines is based upon their dependency on photosynthetically-based food chains.

Living machines are built with parts that are themselves living populations, often extremely diverse and comprised of hundreds of species. A key attribute is that their components replace themselves as they wear out. The life spans of these machines can be extraordinary, lasting for centuries if housed within long-lived materials. They further have abilities to respond and change with variations in inputs. The task may be set by the ecological engineer, but if the living machine is allowed to develop immense complexities it will establish new biotic relationships not seen in nature, thereby expanding options of diversity. The organisms in the ecological digester in my office or at the Center for the Restoration of Waters' ecologically engineered sewage treatment facility at Providence, Rhode Island (Figure 1) have before never experienced the constellation of relationships they currently do as part of living machines. They have the potential to self-design new relationships if they are adapted to the new contained environmental conditions. There are further subsets of self-design. In the Providence facility, for example, there are four parallel treatment systems, each comprised of 14 components that separate the raw sewage from the

FIGURE 1. Solar aquatics facility in Providence, Rhode Island.

clear water discharge. The sub-ecosystems self-design differently at each stage in response to input variations in external factors such as light or to internal variations in the strength of the waste stream. These self-design combinations are dynamic and highly varied.

Another important attribute of living systems is that they are pulse driven. Daily, seasonal, and sporadic variations stamp themselves deeply on the ecology. The background of pulse creates the ability and vigor of the systems to recover from external shocks which is impossible for conventional machines. A living machine can be overwhelmed by overloading or by light deprivation, and in the process lose critical organisms and ultimately the ability to carry out the assigned task efficiently. But, even under severe stress, some of the internal circuits will continue to function, perhaps requiring reseeding to bring the system back to its original task.

Control species within the systems orchestrate the overall ecology. The ecological engineering is in such an early stage of development, however, that we do not yet have a clear idea of whether there are control hierarchies, or if a threshold number of living parts is the basis of control. I have experienced organisms that upset control functions. In the pulp sludge digester, one very active species of fish was keeping the pulp particles in suspension and making the material unavailable for detritus feeding organisms. Removal of the fish species in question fixed the problem. The life histories of organisms are the building blocks of ecological design. It will be essential to graft ecological engineering to a sophisticated knowledge of natural history.

Many years ago Bill McLarney* and I found large fish of the genus *Brycon* in small upland streams in Costa Rica. These streams had little in the ordinary way of foods for the fish in them which, nevertheless grew to sizable fish. We then discovered that *Brycon* could capture and then digest hard and inedible-seeming fruits that fell into the stream. Closer investigation of the anatomy of the fish revealed terrifying-looking teeth capable of shredding hard materials and a long serpentine intestine that was able to digest tough materials. It dawned on me then that the world is a vast repository of unappreciated or unknown biological strategies that have immense importance for humans, if we could develop a science of integrating the stories embedded in nature into the basic systems that sustain us. Human destiny is linked to the natural world. Conservationists and preservationists validly honor nature and work to protect those pristine natural places that remain. But the survival of civilization may well require another fundamental step. It may become essential that we enter into the natural world and use its teachings to reshape and redefine our tools and technologies. Good farmers and gardeners have always had this kind of relationship with nature. With the unfolding and application of ecology, it is now possible to extend this relationship into a new dimension.

TECHNOLOGICAL IMPROVEMENTS

The idea of living machines had to wait for ecology. It also had to wait for the materials sciences to evolve to the point where energy and environmentally responsive materials could be made and manufactured cost-effectively. The containing vessels for the majority of the living machines I have developed need to be fabricated from lightweight, high light-transmitting, flexible materials which can be bonded and waterproofed. They must be capable of handling a variety of stresses including high pressure and ultraviolet radiation. These materials started to become available in the 1970s. By the 1980s new transparent materials like heat mirrors were created to help reduce radiant heat losses. The next step will be materials capable of changing properties based upon internal and external conditions. Such materials will borrow

* Author of *The Freshwater Aquaculture Book.* (1984/1987). Hartley and Marks, Point Roberts, Washington. — Eds.

their properties from strategies employed in nature, perhaps as dramatically as the skin of some lizards. Materials that are "intelligent" are on the verge of being produced commercially. One such example is Cloud Gel, a film material invented by Day Chahroudi, that becomes opaque when its interior surface overheats and transparent when it cools.* During the day, under sunny skies, it prevents overheating without artificial cooling. Other new light-transmitting materials trap internal heat under clear nighttime skies. In combination these materials can regulate and optimize the climates needed by living machines to maintain climatic homeostasis. Structure and containment vessels of beautiful shapes that cover many acres can be developed through this type of gossamer engineering. The term bioshelter was coined to define structures that link these materials with ecology.

The fuel-dependent and energy-intensive processes we know today will be gradually replaced by responsive materials that house and contain various living processes. Widespread development of living machines had to wait for this convergence of biology and materials science. When Buckminster Fuller** opened the radically new pillow-dome at the New Alchemy Institute in 1980, he saw for the first time a living machine housed in one of his geodesic structures. Turning to the principle architect, J. Baldwin, he explained that the pillow-dome was the culmination of his own work, namely the marriage of his engineering and ecology. The pillow dome was glazed with three inflated layers of Tefzel, an inert high light-transmitting material developed by DuPont. The glazing was inflated with Argon gas which functions as invisible insulation. Buckminster Fuller understood the symbiotic relationship between ecological engineering and the materials sciences immediately and knew that it signaled a technological shift of real significance. He urged the DuPont executives to pay attention.

LIVING MACHINES AND SOCIETY

The question of whether living machines will be accepted is as yet not clear. Economically and energetically they make enormous sense. They are already competitive for raising certain foods and purifying concentrated wastes. A commercial facility to handle the latter has been built by Ecological Engineering Associates in Harwich, Massachusetts. Because living machines avoid using hazardous chemicals and are pollution-free in operation they are appealing to environmentalists. The greatest stumbling block to their adoption may be the inertia inherent in conventional engineering and the wide gap between applied ecology and civil and mechanical engineering. Some engineers however are beginning to train themselves in ecology and vice-versa.

The biggest blockade to the emergence of living technologies could be the very phenomena living machines are intended to solve, namely the estrangement of modern cultures from the natural world. Nature is "invisible" to many people in our culture. It is my hope that the aesthetic and emotional feeling that living machines can generate in us will yet carry the day. They can be made beautiful and evocative of a deep harmony that is nature. New economies wrapped in the wisdom of the

* Day Chahroudi, Suntek, Inc., 6817A Academy Parkway, Albuquerque, New Mexico 87109.

** Author of many books on humans, geometry, and nature, including *Operating Manual for Spaceship Earth*, E.P. Dutton, New York, 1978. — Eds.

natural world are capable of creating the future we all desire. I look forward to seeing the results of our efforts to integrate ecological engineering into designed systems here in Sweden.

REFERENCES

Mitsch, William J. and S.E. Jørgensen (eds.). 1989. *Ecological engineering: An introduction to ecotechnology.* New York: John Wiley and Sons.

Odum, H.T. 1971. *Environment, power, and society.* New York: John Wiley and Sons.

Todd, John and N.J. Todd. 1980. *Tomorrow is our permanent address: The search for an ecological science of design as embodied in the bioshelter.* New York: Harper and Row.

Todd, John and N.J. Todd. 1984. *Bioshelters, ocean arks, and city farming: Ecology as the basis of design.* San Francisco: Sierra Club Books.

Chapter 10

SOLAR WASTEWATER TREATMENT IN DENMARK: DEMONSTRATION PROJECT AT THE DANISH FOLKECENTER FOR RENEWABLE ENERGY

Jørgen Hinge and Hamish Stewart

ABSTRACT

During the autumn and winter of 1988 to 1989 a demonstration plant for treatment of domestic wastewater from the Folkecenter was built. The plant began functioning in April 1989 and in 1991 treats 3 to 4 m^3/day. Measurements are regularly carried out to test whether the treated water meets the Danish overall standards of 1.5 mg/l total-P, 8 mg/l total-N and 15 mg/l BOD_5. The results have shown that the plant is able to meet these standards when in stable operation for longer periods. There are, however, still occasional problems resulting in insufficient treatment.

BACKGROUND

The Folkecenter for Renewable Energy is a non-profit organization dealing with research, information, and development of renewable energy. Besides the "traditional" renewable energy sources like wind, solar, and biomass energy, the Folkecenter also has a department for green technologies. This department works with so-called green solutions with regard to food production, waste treatment, recycling, etc.

The Folkecenter approach to wastewater treatment has been to look at its energetic aspects. In Denmark a lot of smaller villages are to be connected to large central wastewater treatment plants, where the wastewater must often be pumped over great distances. This method of treating small quantities of wastewater involves a large electricity consumption. The electricity is most usually generated in coal-fired power plants, and thus the side-effect of treating the wastewater is increased air pollution.

The main purpose of the project has been to construct a model for decentralized treatment of domestic wastewater, which involves little or no use of fossil energy. Second, there is a focus on reutilization of the nutrient resources in the wastewater.

DESCRIPTION OF THE PLANT

The plant consists of three different and independent parts (all covered with plastic greenhouses):

1. Solar algae system
2. Ditch system
3. Root-zone beds.

0-87371-990-5/97/$0.00+$.50

These three parts can function independently, together, or be interconnected in several ways.

1. Solar algae system. The solar algae system is the central part in our treatment plant and therefore most attention has been given to this part. It is based on the concepts of the "Sugarbush" plant and consists of four basins in series each 7 × 2 m and 1 m deep. In the first three basins, bacteria, algae, and macrophytes together with various kinds of invertebrates and fish remove the nutrients from the wastewater. In the last basin the water is led through soil in which *Typha* and *Iris* grow.

2. Ditch system. In a system of shallow ditches (20 cm deep), different kinds of macrophytes are grown with their roots directly submerged in the water (hydroponics). This part of the treatment plant is basically for experimental purposes, and the activity here does not contribute significantly to the cleaning.

3. Root-zone beds. The last part of the system is made up of four parallel basins of the same proportions as the solar algae system. The basins are soil filled, with different kinds of plants being tested for their ability to transport oxygen into the soil. These basins act as a polishing for the solar algae system, but can, as previously mentioned, be fed "raw" wastewater.

RESULTS

From April 1989 to January 1990 a measurement program was carried out in cooperation with The Botanical Institute, University of Århus, DK. Following that, the Folkecenter has carried out field tests every month. During the winter of 1990 to 1991, equipment has been bought to establish a laboratory at the Folkecenter. From now on we will be able to carry out precise analyses on site. In Table 1, key results from the measurement program and the field tests are shown.

The station numbers refer to the Roman numerals in Figure 1.

From August to October 1990 the results showed basically the same trends, whereas from late October to January there were often problems, causing us periodically to lead the wastewater to the Folkecenter's underground seepage system. In February 1991 however — despite the lowest temperatures in 4 years — the plant was back in operation again.

CONCLUSIONS

- The plant has considerable capability to clean household wastewater.
- There seems to be no decrease in treatment efficiency during colder periods.
- Changes in design still need to be made to ensure hydraulic stability.
- A back-up aeration facility is necessary.

TABLE 1
Key Parameters in the Solar Wastewater Treatment Plant[a]

Date		Station I	Station V	Station XV
19/4-89	Total-N	80	35	20
	Total-P	13	3	3
	BOD (=BI_5)	151	24	64
18/5-89	Total-N	80	9	20
	Total-P	12	7	2
	BOD	270	149	56
27/6-89	A poisoning of the system caused an almost total stop of all functions.			
1/8-89	Total-N	80	8	(18)
	Total-P	18	<1	(2)
	BOD	208	22	(50)
	A fraction of "raw" water was led directly to the root-zone system, so the water at station XV is obviously more "polluted" than at station V.			
21/10-89	Total-N	106	30	20
	Total-P	14	6	1
	BOD	350	98	—
6/12-89	Total-N	55	32	2
	Total-P	11	11	1
	BOD	185	42	56
27/1-90	Total-N	65	40	20
	Total-P	12	u	u
	BOD	>350	210	40
1/3-90	Total-N	50	25	7.5
	Total-P	8	u	u
	BOD	290	120	10[(1)]
30/3-90	Total-N	60	40	5
	Total-P	15	<1	u
	BOD	300	150	30
27/4-90	Total-N	30	10	10
	Total-P	10	u	u
	BOD	150	90	50
1/6-90	Total-N	65	35	5
	Total-P	15	12	u
	BOD	270	—	15–40
27/6-90	Total-N	40	10	5
	Total-P	12–13	<1	u
	BOD	210	200	45
27/7-90	Total-N	70	55	20
	Total-P	10	5	<1
	BOD		No measurement	
30/7-90	BOD	320	100	70

[a] Station numbers refer to the Roman numerals in Figure 1.

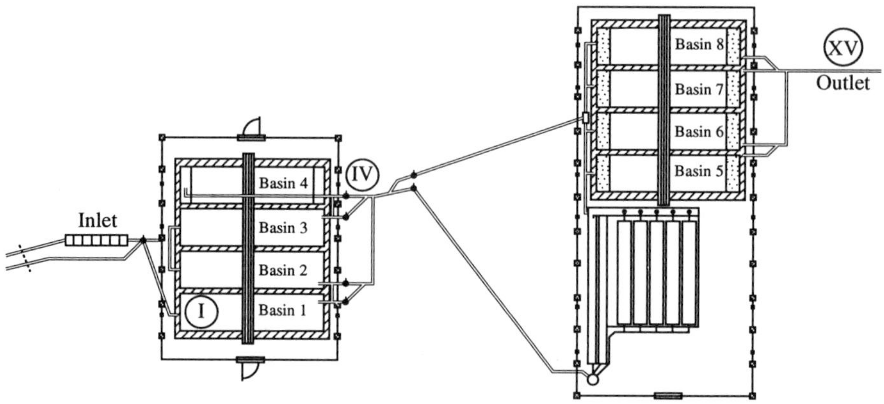

FIGURE 1. Overview of the solar wastewater treatment plant.

Chapter 11

RESOURCE-SAVING AND ECOLOGICAL ASPECTS OF USING WASTEWATER FROM EEL BREEDING AS A SOURCE OF NUTRIENTS, WATER, AND CARBON DIOXIDE FOR PLANT PRODUCTION

Gert Jungersen

ABSTRACT

At DIFTA, The Danish Institute for Fisheries Technology and Aquaculture, formerly The Danish Aquaculture Institute, we have been working on a project called "Green Wastewater Treatment" since 1988. The main aim of the project is to investigate the possibilities of using wastewater as the basis for production of greenhouse crops.

The framework of these experiments is an eel breeding plant, using water recirculation in a 400 m^2 commercial greenhouse. The wastewater from eel production contains high concentrations of nutrients, i.e., about 100 mg/l N and 10 to 15 mg/l P, and therefore it is very suitable as a source of water and nutrients for production of plants. Wastewater from many other sources, especially within the field of animal and vegetable processing, contains similar amounts of N and P.

The potential growth, yield, and nutrient removal rate have been tested for different species of edible and ornamental plants, including tomato (*Lycopersicon*), corn (*Zea mays*), *Ficus benjamina*, *Setaria*, *Cyperus*, *Geranium*, and *Chrysanthemum*. The average annual rates of N and P removal were in the range of 150 to 200 mg N/m^2/day and 20 to 40 mg P/m^2/day for ornamental plants, and in the range of 350 to 450 mg N/m^2/day and 55 to 70 mg P/m^2/day for tomato plants. The average water consumption was about 3 l/m^2/day and 5 l/m^2/day, respectively. Both the quality and the quantity of the plants produced were comparable to results from gardening plants, using commercial fertilizer.

When integrating intensive fish farming and gardening, the environmental load from the fish farm can be almost eliminated, at the same time that valuable nutrient and water resources are recovered. Moreover, the carbon dioxide produced by the biological filters can be reused to raise the concentration of carbon dioxide in the greenhouse, thereby increasing the plant production.

So far, the conclusion of the project is that integration of fish breeding and plant production is profitable from an ecological as well as from an economical point of view.

0-87371-990-5/97/$0.00+$.50

PART I: PLANT PRODUCTION IN PROCESS WATER FROM EEL PRODUCTION

Since 1988, DIFTA has been involved in a project called "Green Wastewater Treatment." The project has been supported economically by The National Agency of Industry and Trade and IFV-energy/NESA, a company producing and delivering energy to Copenhagen and Northern Zealand. The main idea of the project is to examine the possibilities of using wastewater as the main source of water and nutrients for production of valuable plants.

The growth and nutrient removal of different plant species, including *Ficus benjamina*, *Geranium*, *Chrysanthemum*, *Eucalyptus*, *Cyperus*, *Setaria*, willow (*Salix*), elephant grass (*Mischantus sinensis*), corn (*Zea mays*), and tomato (*Lycopersicon*), have been tested during the period of investigation.

SYSTEM DESCRIPTION

The experiments take place in an eel breeding plant and a standard greenhouse of 400 m^2.

Eel Production System

The eel production system is based on the principles of recirculation. It consists of four circular fish ponds, with a total surface area of 28 m^2. The fish ponds are connected in series to a swirl concentrator, which removes larger particles from the water, and two biological upflow filters with a total filter volume of 12 m^3 (Figure 1). The biofilters contain a plastic material with an effective surface of 200 m^2/m^3, to which a microbial film is attached.

The main purposes of the biofilm are breakdown of organic matter, thereby reducing the biological oxygen demand of the water, and oxydizing of ammonia and nitrite in order to eliminate the toxic effects of these substances.

Greenhouse System

The greenhouse contains three different cultivation systems, all based on hydroponic techniques. The added process water from the fish breeding system is recirculated separately within each system.

The only modification of the wastewater is adjustment of the pH level to 5.7 with sulfuric acid, in order to increase the amount of manganese, iron, and phosphorus available to the plants.

RESULTS

Removal of nutrients and water

During 1988 to 1990 the removal rates of nitrogen and phosphorus were examined for different ornamental and edible plant species.

The average annual rate of N and P removal were in the range of 142 to 180 mg N/m^2/day and 17 to 35 mg P/m^2/day for ornamental plants, and in the range of 338 to 429 mg N/m^2/day and 52 to 70 mg P/m^2/day for tomato plants. The average values for corn and elephant grass lie in between the levels for ornamental plants and tomatoes (Table 1). The average water consumption was about 3 l/m^2/day and 5 l/m^2/day, respectively (Table 1).

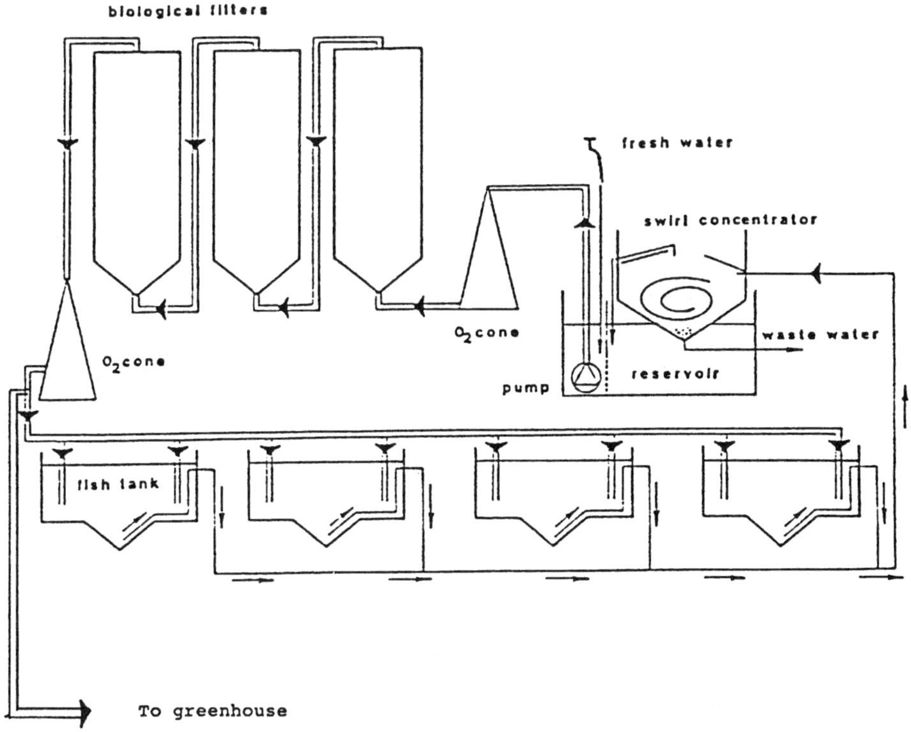

FIGURE 1. Recirculation, eel-breeding system.

TABLE 1
Average Annual Removal Rates of Nitrate-N and Phosphate-P in the Three Cultivation Systems in 1988 to 1989

System	Year	Plant species	N-removal ($mg/m^2/day$)	P-removal ($mg/m^2/day$)
Ebb/flow	1988	*Cyperus* "Zumula" *Setaria chevalieri*	180	22
	1989	*Ficus benjamina*	142	17
NFT	1988	*Cyperus diffusus*	152	35
	1989	Tomato	338	52
Floor	1988	Corn (*Zea mays*) *Cyperus* "Zumula"	234	20
	1989	Elephant grass	221	49

Note: NFT = Nutrient film system.

The seasonal variation in the rates of nutrient removal and consumption of water varies a lot among the examined plant species. Figure 2 shows the seasonal variation in removal of nitrogen and phosphorus and in consumption of water, for elephant grass, *Ficus benjamina,* and vegetative tomato plants. Figure 3 shows the seasonal variation for a production of tomatoes.

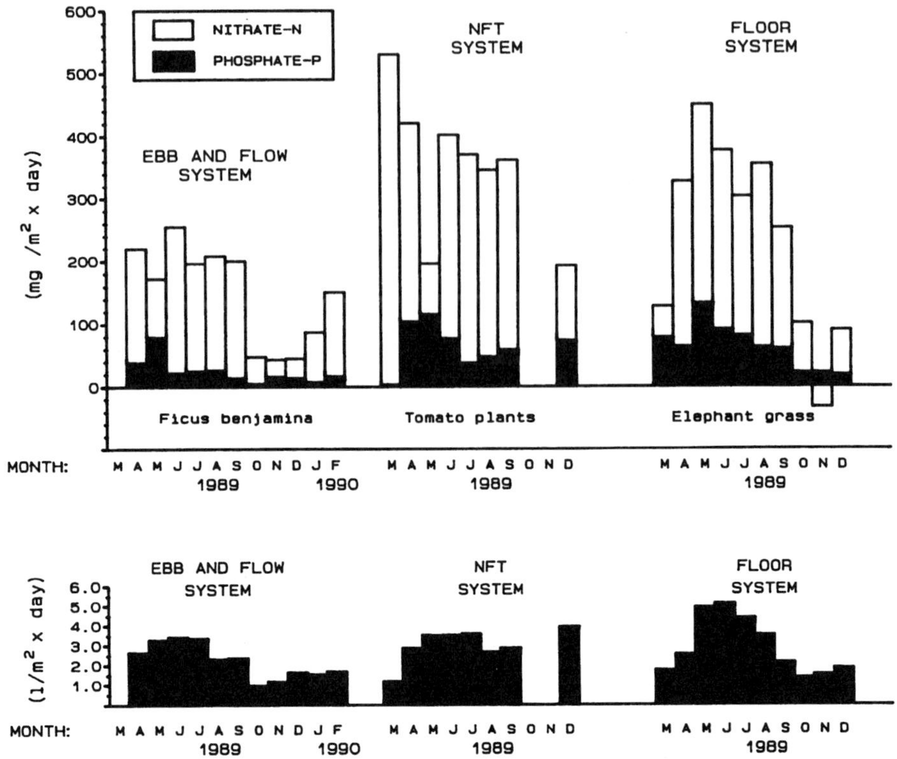

FIGURE 2. Removal of nitrogen and phosphorus and water consumption in the three cultivation systems in 1989/90.

In general, ornamental plants and grasses like corn and elephant grass have the greatest seasonal variation, while the seasonal variation in nutrient removal for tomato plants is much less. The minimum removal rates of nitrogen and phosphorus for ornamental plants generally lie in the range of 15 to 25% of the seasonal average, while the minimum rates for tomato plants are 50 to 75% of the average values.

Quality of the Plant Products

Both the quality and the quantity of the plants produced were comparable to results from plants grown in gardens, using commercial fertilizers.

Most of the ornamental plants have been sold on the market at today's prices, which indicates good quality. The period of growth, from seeds or cuttings to plants ready for sale, has also been comparable to commercial plant production.

The quality of the tomatoes produced was excellent and a grading of the fruit has shown that over 80% were first class, and less than 20% were of poorer quality.

In Denmark however, it is not legal to sell edible products produced in non-treated wastewater, even though it comes from fish breeding plants.

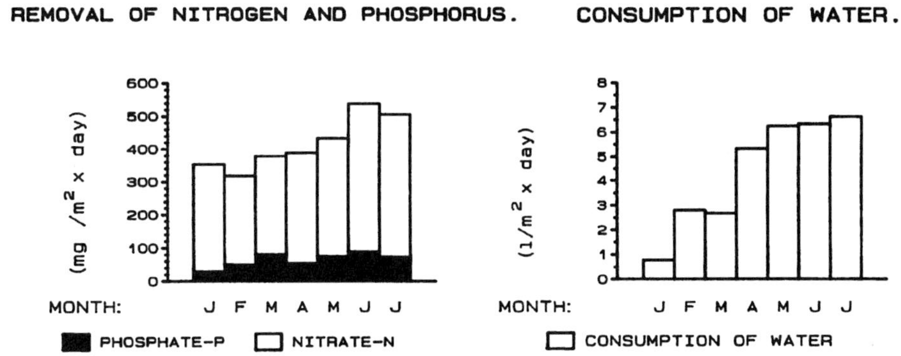

FIGURE 3. Removal of nitrogen and phosphorus and water consumption in relation to production of tomatoes in 1990.

PART II: RESOURCE RECOVERY IN INTEGRATED FISH/PLANT SYSTEMS

From an ecological point of view, the idea of integrating fish breeding with production of plants is obvious, as the two processes are complementary to each other. It means that the waste products from one of the processes form the biological basis of the other.

Fish biomass is built up from the breakdown of fish feed, which releases energy. The waste products from the fish breeding and the activity of the biological filters (water, nutrients, and carbon dioxide) form the basis of the plant production. Plant biomass is built up, using light energy. In practice it is possible to reuse water, nutrients, and carbon dioxide from the fish breeding in the plant production.

The effect of the energy input from the fish production on the total energy budget has not been evaluated yet.

AN EXTRAPOLATION: INTEGRATED EEL BREEDING AND TOMATO PRODUCTION

The following hypothetical example is based on an eel-breeding plant with an annual production of 100 tons, integrated with a production of tomatoes in ordinary greenhouses.

The area of tomato plants is calculated from the average consumption of nitrogen (i.e., 429 mg N/m^2/day) and the total annual output of nitrogen from the eel breeding of 4.5 tons N. The result is a total area of tomato plants of 29,000 m^2, which equals a greenhouse system of 33,000 m^2.

The annual yield of tomatoes is at least 30 kg/m^2, giving a total annual production of tomatoes of 870 tons.

Nitrogen and Phosphorus Flow

Figure 4 shows the resource flow of nitrogen and phosphorus in relation to production of tomatoes. The calculations are based on an average concentration of

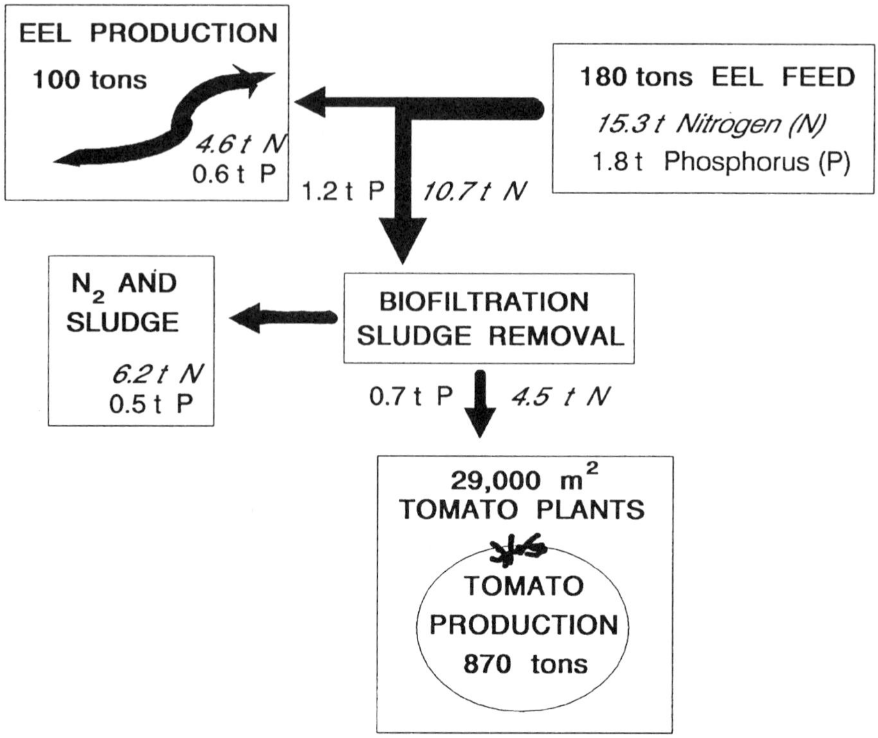

FIGURE 4. Flow of nitrogen and phosphorus in an integrated eel/tomato production system.

nitrogen in the process water of 125 mg N/l, and on the average ratio between nitrogen/phosphorus 6.3:1. The nitrogen concentration in DIFTA's eel-breeding plant has varied a lot during the period of investigations (20 to 137 mg N/l), due to experiments with microfiltration and denitrification in the eel-breeding system. These experiments have shown that it is possible to control the levels and ratio of nitrogen and phosphorus in this system.

The N/P ratio in the process water is very close to the ratio between the removal rates in the plant system, which is 6.1:1. This means that, on average, both N and P are used up in the system.

Water Flow

The consumption of water in a recirculation eel-breeding plant is approximately 0.2 m^3/kg of eel feed used for the fish. This means that 68% of the water demand in the tomato production can be met by the process water from the fish production. The rest of the water demand can be met by collecting rainwater from the greenhouse roofs (Figure 5).

Carbon Dioxide Flow

The amount of carbon dioxide produced by the respiration of fish and biological filters is about 1 kg/kg eel feed used. Most of the carbon dioxide is stripped off in

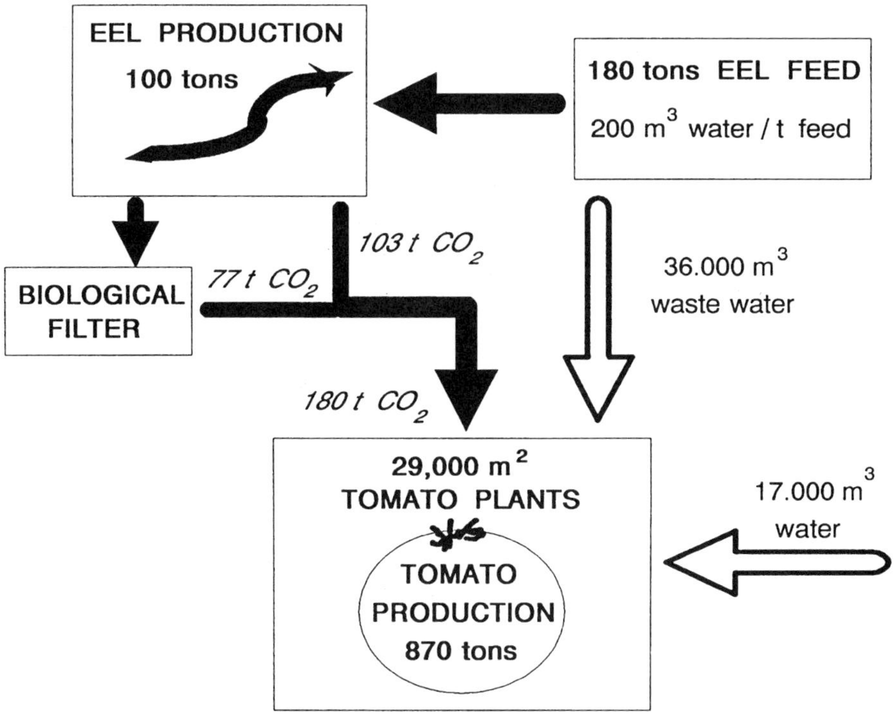

FIGURE 5. Budget of water and carbon dioxide in an integrated eel/tomato production system.

the biological filters or in aeration systems. The total annual amount of carbon dioxide produced is 180 tons or 6.2 kg/m^2, which is about twice the average amount of carbon dioxide normally used for commercial tomato production in Denmark (Figure 5).

The consumption of carbon dioxide in commercial gardening is unevenly distributed over the year. Normally, carbon dioxide is not supplied in the summer period due to high rates of aeration in the greenhouses. The maximum requirement in commercial gardening is 1 to 2 g/m^2/h, compared to an average production rate in the fish/biofilter system of 0.7 g/m^2/h. Therefore, only about 80% of the requirement of carbon dioxide could be met by the fish-breeding system, even though the production is twice the demand of the tomato production.

Other Advantages of Integrated Systems

As mentioned earlier, some energy will be exchanged from the fish system to the plant system together with water and carbon dioxide enriched air. But it is difficult to calculate the exact amount. Anyway, the energy transferred from the fish system will only account for a small part of the total energy consumption in the greenhouse system.

On the other hand, ventilation of the greenhouse air through the biological filters will have a stabilizing effect on the climate in the greenhouse, especially in hot periods.

DISCUSSION

So far, the conclusion of the project is that it is possible to cultivate many ornamental and edible plant species in wastewater from eel breeding, without significant reduction in yield and plant quality. Furthermore, many plants grown in hydroponic systems are able to reduce the concentrations of nitrogen and phosphorus to very low levels, without a reduction in the growth rate.

Effluents from other sources, especially from vegetable and animal production, can with advantage be used as a basis for plant production, too. The use of domestic wastewater for greenhouse production does involve some hygienic problems, which may be overcome with different methods of disinfection.

The most obvious advantage in the fish breeding/plant production system is the possibility to make a very close physical and ecological integration of two productions. In that way, most of the waste products from the eel breeding are reused in the plant production, and the oxygen produced in the plant system is returned to the fish. Moreover, the climate in the plant system is stabilized due to the integration.

Therefore, it seems to be ecologically and environmentally reasonable to integrate these two productions. It seems possible to get an economic benefit from the integration, as well.

Under temperate climate conditions, the reuse of water, nutrients, and carbon dioxide may result in a savings of about 2 to 3% of the total annual running costs in the plant production. In tropical and subtropical areas, where the energy used for heating and lighting are almost eliminated, savings may be much greater. The possible savings in the fish production is 3 to 8% of the total annual running costs, depending on the methods used for disposal of the wastewater.

In the near future, our work will be focused on implementation of full-scale projects using the described techniques. New components may be added to the integration:

- The production of sludge in a 100 ton eel-breeding plan is up to 1000 tons/y, with a dry matter content of about 5%. Biogas fermentation and composting of this sludge with straw or other fiber materials can produce gas for energy production and a compost which can be used as a growth medium for ornamental plants.
- The eel breeding may be combined with production of carp which live off of the excrements and feed waste from the eels, thereby improving the overall utilization of feed.

REFERENCES

The Danish Aquaculture Institute, ATV. 1989. *Pilot project: "Green Wastewater Treatment," 1st year report.* In Danish.

The Danish Aquaculture Institute, ATV. 1989. *Experiments with human wastewater as a source of nutrients for plant production.* In Danish.

The Danish Institute for Fisheries Technology and Aquaculture, ATV (1990). *Pilot project: "Green Wastewater Treatment," 2nd year report.* In Danish, with English summary.

Chapter 12

LIVING WITH YOUR WASTE/WASTEWATER

Jesper Saxgren

INTRODUCTION

During most of this conference, we have been dealing with problems and solutions at one end of the wastewater pipe — where it all ends.

I would like to concentrate on the other end of the pipe — where it all begins — and argue for smaller decentralized units (on people's doorsteps) and the importance of aesthetics in the design of these systems.

But first I want to emphasize that the size of the wastewater problem is closely connected to how much water we actually turn into wastewater. If we reduce the water consumption, we reduce the amount of water we have to treat and we can reduce the size and capacity of the wastewater purification plant.

So when we talk about wastewater treatment methods, we should not forget to mention that everything begins with our own way of using or misusing water.

The more responsibly and carefully we act at the source, the less of a problem we have at the end.

But how can we change our actions in a more responsible way?

By shortening the distance between our actions and the consequences of our actions, by moving the consequences inside our range of vision so that it becomes visible and evident, that it matters whether we act in one way or in another. That different actions give different consequences.

THE SYSTEM IN FLØNG — AN EXAMPLE

At The Production Highschool in Fløng, Denmark, we tried to set up a system dealing with these problems and perspectives (Figure 1 and 2). Under a 150 m^2 pyramid-shaped, attached greenhouse, we have established an integrated circle of fish, crayfish, and plants, where everything recirculates within the system, including all organic waste from the school kitchen (60 people).

The integrated circle of fish and plants is built around two 15 m^3 basins. One is established as a polyculture dam with four different levels of carp fish to consume specific feeds available in various layers in the water.

The first level is the grass carp (*Ctenopharyngodon idellus*), which feed on macrophytes grown in the wastewater from the fish plus vegetable waste from the school kitchen, green plant parts from the greenhouse, comfrey, dandelions, and some grass from the outside.

The common carp (*Cyprinus carpio*) comprises the second level. It forages through much wider and more varied levels. It consumes part of the feces of the grass carp, snails, water fleas, plant seeds, algae, and bread remains from the kitchen.

0-87371-990-5/97/$0.00+$.50

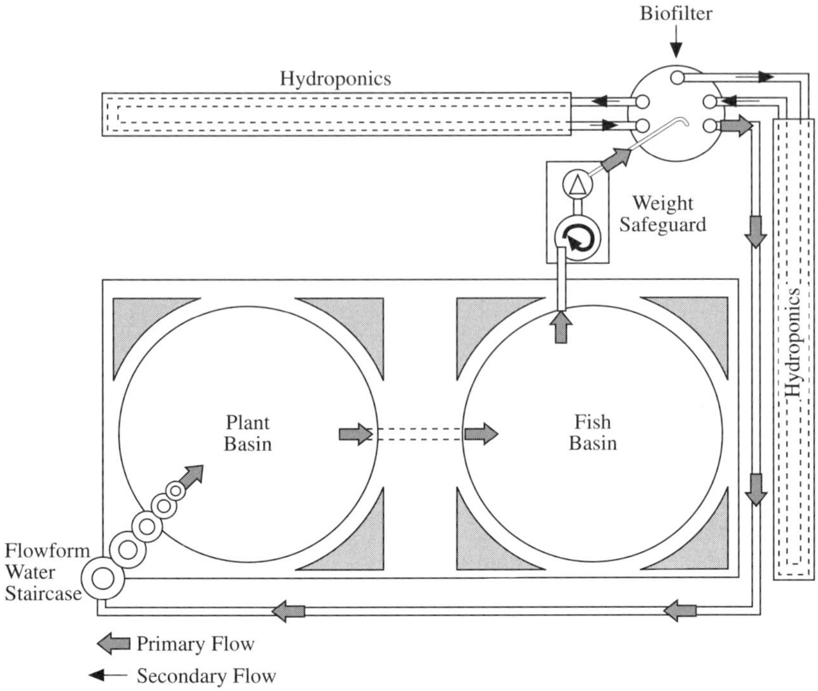

FIGURE 1. An overview of the Fløng wastewater treatment plant.

FIGURE 2. Wastewater treatment plant at Fløng.

At the third level we have the silver carp (*Hypophtalmicthys molitrix*), which live on filtering green algae in the water.

And finally a fourth fish from the carp family — *Leusiscus idus* — the primary function of which is to work as an indicator fish, because it responds to lack of oxygen earlier than the other fish. It feeds on insects at the water surface.

The nutrient-loaded fish water (wastewater) is circulated through a whirl separator, which separates bigger sludge particles, to a biofilter where nitrogen compounds are oxidized to nitrate by means of bacteria. (The sludge particles are fed to compost worms.)

From the biofilter the water runs to the hydroponics section where common vegetables such as lettuce, spinach, silver beet, and watercress, together with more exotic plants such as water chestnuts, are grown directly in the fish water.

Finally, the water runs by a water staircase (Flowform), where it is oxygenated, to basin number two. Here the primary binding of nutrients/purification of water takes place by means of water plants. A floating plant — *Pistia stratiotes* — has done especially well. During the summer season it doubles every 14 days, which means that from a surface area of 16 m^2 we have been harvesting 400 full-grown plants every fortnight.

In this basin we also aim at establishing as complex and diverse a system as possible, in order to obtain the widest possible absorption and conversion of the nutrient load.

That means high populations of water fleas (*Daphnia*), cyclops, gammarids, crayfish, and the like.

Finally, the water is pumped back to the fish basin and the circle is closed, which means that the same water circulates again and again — now in its fourth year.

THE WASTEWATER CYCLE AS AN EDUCATIONAL TOOL

"How often do you renew the water?" is a question often put to us by visitors. And it is of great pedagogical value to be able to answer and show that we do not renew, but only refill the amount of water we lose by evaporation so that in principle it is the same water which is purified again and again.

And the proof is right there and can be read directly in the health of the fish. The principles from the great natural cycle can be read and understood on the basis of the smaller, constructed cycle.

And contrary to the great cycle, where it is often very difficult to predict or see the immediate consequences of our actions because of too many unknown factors or because the distance between our action and its consequence is too great to reveal the link, here we get a concrete and immediately readable response to our actions or lack of actions in the prosperity and behavior of fish and plants.

By *observing* the prosperity and healthiness of fish and plants, we must constantly *consider* whether it is satisfactory or not, and on the basis of that we get a measurement for the quality and suitability of our *actions*.

If we observe malfunctions, we must consider the reason and on account of that come up with more suitable actions. The response to that must then again be observed.

The pedagogical strength of this system lies in the constant feedback we get. We cannot cheat within it, and it matters whether "nobody sees" or not!

The idea (the hope) is that this way of acting and understanding can be used in "real life" as well, as a suitable pattern for a more responsible way of acting in the future.

DECENTRALIZED SOLUTIONS

Through decentralized units close to the source we ensure attention and lucidity (the precondition for participation and responsibility) and the possibility for using solar-based "green" low energy/technology solutions.

Because of the closeness to the source we know much more about the sewage conditions and therefore how it can and should be treated.

A plant-based "living" wastewater treatment unit puts demands on the users at the other end if it is to function well. It demands consideration and responsibility.

But isn't that what we actually need on all levels — a more careful and respectful way of living, where things matter. When you virtually live with your waste/wastewater as an integrated part of your environment, and do not move it as far away as possible or at least out of sight, you have to care.

All of a sudden, questions like how much waste/wastewater we produce and how we treat it become relevant to the individual and not only a problem for experts and society.

And through that we might begin to consider and change our ways of acting and begin to understand that waste/wastewater does not have to be a complicated problem we must get rid of, but could also be a valuable resource — a nutrient capital which should be "reinvested" in the "quicker" and more accessible cycles "on land" instead of being deposited at the bottom of the sea thereby losing accessibility.

But to show and convince people that wastewater treatment is something we could live with as an integrated part of our nearest environment requires that aesthetics play a central role in our designs. It not only can be beautiful, it should be.

Let wastewater blossom!

Chapter 13

ENERGY BUDGET FOR A POLYCULTURE-BASED WASTEWATER TREATMENT SYSTEM

Hamish Stewart, George Chan, and Jørgen Hinge

ABSTRACT

Conventional Danish agriculture operates with an on-farm energy deficit of more than 2:1. The concept behind the described project is that it is possible and economically viable for a farming system in Denmark to produce a net energy surplus. Our calculations have shown that the projected system has not only a positive energy balance of almost 2:1, but also produces protein in excess of the system's needs. There is no demand for artificial fertilizer or pesticides, and fuel use is low, due to the system's compactness. Animal manures are the basis for a production of high protein animal feed and high value edible crops. All plant wastes are used as roughage in animal feeds or recycled through the biogas plant. A pilot project on 0.5 ha will be established in 1991.

INTRODUCTION

The starting point for the project we will describe has been the energetics state of Danish agriculture. Integrated farming systems in China and other parts of the world have supplied the inspiration for establishing this alternative.

Conventional Danish agriculture operates with an on-farm energy deficit of more than 2:1. In other words, in energetic terms Danish agriculture consumes more than twice the amount of resources it produces.

Even with the most environmentally benign farming methods, such as organic dairy farming, the on-farm energy deficit is as much as 1.5:1. The deficit is due not only to the use of electricity in stables and diesel oil for machinery but also to the import to the farm of animal feed.

Danish farmers generally produce crops that are relatively low in protein but which give a high yield. The large quantities of carbohydrate are then supplemented with high-protein feedstocks which can be bought cheaply on the international market. Again, if measured in energetic terms, the importation of protein-rich feedstuffs is approximately double the value of the protein content in the foodstuffs exported from Denmark, mainly meat, butter, cheese, and cereals.

It is estimated that the area (mostly in Asia and America) required to produce the protein that Denmark imports is about equal to the 2.7 million ha that constitute the Danish agricultural area.

There is a clear need for improvement if Denmark is to meet its declared objective of living up to the goals of the Brundtland report. The prime methods by which

0-87371-990-5/97/$0.00+$.50

these goals can be met by agriculture are through a larger home production of protein and a reduced consumption of artificial fertilizers and other energy-expensive farming aids such as pesticides.

METHODS

The integrated farming project is an example of how these goals can be achieved. The concept behind the project is that it is possible and economically viable for a farming system in Denmark to produce a net energy surplus. Our calculations have shown that the projected system has not only a positive energy balance of almost 2:1, but also produces protein in excess of the system's needs. There is no demand for artificial fertilizer or pesticides and fuel use is low due to the system's compactness.

By maximizing the use of solar energy and biogas technology it should be possible under an organic farming strategy to produce 106 GJ of biomass/ha, which is twice the amount that is normally produced on a Danish agricultural hectare.

The 2 ha farm for which the energy calculations have been made is composed of:

Field I	**Area (m^2)**
Greenhouse	760
Algae ponds	270
Polyculture ponds	6160
Aeroponics	1500
Hydroponics	500
Other	810
	10,000

Field II	**Area (m^2)**
Algae ponds	540
Polyculture ponds	6160
Aeroponics	1500
Hydroponics	500
Other	1300
	10,000

The greenhouse system is in effect an indoor model of the outdoor system. Adjoining the greenhouse is animal housing with a biogas digester and a workshop for processing of crops.

Animal manures are the basis for a production of high-protein animal feed and high-value edible crops. All plant wastes are used as roughage in animal feeds or recycled through the biogas plant.

The organic farming rules require a minimum field area of 200 m^2/pig. For geese the allowed area is 10 m^2/bird. These requirements mean that in the self-sufficient and organic strategies there is a need for an additional minimum area of 1.5 ha for animal welfare. Doubling this area to 3 ha enables harvest of a cash crop in alternating years. The energy value of the cash crop will be in the region of 70 GJ/year.

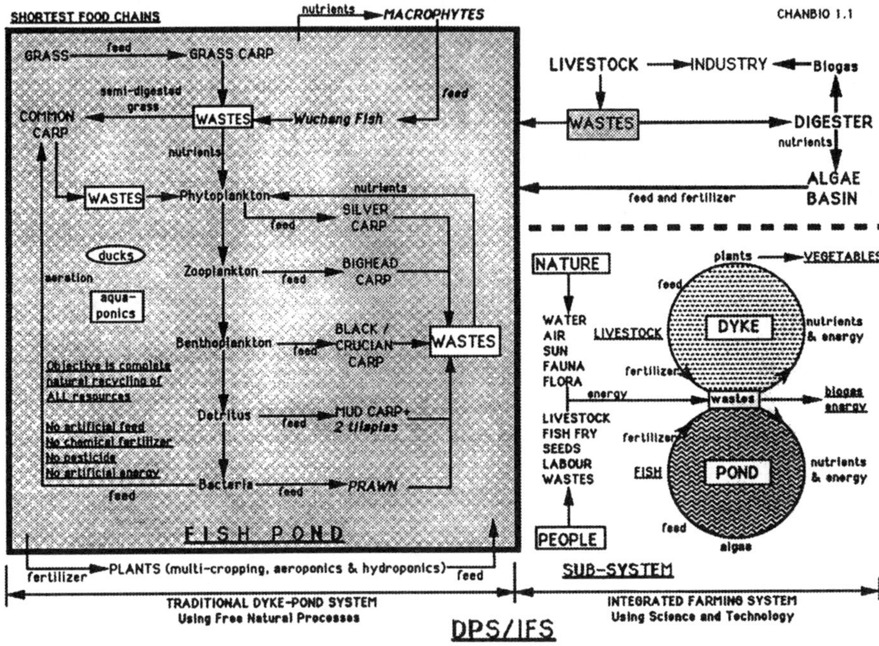

FIGURE 1. The elements of the Chinese dyke-pond system.

The additional field area and cash crop value has not been included in the calculations for ease of comparison of the different strategies.

In practice, the project will follow the organic farming guidelines. The more intensive animal farming of the economic and super economic strategies is regarded as being ethically inappropriate.

THE FLOW OF MATERIALS THROUGH THE SYSTEM

Pig manure is collected for biogas digestion. Under a retention time of 10 to 20 days, approximately 50% of the organic matter is reduced by bacteria and pathogens are broken down. Nutrients are released by the anaerobic process.

The degassed slurry from the digester flows to the algae ponds where organic content is further reduced by aeration and the nutrients are made available to the culture of *Spirulina. Spirulina*, due to its high protein content, is used as protein feed.

The still nutrient-rich water flows from the algae ponds to the 3 m-deep polyculture ponds. Organic content is further reduced by microorganisms in the water and on the surface of plant roots.

The main body of the water is used for a carp polyculture of five species, each feeding at a particular trophic level. The fish will be harvested while small and used as animal feed.

On the surface of the ponds are floating plants such as *Lemna* and *Pistia* which are partly eaten by grass carp and partly harvested as animal feed. Approximately

25% of the pond surface is used for 2 m-high floating aquaponic towers, where marketable produce is grown, e.g., lettuce, strawberries, beans, etc. Nutrient-rich water from the ponds is drawn up by capillary action.

The same water treatment strategy is used in the two other cultivation strategies which are located alongside the ponds.

Water is pumped from the ponds to aeroponic cultivation towers of 2 to 3 m high, and to the 30 cm deep hydroponic troughs where crops such as watercress, tomatoes, cucumber and melons are grown. Parts of the plants that cannot be sold are used as animal feed.

The slurry which now has undergone primary, secondary, and tertiary treatment is now recirculated as treated wastewater to the stable, where it is used to wash manure into the biogas digester.

ENERGY CALCULATIONS

Five different operating strategies, based on varying levels of animal intensiveness and feed import have been reviewed for the 2 ha system. (Table 1).

As can be seen in Tables 2 and 3, the energy balance is worsened with increasing imports of feed, even though the economic result is improved.

CONCLUSION

There is naturally a large degree of uncertainty as to how well the described farming system will live up to the theoretical calculations when tested in a pilot project. There are, in addition, challenges in the cultivation techniques, questions of hygiene and choice of species for a maximum yield.

The positive aspect of the project, if economic viability can be demonstrated, is that it provides a possible solution to the preservation of small farming units.

TABLE 1
Conditions for the Five Strategies (on the basis of a 2 ha system)

Strategy	No. and type of animals	Crops for sale (tons)
Self-sufficiency	84 bacon pigs 1200 geese	16.7
Organic	84 bacon pigs 1200 geese	24.3
Ecological	84 bacon pigs 2400 geese 1500 ducks	28.0
Economic	132 bacon pigs 4500 geese 1500 ducks	38.5
Super economic	360 bacon pigs 4500 geese 2100 ducks	48.0

TABLE 2
Integrated Farm of 2 ha Financial Balance (D.kr) and Energy Balance (GJ) for Five Operating Strategies

Strategy	Self-sufficiency	Organic	Ecological	Economic	Super economic
Import of feed (tons)	0	7	68	161	210
Finan. balance (1000 D.kr)	84	137	372	534	650
Energy balance GJ	+292	+292	+13	–148	–162

TABLE 3
Energy Balances and Income for the Five Strategies

Strategy	Self-sufficiency	Organic	Ecological	Economic	Super economic
All figures in GJ					
Feed needs	401.7	401.7	821.7	1,334.1	1,406.1
Feed production	533.2	533.2	533.2	533.2	607.6
Products sold out of the system (export)					
Pigs	79.9	79.9	79.9	125.5	125.5
Ducks	—	—	36.4	36.4	50.8
Geese	93.6	93.6	187.2	491.4	491.4
Vegetables	131.5	191.7	237.5	469.8	567.0
Energy export	305.0	365.2	541.0	1,123.1	1,234.7
Products bought into the system (import)					
Piglets	11.8	11.8	23.2	23.2	23.2
Poultry young	1.5	1.5	4.9	7.5	8.3
Feed	—	60.2	526.0	1,240.0	1,365.5
Energy import	13.3	73.5	554.1	1,270.7	1,397.0
Energy balance	291.7	291.7	13.1	–147.6	–162.3
Annual income in 1000 D.kr	83.9	136.6	372.5	533.7	649.7

Chapter 14

A PLAN FOR A COMMUNITY GREENHOUSE AND WASTE TREATMENT FACILITY FOR COLD CLIMATE REGIONS

Robert L. Crosby, Jr.

ABSTRACT

A plan for a community-scale integrated Greenhouse/Waste Treatment Facility is proposed. The facility would use biogas produced from the anaerobic digestion of organic wastes to generate electricity, heat, light, and carbon dioxide for enhanced plant growth in a controlled environment. A design procedure and calculations are provided. Assumptions include: (1) Methane yield > .30 standard cubic meters/kg volatile solids added (SCM/kg VS), (2) methane-to-electricity conversion efficiency > 28%, (3) total recoverable energy from combustion of methane > 80%.

INTRODUCTION

Biorealis Systems, Inc. was recently awarded a grant from the Alaska Science & Technology Foundation to build and test a pilot-scale, high-efficiency methane digester. The prototype digester will be a three-phase, automatically controlled, plug-flow reactor designed to emulate the digestive system of a living organism. The first phase will be optimized for growth of acid-forming bacteria, the second for growth of methanogens. Effluent from the second stage will pass through a final aerobic phase for polishing before being stored for reuse (for mixing) or discharged to the environment. System temperature, pH, and gas production rate will be continuously monitored and used to automatically control feed and sludge recirculation rates, temperature, and peristalsis.

The project is intended as a research effort to determine what is the maximum obtainable efficiency to convert a given quantity of organic material to usable fuel. Efficiency is defined in terms of:

1. Methane yield per unit of capital cost: primarily a function of digester tank volume, which is a function of production rate, hence required retention time. Production rate is, in turn, a function of ability to reliably maintain optimum conditions for a variety of species of organisms, which interact with each other in complex ways.
2. Methane yield per quantity of input: primarily a function of temperature. Evidence suggests that, at higher temperatures (in the range of thermophilic organisms) more energy is available to organisms for metabolic activity and the production of cell mass, enabling a more complete conversion of substrate to methane (Cooney & Ackerman 1975, Loll 1976).

0-87371-990-5/97/$0.00+$.50

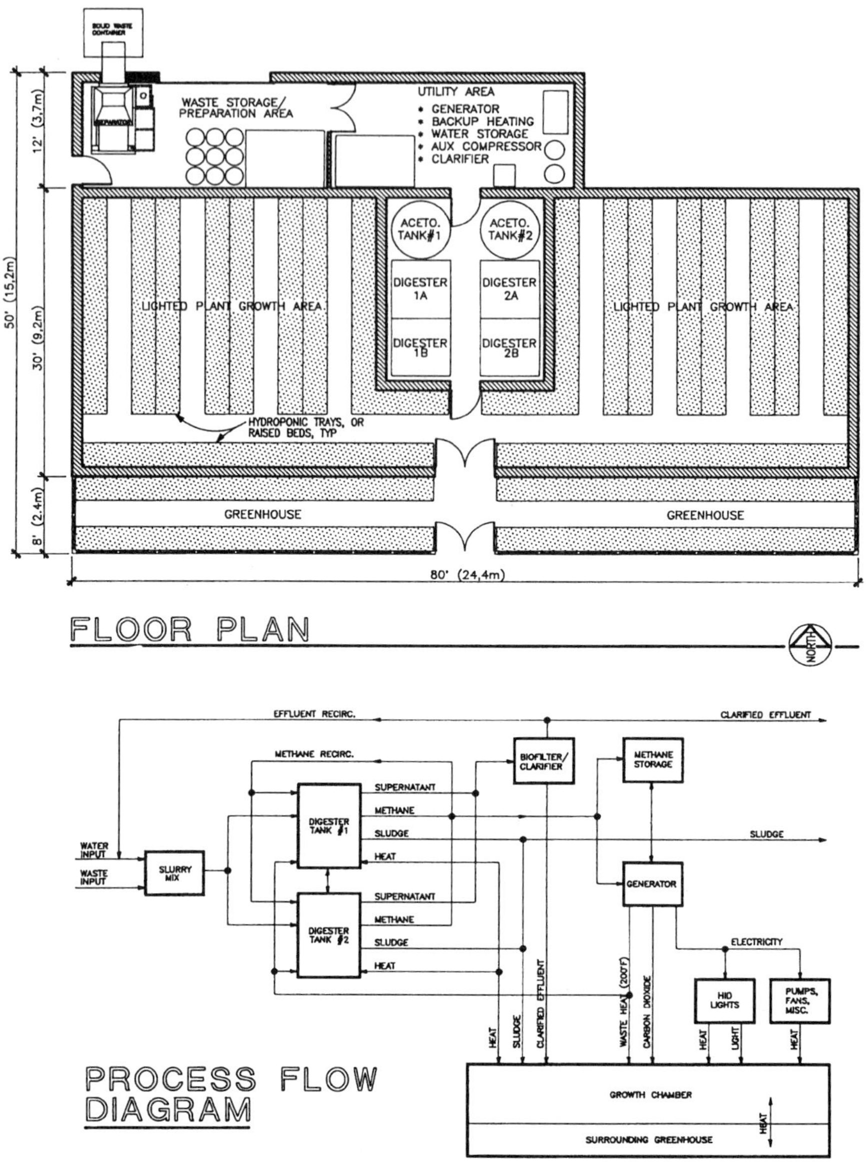

FIGURE 1. Overview of a community-scale greenhouse/waste treatment facility.

3. Methane yield per person-hour of labor: a function of level of automation and system design for ease of maintenance.

The digester is being developed for use as the central component of an integrated plant growth facility which would derive the majority of its heating, lighting, and carbon dioxide needs from the biogas produced. Such a facility could provide a source of fresh produce — and increased local self-reliance — for rural Alaskan communities, while processing waste in an environmentally sensitive manner.

FACILITY DESCRIPTION

The proposed facility will consist of six functional components/areas, including: (1) Waste storage/preparation area; (2) Methane digester; (3) Generator Room; (4) Controlled-Environment Growth Chamber; (5) Surrounding greenhouse; (6) Utility area (see Figure 1).

The various functions will be arranged so that the highest temperature processes (i.e., generator and digester) are located at the center of the building, with the progressively lower temperature processes surrounding them. The "UA" (product of Heat Transfer Coefficient × Area) of partitions between each temperature zone will be designed so that the rate of heat flow through the partition at design temperatures matches the minimum heat input rate from the internal process. The temperature in each space is then regulated by treating the heat source at the core as a cooling load, and dumping excess heat to the surrounding area; in effect, making multiple use of the same heat by matching the output of one process to the input requirements of the next.

A computerized control system will be used to monitor and control individual processes, as well as to regulate heat transfer between the various areas.

WASTE STORAGE/PREPARATION

Different rural villages in Alaska use different waste collection methods; a "honey-bucket" system probably being the most common. This is a bucket lined with a plastic bag, inside a vented plywood enclosure with a hinged seat. Every few days or so, the home-owner must lift the lid, remove the bag and dispose of it. Disposal may mean tossing it in a used 55-gal fuel drum outside the door, for the community truck to pick up; dumping it in a lined, covered pit provided for the purpose by the Public Health Service; or carrying it on the back of a three-wheeler to the nearest convenient body of water. In many villages, waste treatment facilities are inadequate for basic sanitation and health.

The required size and function of the waste storage/preparation area will depend on the collection method used. With a honey-bucket collection system, non-organic solids (plastic bags, beer cans, etc.) must be separated from the digestible wastes, C/N ratios balanced, and water added as required to provide an approximately 8% slurry suitable for introduction to the digester.

An improved collection system, e.g., low-flush toilets, small holding tanks, or heated pumper trucks, could reduce, or even eliminate, the need for a temporary waste storage and preparation area.

METHANE DIGESTION

The process by which anaerobic bacteria decompose organic matter into methane, carbon dioxide, and a nutrient-rich sludge involves a series of reactions requiring the cooperative action of several organisms. The methanogenic bacteria which form only the final stage of the digestion process are obligate anaerobes. These bacteria inhabit a

TABLE 1
Methane Digester Calculations

Location: **Rural Village, AK**
No. people: **300**

Raw material characteristics and quantities: (Leckie, et al. p. 196)

Waste Material	lb/D/ Unit	Units (EA)	Total (LB/D)	M %	TS %	VS %	C %	N %	TS lb/D	VS lb/D	C lb/D	N lb/D	C/N ratio	H_2O lb/D
Human feces	0.5	300	150.0	0.73	0.27	0.92	0.48	0.06	40.5	37.3	19.4	2.4	8.0	109.5
Human urine	1.5	300	450.0	0.94	0.06	0.75	0.14	0.18	27.0	20.3	3.9	4.9	0.8	423.0
Green garbage	1.0	300	300.0	0.50	0.50	0.75	0.55	0.02	150.0	112.5	82.5	3.2	26.2	150.0
Fish scrap	0.0	0	0.0	0.65	0.35	0.90	0.20	0.07	0.0	0.0	0.0	0.0	ERR	0.0
Scrap paper	0.8	300	225.0	0.07	0.93	0.97	0.41	0.00	209.3	203.0	85.8	0.1	820.0	15.8
Sawdust	0.8	300	225.0	0.25	0.75	0.99	0.50	0.00	168.8	167.1	84.4	0.2	500.0	56.3
Chicken manure	0.3	0	0.0	0.65	0.35	0.65	0.50	0.06	0.0	0.0	0.0	0.0	ERR	0.0
Pig manure	6.0	0	0.0	0.87	0.13	0.85	0.58	0.04	0.0	0.0	0.0	0.0	ERR	0.0
Cow manure	36.0	0	0.0	0.86	0.14	0.80	0.31	0.02	0.0	0.0	0.0	0.0	ERR	0.0
Horse manure	30.0	0	0.0	0.84	0.16	0.80	0.58	0.02	0.0	0.0	0.0	0.0	ERR	0.0
Totals			1350.0						595.5	540.0	276.0	10.7	25.8	754.5

Intermediate results:			
Added H_2O required for 8% slurry (lb/d):	6093.8		[92 lb H_2O * (ts/8)] — H_2O already present
Total slurry weight (lb/D):	7443.8		(Total wastes + added H_2O)
Total slurry vol. @ D = 65 lb/cf (cr/d):	114.52		(Total slurry wt/65 lb/CF)
VS concentation of slurry (lb/cf):	4.72		(VS in lb/d/slurry vol in CF/D)
Chemical oxygen demand (lb/day):	405.0		(Assumes COD = 1.5*VS, and 50% of VS is non-biodeg.)
Chemical oxygen demand conc. (ppm):	56589		(VS conc. in lb/cf * 1.5 * .5/16,000 ppm)
Temperature of digestion process:	95.00		(See Lookup table: Leckie et al. p. 204)
k (bacterial comsumption rate):	6.67		(Extrap. from table: use w/caution)
K_c (minimum nutrient req'd for bact. mult):	2235		(Extrap. from tble: use w/caution)
Min solids ret. time (days):	4.35		(Leckie et al. p. 204: note poss. error!)
Safety factor:	12		
Detention time (days):	52		(Min calc. time × safety factor)

Digester size requirements:	**(Meso)**	**(Thermo)**
Total digester volume (cf):	5980.3	1993.42 (Total vol cf/day × SRT in days)
Desired tank diameter (ft):	16.00	12.00
Digester tank height (ft):	29.74	17.63
Cubic tank dimensions (ft × ft × ft):	18.10	12.55 (Cube root of vol.)

Output quantities:	**(Meso)**		**(Thermo)**
Sludge solids init (lb/day):	325.48		
Sludge solids bact (lb/day):	9.085		
Sludge solids total (lb/day):	334.56		
Supernatant total (lb/day):	6093.75		
Methane gas production (cf/day):	1365.8		2048.66
Total biogas production (cf/day):	2048.7		3072.99
Heat value (btu, @900 btu/cf):	1.2E+06		1.8E+06
Equivalent #2 fuel oil (gal/day):	8.78		13.17
Usable electricity @28.0% efficiency:	100.84	KW/day	151.26 KW/day

very narrow range of temperature and pH and depend on highly specific but simple substrates, primarily acetate and other volatile acids produced by other organisms.

The various organisms which first break down the raw wastes into the simpler fatty acids grow much more quickly than the methanogens and are more tolerant of pH and temperature fluctuations. As a result of these dependencies, a single-stage digestion process is inherently unstable, and an imbalance can create a positive feedback situation which will quickly bring the entire process to a halt. That is, a slight drop in methane production rate without a corresponding reduction in substrate production will cause the pH to drop, causing a further reduction in methane production, causing still lower pH and so on, until the system crashes.

Recent research indicates that a two-phase system, in which the first is optimized for acidogenesis and the second for methanogenesis, yields considerable gains in both conversion efficiency and system stability over single-stage systems. In one experiment, a mesophilic (35°C) two-phase system exhibited the same methane yield and solids reduction at a 3-day hydraulic residence time (HRT) as those of single-stage, high-rate digestion at 15- and 17-day HRTs. Yields of better than 0.30 SCM methane/kg VS added were obtained at a loading rate of 7 kg VS/m^3-day (Ghosh 1987).

ENGINE/GENERATOR

The electric generator will be the primary energy source for the facility, converting methane produced in the digester to electricity, heat, light, and carbon dioxide (used by plants in photosynthesis). Generator size will be determined by the amount of fuel available, which is determined, in turn, by the quantity and characteristics of the local wastes available for digestion.

Much promising work has been done in the last decade in the area of engine design. With the availability of new materials (e.g., superalloys and ceramics), there is renewed interest in Stirling cycle heat engines, which are inherently more efficient than Otto or Diesel cycle engines, but which have traditionally suffered from a lack of power.

Mechanical Technology, Inc. (MTI), a U.S. research and development firm, has developed, under contract to the federal Department of Energy and the National Aeronautics and Space Administration (NASA), a Stirling cycle free-piston linear alternator capable of 40% efficiency (heat in to 60 Hz AC out). Swedish and Japanese firms are also involved in the research and development of high-efficiency heat engines capable of burning multiple fuels.

The MTI alternator, though still in development, would be ideal for this application, providing the requisite conversion efficiencies while eliminating the need to condition the gas.

CONTROLLED ENVIRONMENT GROWTH CHAMBER

A growth chamber is a heated, artificially lighted room designed to provide optimum conditions for plant growth. Controlled environmental factors include soil temperature, moisture, and pH; air velocity, temperature, humidity, and CO_2 concentration; and lighting (intensity, photoperiod, and spectral emission).

The relation between plant and environment is extremely complex. In general, however, to speed plant growth we want to increase the (1) photosynthesis rate and (2) nutrient uptake:

TABLE 2
Total Energy Input

Elec Cost $0.38 /KWH
Heat Cos $0.05 /KWH (#1 F.O. @ $1.80/GAL)

Load Component	KW/Hr	Hr/Day	KW/Day	$/DAY	$/YR
LIGHTING:					
Growth Chamber Lights	8.28	12	117.00	$44.46	$16,228
Utility Space lights	0.55	8	4.42	$1.68	$612
	SUBTOTAL:		121.42	$46.14	$16,840
POWER					
Compressor	0.75	1	0.75	$0.28	$103
Biofilter Pump	0.22	24	5.37	$2.04	$745
Digester tank stirring	0.22	4	0.90	$0.34	$124
Generator Cooling fan	0.37	12	4.48	$1.70	$621
Generator Cooling Pump	0.22	12	2.69	$1.02	$372
Growth Chamber Fan	0.37	8	2.98	$1.13	$414
Greenhouse Cooling Fan	0.37	8	2.98	$1.13	$414
Controls	0.22	24	5.22	$1.98	$724
Shredding/Waste Prep	8.95	0.5	4.48	$1.70	$621
	SUBTOTAL:		29.84	$11.34	$4,139
	ELEC. TOTAL:		151.26	$57.48	$20,979
HEAT: (If it were additive)					
Digester tank heat			31.50	$1.54	$561
Growth Chamber heat			97.57	$4.76	$1,738
Greenhouse heat			117.20	$5.72	$2,088
	HEAT TOTAL:		246.27	$12.02	$4,387
	TOTAL ENERGY:		397.53	$69.50	$25,366

1. During photosynthesis, a plant uses light energy to break down molecules of atmospheric carbon dioxide, expelling oxygen and using the carbon to build tissue. This process may be accelerated by providing the plant with abundant CO_2 in the presence of high intensity light of the appropriate wavelength (photosynthetically active radiation (PAR), 400 to 700 nm).
2. Most terrestrial plants absorb nutrients through their roots. All other things being equal (i.e., metabolic demand, luxury consumption of nutrients, etc.), the rate at which a plant can take up waterborne nutrients is ultimately limited by the rate of evapotranspiration from the leaf surface. This rate, in turn, is a function of the relative humidity and movement of the air at the leaf surface.

Under optimum conditions, food crops such as lettuce, tomatoes, peppers, cucumbers, eggplant, etc. have been produced in controlled-environment growth chambers with annual yields as much as five times higher per unit area than in conventional greenhouses.

GREENHOUSE

The greenhouse area will wrap around the growth chamber on three sides, providing additional enclosed plant growth space in summer and a thermal buffer in winter. Preliminary heat load calculations on the prototype design indicate that greenhouse temperature can be maintained above freezing at outdoor temperatures down to –18°C.

Heat vented outward from the core, as described above, will be routed through ducts buried in raised growing beds in both the growth chamber and greenhouse. The thermal mass of each component will be sized to provide a minimum of 12 hours of carry-over at design temperatures, to allow for generator "off" time.

CALCULATION METHODOLOGY AND NOTES

Calculations were performed using computer spreadsheet models created in Quattro Pro®, an MS-DOS spreadsheet program similar to, and compatible with, Lotus 1-2-3®, Microsoft Excel®, etc. Methodology used is as follows:

1. Input estimated quantity and characteristics of wastes available; calculate digester size, daily biogas production, and total daily electric energy production. Methodology from "Other Homes and Garbage" by Leckie, Masters, Whitehouse, and Young, 1975. (Note: Gas production estimates modified from original to account for higher temperatures and multi-stage design. See discussion in introduction regarding performance at higher temperatures and under "Methane Digestion" section regarding improved performance of two-phase systems.)
2. Estimate total energy requirements of facility, including process requirements and overhead (general illumination, fans, pumps, etc.). Subtract process energy requirements from total daily electric energy available. Assume remaining quantity of total electric energy is available for plant illumination (see Table 2).
3. Based on illuminance levels required for optimum plant growth (from literature), calculate amount of area which can be adequately lighted for plant growth purposes. Design and lay out facility based around estimated digester tank size and lighted plant growth space (see Table 3).
4. Calculate maximum UA (product of surface area/heat transfer coefficients of individual building components) to match internal processes.

TABLE 3
Growth Chamber Lighting Requirements

Available electricity (kw/day):	**117.00 kwh/day (From methane calcs)**
Desired photoperiod (hr/day) =	**12.0 hr/day**
Required illuminance (w/m^2)	**50.0 w/m^2 (400–850 nm)**
Required illuminance (w/ft^2)	**4.6 w/ft^2 (400–850 nm)**

Light type	Total input (watts)	Lamps (ea)	% PAR (400–850)	Illum area (s.f.)	W/refl. (s.f.)	Footcandles
Fluorescent	46	212	21%	441	705	1670
Metal halide	460	21	30%	629	1007	1460
Hi press sodiu	470	21	36%	755	1209	1480
Mh + hps	>>>>	>>>>	>>>>	>>>>	1108	

Food crop	Req'd area (SF/plant)	Area (SF)	Yield (lb/plant)	No. of crops/yr	Total yield (lb/yr)	Mkt val> ($/lb)	Gross income	Crop value $/SF
Cucumber	3.00	50	20.0	3.0	1000	$1.47	$1,468	$29.37
Tomato	4.00	350	25.0	3.0	6563	$3.98	$26,097	$74.56
Lettuce	1.09	150	1.2	52.0	8581	$2.65	$22,710	$151.40
Bell pepper	3.00	50	20.0	3.0	1000	$1.32	$1,317	$26.33
Fancy peppers	3.00	50	20.0	3.0	1000	$3.84	$3,844	$76.87
Fancy herbs	0.50	150	0.06	52.0	975	$53.92	$52,570	$350.47
Mushrooms	0.05	150	0.08	26.0	6500	$3.84	$24,984	$166.56
Snow peas	3.00	0	10.0	3.0	0	$5.31	$0	ERR
Asparagus	2.00	100	20.0	3.0	3000	$5.31	$15,920	$159.20
Ornamentals	1.00	50	1.0	12.0	600	$3.99	$2,394	$47.88
Totals		1100			29,218		$151,304	

TABLE 4
Estimated Budget for a Community-Owned Methane Digester/Greenhouse

Monthly Income	
Utility fee/member dues:	\$900 (\$15/month per household × 60)
Plant and crop sales	\$12,609 (from lighting calcs: \$/yr/12)
Profit from misc retail:	\$1,000
Monthly total:	\$14,509
Monthly Costs	
Debt service	\$6,960 (\$580,000, 12%, 15 yr)
Misc overhead:	\$1,050 (Supplies, insurance, utilities, maintenance, etc.)
Staff/Salaries	
1 full-time manager:	\$4,000 (Biol, Hort, @ \$48,000/yr)
1 assistant/gen. maint	\$1,666 (Part-time @ \$20,000/yr)
1 bookkeeper	\$833 (Part-time @ \$10,000/yr)
2 labor/retail sales:	\$0 (Coop member supported, at 2 day/mo/household)
2 retail sales	\$0
Monthly total:	\$14,509

REFERENCES

Cooney, C.L., and Ackerman, R.A. 1975. "Thermophilic Anaerobic Digestion of Cellulosic Waste." *European Journal of Applied Microbiology* 2: 65-72.

Ghosh, S. 1987. "Improved Sludge Gasification by Two-Phase Anaerobic Digestion." *Journal of Environmental Engineering*, Vol. 113 No. 6, December, 1987.

Loll, U. 1976 "Engineering Operation and Economics of Biodigestion." In: H.G. Schlegel and J. Barnea (eds.). *Microbial Energy Conversion.* Göttingen, W. Germany: Erich Goltze KG. pp. 361-78. As cited in Anderson, Russell E., *Biological Paths to Energy Self-Reliance. p.* 262.

Chapter 15

A DIAGNOSTIC AND ECOLOGICAL APPROACH TO THE PURIFICATION OF SEWAGE, TOXIC SUBSTANCES, AND WATER BODIES

Karl F. Ehrlich, Marie-Claude Cantin, and Andrée Turcotte

ABSTRACT

Sewage, toxic organic substances, and lake pollutants have a common denominator: the ability of microorganisms to use these substrates as a food source. Efficient water purification is, in large part, dependent upon the presence and activity of appropriate communities of water purifying microorganisms. Not only is the presence of the correct organisms necessary to achieve efficient water purification, but the bacteria must also be in the right physiological condition and in an environment with appropriate physico-chemical conditions to accomplish the desired purification task. Traditional wastewater treatment techniques too often ignore certain underlying biological realities, thus not optimizing wastewater purification.

Dramatic improvements in cost-effective water purification, in existing facilities, have been achieved by ecological engineering based on the use of three components:

- Process evaluation in terms of physico-chemical and biological activities and rates.
- Physiological engineering and seeding of communities of water purifying bacteria.
- Environmental control of treatment processes.

This approach is being used in sewage plants with flows as large as 640,000 m^3/d to reduce sludge production by 30%, remove sludge accumulations from lagoons, increase energetic efficiencies, and/or simply control odors. Similarly, the biodegradation of hydrocarbons can be efficiently handled. Phenol in water has been biodegraded from 220 ppm to undetectable levels within 24 h. A combined concentration of 400 ppm benzene, toluene, and xylene has been decreased to 0.5 ppm within 12 h. Refinery soil contaminated with 460 ppm of polyaromatic hydrocarbons (PAH) has been biodegraded to about 50 ppm in 6 weeks.

This form of ecological engineering is not limited only to wastewater treatment facilities but is also being used in aquaculture and lake restoration to reduce soluble phosphorus and control aquatic weeds. Growth of Eurasian milfoil (*Myriophyllum spicatum*) was reduced by 90% as compared to a control after 4 months of seeding lake sediments with physiologically engineered communities of water-purifying bacteria. This bacterial augmentation

0-873713-990-5/97/$0.00+$.50

resulted in increased production of bacterial, invertebrate and fish populations while decreasing plant biomass. Small shifts in population dynamics at the decomposer trophic level affected higher levels within the food web. In this case, the result was the conversion of organic wastes into bacterial biomass and increased channeling of nutrients up the animal side of the food web, thus limiting nutrients available for the excessive proliferation of Eurasian milfoil.

INTRODUCTION

Even though many phyla of organisms and diverse technologies can and do contribute to water purification, bacteria play a primary role in most systems. Sewage, organic toxic substances, and lake pollutants have a common denominator: the ability of microorganisms to use these substrates as a food source. Efficient water purification is, in large part, dependent upon the presence and activity of appropriate communities of water purifying microorganisms (Sonea & Panniset, 1980). Chrost et al. (1986) have demonstrated not only that bacterial populations change quantitatively and qualitatively both temporally and spatially within a lake but that the enzymatic activity and availability, which are critical to decomposition, also fluctuate widely. All bacteria are obviously not the same, and all essential strains are not always present when needed. This explains why total bacteria counts, so often documented in the scientific literature, do not give adequate indication of bacterial purification efficiency. Not only is the presence of the correct organisms necessary to achieve efficient water purification, but the bacteria must also be in the right physiological condition and in an environment with appropriate physico-chemical parameters to accomplish the desired purification task. Traditional wastewater treatment techniques too often ignore certain underlying biological realities, thus not optimizing wastewater purification. One only has to read the newspapers or listen to the news to be aware of these technological limitations: whether it be poorly operating sewage treatment facilities, inept attempts to remedy contaminated soils, sludge accumulation from aquaculture or accelerated eutrophication of our water bodies. Better understanding of the purification processes can help to cost-effectively mitigate many of these problems. It is not the intent of this article to provide a comprehensive review of bacterial purification processes but rather an overview of certain microbial realities and a diagnostic approach to the application of these realities to achieve specific goals. Results achieved with this methodology are also presented.

FACTORS AFFECTING WASTEWATER TREATMENT EFFICIENCY

There are four major factors affecting the efficiency of wastewater purification: the presence of appropriate communities of microorganisms, the physiological condition of these organisms, the physico-chemical conditions within the treatment facility or area to be treated, and the time available to accomplish the desired task (i.e., the hydraulic retention time).

APPROPRIATE STRAINS

Water purification is a function of teams of microorganisms, where the byproduct of the activity of one strain can serve as a nutrient for others. Lack of a member of the essential bacterial community weakens the purification team. It is common knowledge that bacteria can grow in most environments. This does not mean that all necessary strains will always be present. The ubiquity principle states that bacteria can grow in most environments; it does not state that all strains are everywhere all the time. The lack of ubiquitous presence of bacteria is, in fact, what Chrost et al. (1986) have shown in natural water bodies. Bacteria from our digestive systems are adapted to growing under specific conditions, which are very different from those found in wastewater treatment facilities. Specific bacterial communities are required to accomplish water purification. If all essential strains are not always available even in natural water bodies and are not in our bodies, how can one expect to find and maintain them in a wastewater treatment facility, particularly when the water used in domestic consumption is sterilized to begin with? The same logic applies to industrial processes many of which are sterile or use very specific strains (dairy plants). Ecological engineering is based on the concept of bioaugmentation, which can be defined as the seeding of water-purifying organisms to achieve specific goals. Bacterial augmentation is a subset of bioaugmentation and is used to assure the presence of essential bacterial communities. This approach has been used with success, to accelerate degradation of organic wastes, accelerate nitrification, reduce soluble phosphorus and control algal blooms (Hung et al. 1986; Bonnett 1987; Zachopoulos 1987; Ehrlich et al. 1989). Seeding bacteria alone, however, cannot guarantee the desired result, if the process to which it is applied is not understood.

Results of studies using four biotechnological products, in particular, are presented: Bacta-Pur™ E-2000+, S-2000+, G-2000+, and N-2000+. The four mixtures contain communities of water purifying bacteria balanced or grown in a particular manner to achieve specific results. E-2000+ contains *Aerobacter aerogene*, *Bacillus subtilis*, *Cellulomonas biazotea*, *Pseudomonas denitrificans*, *P. stutzeri*, *Rhodopseudomonas palustris*, *Nitrosomonas* sp., and *Nitrobacter winogradskyi*. The role of each of these strains has been discussed by Ehrlich et al. (1989). This community is grown to accelerate removal of BOD. S-2000+ contains the same strains but grown to increase production of exoenzymes; this mixture is used primarily for sludge reduction. G-2000+ contains an additional strain (*Acinetobacter*) for fat breakdown. N-2000+ contains *Nitrosomonas* sp., *Nitrobacter winogradskyi* and two *Bacillus*; this mixture is used principally for nitrification.

PHYSIOLOGICAL CONDITION

The ability of microorganisms to accomplish specific tasks is dependent, in large part, on their physiological condition during the purification process. The growth of bacterial populations can be divided into four phases (Figure 1):

- The lag phase when bacteria are adapting to a new environment or new food (pollutant to degrade). This phase can involve the synthesis of internal enzymes to handle the target food.
- The exponential phase when bacteria are rapidly growing on the available soluble food (e.g., BOD reduction or nitrification).

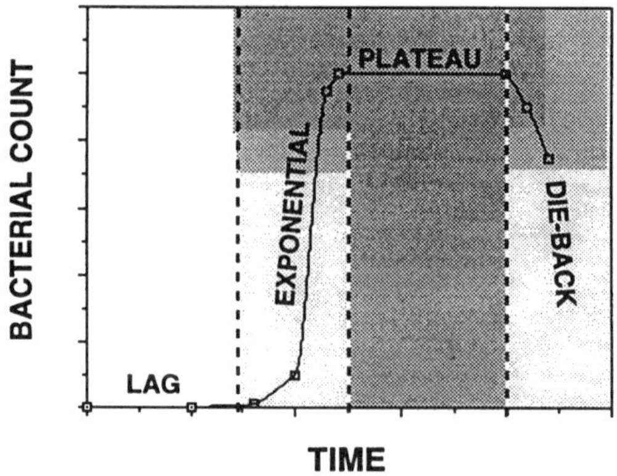

FIGURE 1. Phases of bacterial growth.

- The plateau or stationary phase when growth is reduced because of food limitation. Bacteria secrete exoenzymes and extracellular enzymes to solubilize solids and produce usable food.
- The die-back phase when bacteria die and lyse due to starvation. Cell lysis liberates extracellular enzymes and reduces bacterial biomass, converting it into soluble food for surviving bacteria.

PHYSICO-CHEMICAL CONDITIONS

The efficiency with which bacterial populations degrade target pollutants is dependent, in large part, on factors such as pH, temperature, oxygen concentration, and/or availability of alternative electron receptors, availability of essential nutrients, lack of inhibitory substances, water movement, and the stability of these conditions. Organic pollutants, whether sewage or hydrocarbons or agricultural residues, are carbon sources for heterotrophic bacteria. In addition to carbon, bacteria require other major and minor nutrients. Nitrogen and phosphorus are the other major nutrients, which should be supplied in a ratio of 100:7:1 for BOD:N:P (Hänel, 1988). Bacterial nutrients must be soluble to be usable by the bacteria. Total phosphorus measurements include soluble and insoluble forms and do not assess what is available for bacterial activity. Soluble orthophosphate is what is usable. Domestic sewage generally contains, at least initially, adequate levels of N and P. One cannot assume, however, that even if nutrients are adequate in the influent that this will be the case throughout the treatment process. Excessive use of flocculants (i.e., alum, positively charged polymers, iron salts, etc.) or the use of such products in an inappropriate place in the treatment process can convert soluble phosphorus needed for bacterial growth to an insoluble form, limiting bacterial activity. Industrial wastes generally require nutrient addition.

AVAILABLE TIME

Accomplishing the desired task within the available time requires assuring the presence of the essential organisms, getting the organisms in the appropriate physiological condition(s) and then providing the organisms with conditions for rapid growth. Bacterial augmentation can provide the appropriate bacterial communities,

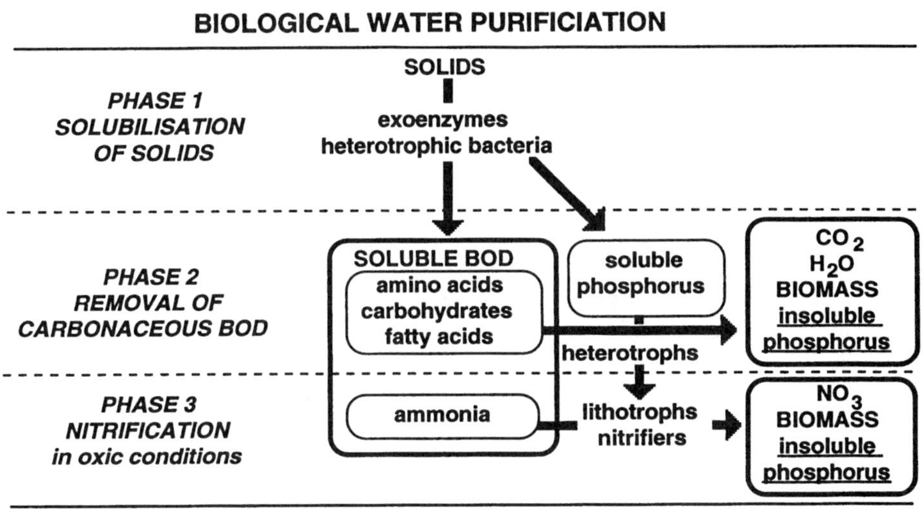

FIGURE 2. Phases of bacterial purification.

but this technique alone does not assure that the organisms will be in the appropriate physiological condition to degrade the target pollutant(s) in the available time. This can be accomplished by physiological engineering, which is discussed below.

PROCESSES OF BACTERIAL WATER PURIFICATION

There are four essential processes of bacterial purification of water: solubilization of solids, removal of carbonaceous BOD, nitrification, and denitrification (Figure 2). Each of these processes requires specific conditions, some of which exclude the simultaneous occurrence of the other processes. Understanding requirements for each of the processes is the first step in controlling them.

SOLUBILIZATION OF SOLIDS

Bacteria can only utilize soluble food, which is able to pass through their cell wall. Exoenzymes or ectoenzymes are produced by bacteria and are attached to cell walls; extracellular enzymes are ones secreted into external environment to solubilize solids (Hänel, 1988). Production of enzymes to solubilize solids largely occurs when soluble food is limiting. Solubilization of solids produces carbonaceous BOD, soluble phosphorus, and/or ammonia, depending upon the nature of the solids. Solubilization also occurs from the lysis of cells which are either excessively nutrient deprived or placed under unfavorable conditions, such as when aerobic cells are kept under anoxic conditions. Solubilization under these conditions releases the internal intracellular enzymes of the cells, which solubilizes the cells as well as surrounding solids. Excessive retention of solids, either in a clarifier or on tank bottoms, results in release of BOD, ammonia, and phosphorus. This can be particularly problematic in aquacultural production because of ammonia/nitrite toxicity and/or solubilized phosphorus leading to unwanted algal or macrophyte proliferation.

REMOVAL OF CARBONACEOUS BOD

Bacteria can be divided into two groups (heterotrophs and lithotrophs) based on their feeding ecology. Heterotrophs use solubilized organic sources of carbon from proteins, fats, carbohydrates, or hydrocarbons to build their body components. Lithotrophs use inorganic carbon — carbon dioxide — for the same purpose. Removal of carbonaceous BOD is largely accomplished by heterotrophic bacteria. Nitrification is a function of lithotrophic bacteria. Removal of carbonaceous BOD results in production of bacterial biomass, carbon dioxide, water, and insolubilization of phosphorus as it is incorporated into cells. Biological inefficiencies result in losses, as carbon dioxide, of 50 to 80% of the carbon in the pollutant as it is converted to bacterial biomass (Blackburn, 1987). Cell carbon incorporation efficiency decreases as one goes from aerobic to fermentation processes. Manipulation of bacterial populations to take advantage of this inefficiency of carbon transfer is a means to reduce the sludge production of wastewater treatment facilities. Results of application of these processes are presented below.

NITRIFICATION

Nitrification not only converts ammonia via nitrite to nitrate but also reduces soluble phosphorus due to cellular incorporation. Nitrification requires the presence of oxygen as well as a source of inorganic carbon. Removal of carbonaceous BOD is a precursor to nitrification (WPCF, 1983). High levels of soluble organic products can inhibit nitrification. Nitrate synthesis generally occurs when soluble carbonaceous BOD is below 20 mg/l. Thus nitrate synthesis across a treatment basin provides a rapid method to assess BOD removal.

DENITRIFICATION

Denitrification requires anoxic conditions and an organic carbon source for the heterotrophic bacteria. The carbon source can come from sludge biomass or raw influent or from additions of products such as methanol, corn starch, methane, or acetate. It should be kept in mind that breakdown of sludge biomass can also lead to ammonia synthesis; similarly certain microorganisms can reduce oxidized nitrogen to ammonia (Hänel, 1988). Temporal or spatial cycling of anoxic and oxic processes can facilitate removal of carbonaceous BOD and nitrogen while reducing energy consumption.

APPROACH AND METHODOLOGY

PHYSIOLOGICAL ENGINEERING

The ability to achieve certain water purification processes, such as sludge reduction or rapid degradation of hydrocarbons, is dependent on getting the bacteria in or through certain physiological conditions in a short period of time (less than the hydraulic retention time). Two physiological engineering techniques have been developed to attain specific physiological conditions of bacterial populations prior to their introduction into the water to be treated:

- Preactivation — optimizes production of exoenzymes and extracellular enzymes to reduce sludge.

- Preconditioning — induces bacterial populations to recognize refractory products such as hydrocarbons and to develop the internal enzymes necessary to degrade them.

Both methods exploit the synergistic benefits of combining bacterial augmentation with existing strains.

PREACTIVATION

Sludge reduction requires the production of exoenzymes or extracellular enzymes to solubilize solids, whose consequent oxidation results in a net reduction of solids to be handled, due to loss of CO_2 and water with each transformation. High levels of carbonaceous BOD within wastewater treatment facilities provide heterotrophic bacteria with food for growth and limit their need to produce exoenzymes. Limited retention times and washout of bacteria in sewage treatment plants mean that significant amounts of bacteria and exoenzymes do not come in contact with the solids to be solubilized. Preactivation is the technique of on-site growing of endemic and seeded bacteria to stimulate production of sludge solubilizing enzymes. Two different basic recipes of preactivation are used, depending whether the goal is to reduce sludge or grease; both methods use a proprietary nutrient mix (activator):

- To reduce sludge, Bacta-Pur™ S-2000+/Activator ratio of 6:1 is used.
- To reduce grease, Bacta-Pur™ G-2000+/Activator ratio of 12:1 is used.

On day 1, the bacteria and activator are placed in a container with non-chlorinated effluent equal to ten times the prescribed volume of bacteria. The temperature is adjusted to 35°C and the container is aerated over 24 h. The second day the heat is turned off and the mixture from day 1 is drained into the system over the next 24 h. A second system is started on day 2 as on day 1, and the cycle is repeated. There can be site-specific adjustments to this basic method.

The controlled environment used for the preactivation allows rapid bacterial growth and production of exoenzymes, independent of ambient conditions. This is particularly advantageous when working in cold water. Preactivation is applicable not only to wastewater treatment plants but also to aquaculture and lakes. Sludge reduction is most effective when the preactivated bacterial community is placed in contact with the sludge to be solubilized. In activated sludge facilities this is in the return activated sludge (RAS). In ponds or lagoons this is directly into the accumulated sludge. Preactivated bacteria can be injected directly into sludge accumulations under aquacultural cages. Weighted manifolds of slotted pipe with a vertical feeder line make application fast and efficient.

PRECONDITIONING

Degradation of refractory substrates, such as hydrocarbons, requires that the bacteria develop the internal enzyme(s) to utilize the substrates for growth. Bacteria often do not develop the needed enzyme(s) until they are in the presence of the target substrate. This can result in a lengthy lag phase, which delays the onset of degradation and is a problem in wastewater treatment plants with short retention times, due to the continual loss of adapted bacteria in the effluent. It is also a problem in the start-up of wastewater treatment, either for batch processes or following plant shutdowns because of toxic shocks. It is sometimes possible to induce the necessary enzyme(s) to degrade recalcitrant substrates by providing the bacteria with other hydrocarbons; this has been referred to as fortuitous metabolism (Crawford and O'Reilly, 1989). Preconditioning is a technique which induces communities of seeded and endemic bacteria to recognize the substrate(s), produce the necessary enzyme(s), and to utilize and degrade the desired substrate(s).

Preconditioning is based on the use of the bacterial mixture Bacta-Pur™ E-2000+ combined with indigenous strains and proprietary nutrient packages, which provide an adequate source of nitrogen and trace elements (PC1), phosphorus, buffering and bacterial carrier (PC2). PC1 and PC2 are used at the respective rates of 0.5 g and 12.5 g/g soluble carbonaceous BOD. The general technique to precondition bacteria for a water application consists of the following procedure (there can be site-specific modifications of these techniques): Day 1 the bacteria and nutrient packages are placed in a container with non-chlorinated effluent equal to 8 times the prescribed volume of bacteria and influent to the wastewater treatment equal to twice the volume of prescribed bacteria. The temperature is adjusted to 35°C and the system is aerated over 48 h. Of this mixture 90% is then introduced into the treatment area. The procedure is then repeated on a 24-h cycle including the 10% of the volume from the previous day.

CASE HISTORIES

Case histories are divided into three sections: sewage, hydrocarbons, and lakes.

SEWAGE

Sludge disposal is becoming increasingly expensive and problematic. Nor is it feasible or desirable to install extended aeration, incineration, or other facilities. The role of preactivation as a tool to reduce sludge production in existing facilities is shown for the city of Pittsburgh, PA. The activated sludge plant receives municipal wastes at a flow rate of 680,400 m^3/d (180,000,000 gal/day). The plant is divided into east and west sections, each of which receives half the flow. A side-by-side comparison permitted demonstration of the benefits of preactivating 1.5 ppm (flow based) of bacterial inoculant for sludge reduction.

Traditional wastewater treatment often relies on the use of large amounts of chemicals, largely flocculants. These are expensive, can increase the difficulty of dewatering sludge, and long-term environmental effects need definition. Bacterial augmentation by preactivation is a technique which can decrease polymer use and improve sludge dewatering. Data are shown for an activated sludge plant, with a flow rate of 5670 m^3/d (1.5 million gal/d) from the city of Lake Megantic, Quebec, Canada. This plant was treated by addition of 1.5 ppm (flow based) of preactivated Bacta-Pur™ S-2000+.

HYDROCARBONS

Environmental contamination of hydrocarbons can be in water or on the land. The acceleration of preconditioning to degrade benzene, toluene, xylene, and phenol has been demonstrated with industrial byproducts from refineries around Montreal,

Quebec, Canada. Soil decontamination is much more complex than water treatment. Not only are appropriate bacterial communities and nutrients required but so are environmental control technologies, particularly with respect to pH and an electron receptor for oxidation processes. Site-specific treatments must be planned based on factors such as the biodegradability of the target substrate, physical nature of the contaminated and underlying soil (e.g., particle size), depth of the contamination, space and equipment available for treatment, time available for treatment, quantity of contaminated soil, climatic conditions, environmental constraints, and land-use goals. The degradation of polyaromatic hydrocarbons in refinery soil was demonstrated using nitrate as the electron receptor. Preconditioners 1 and 2 were used in the preconditioning of indigenous bacteria augmented with Bacta-Pur™ E-2000+, as well as directly in the soil, to provide the appropriate physico-chemical conditions.

LAKES

Nonpoint source pollution is often a major cause of lake eutrophication (Emory et al. 1973, Bengtsson & Gelin 1975). The aging of lakes is related to rates of sediment accumulation exceeding those of decomposition (Wetzel, 1975). Organic sediments, particularly under anoxic conditions, can liberate important quantities of phosphorus (Goldman & Horne, 1983; Okada & Aiba, 1986), which then contribute to proliferation of algae and weeds. Bacterial augmentation was initially examined as a means to remove organic sediments. Many fish farmers have also reported the use of bacterial augmentation to control aquatic plants. Subsequent work looked at bacterial augmentation to control the excessive proliferation of the aquatic weed Eurasian water milfoil (*Myriophyllum spicatum*) through the food web structure. Excerpts of these two studies are presented here.

Lake sediments can contain low concentrations of organic material (10 to 20%) as compared to sewage solids (70 to 80%). It is generally more difficult to remove lower concentrations of biodegradable substrates, which results in typical asymptotic curves of substrate removal. Lake Magog, in the eastern townships of Quebec, Canada (latitude 45° N) is an eutrophic lake (10.8 km^2, average depth of 9.8 m and an average renewal time of 22 days (Gouv. Qc, 1977)). The lake acted as a sewage settling lagoon for the town of Magog until the early 1980s when a sewage treatment plant was brought on line. The lake suffers from severe algal blooms, thick sludge deposits and hypolimnetic anoxia (Gouv. Qc, 1977). An *in situ* study was conducted to evaluate the ability of bacterial augmentation to remove organic deposits. Four experimental parcels of 4 ha (200 m × 200 m) were selected in water of 1 to 6 m deep based on the homogeneity of the sediments. All parcels were located above the thermocline in oxygenated water and contained black muck, gytta type sediments. The parcels received average concentrations of Bacta-Pur™ S-2000+ of 0, 1.42, 2.83, and 5.67 l/ha/wk from May to September. The bacterial suspensions were applied directly to the sediments by pumping through a weighted flexible hose from a small boat, which was moving in a zigzag pattern across each parcel. Total volatile solids were followed through the experimental period.

The second study was conducted in two pairs of 200 l microcosms containing 20 l of sediments from Lake Massawippi (17.9 km^2, average depth of 41.6 m and an average renewal time of 1.5 years, Gouv. Qc, 1978) situated in the eastern townships of Quebec, Canada. The lake is mesotrophic; the southern portion suffers from proliferation of Eurasian water milfoil to the point of limiting water sports. The two pairs of microcosms were planted with fifteen 8- to 10-cm rooted segments of the plant, all coming from the same parent stock to eliminate genetic variability. One pair of microcosms included 10 goldfish (*Carassius auratus*) with an initial average weight 0.3 g in each of the microcosms. The sediments of one microcosm in each of the pairs was seeded with Bacta-Pur™ S-2000+ at a rate of 10 l/ha/wk (0.5 ml/microcosm/wk). Total length of the milfoil, total heterotrophic bacterial and snail counts, and fish biomass were followed as appropriate over 3 to 4 months. Distribution of total phosphorus in the biota, sediments, and water was also followed, but will be reported elsewhere.

RESULTS AND DISCUSSION

SEWAGE

It is theoretically possible to estimate the percentage of solids entering a wastewater treatment plant which is solubilized and thus assess the potential benefits of using the preactivation technology. On a weight-for-weight basis particulate BOD produces two times more sludge than soluble BOD (Benedict et al., 1979). This is due to the loss of about 50% of the BOD as CO_2 and water when the bacteria use soluble BOD for growth. Sludge production is a function of the total suspended solids (TSS) removed plus 0.5 × soluble BOD removed. TSS can be removed either by direct sedimentation or by solubilization (conversion to soluble BOD), transformation into bacteria and subsequent sedimentation of flocculated bacterial biomass.

If one assumes that 100% of the suspended solids entering the plant are solubilized, then the daily sludge production should be (15,227 ÷ 2) kg + (6818 ÷ 2) kg = 11,022 kg (Table 1). Alternatively, if 0% of the suspended solids entering the plant are solubilized then the daily sludge production should be 15,227 kg + (6818 ÷ 2) kg = 18,636 kg. The observed value was 18,409 kg, showing that very little of the entering solids were solubilized, and that sludge reduction should be possible by seeding exoenzyme producing bacteria into the RAS.

Bacterial augmentation increased the total heterotrophic plate count of the mixed liquor in the aeration tank by 25 to 100% (Figure 3). The increase in bacterial inventory within the treatment facility was not merely due to the quantity of bacteria added but rather a result of synergistic effects of the bacterial augmentation. The heterotrophic bacterial concentration of the untreated side of the plant averaged 8.3×10^6 cells/ml. The total volume of the aeration basin is 45,360 m^3, giving a total inventory of 3.8×10^{17} cells. The preactivation for the test side of the treatment plant was carried out in a total volume of 2268 l (600 gal). A final cell concentration in the mixture was 2×10^8 cell/ml, or a cell count of 4.5×10^{14} cells. This concentration represents an addition of only about 0.1% of the background level and, clearly, the observed increases in cell counts were not simply an arithmetical phenomenon.

Sludge production was compared against historical data due to a calibration problem with the sludge pumps. Bacterial augmentation using the preactivation technique reduced sludge production by 550 dry tons/month or about 16% (Figure 4). The benefits of bacterial augmentation disappeared within 30 days of the termination of the bacterial additions. Similar results have been found in other U.S. WWTP in terms of sludge yield (solids produced/BOD removed): N. Charleston, SC 1.34 to

TABLE 1
Pretreatment Daily Removal of Soluble BOD and TSS at Pittsburgh[a]

	Soluble BOD	TSS
In	7,273 kg	16,136 kg
Out	454 kg	909 kg
Removal	6,818 kg	15,227 kg

[a] Q = 680,400 m^3/d, 180 million gal/day.

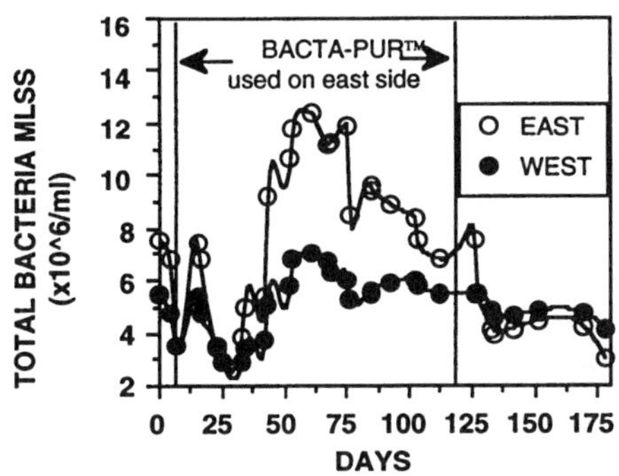

FIGURE 3. Total heterotrophic plate counts in Pittsburgh WWTP with and without bacterial augmentation (redrawn from Flamang, Wong, & Lowe, presentation to WPCF, Oct. 1988).

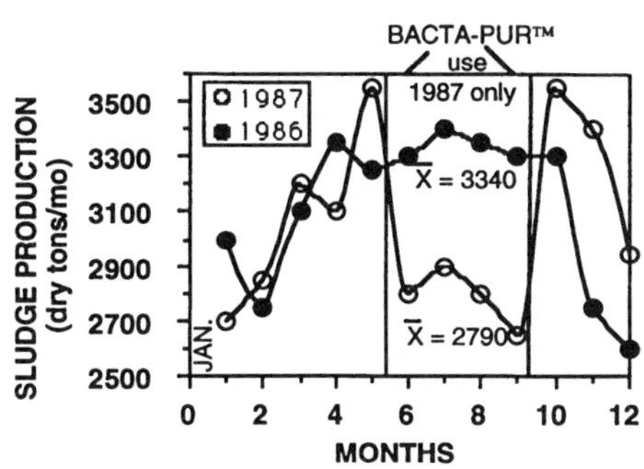

FIGURE 4. Sludge production in Pittsburgh WWTP with and without bacterial augmentation (redrawn from Flamang, Wong, & Lowe, presentation to WPCF, Oct. 1988).

0.86 (36% reduction) and Erie, PA 0.38 to 0.21 (45% reduction) (T. Lowe and J. Wong, pers. comm.)

Bacterial augmentation and preactivation at Lake Megantic, Quebec reduced use of polymer from about 9 kg/dry ton of solids to about 2 kg/dry ton solids (Figure 5). The decrease occurred over 5 months. Bacterial augmentation reduced use of polymer while increasing the sludge dryness. Dewatered sludge solids increased from 13.8 to 15.0% (Figure 6).

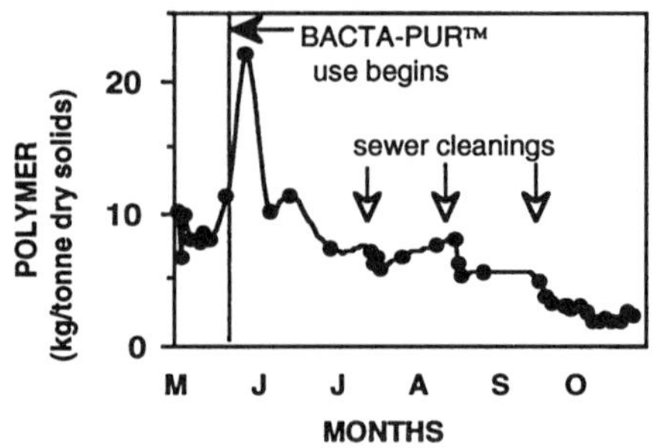

FIGURE 5. Polymer use at Lake Megantic WWTP with and without bacterial augmentation.

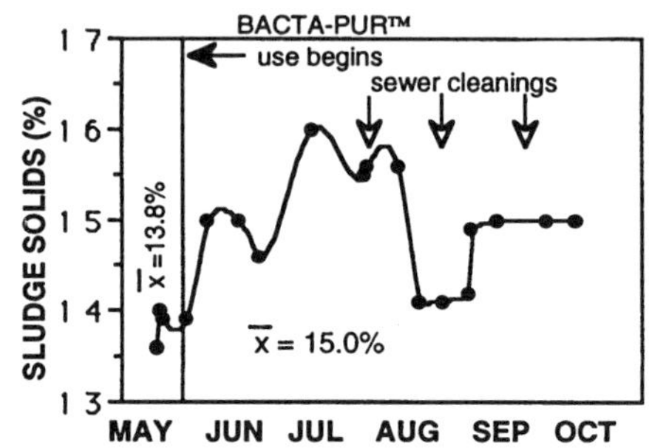

FIGURE 6. Dewatered sludge solids at Lake Megantic WWTP with and without bacterial augmentation.

HYDROCARBONS

Preconditioning of bacterial seedstocks resulted in the ability to degrade rapidly several hundred ppm of total BTX (benzene, toluene, and xylene) (Figure 7). An initial concentration of 400 ppm was decreased to < 5 ppb within 24 h. A subsequent addition of BTX brought the concentration once again to approximately 400 ppm; 24 h later the concentration was reduced to 10 ppb. Concentration of the controls, without the preconditioned bacterial populations, showed no change over the 48 h (Figure 8). Similar results have been achieved with phenol (Figure 9). BTX are very common in our environment as they are components of gasoline. Use of preconditioning techniques in bioreactors offers a cost-effective means to degrade these contaminants.

Total polyaromatic hydrocarbons (PAHs) in soil treated with the Bacta-Pur™ System showed a classical degradation curve; the rate was proportional to substrate concentration (Figure 10). The target of the degradation was ≤ 50 ppm. More than 400 ppm (89%) were degraded during the 6-week trial. The individual PAHs included: 3-methylcholanthrene, acenaphthalene, acenaphthene, anthracene, benzo(a)anthracene, napthalene, benzo(a)pyrene, benzo(b)fluoranthene, benzo(e)pyrene, benzo(ghi)perylene, benzo(k)fluoroanthene, crysene, dibenzo(ah)anthracene, fluoranthene, fluorene, indeno(1,2,3cd)pyrene, phenathrene, and pyrene.

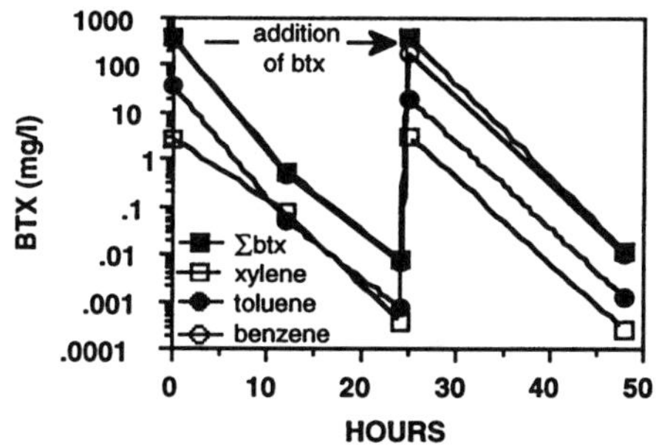

FIGURE 7. Degradation of BTX with preconditioned bacterial populations.

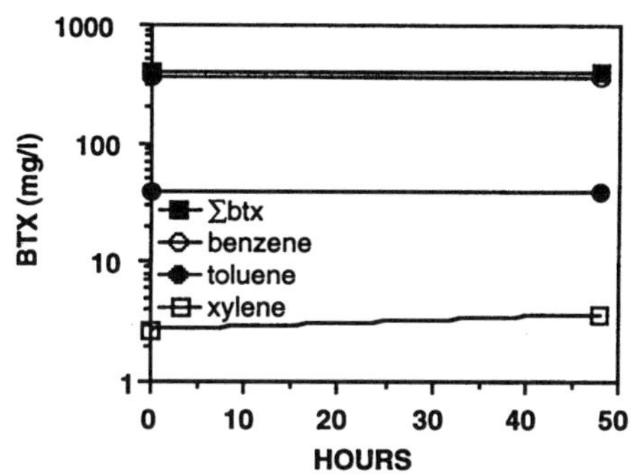

FIGURE 8. Degradation attempts of BTX without preconditioned bacterial populations.

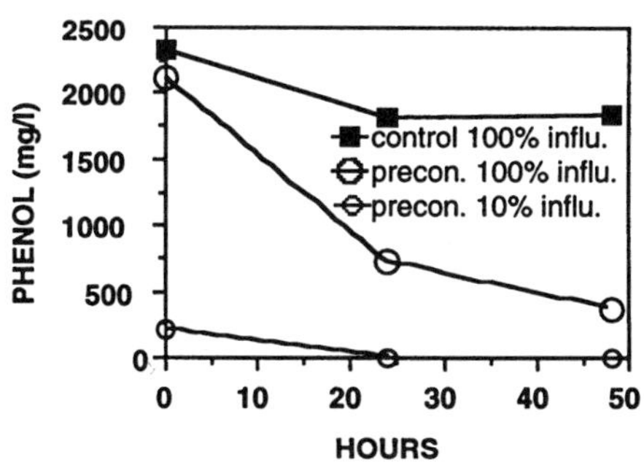

FIGURE 9. Degradation of phenol with preconditioned bacterial populations.

SEQUENCE OF BIODEGRADATION

Biodegradation of hydrocarbons is accomplished in steps, and the complexity of the processes, from chemical and biochemical points of view, is a function of the complexity of the substrates to be biodegraded (Figure 11). Bacteria are selective with respect to their preference for different substances. The degradation of one polyaromatic hydrocarbon can generate other PAHs and the life expectancy of these

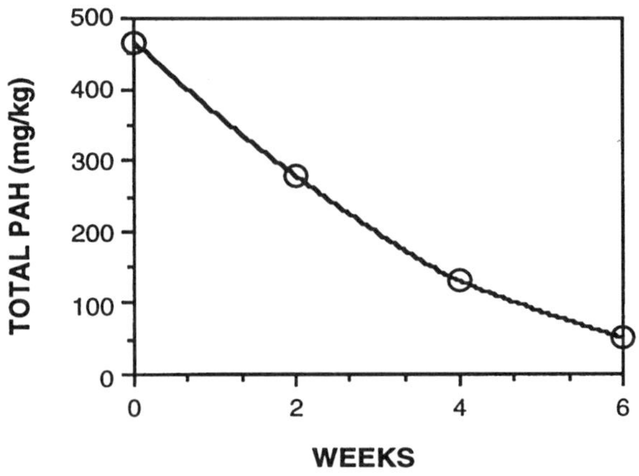

FIGURE 10. Degradation of polyaromatic hydrocarbons in soil with preconditioned bacterial popultions.

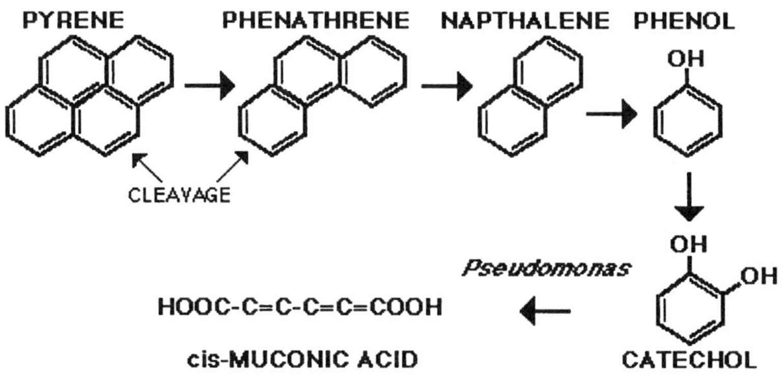

FIGURE 11. Sequence of biodegradation.

synthesized PAHs will be dependent upon many factors. For example, the biodegradation of pyrene by bacteria attacking one double bond generates phenanthrene. It is even possible to imagine the control of the bacterial processes in a bioreactor to produce certain intermediaries for recovery.

LAKES

Total volatile solids (TVS), the biodegradable component of the sediments, decreased with increasing rates of bacterial augmentation (Figure 12). Except for the initial sampling period, 19 days after the start of the bacterial augmentation, the station receiving 1.42 l Bacta-Pur™ S-2000+/ha/wk contained less TVS than the control, but the difference did not increase with time (Figure 12). The original intent of the study was to measure changes in sediment depth by means of permanently fixed stakes. Localized currents around the stakes caused erosion, making this technique unusable. Two weeks after the start of the bacterial augmentation, the assessment was changed to TVS, but the time zero data point was missed. The ability

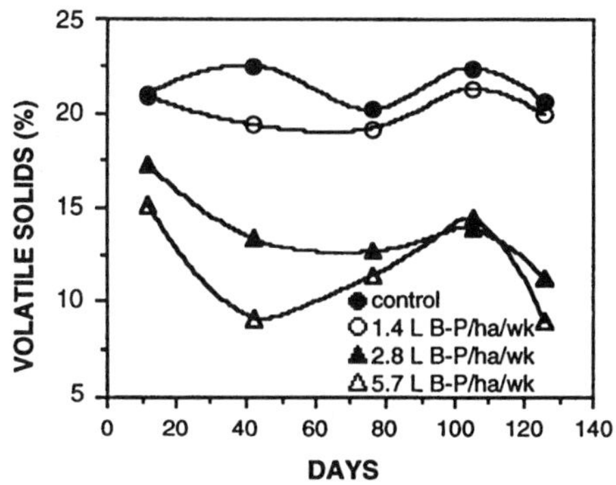

FIGURE 12. Relationship between total volatile solids in the sediments of Lake Magog and bacterial augmentation.

of bacterial augmentation to degrade organic sediments was expected, as this technique had already been demonstrated in sewage at Pittsburgh and Lake Megantic as well as by Hung et al. (1986) and in aquaculture (Ehrlich et al. 1989).

Apparently, naturally occurring numbers and/or species of microflora were inadequate, at least in these cases, for efficient decomposition of organic solids. Wetzel (1975) pointed out that microbial growth in aquatic ecosystems can be extremely variable; environmental fluctuations, biological generation of organic growth inhibitors, microbially formed inorganic growth inhibitors, high salt concentrations, and inappropriate temperatures, pH, or other conditions can reduce biological activity. Furthermore, microbial communities lacking appropriate enzyme systems due to lack of inefficient functioning of certain members may cause reduced decomposition abilities. The more efficient sediment removal with bacterial augmentation may, in part, be explained by species supplementation under marginal conditions, provision of specific exoenzyme producers, and (perhaps most important) seeding of mixtures of bacterial populations selected for their ability to work together in the process of water purification. In addition, the repeated seeding of selected strains could favor their growth.

The microcosm experiments showed that bacterial augmentation reduced the growth of Eurasian water milfoil (*Myriophyllum spicatum*; Figure 13). Total plant length is the sum of the principal and side branches of all plants in the microcosms. Little difference was detected between the control and experimental conditions up to about 70 days; after this time the control plants grew rapidly while in the bacterial augmented microcosm there was die-back. Clearly, manipulation of microbial communities can impact other trophic levels. To try to shed additional light on this subject, the effect of bacterial augmentation was examined in terms of floral and faunal biomasses. The lack of good growth of the plant in the presence of the fish was due, in large part, to increased turbidity and light attenuation caused by the fish.

Bacterial counts in the water column were consistently higher in the microcosms receiving bacterial augmentation (Figure 14). Average cell density in the controls, with or without fish was about 7×10^6/ml, giving at total inventory, in 150 l of water, of 1×10^{12} cells. The weekly addition of 0.5 ml of Bacta-Pur™ S-2000+ added only approximately 2×10^8 cells. Clearly, as in the case of the sewage treatment plant at

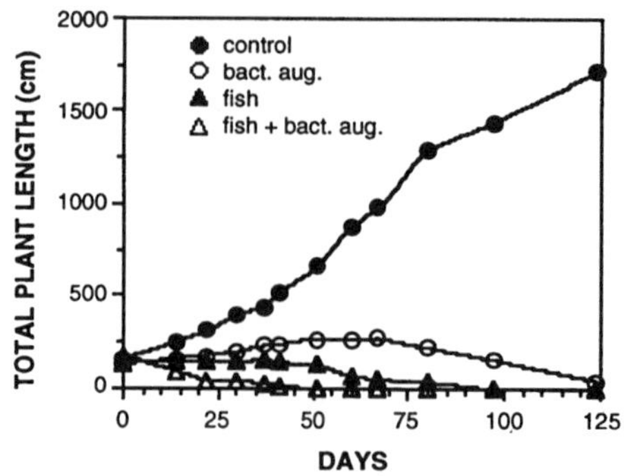

FIGURE 13. Effect of bacterial augmentation on the growth of Eurasian water milfoil.

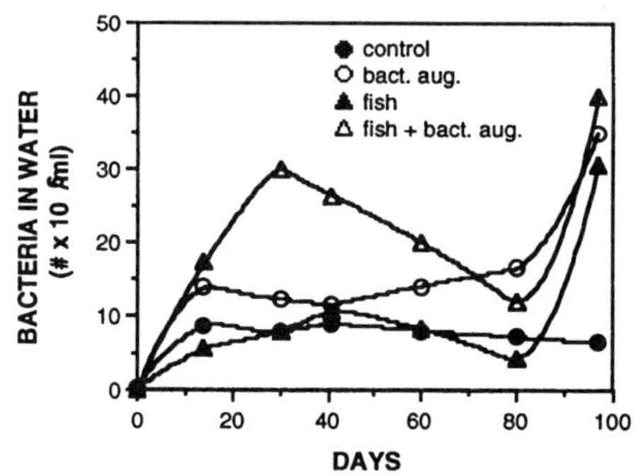

FIGURE 14. Effects of bacterial augmentation bacterial counts in water of microcosms.

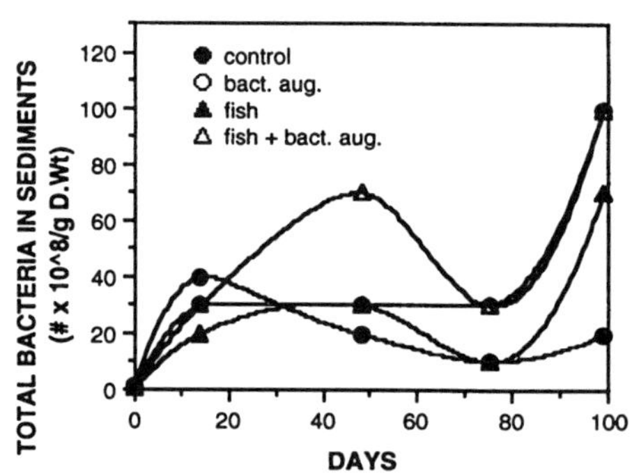

FIGURE 15. Effects of bacterial augmentation bacterial counts in sediments of microcosms.

Pittsburgh, the increase in bacteria cannot be accounted for by a strictly arithmetic addition of cells. The bacterial augmentation apparently stimulated bacterial growth and/or activity. Similar results were found for bacteria in the sediments (Figure 15).

The quantity of snails (*Gyraulus* sp.) in the microcosm without fish increased under the regime of bacterial augmentation (Figure 16). Bacterial generated floc can

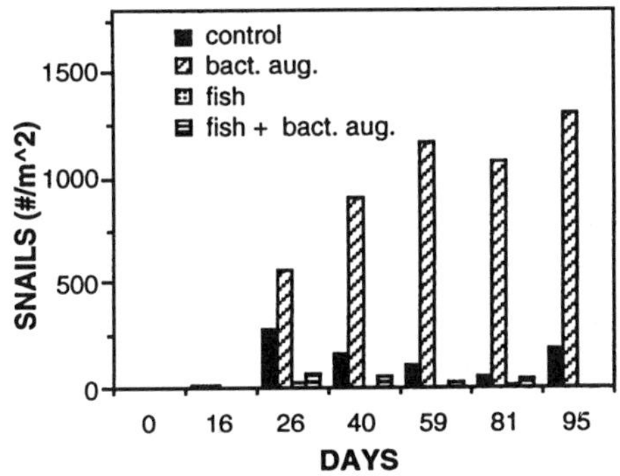

FIGURE 16. Effect of bacterial augmentation and presence of goldfish on number of snail density in microcosms.

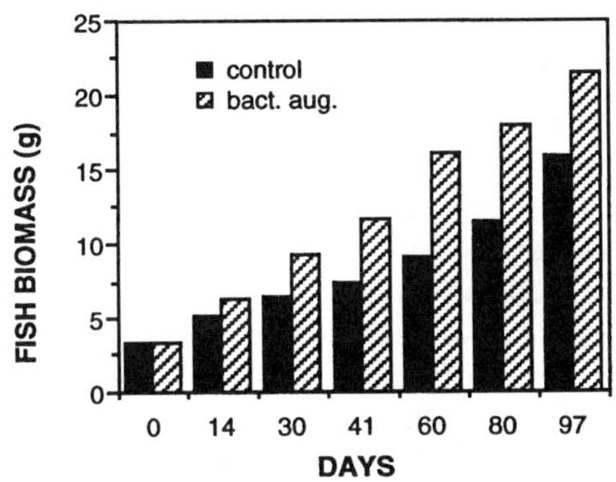

FIGURE 17. Effect of bacterial augmentation on biomass of goldfish in microcosms.

serve as food for snails and could explain the observed increase in the absence of fish. Goldfish like the common carp can feed not only on bacterial floc (Moriarty and Pullin, 1987), but they also readily consume snail egg masses. This could explain the lack of snails despite the bacterial augmentation in the microcosm with fish. The fish, however, appeared to respond to the bacterial augmentation with increased growth (Figure 17).

The observed increases in bacterial and faunal parts of the food web with a simultaneous decrease in flora, changes associated with bacterial augmentation with Bacta-Pur™ S-2000+, could be explained by several phenomena. Part of the increase in bacterial production could be explained by increased bacterial efficiencies due to synergy of a more diverse bacterial community. Simultaneously, solubilization of organic solids, which Bacta-Pur™ S-2000+ is designed to do, produces more food for the bacteria. Fuhs et al. (1972) and Rhee (1972) have show that *Bacillus subtilis*, one of the strains in the augmentation mixture, can absorb orthophosphate more efficiently and to the detriment of algae. This preferential uptake by bacteria over plants may have been even more accentuated by the presence of a bacterial community selected for the complementary nature of the strains and their ability to degrade organic matter rapidly. Furthermore, it is not unreasonable to expect that Eurasian milfoil could be even more disadvantaged than algae with respect to competition for nutrients against bacteria because of the lower surface-to-volume ratio of the macrophyte as compared to the algae. Bacterial augmentation, particularly with physiologically engineered consortia of complementary water purifying strains, offers the potential to slightly shift the balance in food webs by channeling more nutrients through the faunal part of the web. Current research is examining phosphorus flow through the food web, sediments, and water.

CONCLUSIONS

Bacterial augmentation, and particularly physiological engineering of bacterial consortia, is a relatively young science. It appears, however, to have potential in all realms of environmental management, from traditional wastewater treatment plants to advanced ecologically engineered systems to soil decontamination to water body restoration. All of these systems or processes depend, in large part, on growth and activity of bacterial communities. Bacterial augmentation, especially when linked to ecological engineering offers a potential and tool to help transform organic wastes into valued resources such as fish for restocking. It is also more difficult for people to complain that a wastewater treatment plant is not treating water adequately when the effluent is adequate for the production of valued sport fish.

It is, hopefully, clear to all readers that environmental management entering the 21st century should drop some unfounded beliefs which seem to predate the 19th century, such as the spontaneous generation of all necessary strains in wastewater treatment plants. Even if bacteria can be found in most environments and if they can be transported by the winds, it seems rather incongruous that in this day and age of wastewater treatment facilities costing hundreds of millions of dollars that the arrival of the most essential part of the water purification community is usually dependent upon which way the wind blows. Biotechnology can help stabilize treatment efficiency by assuring the presence of specific bacterial communities in the appropriate physiological condition.

REFERENCES

Benedict, A.H., M.S. Merrill & G.S. Mauseth. 1979. "Sludge production, waste composition and BOD loading effects for activated sludge systems." *J. Water Poll. Contr. Fed.* 51: 2898.

Bengtsson, L., and C. Gelin. 1975. "Artificial aeration and suction dredging methods for controlling water quality." In *The Effects of Storage on Water Quality*, pp. 313–342, Marlow, Bucks, England: Water Res. Cntr. Medmemham.

Blackburn, T.H. 1987. "Microbial food webs in sediments." In M.A. Sleigh, ed., *Microbes in the Sea*, pp 39–58. Chichester, England: Ellis Horwood, Ltd.

Bonnett, R.M. 1987. Phosphorus Removal in Municipal Wastewater Treatment. M.Sc. Thesis, Cleveland State Univ., OH, 214 pp.

Chrost, R.J., R. Wcislo, and G.Z. Halemejko. 1986. "Enzymatic decomposition of organic matter by bacteria in an eutrophic lake." *Arch. Hydrobiol.* 107: 145–165.

Crawford, R.L., and K.T. O'Reilly. 1989. "Bacterial decontamination of agricultural wastewaters." In M.E. Huntley (ed), *Biotreatment of Agricultural Wastewater*, pp. 73–89. Boca Raton, Fl: CRC Press.

Ehrlich, K.F., M.-C. Cantin, and F.L. Horsfall III. 1989. "Bioaugmentation: Biotechnology for improved aquacultural production and environmental protection." In K. Murray (ed), *Aquaculture Engineering Technologies for the Future.* Inst. Chem Eng. U.K., Symp. Ser. No. 111: 329–341.

Emory, R.M., C.E. Moon, and E.B. Welch. 1973. "Delayed recovery of a mesotrophic lake after nutrient diversion." *J. Water Polln Contr. Fed.* 45: 913–925.

Fuhs, G.W., S.D. Demmerle, E. Canelli, and M. Chen. 1972. "Characterization of phosphorus limited plankton algae (with reflections on the limiting nutrient concept)." In G.E. Likens (ed.), *Nutrients and Eutrophication: The Limiting Nutrient Controversy.* Spec. symp. Amer. Soc. Limnol. & Oceanogr. 11: 505–514.

Goldman, C.R., and A.J. Horne. 1983. *Limnology.* San Francisco, CA: McGraw-Hill Book Co. 464 pp.

Gouv. Qc, Minist. Richesses Nat. 1977. *Carte de classification, Lac Magog.*

Gouv. Qc, Minist. Richesses Nat. 1978. *Etude Limnologique, Lac Massawippi.*

Hänel, K. 1988. *Biological Treatment of Sewage by the Activated Sludge Process.* Chichester, England, Ellis Horwood Ltd. 299 pp.

Hung, Y.-L., F.L. Horsfall III, D.B. Shah, and S.A. Zachopoulos. 1986. "Bio-conversion of accumulated sludge with bacterial augmentation process in aerated lagoons for municipal wastewater treatment." *Intl J. Environ. Studies* 28:41–56.

Moriarty, D.J.W., and R.S.V. Pullin, eds. 1987. *Detritus and Microbial Ecology in Aquaculture.* ICLARM Conf. Proceed. 14, 420 pp.

Okada, M., and S. Aiba. 1986. "Simulation of water bloom in an eutrophic lake — IV. Modeling the vertical migration in a population of *Microcystis aeruginosa*." *Water Res.* 20: 485–490.

Rhee, G.-Y. 1972. "Competition between an alga and an aquatic bacterium for phosphate." *Limnol. & Oceanogr.* 11:505–514.

Sonea, S., and M. Panniset. 1980. *A New Bacteriology.* Boston, MA: Jones & Barlett Publ., Inc., 140 pp.

Wetzel, R.G. 1975. *Limnology.* Philadelphia, PA: Saunders College Publ. 743 pp.

WPCF. 1983. *Nutrient Control, Manual of Practice FD-7* Facilities Design. Water Polln Contr. Fed., Washington, D.C. 205 pp.

Zachopoulos, S.A. 1987. "Bioaugmentation process for municipal wastewater treatment." Doc. Eng. Thesis, Cleveland State Univ., OH, 586 pp.

Chapter 16

CONTRIBUTIONS OF MARSHES AND SALT MARSHES TO ECOLOGICAL ENGINEERING

John M. Teal

ABSTRACT

Natural marshes exhibit properties and processes useful for (or harmful to) the treatment of wastewater. These include: denitrification, organic matter mineralization, sorption of materials on sediments, burial within sediments, production and consumption of plants and animals, precipitation of chemicals, transfer of volatiles to the atmosphere and dehalogenation reactions. Ecological engineering attempts to replicate and/or enhance these processes for human benefit. To this end we must understand how they work in order to know what are the limits to their functions and to what extent they have to be managed to maintain that function.

Processes such as denitrification and organic mineralization need only the conditions to maintain the marsh for their maintenance. Sorption of materials on sediments and burial, on the other hand, can be saturated. The result in the case of the latter processes is either a cessation of function in that regard, or the necessity of active interference with the marsh to maintain the function. Some examples illustrating these conclusions will be given.

INTRODUCTION

My idea of ecological engineering is to use the ecological processes within natural or constructed imitations of natural systems to achieve engineering goals. In the case of marshes, goals can range from aquaculture to hazardous waste treatment, but here we will concentrate on wastewater treatment. So the question we are to address is: what properties of marshes can be used in wastewater treatment? Once they are identified, we want to know such things as whether or not the identified properties can be replicated and/or enhanced; what are the limits to their functions; and whether they have to be managed to continue to function.

I will list the principal marsh processes of value as: sorption and chemical precipitation, organic production and decomposition, nitrification and denitrification, dehalogenation, and burial. For wastewater treatment we can group the materials to be treated into nutrients, organic contaminants, inorganic contaminants, pathogens, and inert components. I will discuss wetland contributions in terms of the treated substances.

0-87371-990-5/97/$0.00+$.50

NUTRIENTS

We are typically concerned with only two elements: nitrogen and phosphorus. Both frequently are limiting to marsh production, are chemically and biologically active, and are important components of wastewater effluents. These properties lead us to expect uptake of these nutrients by marshes, which we regularly find.

NITROGEN

In a natural marsh the nitrogen cycle is composed of uptake and loss by plants and microbes, solubilization as dissolved inorganic nitrogen (DIN) and release from organic matter in the course of microbial degradation, oxidation of released ammonium by nitrifying bacteria, denitrification of nitrate in anoxic sediments or other anoxic microsites with release of N_2, absorption of ammonium onto fine sediments, sedimentation of organic matter resistant to degradation, and loss of nitrogen compounds from the marsh mostly by downstream flow. If the marsh is accumulating sediment, there will be a net deposition of nitrogen. There could be a net removal by grazing by large herbivores or by insect emigration. In a typical unmodified marsh there is uptake of nitrogen from the sediments and waters in the spring and summer while plant biomass is increasing. This is followed in the autumn by loss of nitrogen as the leaves and stems of the plants become senescent, permeable, and subject to microbial attack (Valiela and Teal, 1979). This pattern is evident even in rather artificial situations such as that in the cranberry bog studied by Howes and Teal (in press). The bog was created from a natural red maple swamp and is highly managed by water control, fertilization, and harvest for the production of cranberries, the fruit of *Vaccinium macrocarpus*. Although the largest inputs to the bog are in the form of fertilizers (indicated in the fluxes seen in early summer), and the outflow is interrupted for frost protection, insect control, and harvest (seen most notably in the interruption of flux during autumn), the annual cycle referred to for an unmodified, natural marsh is clear (Figure 1).

The aquatic part of this cycle is what one sees when looking at the marsh as a "black box," as one would examine a marsh to see how effective it was in removing nitrogen from wastewater. But in spite of the size and periodicity in "input-output," most of the nitrogen cycling occurs within the typical marsh, consisting of transfers among sediments, roots, and plant tops (Howes et al., 1983). But if the system is subject to serious disturbance, this stored and recycled nitrogen becomes available for export. Even in the undisturbed marsh, much of the export of nitrogen can come from water from sediment pore spaces leaking into outflowing waters (Jordan and Correll, 1985).

But even if the marsh is "full" of sediment and there are no significant losses to emigration or grazing, there can still be net removal of nitrogen via denitrification. For example, Bowden (1986) indicated the probable importance of denitrification in a freshwater tidal marsh, and Smith et al. (1985) estimated that 50% of the NO_3-N entering the estuary of one of the distributaries of the Mississippi River was lost through denitrification. In the salt marsh we have studied, denitrification accounts for an amount of nitrogen equal to that added by precipitation and freshwater (groundwater) flows (Valiela and Teal, 1979).

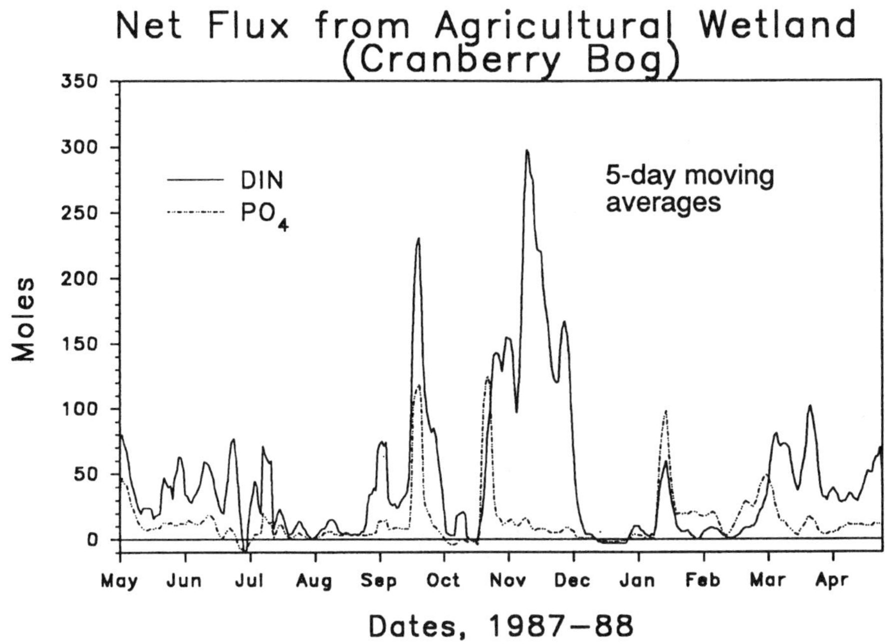

FIGURE 1. DIN = dissolved inorganic nitrogen.

In natural marsh systems to which nitrogen is added, denitrification remains an important avenue of loss. In the Massachusetts cranberry bog, about 40% of the total nitrogen inputs from all sources are lost through denitrification (Howes & Teal, 1991). We have been fertilizing experimental plots on the salt marsh referred to above for some 20 years. We made a study of the fate of nitrogen added over various periods, in two forms and at different application rates (Figure 2). At the low levels of addition of nitrogen in the form of nutrient-enriched sewage sludge (LF), all the nitrogen was either stored in the sediments or taken up by the plants and lost from the plots when the tops senesced, died, and were washed away by the tides. In other plots fertilized with urea (U+P) or with sludge but at a rate nine times greater than that in LF (XF), only a fraction of the nitrogen added over the years could be accounted for either in the sediments or as taken up and lost via the plants. Since previous research (Valiela et al., 1973) showed that only a minor fraction of the added fertilizer was washed out by the tides directly after addition and that denitrification occurs rapidly in the marsh (Kaplan et al.,1979), the most logical conclusion is that the nitrogen labelled "lost" in Figure 2 has been lost by denitrification. This amounts to as much as 70% of added nitrogen added over a period of 17 years (Howes and Teal, unpublished data).

Nitrogen may also be removed from marshes by the non-biological harvest mechanism, flooding. Heckman (1986) found significant removal of both nitrogen and phosphorus from the freshwater portion of the Elbe estuary by winter storm tides that strand nutrient-rich plant remains on the surrounding uplands.

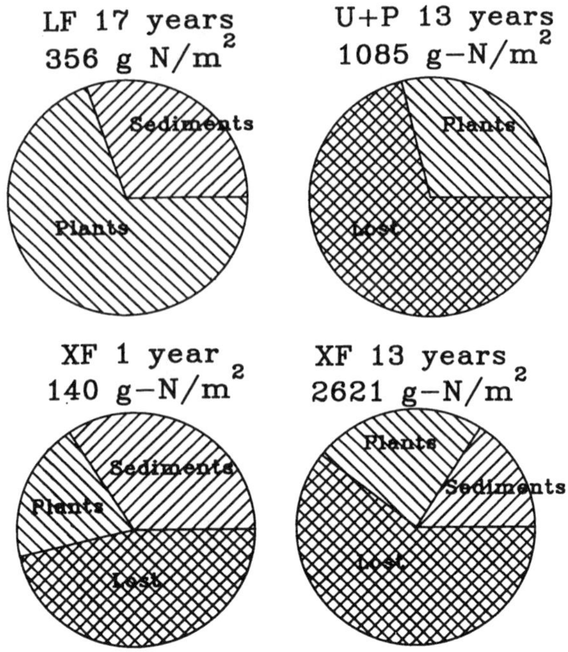

FIGURE 2. The fate of nitrogen added to a salt marsh. Codes for soil amendments: LF = sewage sludge; XF = sewage sludge at a rate nine times greater than in LF; U + P = urea.

PHOSPHORUS

Phosphorus does not "disappear" into the atmosphere from marshes under normal circumstances. It can be removed from the system by harvest as in the last example. But usually the most significant mechanism for removal of phosphorus from waters flowing through a marsh is precipitation in sediments as aluminum or iron salts. This system can be saturated in a period that can be predicted from the concentrations of these cations (Richardson, 1985). When that occurs the removal of phosphorus will be limited to that exported in biomass or stored in biomass accumulation as organic sediments.

ORGANIC CONTAMINANTS

We are concerned here with those compounds that are hard to decompose, are toxic, or both. Compounds such as lignins are hard to decompose, especially under anoxic conditions, and contribute to peat formation in natural marshes. Of more concern are compounds like polyaromatic hydrocarbons (PAH) in which the aromatic ring structures are attacked only with difficulty and for which the intermediate decomposition products can be toxic or mutagenic. These are natural compounds, and organisms have mechanisms to attack the molecules, but exposure to them has probably been at only relatively low concentrations, for the most part on an evolutionary time scale. They become a problem when concentrated in marshes by our activities, such as in oil spills or in marshes used as road runoff treatment systems.

Compounds like PCBs and other halogenated compounds, the microbial degradation of which is hindered by the halogen, are even more difficult to deal with.

Marshes can deal with these compounds to a considerable extent if given sufficient time. The West Falmouth Oil Spill, for example, was perhaps 90% recovered from the severe, though local, effects of a fuel oil spill in 1969 after about 5 years. However, in the most heavily oiled intertidal sediments there were still large amounts of oil at depths of 10 to 15 cm in anoxic sediments in 1989, 20 years after the event (Farrington et al., MS.). Although anoxic conditions in sediments slow decomposition, they also allow anaerobic microbes to dehalogenate molecules. Chlorinated pesticides contained in the sludge placed on salt marsh plots were closely bound to the area of application, and degraded within 2 years (Krebs et al., 1974). I believe the alternation of oxic with anoxic conditions as occurs in a tidal marsh contributes to organic degradation by allowing both types of microbial systems to exist side by side.

INORGANIC CONTAMINANTS

Metals are of most concern in this category. They are sequestered in marshes both by sorption on sediments and by chemical precipitation. In salt marshes with high concentrations of sulfide, many heavy metals (e.g., mercury and lead) are bound as sulfides with very low solubilities and are effectively removed from the biological part of the system as long as they remain in the anoxic zone. But other metals were more mobile (e.g., cadmium and zinc), and not retained very effectively (Valiela et al. 1974). In freshwater systems, I expect similar behavior over the long term, though a 1-year loading of an artificial sewage treatment wetland in California showed over 97% removal of cadmium, zinc, and copper (Gersberg et al. 1984).

PATHOGENS

There is little information on natural marshes with regard to their treatment of pathogens. Normal die-off and filtering out are expected and found where looked for, though most of the data comes from wastewater treatment systems (e.g., Bauerfeind, 1986). Even less is known of the fate of viruses, though again there are studies indicating less than 1% survival rates in artificial wetlands used for treatment (Gersberg et al., 1987).

INERT CONTAMINANTS

I include sand, silt, etc. here and list if only for completeness. These materials are "treated," removed by marshes by the filtering actions of their plant covers and the fact that velocity of water movement is slowed by the area of the marsh in comparison to the channels into the wetland. Sediment accumulation is controlled by the balance between deposition and erosion, and wetlands can be filled with sediments just as with the things associated or sorbed to them. In marine marshes, continued slow sedimentation is possible nearly throughout the world at present, as marshes grow in height to maintain their position in relation to rising sea level. This amounts to 1 to 2 mm/yr on average, sufficient to permit significant accumulation of both sediments and their associated pollutants. Banus et al. (1974) give a picture of the accumulation of lead in salt marsh sediments over time since the European settlement of North America.

INTERACTIONS WITH MARSH PROCESSES

From the above considerations, we seem to know a fair amount with regard to natural wetlands about the behavior of, at least, the more typically regulated substances in wastewater treatment. Nitrogen processing is very active, with denitrification a continuous sink. Phosphate is removed from inflowing water mostly by sorption and precipitation in sediments; if these processes are saturated, there is little removal. Both nutrients can stimulate productivity of wetlands and thus the associated and usually minor removal of nutrients by grazing or flood transport. The breakdown of organic compounds is a continuing process, stimulated in the case of some classes of compounds by alternation of oxic and anoxic conditions in the marsh sediments. Metal and sediment sequestration is limited, like that of phosphorus, by the space and conditions within the marsh that favor accumulation of these materials.

The nature of the marsh can be changed by the additions of wastewater, in the most severe cases making it no longer a marsh. This is most likely to occur as a result of the additions of large amounts of inorganic sediments which raise the level of the marsh and promote drainage. Marsh deterioration may also result from excessive loading of other materials. In our experience, a Massachusetts salt marsh can be loaded very heavily without obvious deterioration. The application of sewage sludge at the rate of 195 g/m^2 yr (the XF plots of Figure 2), equivalent to 1.95 T–N/ha yr, has not greatly changed the macro-flora and fauna of the Great Sippewissett Salt Marsh although it has increased the productivity markedly (Teal, 1986) and affected relative abundance and distribution of organisms. This sludge contains significant quantities of heavy metals and other pollutants, so one can only conclude that whatever else may be affected, the macroscopic salt marsh ecosystem itself is quite resistant to loading with these pollutants (Teal, 1985). The microbial community did show an increase in metal and antibiotic resistant bacteria and a decrease in diversity in the plots receiving sludge (Hamlett, 1986).

But smaller changes can have significance for marsh processing of materials. For example, enhanced grass productivity in salt marshes results in an increase in oxidation reduction potential in the sediments through a positive feedback between plants, redox, and nutrients (Howes et al., 1986). This in turn affects other processes, especially those that depend upon productivity and/or sediment redox. In larger plants resulting from increased nutrients, there are larger aerenchymal spaces which could result in reduced resistance to diffusion of gases from the sediments to the atmosphere. Figure 3 shows transport of various gases through a *Spartina* stem (Howes and Goode, 1991). On the one side are the gases which are active in plant metabolism. They exchange with the atmosphere through the stomata. As shown so dramatically by Hofmann (this volume) they also exchange through the roots, though we find that exchange of oxygen is greatly reduced when the roots are in water compared to when they are in air (Howes and Teal, in press). On the other side are

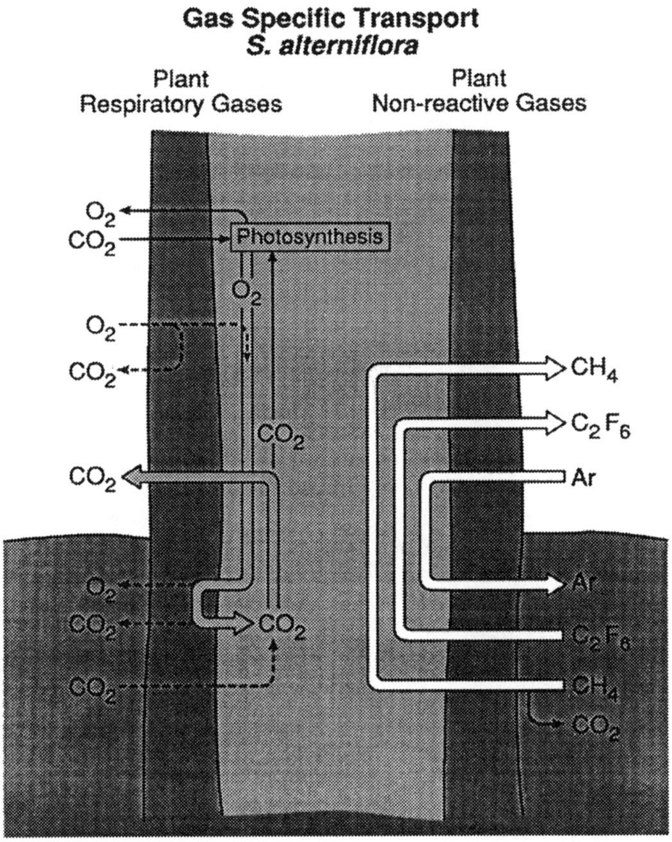

FIGURE 3. Transport of various gases through a *Spartina* stem.

shown some gases which are not active in the plant's metabolism. Since some of these gases could be toxic contaminants of wastewater, their residence time in sediments would be reduced. This could be beneficial if air quality was not adversely affected or detrimental if air quality were a concern and it was desirable to keep the volatile compounds within the sediment for as long as possible to enhance their degradation.

It is obvious from the very existence of this conference that the properties of marshes and other wetlands have value for the treatment of wastewaters. It is also clear that for many of the materials to be treated, wetlands have to be managed if their desirable function is to be maintained. For their most efficient use, we need the fullest possible understanding of marsh functions under natural conditions as well as under the stresses that will result from watering them with our wastes.

ACKNOWLEDGMENT

This work was supported by the C.S. Mott Foundation.

REFERENCES

Banus, M., I. Valiela and J.M. Teal. 1974. "Exports of lead from salt marshes." *Mar. Pollut. Bull.* 5:6-9.

Bauerfeind, S. 1986. "Microbiological and chemical investigations in a wastewater treatment plant with a wastewater pond planted with marsh plants during winter 1983/84." *Zentralbl. Bakteriol. Mikrobiol. Hyg.*, Abt. B., 182:177-192.

Bowden, W.B. 1986. "Nitrification, nitrate reduction, and nitrogen immobilization in a tidal freshwater marsh sediment." *Ecology* 67:88-99.

Farrington, J., J. Teal, B. Tripp and C. Phiney. MS. "The West Falmouth Oil Spill, twenty years later."

Gersberg, R.M., S.R. Lyon, B.V. Elkins and C.R.Goldman. 1984. *Removal of heavy metals by artificial wetlands.* 13 pp; Ecol. Res. Ser. U.S. Environ. Protect. Agency. EPA/600/D-84/258.

Gersberg, R.M., S.R. Lyon, R. Brenner and B.V. Elkins. 1987. "Fate of viruses in artificial wetlands." *Appl. Environ. Microbiol.* 5:731-736.

Hamlett, N.V. 1986. "Alteration of a salt marsh bacterial community by fertilization with sewage sludge." *Appl. Environ. Microbiol.* 52:915-923.

Heckman, C.W. 1986. "The role of marsh plants in the transport of nutrients as shown by a quantitative model for the freshwater section of the Elbe Estuary." *Aquat. Bot.* 25:139-151.

Howes, B.L., I. Valiela and J.M. Teal. 1983. "Transport and cycles of nitrogen and carbon in New England salt marshes." *Amer. Soc. Limnol. Oceanogr. Abst.*

Howes, B.L. and J.M. Teal. Submitted. "Nitrogen balance in a Massachusetts cranberry bog." *Ecological Applications.*

Howes, B.L. and R.E. Good. 1991. "Factors controlling gas exchange in salt water wetlands: Plant vs. sediment exchanges and the role of hydrology." Abstracts, Wetland Biogeochemistry Symposium, Louisiana State University, Baton Rouge, Feb. 1991.

Howes, B.L., J.W.H. Dacey and D.D. Goehringer. 1986. "Factors controlling the growth form of *Spartina alterniflora*: Feedbacks between above-ground production, sediment oxidation, nitrogen and salinity." *J. Ecol.* 74:881-898.

Howes, B.L. and J.M. Teal. Submitted. "Oxygen loss from *Spartina alterniflora* and its relation to rhizosphere waterlogging." *J. Exper. Mar. Biol.*

Jordan, T.E. and D.L. Correll. 1985. "Nutrient chemistry and hydrology of interstitial water in brackish tidal marshes of Chesapeake Bay." *Estuar. Coast. Shelf Sci.* 21:45-55.

Kaplan, W., I. Valiela and J.M. Teal. 1979. "Denitrification in a salt marsh ecosystem." *Limnol. Oceanogr.* 24:726-734.

Krebs, C.T., I. Valiela, G. Harvey and J.M. Teal. 1974. "Reduction of field populations of fiddler crabs by uptake of chlorinated hydrocarbons." *Mar. Pollut. Bull.* 5:140-142.

Richardson, C.J. 1985. "Mechanisms controlling phosphorus retention capacity in freshwater wetlands." *Science* 228:1424-1426.

Smith, C.J., R.D. DeLaune and W.H. Patrick, Jr. 1985. "Fate of riverine nitrate entering an estuary: 1. Denitrification and nitrogen burial." *Estuaries* 8:15-21.

Teal, J.M. 1985. "The role of one salt marsh in coastal productivity." pp.241-258. In *International Symposium on Utilization of Coastal Ecosystems: Planning, Pollution and Productivity.* Vol I. N.L.Chao & W.Kirby-Smith, Eds. Fundacao Univer. do Rio Grande & Duke Univer. Mar. Lab. 496 pages.

Teal, J.M. 1986. *The ecology of regularly flooded salt marshes of New England: A community profile.* Fish and Wildlife Service, Division of Biological Services, Washington, D.C. Biological Repon 85(7.4).

Valiela, I. and J.M. Teal. 1979. "The nitrogen budget of a salt marsh ecosystem." *Nature* 280:652-656.

Valiela, I., M. Banus and J.M. Teal. 1974. "Metal retention by salt marsh sediments and the response of marsh bivalves to enrichment with metal-containing fertilizers." *Environ. Pollut.* 7:149-157.

Valiela, I., J.M. Teal and W. Sass. 1973. "Nutrient retention in marsh plots experimentally fertilized with sewage sludge." *Estuar. Coast. Mar. Sci.* 1:261-269.

Chapter 17

THE ROLE OF PLANTS IN SUBSURFACE FLOW CONSTRUCTED WETLANDS

Karlfriedrich Hofmann

ABSTRACT

The role of plants in subsurface flow constructed wetlands is evaluated from studies of the use of *Phragmites* in sewage sludge treatment. The plants do much more than simply supplying a large surface area for microorganisms. *Phragmites* is particularly good in stabilizing accumulating sediments on the surface, thereby providing infiltration pathways through the upper layer. The continually extending root system maintains the initial permeability of the filter. The plant litter introduces biopolymers like cellulose, and therefore enhances humus formation. The oxygen supply of the constructed wetland is improved in two ways. First, mechanical effects of extending roots as well as the plant's great demand for water enable the formation of air-filled pores. Second, oxygen leaks out from the roots, particularly between the epidermal cells of young secondary roots and laterals. Therefore, the rhizosphere becomes an oxygen-rich microzone in a substratum which is otherwise anoxic or anaerobic. The rate at which oxygen is supplied directly by *Phragmites* is likely to be between 5 and 6 g O_2/m^2/day. The plant-mediated aerobic, anoxic, and anaerobic zones within the soil profile facilitate a sequence of mineralization, nitrification, and denitrification. In subsurface flow constructed wetlands with an intermittent flow regime, the plants combine the benefits of an aerobic biological filter with the traditional concept of helophyte wetland.

INTRODUCTION

Constructed wetlands have been shown to have a significant capacity for water treatment and resource recovery in many working examples worldwide. Reports of practical experiences and the latest research results are presented by Cooper and Findlater (1990).

One of the principal plants involved in constructed wetlands is the common reed (*Phragmites australis*). Reed is among the most widespread and variable plant species in the world. Subsurface flow constructed wetlands employ marsh plants like reed, with specially evolved air conducting tissues (aerenchyma), to provide a passage of oxygen to the microorganisms surrounding their roots and rhizomes. Carefully prepared mixtures of gravel, sand, and soil are used as rooting-filtration-adsorption media and undersealed with an impervious liner (Figure 1). The flow is engineered by suitable design and construction of the artificial wetland.

0-8493-990-5/97/$0.00+$.50

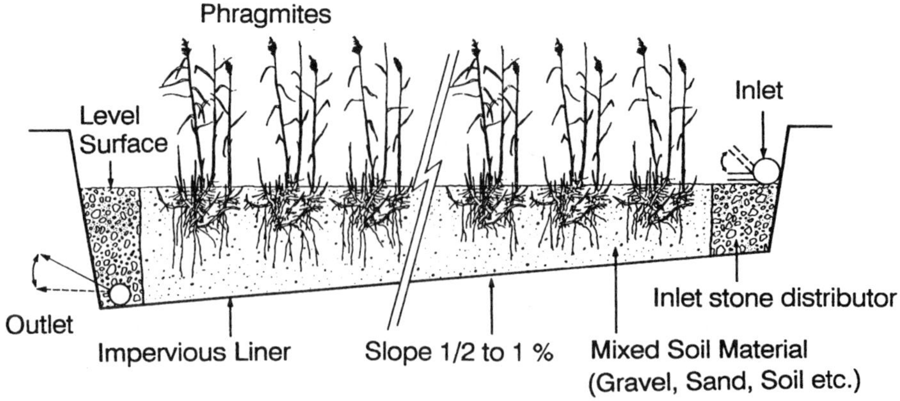

FIGURE 1. Typical arrangement of a constructed wetland for water treatment (after Cooper and Findlater, 1990).

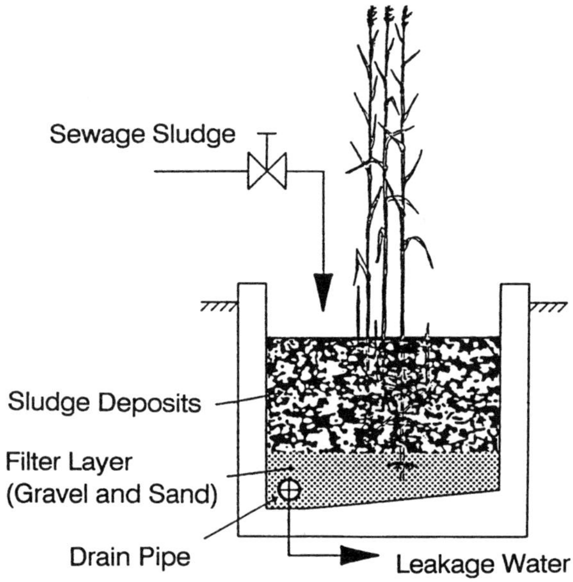

FIGURE 2. Sewage sludge treatment in reed beds (schematic cross-section view).

Considerable debate still exists as to how the reactions between pollutants, substratum, plants, and microorganisms occur and to what extent, with many conflicting opinions. This paper focuses on the direct and indirect effects of plants in such systems, based upon results of studies on the use of *Phragmites* in sewage sludge treatment (Figure 2; for details see Klee and Hofmann 1987; Hofmann 1990). Reasons for the effectiveness of *Phragmites* in constructed wetlands, and clues to a more successful design and construction of the systems, can be gained from a better understanding of the plant's role.

PLANT BIOMASS

The above-ground productivity of reed has been assessed in many studies, leading to the conclusion that *Phragmites* is one of the most productive species in temperate latitudes. In reed beds loaded with sewage sludge, particularly high yields are obtained, provided that sludge type and loading rate are tolerated by the plants. A 3-year-old *Phragmites* clone in a sewage sludge drying bed produced up to 4 $kg/m^2/year$ dry weight (Hofmann 1986). The experiments indicate that growth is impeded at loading rates beyond 30 $kg/m^2/year$ dry weight, that is at extremely high nutrient concentrations. If any damage occurs to the stems, reducing their ability to aerate the rhizomes, there is obviously a danger that anaerobic conditions of such severity take place that regrowth is completely stopped.

The below-ground productivity of reed in a sewage sludge drying bed reached about 2.3 $kg/m^2/year$ dry weight. The underground to above-ground ratio was 0.7 (Hofmann and Klee 1988). Such low values are typical of hypertrophic sites.

A dense reed stand tends to shade out photosynthetic eucaryotes on the soil surface, and secondary pollution of the effluent is reduced. Moreover, plant cover and leaf litter on the surface have an insulating effect, thereby ensuring winter performance of the constructed wetland. Winter harvesting every 3 or 4 years can improve light penetration into the reed stand and diminishes the rate of surface accretion.

WATER BUDGET

Annual evapotranspiration in dense reed stands has been measured as 1 to 2 m/year in several European countries. A sewage sludge drying bed is loaded at fortnightly intervals. Water loss is lowered by the intermittent flow regime to 575 mm/year but remains as high as above free water. In contrast to this, only 390 mm/year water are lost from the unplanted control (Figure 3).

The plant's great demand for water caused the applied sewage sludge to gradually lose its water. After a 1-year drying period without sludge loading, it was finally converted into a "friable soil" containing 40% dry matter (Figure 4), whereas the unplanted control contained only 32% dry matter.

EFFECTS OF ROOTS ON SOIL STRUCTURE

The sludge deposits are subjected to a differentiation. Two distinct zones were formed; an anaerobic, black upper layer, and a more oxidized, brown lower layer (Figure 5). After a 1-year drying period without sludge applications the whole sludge profile was oxidized (Figure 4). In contrast to this, both layers remained black in the unplanted control.

The lower layer is densely rooted with horizontal rhizomes (Figure 5 and 6). Roots, if present, are unbranched or only slightly branched. Vertical rhizomes grow upwards from the end of horizontal rhizomes or from side branches (Figure 5 and 6). Their roots are more branched (Figure 6). Buds on vertical rhizomes can produce either vertical or horizontal rhizomes. The rhizomes in the upper layer are mainly vertical (Figure 5).

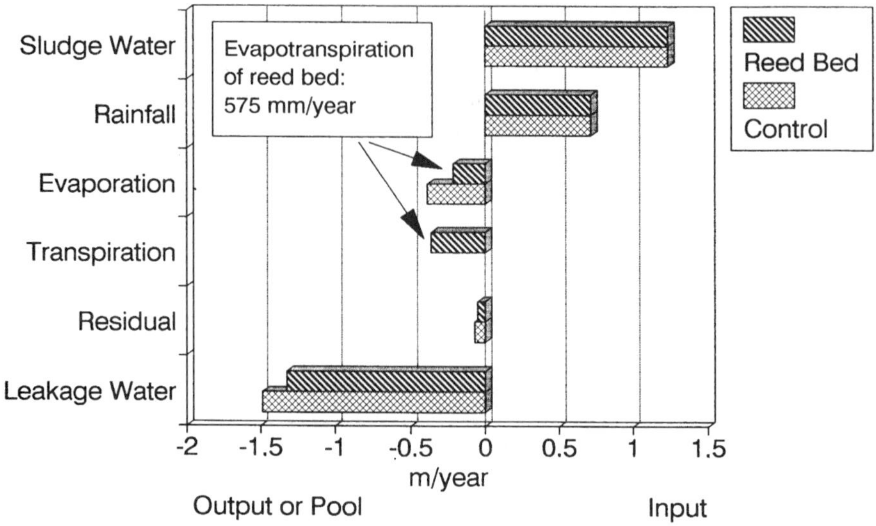

FIGURE 3. Water budget for the reed bed and the unplanted control (calculated from daily measurements of rainfall and leakage water and monthly measurements of residual).

FIGURE 4. The sewage sludge in the reed bed is converted to a 30 cm thick layer of "friable soil."

On reaching the surface, a vertical rhizome will continue growing as an aerial shoot. Reed continually extends its rhizome system into the sludge deposits and occupies up to 5% of the total volume (Figure 7).

Mechanical effects of extending rhizomes and roots in conjunction with plant transpiration removing water from the surrounding substratum enable the formation

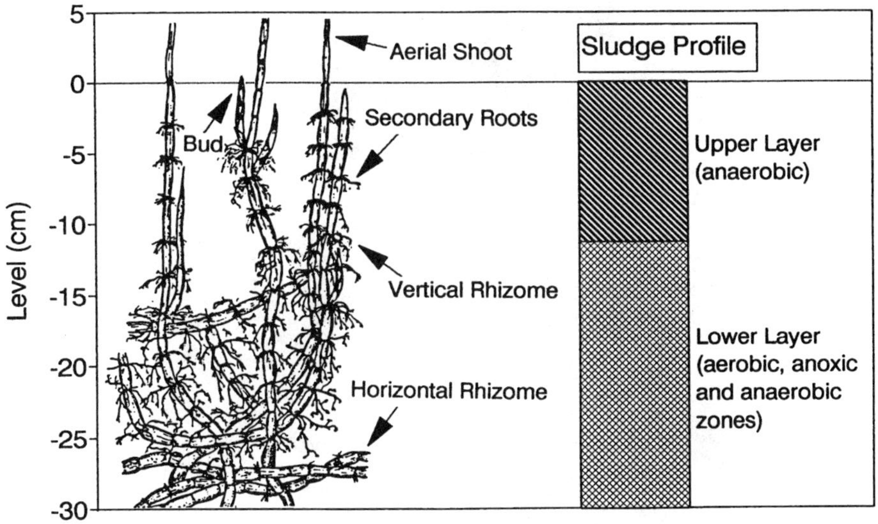

FIGURE 5. Rooting system of reed and soil profile in a sludge drying bed (Drawing: S. Weisshaar).

of air-filled pores. The amount of air pore space in a sewage sludge drying bed was nearly doubled by the plant-mediated processes (Figure 7).

ORGANIC MATTER DECOMPOSITION

The decomposition rate of organic matter increased from 30% (unplanted control) to 42% and reached 7.4 kg/m^2/year. The input of organic matter in the reed bed was about 30% higher because of the plant litter (Figure 8). The enhancement of mineralization processes in the reed bed results in significantly lower BOD and COD values of the leakage water compared to the unplanted control (Hofmann 1990).

In all probability, organic matter decomposition in the unplanted control is exclusively anaerobic. The comparative enhancement of decomposition in the reed bed can be interpreted, indeed, as a result of aerobic processes.

NITROGEN DYNAMICS

The nitrogen mass for the reed bed shows that there is a nitrogen removal from the system of 235 g N/m^2/year (Figure 9). However, in the unplanted control nitrogen removal is insignificant (15 g N/m^2/year). The reed bed can provide the necessary conditions for both nitrification and denitrification. In contrast to this, no nitrification at all takes place in the unplanted control.

Indeed, viable count of the autotrophic nitrifiers *Nitrosomonas* and *Nitrobacter* in the rhizosphere of the investigated sewage sludge drying bed was about 2 to 15 times higher than in the unplanted control (Figure 10). The number of heterotrophic nitrifiers was about 14 times higher. Heterotrophic nitrifiers, particularly cytophagas, coryneform bacteria, and pseudomonads, are likely contributing to nitrification processes.

FIGURE 6. Dense system of horizontal rhizomes in the lower layer of a sewage sludge drying bed after removing the sludge deposits (top) and a vertical rhizome from the upper layer with branched adventitious roots (bottom). The vertical rhizome continues growing as an aerial shoot (bottom, above).

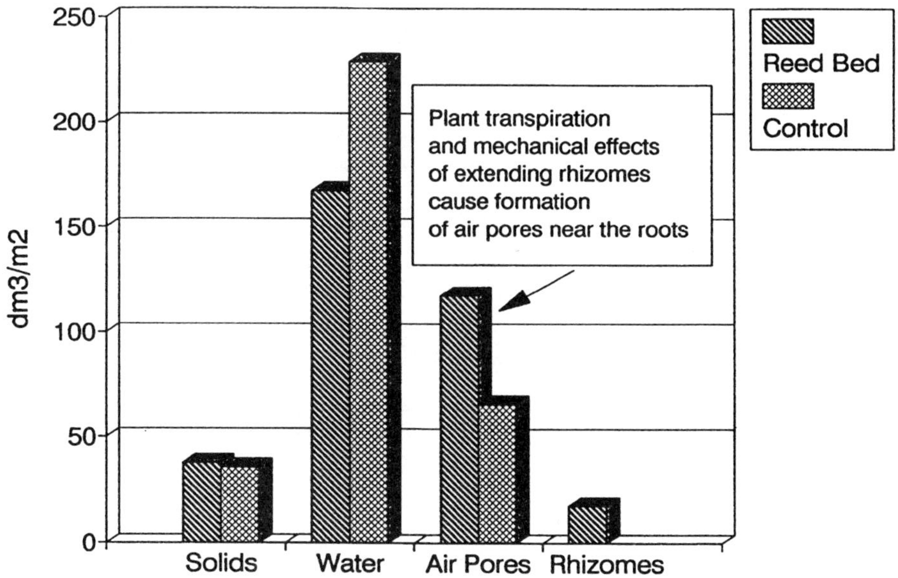

FIGURE 7. Volume composition of soil profile in sewage sludge drying beds.

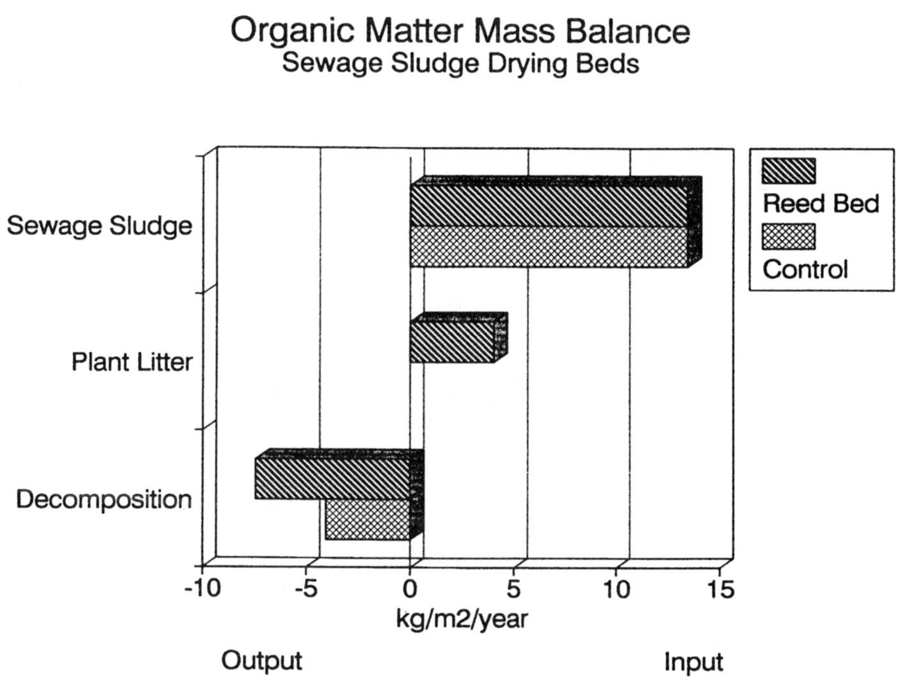

FIGURE 8. Organic matter mass balance of the sewage sludge drying beds (calculated from measurements of organic matter input via sewage sludge or plant litter, output via leakage water, and residual in the beds).

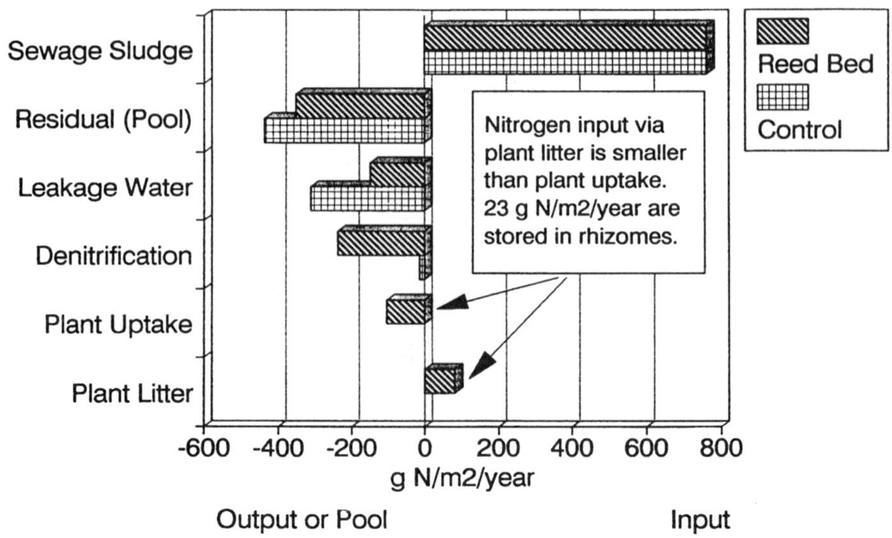

FIGURE 9. Nitrogen mass balance of the sewage sludge drying beds (calculated from measurements of nitrogen input via sewage sludge and plant litter, output via leakage water and plant uptake, and residual in the beds).

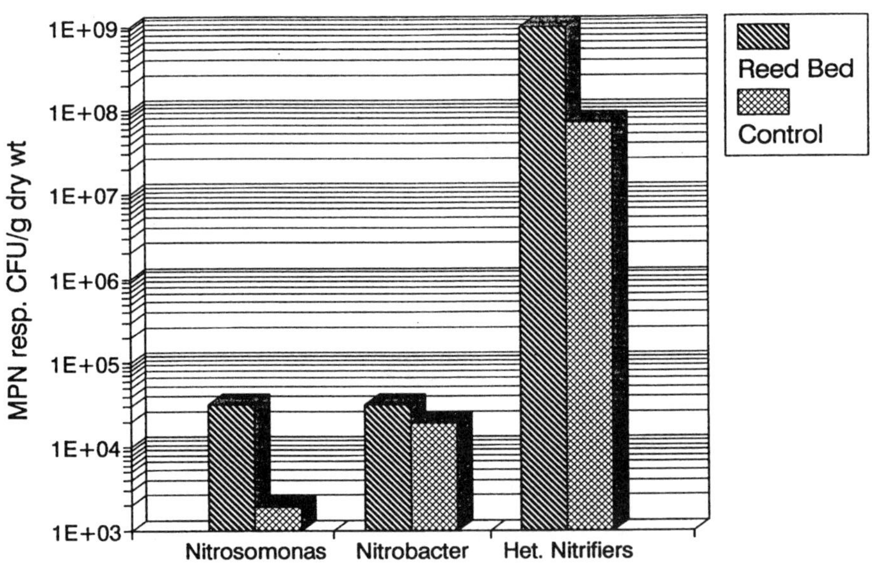

FIGURE 10. Concentrations of autotrophic and heterotrophic nitrifying bacteria in the reed bed's rhizosphere and the lower layer of the unplanted control.

The numbers of denitrifying bacteria increased from 2.6×10^8 (unplanted control, lower layer) to 2.3×10^9/g of sludge dry weight (reed bed, rhizosphere). *Pseudomonas*, coryneform bacteria (e.g., *Arthrobacter*), and *Cytophaga* were the predominant groups.

NUTRIENT STORAGE IN PLANTS

Reed stature and density depend obviously on nutrition. Part of the nutrients assimilated can be removed from the wetland system by winter harvesting. The effect of this on total nutrient removal is limited because the recharging of rhizomes with carbohydrates and plant nutrients is completed in summer and early autumn.

Productivity studies in reed-planted sewage sludge drying beds suggest that the growth of new rhizomes and the storage of reserve material in old rhizomes commences in late July (Hofmann 1986). This has consequences for the timing of the period in which the water table of a constructed wetland should be lowered to induce deepening of the actively growing rhizomes.

OXYGEN SUPPLY

It is well-known that oxygen passes from and through the emergent shoots of wetland plants into the submerged and buried organs. Oxygen has a two-fold importance. First, it is necessary for root respiration. The second requirement for an oxygen transport system in marsh plants is to oxidize the many toxic compounds in the rhizosphere. The soil chemistry of water-saturated solids includes a mixture of compounds like reduced iron or manganese, hydrogen sulphide, lower fatty acids, etc. Many or all of these substances can be toxic to plants.

Measurements of redox potential, nitrate, and other parameters in a reed-planted sewage sludge drying bed and its leakage water clearly demonstrate that the rhizosphere becomes an oxygen-rich microzone in a substratum which is otherwise anoxic (no free oxygen, but with nitrate present) or anaerobic (no free oxygen or nitrate). The leakage water of the reed bed contained significantly less ammonia and more nitrate that the unplanted control (Hofmann 1990). Measurements of redox potential indicate that the upper layer of sludge deposits is anoxic to anaerobic both in the reed bed and the unplanted control (–20 to –100 mV; Figure 11). Redox potentials of the reed bed's densely rooted lower layer reached +95 to +170 mV. In the lower layer of the control, however, we measured only –60 to +60 mV (Figure 11).

Microscopical analysis of bacterial flora shows that young secondary roots and fine laterals are likely to be the most important oxygen sources (Figure 12). Oxygen leaks out particularly from intercellular spaces between epidermal root cells (Figure 13).

Based on budgets of carbon and nitrogen, an oxygen mass balance was tentatively calculated for the reed-planted sewage sludge drying bed (Figure 14). The rate at which oxygen is supplied by *Phragmites* seems to be between 5 and 6 g O_2/m^2/day, on a yearly average.

MICROORGANISMS

The rhizosphere of the reed bed is characterized by increased concentrations of heterotrophic bacteria, diatoms, testate amoeba, and other microorganisms (Klee and Hofmann 1987). Bacterial numbers have been clearly shown to increase by up to 9.6×10^9 aerobic heterotrophic bacteria/g dry wt (Figure 15). Pseudomonads, coryneform bacteria, flavobacteria, and cytophagas were the most prevalent bacteria

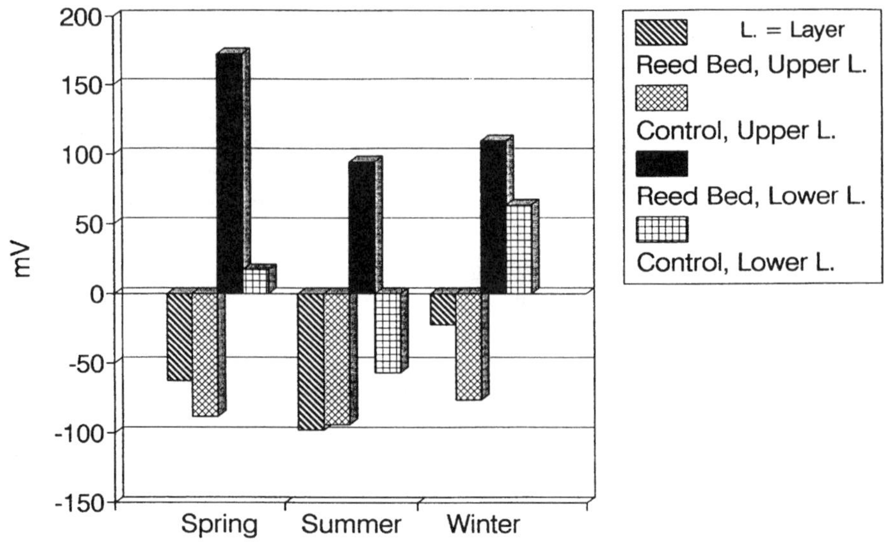

FIGURE 11. Redox potentials in the reed bed and the unplanted control (mean values).

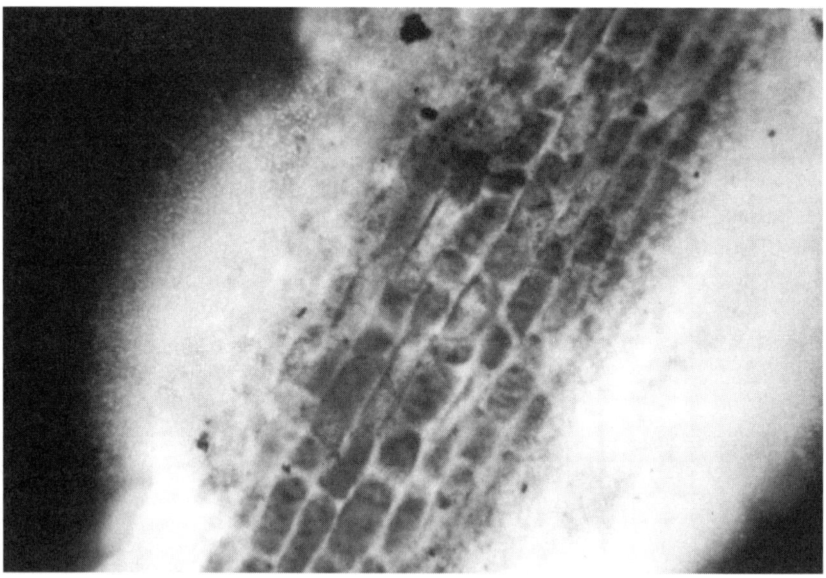

FIGURE 12. Dense "bacterial clouds" on a young secondary root (acridinorange stain).

found in the rhizosphere. Pseudomonads are widespread in nature. Because of their catabolic versatility, they play an important role in the mineralization of organic matter in nature. Bacteria of the genus *Arthrobacter* form a numerically important fraction of the indigenous bacterial flora of soils. They can use a wide and diverse

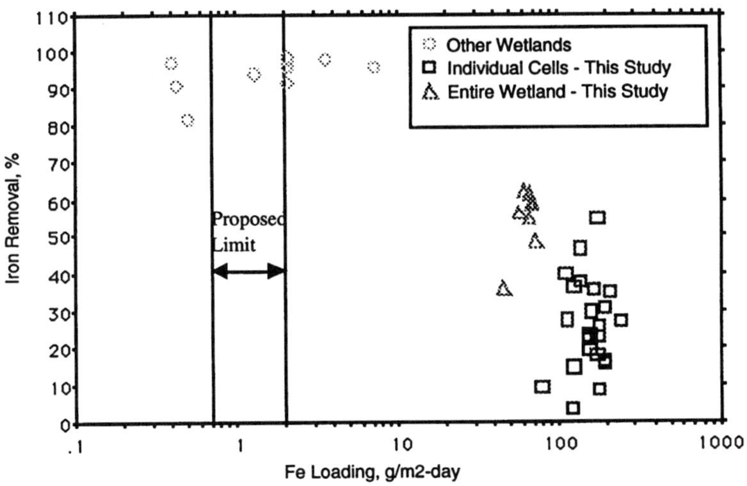

FIGURE 2. Iron removal as a function of iron loading for mine drainage wetlands (Fennessy and Mitsch 1989b). Data from Coshocton County, Ohio wetland and other wetland studies in Tennessee and Alabama.

loading rates (Figure 2). Preliminary ecological engineering design criteria for construction of these types of *Typha* wetlands for removal of iron have been developed (Fennessy and Mitsch, 1989b).

RIVER RESTORATION

The restoration of entire rivers has been shown to be an elusive goal in many parts of the world because of years of drainage and channel "improvement," floodplain development, and significant loads of sediments and other nonpoint pollutants. Even in pollution control approaches, we have paid too much attention to the stream itself and not enough to its interactions with its floodplain. One ecological engineering project on the Des Plaines River north of Chicago in Lake County, Illinois involves restoration of a length of a river floodplain and establishing experimental wetland basins (Figure 3). This project, begun in 1982 as the "Des Plaines River Wetland Demonstration Project" has as its goals "to demonstrate how wetlands can benefit society both environmentally and economically, and to establish design procedures, construction techniques, and management programs for restored wetlands" in a riverine setting (Hey et al., 1989). On the entire 182-ha site, woody and scrub vegetation has been replanted with native prairie species and oak savannas. Abandoned quarry lakes are connected to the river to give additional sediment trap efficiency and have been connected to the river as backwater habitats for fish and shorebirds. Some wetlands have been constructed with flap-gates to allow flooding by the river and trapping of water and sediments as the river recedes. Four experimental wetland basins (1.6 to 4.7 ha) have been constructed and instrumented at the northern half of the site for precise hydrologic control to achieve ecosystem experimentation conditions to investigate hydrologic design of wetlands subjected to high sediment loads. Early indications of the experimental wetlands after 1 year of study

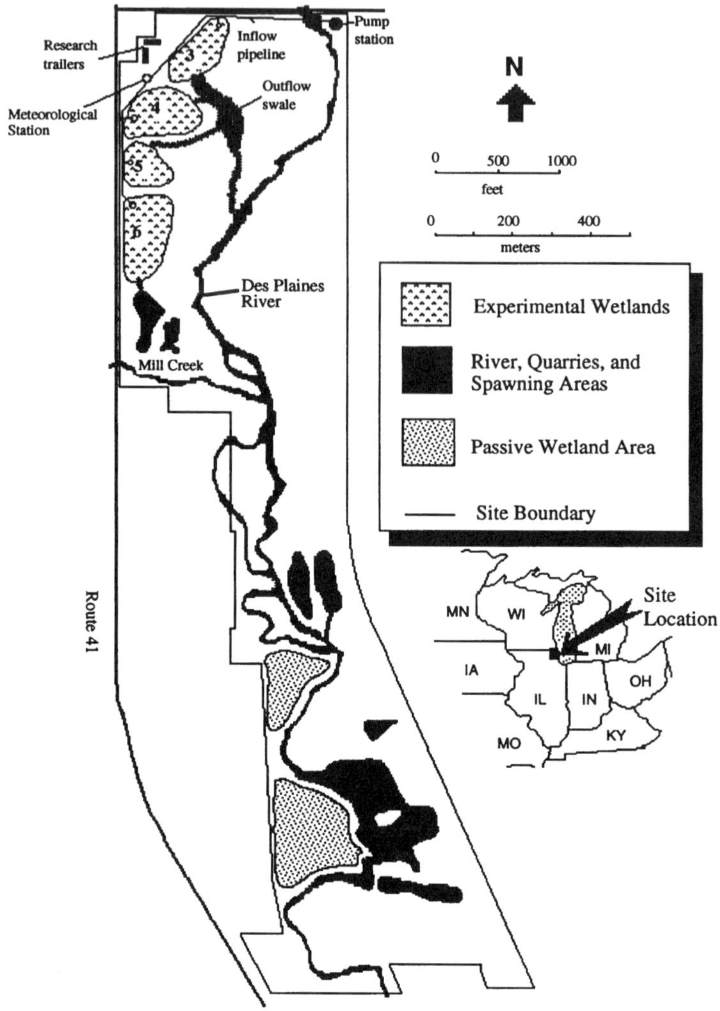

FIGURE 3. Des Plaines River Demonstration Project in Lake County, Illinois, showing four experimental wetland basins in the northern part of the site and passive wetlands and reclaimed quarries in the lower part of the site (Hey et al. 1989).

is that there is substantial productivity in all of the wetlands and most of the phosphorus and sediments are retained (Mitsch et al., 1990).

FISH PRODUCTION IN WETLAND SYSTEMS

The Lake Go Reed Wetland and Fish Farm, located in Yixing County, China integrates a *Phragmites* wetland with fisheries production. Water levels are controlled in such a way as to maximize production of herbivorous grass carp (*Ctenopharyngodon idella*) and Wuchang fish (*Megalobrama amblycephala*) as well as harvest of the reed grass (*Phragmites* sp.) for various uses, including fuel. This application

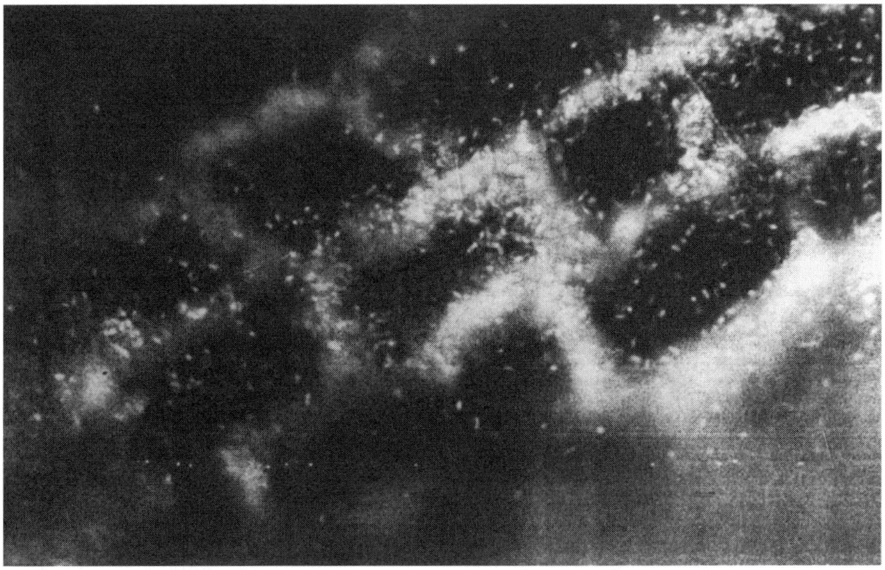

FIGURE 13. The pattern of bacterial population reflects the arrangement of epidermal root cells. Oxygen is likely leaking out between the cell walls (acridinorange stain).

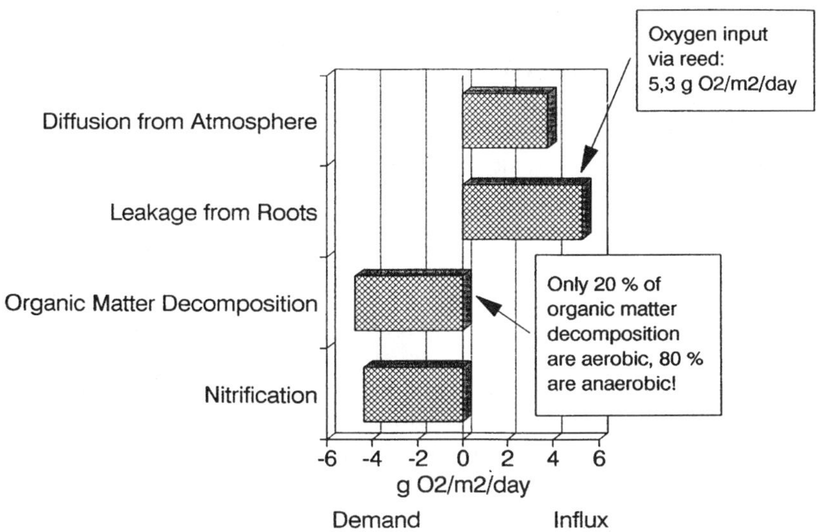

FIGURE 14. Tentative calculation of an oxygen mass balance for the reed bed.

range of organic substrates. Flavobacteria are widely distributed in nature and especially common in water. Cytophagas are responsible for the aerobic breakdown of cellulose in aquatic habitats. Quantitative and qualitative changes in the reed bed's bacterial flora may be largely a consequence of oxygen leakage rather than of root exudation.

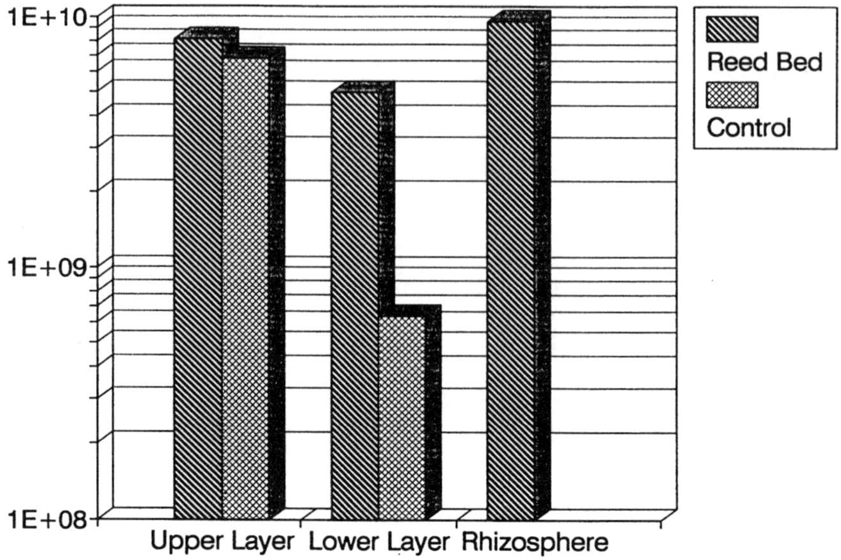

FIGURE 15. Concentrations of aerobic heterotrophic bacteria in the sludge profile (mean values from plate counts; see Hofmann 1990).

It is well known that wetlands offer a unique combination of physical, chemical, and biological factors which contribute to inactivation and removal of pathogenic bacteria. The increased removal of heterotrophic bacteria and enterobacteria in the leakage water of the reed bed (Figure 16) is mainly explained by the plant-mediated improvement of the physical filter. Enterobacteria and pseudomonads probably display an antagonistic behavior in the rhizosphere (Hofmann and Klee 1988). There was no evidence for a bactericidal effect of root exudates.

CONCLUSIONS

The described plant-mediated processes clearly demonstrate that the plants serve much more of a function than simply supplying a large surface area for microorganisms.

The roots certainly do promote sludge and water treatment beyond that expected of physical settlement. *Phragmites* is particularly good in stabilizing accumulating sediments on the surface, thereby providing infiltration pathways through the upper layers. Usually, the development of a dense and deep rhizome system does not create a high hydraulic conductivity as suggested by Kickuth (1981). The initial permeability of the filter, however, is maintained. Solids contained in the sewage gradually reduce the specific surface area in a constructed wetland. This is compensated by the extending root system which creates new surfaces.

Sand with a low clay content often is used in subsurface flow constructed wetlands to ensure good initial conductivity. Such media exhibit little or no aggregation. Now the plant litter introduces more stable organic compounds (e.g., cellulose),

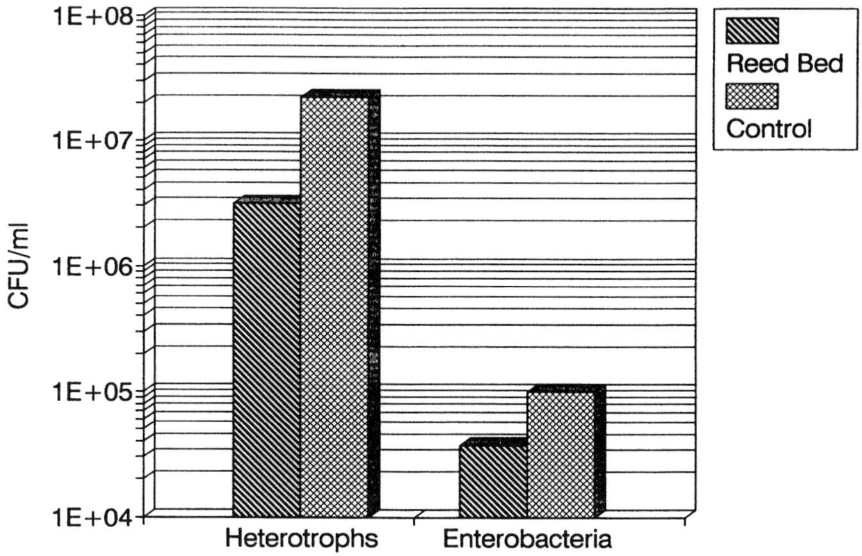

FIGURE 16. Concentrations of aerobic heterotrophic bacteria and enterobacteria in the leakage water of sewage sludge drying beds.

and therefore enhances humus formation. Cellulose and starch decomposing bacteria (*Cytophaga* spp.) have been found in large numbers in the rhizosphere of a sewage sludge drying bed planted with reed. It is possible that the cytophagas contribute to the stabilization of substratum structure due to their ability to produce slime which enables some kind of aggregation. Therefore, a yearly harvesting in constructed wetlands is not recommended.

The plant's ability to create oxidized zones in the substratum is the key mechanism for the success of water treatment in constructed wetlands. But its efficiency has been challenged by some scientists and engineers. The fact is that plant-mediated aerobic, anoxic, and anaerobic zones within the soil profile facilitate a sequence of mineralization, nitrification, and denitrification. In subsurface flow constructed wetlands with an intermittent flow regime, the plants are a link which makes it possible to combine the benefits of an aerobic biological filter with the traditional concept of helophyte wetland. Therefore, constructed wetlands of this type do not mimic natural wetlands. They are, indeed, hybrid systems.

Conventional sewage purification plants are being employed in the industrialized countries to help control municipal and industrial sources of water pollution. In the future it is clear that additional technologies will be needed to upgrade many of these treatment facilities while more attention will need to be given to controlling many of the non-point sources of water pollution as well, if any drastic improvements in water quality are to be realized. Wetlands act as a buffer between terrestrial and aquatic ecosystems. Therefore, the use of constructed wetlands in water pollution control can contribute in a significant manner to a more ecological approach to wastewater treatment.

REFERENCES

Cooper, P.F. and B.C. Findlater. (eds.). 1990. *Constructed wetlands in water pollution control.* Proceedings of the international conference on the use of constructed wetlands in water pollution control, held in Cambridge, UK, 24-28 September 1990. Oxford: Pergamon Press.

Hofmann, K. 1986. "Wachstumsverhalten von Schilf (*Phragmites australis* [Cav.] TRIN. ex STEUDEL) in klärschlammbeschickten Filterbeeten." *Arch. Hydrobiol.* 107:385-409.

Hofmann, K. 1990. "Use of *Phragmites* in sewage sludge treatment." In: Cooper and Findlater (1990), 269-277.

Hofmann, K. and O. Klee. 1988. *Untersuchungen über den Einsatz von Pflanzen zur Klärschlamm-Entwässerung. Teilvorhaben 3: Biologisch-bakteriologische Untersuchungen.* UFOPLAN Abfall-wirtschaft, Forschungsbericht 1031-1227.

Kickuth, R. 1981. Abwasserreinigung in Mosaikmatrizen aus aeroben und anaeroben Teilbezirken. In: F. Moser, Hg., *Grundlagen der Abwasserreinigung,* Oldenbourg, München Wien, pp. 639-665.

Klee, O. and K. Hofmann. 1987. "Untersuchungen über den Einsatz von Pflanzen zur Klärschlamm-Entwässerung. Teil II: Biologisch-bakteriologische Untersuchungen." *Korr. Abw.* 34:1197-1203.

Chapter 18

WASTEWATER NUTRIENT REMOVAL AND RECOVERY IN AN INFILTRATION WETLAND

Karin Sundblad and Hans-B. Wittgren

ABSTRACT

The efficiency of an infiltration wetland in removing nitrogen, phosphorus, and organic matter from wastewater was studied for a period of 3 years in field lysimeters (3 × 4 × 1 m) with reed sweetgrass, *Glyceria maxima*. The potential for recovering wastewater nutrients in harvested biomass was also investigated. The focus of the experiments was the effects of various application frequencies, with a constant load of 140 mm/week, on nutrient recovery and treatment efficiency.

The maximum amount of P and N removed with the harvested biomass was 4.8 and 32 g/m^2, respectively. The relative removal during one growing season varied from 12 to 28% of the amount applied for P and from 26 to 55% for N. Over the whole study period, the removal efficiency for P was more than 99%, and 86 to 88% of this was retained in the soil. For N, the removal amounted to 57 to 76% and was related to observed decreases in soil permeability and to temperature variations. Plant uptake and gaseous losses were found to be of about the same magnitude. It was concluded that alteration of application frequency is an important means of manipulating treatment efficiency, and that the treatment efficiency with regard to P met the Swedish requirements for tertiary treatment.

INTRODUCTION

Constructed wetlands with emergent aquatic macrophytes have been suggested to be suitable for low-cost wastewater treatment in regions with a temperate climate (Lakshman 1979; Seidel 1976; Wolverton 1987). In recent years, many reports have presented results from systems with a horizontal flow of wastewater through sand or gravel, or in a few cases even fine soils (e.g., Brix 1987; Bucksteeg et al. 1985, Cooper and Findlater 1990, Hammer 1989). In contrast, the results summarized in the present paper were obtained in field experiments with a constructed wetland designed to allow *infiltration* of the applied wastewater (Wittgren and Sundblad 1987). This system can be viewed as an intermediate between a wetland and a slow-rate land-treatment system, i.e., an infiltration wetland. The experiments had two equally important aims: to study the potential for recovering wastewater nutrients in plant biomass and to investigate the capacity of the system to treat wastewater.

In order to produce a potentially valuable crop, reed sweetgrass *Glyceria maxima* (Hartm.) Holmb., which has been used as fodder in several countries, was grown in

0-87371-990-5/97/$0.00+$.50

the infiltration wetland (Lambert 1947). The concept of recycling wastewater nutrients is promising because it includes the potential of producing high yields of biomass. However, in a temperate region like Sweden the mean temperature is close to or below 0°C for several months of the year. During this colder period, there will be no plant growth, and all nutrient removal will be a result of physical, chemical, and microbiological processes in the soil substrate.

Soil redox status is of considerable importance for plant growth and nutrient availability, as well as for other soil processes that affect wastewater treatment. For example, alternating submerged and dry conditions have been shown to be important for the extent of N losses through denitrification from soils (Patrick and Wyatt 1964, Reddy and Patrick 1984). To evaluate the potential to, in particular, increase the N-removal capacity in the infiltration wetland, we studied the effects of different application frequencies. It was hypothesized that a low frequency would promote alternating aerobic and anaerobic conditions in the soil, and thereby increase gaseous losses of N. Furthermore, the treatment result with regard to P is also reported.

METHODS

EXPERIMENTAL DESIGN

This study was performed in four field lysimeters (3 by 4 m) near Linköping, Sweden (15°40' E, 58°25' N). At this location, the mean annual temperature is 6°C, and the mean temperatures of the coldest (February) and warmest (July) months are –4°C and 17°C, respectively. The mean annual precipitation is 510 mm, and the mean annual potential evapotranspiration is 580 mm (statistics from Swedish Meteorological and Hydrological Institute, Norrköping, Sweden). The field lysimeters, previously described by Wittgren and Sundblad (1987), were filled to a depth of 0.75 m with loamy sand (3% clay, 13% silt, 59% sand and coarser material). Reed sweetgrass, *Glyceria maxima*, was planted in the lysimeters in the summer of 1984. Starting in June 1985, the grass was harvested twice per growing season, and each time the harvested biomass and its nutrient content was measured. Starting in March 1985, domestic wastewater was applied through pipes to the soil surface; the wastewater was pretreated in a facultative pond during 1985. From 1986, a smaller pond was used for pretreatment, which resulted in higher N concentrations in the applied wastewater (Table 1).

The lysimeters received wastewater (140 mm/wk), with different application frequencies, from the middle of May to October in 1985 and from the middle of April to August in 1986 (Table 2). This schedule resulted in ponding during part of each week in lysimeters III and IV. During periods of low temperature, the load was reduced to ensure that all the wastewater applied would infiltrate. In 1987, the application frequency for lysimeter I was changed to 70 mm twice a week to study how this would influence soil permeability and treatment results. In the same year, due to a reduced soil permeability, lysimeters II to IV received a reduced load (Table 2). From April to August 1986, lysimeter IV was not used because of a leak, which was later repaired.

Details of water flow measurements, water and soil sampling, harvesting, and analytical methods have been described elsewhere (Wittgren and Sundblad 1987,

TABLE 1
Average Composition of Wastewater Applied to Field Lysimeters with *G. maxima* during 1985 (April to October), 1986 (April to August), and 1987 (June to October)[a]

	1985	1986	1987
pH	7.9 (0.06)	7.4 (0.09)	7.3 (0.03)
Conductivity (mS cm^{-1})	0.94 (0.02)	1.2 (0.05)	1.2 (0.05)
COD (mg 0_2 L^{-1})	274 (14)	305 (22)	222 (9.4)
BOD_7 (mg 0_2 L^{-1})	130 (22)	120 (24)	108 (6.2)
Total–N (mg L^{-1})	29 (1.2)	41 (1.6)	39 (1.6)
NH_4–N (mg L^{-1})	17 (1.5)	32 (1.4)	32 (1.6)
NO3–N (mg L^{-1})	<0.5	<0.5	<0.5
Total–P (mg L^{-1})	10 (0.36)	12 (0.44)	11 (0.85)

[a] S.E. are shown in brackets.

TABLE 2
Wastewater Application Schedules for Field Lysimeters with *G. maxima*[a]

	Lysimeter Load mm wk^{-1}			
Time period	**I**	**II**	**III**	**IV**
		(d)		
Apr. 85–Oct. 85	14 × 10	7 × 20	2 × 70	1 × 140[b]
Apr. 86–Aug. 86	(0)	(0)	(1–3)	(2–3)
Jan. 86–Mar. 86	7 × 10	7 × 10	2 × 35	2 × 35
Nov. 86–Dec. 96				
Jun. 87–Oct. 87	2 × 70	5 × 20	1 × 60	1 × 60
	(0.5)	(0.5–1)	(3–4)	(3–4)

[a] Values represent the number of applications times amount applied and (in parentheses) approximate duration of ponding after each application, in days.
[b] Reduced application during Apr. 85 to Oct. 85 and no application during Apr. 86 to Aug. 86 due to leak.

Sundblad and Wittgren 1989). The general sampling programs for water and soil analyses are outlined below.

WATER ANALYSES

Throughout the study period, weekly wastewater and leachate samples were analysed for pH, conductivity, total N, NH_4-N, NO_3-N, NO_2-N, and BOD_7 (only in wastewater). Starting in 1986, the sampling frequency was reduced to every second week for total-P and BOD_7 in leachate and COD in wastewater. Weekly samples of leachate were pooled in proportion to the leachate flow to obtain one sample per month for analysis of COD during 1987. Weeks without analyses were assigned mean values of the concentrations in the previous and following week. In the present paper, only results concerning N and P are reported.

SOIL ANALYSES

To study accumulation of organic matter and N, soil samples were collected from each lysimeter in May and November 1985, April and November 1986, and November 1987. Five soil samples from each level (0 to 15, 15 to 30 and 30 to 45 cm) were analyzed for Kjeldahl-N and loss on ignition (600°C, 2 h). In 1986 and 1987 the upper 15 cm were analysed in 5-cm sections.

RESULTS AND DISCUSSION

NUTRIENT RECOVERY AND BIOMASS PRODUCTION

The following is a summary of the results of biomass production and nutrient uptake previously presented by Sundblad and Wittgren (1989). A considerable number of terrestrial plant species competed successfully with *G. maxima* in the non-ponded lysimeters I and II, whereas pure stands of *G. maxima* were formed in the temporarily flooded lysimeters III and IV. In the latter, the yearly total yields were 870 to 1165 g/m^2, and we concluded that the highest yields were about optimal for the prevailing climatic conditions.

Uptake of N and P by *G. maxima* in the ponded lysimeters amounted to a yearly maximum of 32.1 g/m^2 of N and 4.8 g/m^2 of P for the ponded lysimeters (Table 3). Plant uptake was lower in 1987, when the load had been reduced, whereas the considerably increased N-P load in lysimeter III during 1986 had no significant effect on the amount removed with the crop. Hence, it can be concluded that the highest nutrient removal by the harvested biomass was about maximum for this climatic region. Uptake of N was lower in lysimeter IV than in lysimeter III during the first year, and we assume that this reflects the effects of ponding on soil redox conditions and thereby on nutrient availability to plants (Wittgren and Sundblad 1987). Thus, to obtain maximum yields and nutrient recovery in biomass, it is important to consider both total load of nutrients and application frequency.

By relating the amount of N and P applied during the vegetation period to the amount of nutrients removed with the crop, the efficiency exhibited by the system in recirculating nutrients for human utilization can be estimated. A maximum of 55% of N and 28% of P were removed with the crop (Table 3). Crop uptake of N was highest in 1986, when the load was increased, but the relative removal during this year was comparatively much lower than during the other years. The proportion of P removed with the crop might be increased by decreasing the load, in which case, depending on the N:P ratio of the wastewater, an additional amount of N may have to be added to achieve maximum biomass yield and nutrient uptake. Applying wastewater during the entire year will further reduce the relative amount of N and P recycled, since there is no uptake by plants during several months, as will be discussed below.

TABLE 3
Nitrogen and Phosphorus Uptake of *G. maxima* in Field Lysimeters Relative to Amounts Applied with Wastewater from April 15 to October 1, 1985 to 1987

	Nitrogen			Phophorus		
Lysimeter and year	Applied ($g\ m^{-2}$)	Crop uptake ($g\ m^{-2}$)	Relative removal (%)	Applied ($g\ m^{-2}$)	Crop uptake ($g\ m^{-2}$)	Relative removal (%)
Lysimeter III						
Year 1	52.6	29.4	55	20.8	4.8	23
Year 2	122	32.1	26	36.9	4.6	12
Year 3[a]	37.2	20.5	55	11.3	3.0	26
Lysimeter IV[b]						
Year 1	47.0	21.5	46	17.8	3.2	18
Year 2	78.3	27.6	35	24.6	4.2	17
Year 3	36.8	19.8	54	11.2	3.1	28

Note: Wastewater was applied on 2 (III) and 1 (IV) occasion(s) per week. Nutrients applied after the second harvest have been considered part of the next growing season.

[a] One application per week in year 3 due to decreased total load.
[b] Due to a leak in years 1 and 2, this lysimeter received a lower total load. Estimated from data on water balance (Wittgren and Sundblad. 1987).

From Sundblad and Wittgren, 1989.

NITROGEN REMOVAL AND TREATMENT EFFICIENCY

The N removal efficiency for the whole experimental period (1985 to 1987) was 57, 69, and 76% for lysimeters I, II, and III respectively (Table 4). No balance could be calculated for lysimeter IV due to lack of data for 1986 (a leak was discovered and repaired).

In 1985 the concentrations of total-N in leachate were closely related to the application frequency (Wittgren and Sundblad 1987): lower concentrations accompanied a lower number of applications per week. The present results show that this observed relationship persisted during the following years (Table 4). The cause of this was the interactive effect of application frequency and decreasing soil permeability on N transformation processes, as will be discussed in a forthcoming paper. Decreasing soil permeability gradually increased the period of soil submergence after each application of wastewater (Table 2). Consequently, more reduced conditions were created in the soil, and this would have favoured gaseous losses of N by denitrification. Rapid nitrification of the applied NH^+_4-N could occur in the surface soil during non-ponded conditions, as indicated by the potential ammonium oxidation rate in lysimeter III, which was as high as in the predominantly oxidized soil of lysimeter I (unpublished result). A discussion of the nitrogen transformations in the water and topsoil of the lysimeters has been presented elsewhere (Wittgren and Sundblad 1990).

TABLE 4
Total Amount of N Applied, Leached, Removed with Harvested Biomass, and Accumulated in Soil in Lysimeters I to III after Wastewater Application during a Total of 22 Months from April 1985 to October 1987

	Lysimeter		
	I	II	III
		$g\ m^{-2}$	
Applied	374	350	313
Harvested			
G. maxima	53	62	82
Other species	34	29	3
Accumulated in soil	44	13	26
Leached	163	110	76
		% of applied	
Removal efficiency	57	69	76
Not accounted for	21	38	40

Nitrogen not accounted for by harvest, soil accumulation or leaching amounted to 77, 130 and 121 g/m^2 in lysimeters I, II, and III, respectively (Table 4), or, in other words, between 21 and 40% of the amount of N applied. In calculating these values, soil accumulation of N was estimated for the top 15 cm, using data from May 1985 and November 1987. Some of the N not accounted for can be attributed to storage in below-ground plant biomass. Estimates of such biomass usually vary considerably, but in a natural stand of *G. maxima* the permanent pool of N in below-ground biomass was estimated to be 33 g/m^2 (Ulehlová et al. 1976). Assuming this amount was stored, 43, 97, and 88 g/m^2 in lysimeters I–III, respectively, were attributed to ammonia volatilization or denitrification (Wittgren and Sundblad 1990). These latter values are of the same magnitude as amounts removed by the harvested plant biomass.

The total N-removal efficiency of 57 to 76% (Table 4) agrees with reports from Danish full-scale horizontal-flow constructed wetlands receiving wastewater load similar to that in our experiments (Brix 1987). This load corresponded to an areal demand of about 20 m^2/p.e. Danish plants receiving 3 to 10 times this load had N removal efficiencies ranging from 10 to 30%, mainly because a large part of the wastewater spread on the surface instead of passing through the root-zone. However, it must be pointed out that the yearly N removal efficiency in the infiltration wetland was somewhat lower than that presented in Table 4, since only one period with low temperatures was included in our calculation.

During the period when the average temperature of the leachate was 2°C (Jan to March 1986), 50% of the amount of N applied was removed (Table 5). High concentrations of NO_3^--N were found in the leachates from all the lysimeters, whereas concentrations of NH_4^+-N were low. Consequently, most of the applied NH_4^+-N must have been nitrified. The rate of nitrification was apparently higher than the denitrification rate. A possible explanation for this is the low temperature, even though the effect of low temperatures on these processes is not fully understood. Focht and

TABLE 5
Average Concentrations (S.E. in Brackets) of BOD_7, N, and P in Wastewater Applied and in Leachate from Lysimeters I–IV during the Winter Season (January–March) of 1986. Removal Efficiencies Were Calculated from the Accumulated Amounts Applied and Leached

	Wastewater		Lysimeter leachate		Removal efficiency
Parameter	n	value	n	value	(%)
BOD_7 (mg O_2 L^{-1})	7	131 (9.4)	28	3.1 (0.24)	98
Total–N (mg L^{-1})	11	51 (1.4)	24	25 (1.3)	50
NH_4–N (mg L^{-1})	11	48 (1.3)	44	1.4 (0.10)	
NO_3 + NO2–N (mg L^{-1})	11	<0.5	44	22 (0.93)	
Total–P (mg L^{-1})	8	13 (0.46)	20	0.04 (0.01)	100

Verstraete (1977) cited reports showing that autotrophic nitrifying bacteria become acclimated to the temperature regime of their habitat. A recent investigation of nitrification in Swedish agricultural soils showed that a Q_{10} of about 1.7 was valid down to 2°C (Berg 1986). Malhi and McGill (1984) suggested that nitrification may proceed even under snow cover, which is supported by our results. On the other hand, for denitrification most reports show a profound break in the rate curve below 12 to 15°C with higher Q_{10} values at lower temperatures (Focht and Verstraete 1977). Reported minimum temperatures for the occurrence of denitrificaton in soil range from 2.7 to 10°C (Firestone 1982). Mean N-removal efficiencies ranging from 15 to 85% have been reported for a winter period in constructed wetlands in Denmark (Brix 1987). This was, however, attributed to adsorption to soil particles rather than to nitrification/denitrification.

PHOSPHORUS REMOVAL AND TREATMENT EFFICIENCY

Leachate concentrations of total-P were generally below 0.1 mg/l, even during the cold season (Table 5). Considering the entire experimental period, more than 99% of the amount of P applied was removed in lysimeters I to III (Table 6).

Phosphorus in harvested biomass accounted for 11 to 13% of that applied, which leaves 86 to 88% that must have been retained in the soil in organic or inorganic form. This means that soil organic and inorganic P content had increased by 80 to 100 g/m^2. The importance of soil processes for P removal has also been demonstrated in constructed wetlands with horizontal water flow: removal efficiencies were as low as 11% when wastewater did not move through the soil (Brix 1987).

Most soils have rather high capacities to retain P; these are determined by the net effect of physical, chemical and microbial processes involved in P transformations. The retention generally involves rapid adsorption processes and slower chemical reactions leading to the formation of a solid phosphate phase (Beek and van Riemsdijk 1979). Soil P retention capacity is of crucial importance for the longevity of a constructed wetland for wastewater treatment and is influenced by several factors, such as chemical composition of the soil, pH, oxidation-reduction reactions, and the characteristics of water flow.

TABLE 6
Total Amount of P Applied, Leached, and Removed with Harvested Biomass in Lysimeters I to III after Wastewater Application from April 1985 to October 1987

	Lysimeter		
	I	II	III
		$g\ m^{-2}$	
Applied	115	109	97
Harvested			
G. maxima	8	10	12
Other species	5	5	0.6
Leached	1	0.5	0.8
		% of applied	
Removal efficiency	99	100	99

CONCLUSIONS

The present study shows that an infiltration wetland represents an efficient method for removing and recycling nutrients from wastewater. Using this method, the Swedish tertiary requirement for phosphorus removal was reached with no dependence on temperature variations. Significant removal (50%) of nitrogen occurred even when water temperature was close to zero. During periods of low temperatures, the nitrogen leached out as nitrate, which implies that the rate of nitrification did not limit the nitrogen removal capacity of the infiltration wetland.

On a yearly basis, uptake of nitrogen by plants may be of the same magnitude as gaseous losses of nitrogen. However, for the entire experimental period plant uptake accounted for less than 15% of the applied P and 23 to 27% of the applied N, which emphasizes the crucial role of soil chemical and microbiological reactions in the treatment result. Scheduling the application of wastewater provides a means of increasing the efficiency of such processes, as was demonstrated by the differences between lysimeters with regard to N-removal: 57 to 76%.

REFERENCES

Beek, J. and W. H. van Riemsdijk. 1979. "Interaction of ortophosphate ions with soil." In: G. H. Bolt (ed.). *Soil chemistry.* B. Physico-chemical models, Devel. Soil Science 5B. Elsevier Scient. Publ., Amsterdam, pp. 259-284.

Berg, P. 1986. "Nitrifier populations and nitrification rates in agricultural soil." Ph.D. dissertation, Report 34, Dep. of Microb., Swedish Univ. Agric. Sci., Uppsala.

Brix, H. 1987. "Treatment of wastewater in the rhizosphere of wetland plants: The root-zone method." *Water Sci. Tech.*, 19(Rio):107-118.

Bucksteeg, K., H.H. Müller, R. Mörtl, H. Adelt, G. Scheer, W. Labitzky, H. Sendl, O. Hörmann, and P. Kästner. 1985. "Erste Erfahrungen mit 12 Sumpfpflanzenkläranlagen. *Korrespondenz Abwasser*, 5:376-385.

Cooper, P. F. and B.C. Findlater (eds.) 1990. *Constructed wetlands in water pollution control.* Proceedings from the international conference in Cambridge, UK, 24-28 September 1990. Advances in Water Pollution Control, Pergamon Press, Oxford. 605 pp.

Firestone, M. K. 1982. "Biological Denitrification." In: F. J. Stevenson (ed.). *Nitrogen in agricultural soils.* Agronomy, 22. ASA, CSSA, SSSA, Madison, WI, pp. 289-326.

Focht, D. D. and W. Verstraete. 1977. "Biochemical ecology of nitrification and denitrification." *Adv. Microbial Ecol.* 1:135-214.

Hammer, D. A. (ed.) 1989. *Constructed wetlands for wastewater treatment.* Proceedings from the first international conference in Chattanooga, Tennessee, 13-17 June, 1988. Lewis Publishers, Inc., Michigan. 831 pp.

Lambert, J. M. 1947. "Biological flora of the British Isles. *Glyceria maxima.*" *J. Ecol.* 34:310-344.

Lakshman, G. 1979. "An ecosystem approach to the treatment of wastewater." *J. Environ. Qual.* 8:353-361.

Malhi, S. S. and W.B. McGill. 1982. "Nitrification in three Alberta soils: Effect of temperature, moisture and substrate concentration." *Soil Biol. Biochem.* 14:393-399.

Patrick, W.H. and R. Wyatt. 1964. "Soil nitrogen loss as a result of alternate submergence and drying." *Soil Sci. Soc. Am. Proc.* 28:647-653.

Reddy, K. R. and and W.H. Patrick Jr. 1984. "Nitrogen transformations and loss in flooded soils and sediment." *CRC Critical Reviews in Environmental Control* 13(4):273-309.

Seidel, K. 1976. "Macrophytes and water purification." In: J. Tourbier and R. W. Pierson, Jr. (eds.). *Biological control of water pollution.* University of Pennsylvania Press, Philadelphia, PA, pp. 109-121.

Sundblad, K. and H.-B. Wittgren. 1989. "*Glyceria maxima* for wastewater nutrient removal and forage production." *Biol. Wastes* 27:29-42.

Ulehlová, B, E. Klimo, and J. Jakrlová. 1976. "Mineral cycling in alluvial forest and meadow ecosystems in southern Moravia, Czechoslovakia." *Int. J. Ecol. Environ. Sci.* 2:15-25.

Wittgren, H. B. and K. Sundblad. 1990. "Removal of wastewater nitrogen in an in-filtration wetland with *Glyceria maxima.*" In: Cooper, P. F. and Findlater, B. C. (eds.). *Constructed wetlands in water pollution control.* Proceedings from the international conference in Cambridge, UK, 24-28 September 1990. Advances in Water Pollution Control, Pergamon Press, Oxford. pp. 85-96.

Wittgren, H.-B. and K. Sundblad. 1987. "Wastewater treatment in a wetland filter: Effects of varying application frequency on nitrogen removal." In: K. R. Reddy and W. H. Smith (eds.). *Aquatic plants for water treatment and resource recovery.* Magnolia Publ. Inc., Orlando, FL, pp. 513-523.

Wolverton, B. C. 1987. "Aquatic plants for wastewater treatment: An overview." In: K. R. Reddy and W. H. Smith (eds.). *Aquatic plants for water treatment and resource recovery.* Magnolia Publ. Inc., Orlando, FL, pp. 3-15.

Chapter 19

SEASONAL VARIATIONS IN THE EFFICIENCY OF NUTRIENT REMOVAL FROM DOMESTIC EFFLUENT IN A QUASI-NATURAL FIELD OF REEDS (*PHRAGMITES AUSTRALIS*)

Hanna Obarska-Pempkowiak

ABSTRACT

Municipal liquid wastes from the small seaside resort Frombork (in northeast Poland) after primary treatment in a storage Imhoff tank are treated in a natural reed wetland and then directed to the sea. In 1990 measurements were carried out aimed at establishing the efficiency of organic matter and nutrient removal in the reed field. Average percentages of removal were as follows:

• BOD_5	Average	57.5	Winter	35.9	Summer	81.0
• COD	Average	59.7	Winter	39.6	Summer	76.1
• N (total)	Average	26.3	Winter	20.6	Summer	32.0
• P (phosphate)	Average	2.7	Winter	–2.1	Summer	8.4

A clear seasonal dependence was established for all four listed contaminants.

The mass balance of nutrients indicated that just over 9% N and 3% P were converted into reed biomass. Moreover, these amounts of nutrients are returned to the system in fall and winter. Therefore, removal of nutrients due to assimilation by reeds can be deemed insignificant. The reduction of nitrogen is most likely caused by denitrification, sorption playing minor role because of great average depth (~ 25 cm) of sewage in the field.

Effective removal of organic matter is attributed to long storage time of sewage in the system. However, even this can be improved if sewage is stopped from winding its way across the system and spread more evenly.

INTRODUCTION

Liquid wastes produced by small plants located in villages and small towns as well as domestic sewage generated there constitute a small proportion of the total amount of wastewater produced in Poland. For instance, in the Gdansk Region small industrial enterprises produce 0.8% of the total amount of wastewater, although loads of pollutants they introduce to surface waters are equal to 2.3% BOD, 2.6% COD, and 3.9% suspended solids (Obarska-Pempkowiak and Kowalik, 1989). It follows from the figures that contaminants in this type of sewage are more concentrated than

0-87371-990-5/97/$0.00+$.50

in other ones. Thus, the sewage is potentially more dangerous to the environment. The situation is caused by inadequate purification resulting from the poor maintenance, or lack thereof, at purification plants. Unsteady inflow of sewage, causing frequent hydraulic overloading of facilities, contributes to the problem. Moreover, there exists a shortage of specialists capable of proper maintenance of conventional sewage purification plants. It seems that the so-called alternative methods of purification, which have gained popularity around the world (Brix 1987, Bucksteeg 1989, Cooper and Boon 1987, Nichols 1983), often could be used instead of the conventional ones. Due to reduced requirements of manpower and energy as well as ease of installation, they are best suited for the countryside. Unfortunately very few of such facilities have been in operation in Poland yet, mostly due to the lack of both basic knowledge on the subject and proper guidelines how to start them.

The purpose of this work is an analysis of a year-round operation of a quasi-natural field of reeds irrigated with domestic wastewater after primary treatment. The facility is located in Frombork on the coast of the Vistula Lagoon. The analysis is based on year-round measurements of efficiencies of pollutant removal from wastewater flooding the field.

OUTLINE OF THE SEWAGE PURIFICATION SYSTEM IN FROMBORK

Municipal wastewater from the town of Frombork is collected in sewerage and brought to an Imhoff tank situated at the outskirts of the town some 30 m from the shore of the Vistula Lagoon (Figure 1). The flow of sewage averages 850 m^3/day, and the retention time in the tank is some 2 hours. The collected sewage is domestic, characterized by concentrations of organic substances and nutrients in the following ranges: BOD_5 from 147 to 246 g O_2/m^3, COD from 388 to 459 g O_2/m^3, suspended solids from 201 to 513 g/m^3, pH from 6.8 to 8.0. The removal of organic substances from sewage in the tank is in the range from 20% to 30%. Therefore, there is no doubt that inflow of poorly purified sewage was contributing substantially to pollution of the Vistula Lagoon.

Starting in 1985, sewage has been directed from the Imhoff tank to a field of reeds separated with artificial dams from reeds naturally growing along the coast. The sewage is drained to the field through a 200-mm pipe at a point marked in Figure 1. Then it flows across the field. After passing through the field, the sewage is directed to the Vistula Lagoon via a ditch running along the northeast boundary of the field. Near the outlet of the pipe, the sewage is about 40 cm deep, while at the vicinity of the entrance to the ditch the depth is about 15 cm. The total area of the field is 22,100 m^2. Assuming an average depth of 25 cm, the total volume of sewage in the field is 5525 m^3. Keeping in mind that the inflow of sewage is 850 m^3/day, the retention time is calculated to be 6.5 days. In the first 4 years of operation, the system was found to reduce the concentration of organic substances by 70% (Kowalik and Obarska-Pempkowiak, in press). Unfortunately no measurements of nitrogen and phosphorus were carried out then. In the period from January 1990 to December 1990, measurements of nitrogen and phosphorus concentrations were

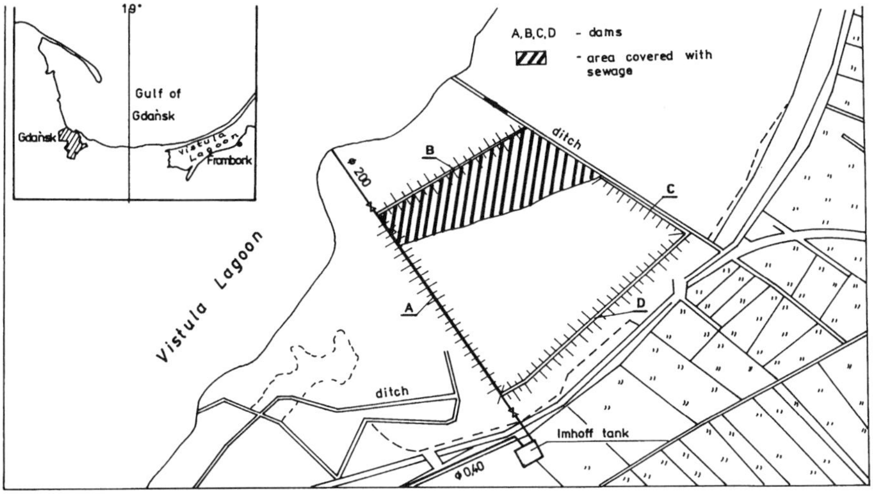

FIGURE 1. Location of the system.

carried out in order to establish the performance of the system with regard to nutrients and the results are reported in the present paper.

METHODS

Concentrations of the basic forms of nitrogen, e.g., ammonium (NH_4^+), nitrate (NO_3^-), nitrite (NO_2^-), and organic nitrogen (N_{org}) as well as phosphates (PO_4^{3-}), organic substances (COD and BOD_7), and suspended solids were measured. Samples were collected just after the Imhoff tank and after the field of reeds. Two modes of sampling were carried out: throughout a year samples were collected once a week (4 to 5 times a month), and, for 4 weeks during the summer, samples were collected daily, so that stability of the effluent composition could be assessed. After collectiong, samples were preserved, brought to the laboratory, and analyzed according to standard methods (Hermanowicz et al. 1976).

Additionally, green parts of reeds collected from both irrigated and natural fields were analyzed for both nitrogen and phosphorus. This was carried out so that accumulation of both elements in the plants could be assessed. Results were calculated on a dry weight basis.

RESULTS

The average monthly concentrations of the measured contaminants are presented in Figure 2 and Figure 3. The permissible concentrations of contaminants in effluents, according to Polish standards, are also presented in the figures. Total nitrogen — the sum of ammonium, organic, nitrate and nitrite nitrogen — is denoted by N_{tot}.

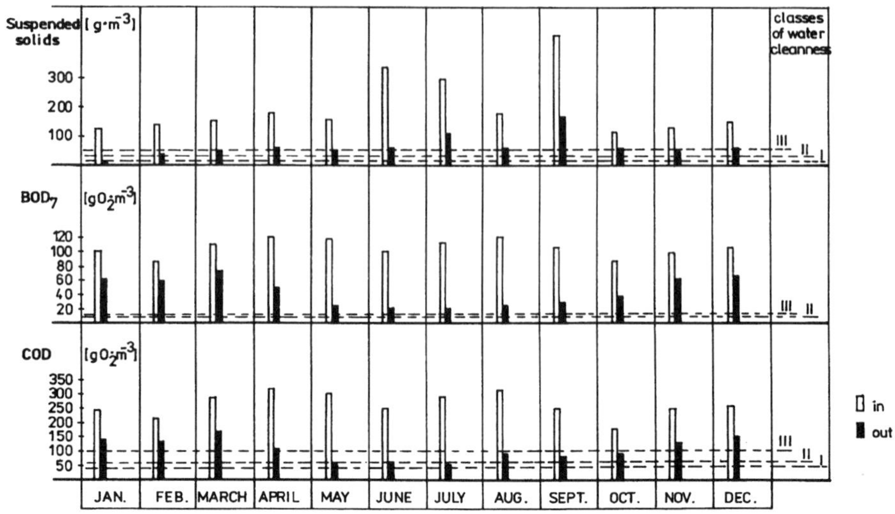

FIGURE 2. Average monthly concentrations of organic substances in inflowing and outflowing sewage.

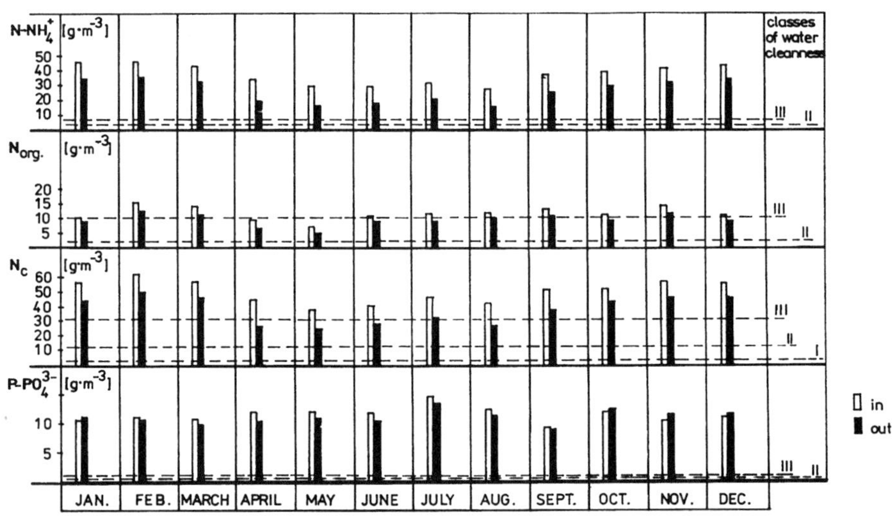

FIGURE 3. Average monthly concentrations of nutrients (N and P) in inflowing and outflowing sewage.

The average yearly concentrations of contaminants in inflowing and outflowing sewage were as follows: COD 268 and 106 g O_2/m^3, BOD_7 105 and 44 g O_2/m^3, suspended solids 185 and 65 g/m^3, N_{tot} 50.3 and 37.5 g N_{tot}/m^3, P 11.6 and 11.2 g PO_4^{3-}/m^3.

Average monthly efficiencies of the system are presented in Figure 4. Mass balances for organic matter as well as bioutilization of nitrogen and phosphorus in both irrigated and natural fields of reeds are presented in Figure 5.

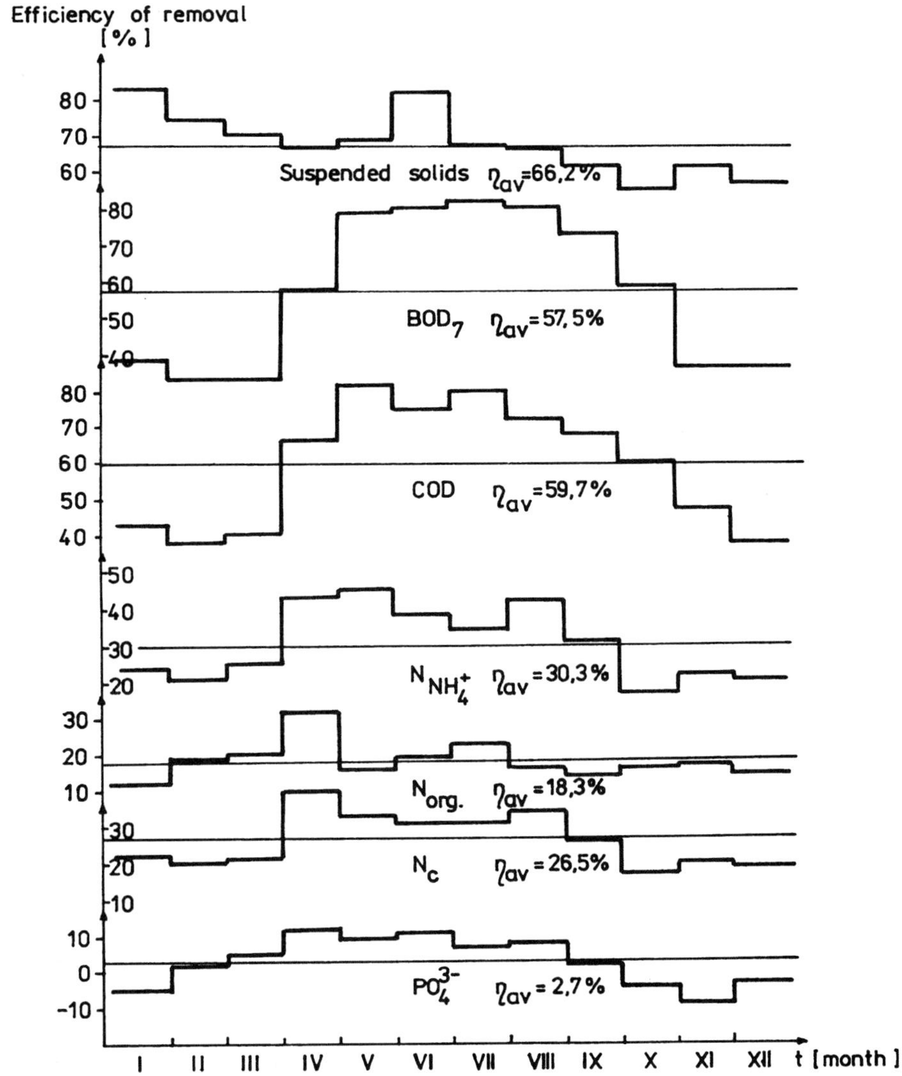

FIGURE 4. Average monthly efficiencies of removal of selected contaminants in the quasi-natural field of reeds in Frombork. N_c = total nitrogen. Horizontal lines indicate the year-round average removal rates, as do the percentages given in the middle of the diagram.

DISCUSSION

Daily measurements of concentrations indicated a rather stable composition for the investigated sewage. It is safe, therefore, to calculate average monthly concentrations of various pollutants on the basis of four to five analyses performed on samples collected once a week (Obarska-Pempkowiak 1991).

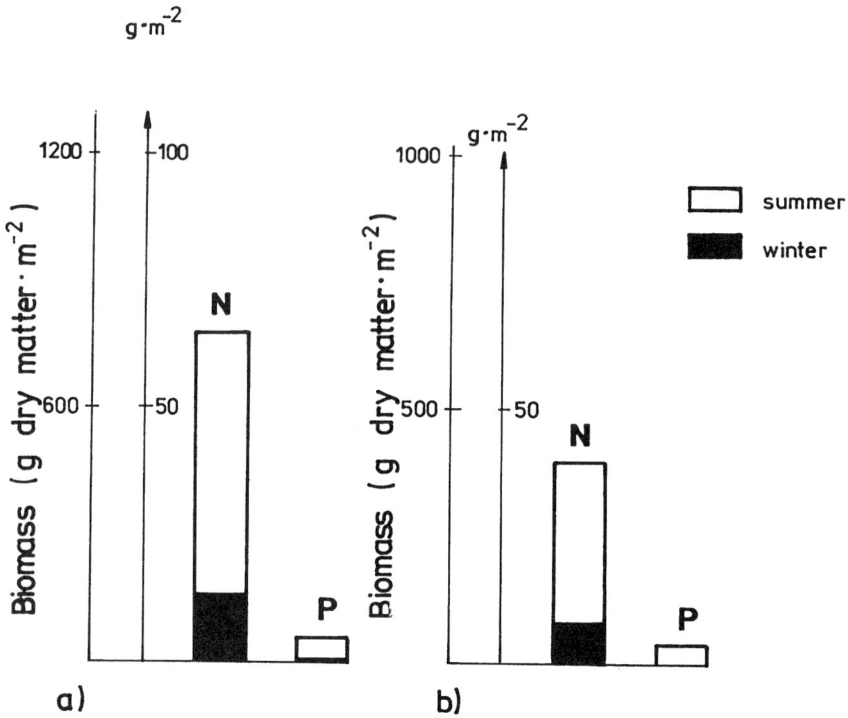

FIGURE 5. Biomass production (left scale) and content of N and P in reeds growing in a) a field irrigated with sludge and b) a natural field.

Suspended solids concentrations are the least stable. This may be caused by the lack of a sand trap, periods of poor operation of the tank, and improper situation of the sampling site after the field of reeds. Another contaminant characterized by a broad and irregular changeability is total nitrogen (N_{tot}). This is most likely caused by natural fluctuation of ammonium nitrogen (NH_4^+) intensified by very irregular concentrations of organic nitrogen (N_{org}). Much smaller standard deviations of N_{org} in sewage after the field, compared with inflowing sewage, indicate the averaging function of the field.

No regular fluctuations of sewage composition were found in samples of sewage flowing to the field. This is in accordance with the lack of industry and the stable water consumption in the town.

Calculated reduction of concentrations in the field, listed in Figure 2 and Figure 3, differ substantially for various contaminants. The efficiency is most pronounced for suspended solids. It is also most stable throughout the year, although it is rather small (as compared with literature data). This may be caused by improper sampling, however.

Substantial reduction of both COD and BOD_7 concentrations in sewage after passing the field is also apparent. Rather large relative standard deviation, exceeding 30%, might indicate changeability of the parameters. However, a glimpse at Figure 2 is enough to conclude that the large standard deviation is caused by regular fluctuation, manifesting itself with large efficiencies in summer (80%) and much smaller in winter (37%). The feature is true for both parameters. This is caused by the obvious dependence of the intensity of biochemical degradation of organic substances on temperature. The fact that even in winter there is a 30% reduction of organic substances indicates that processes of sorption and sedimentation are quite intensive also in winter.

A clear seasonal fluctuation of the efficiency of the NH_4^+ removal can also be noticed. Since sorption of NH_4^+ is independent of temperature, larger efficiency in summer must be seen as a result of denitrification as well as assimilation by reeds. Seasonal changes of the N_{tot} removal follow that of NH_4^+, however rather irregular changes of the N_{org} efficiency (relative standard deviation 25%) and the lack of seasonal changeability of N_{org}, which sum up with NH_4^+ to give N_{tot}, make the feature less clear (Figure 3).

Efficiency of the PO_4^{3-} removal shows signs of seasonal changes too. Surprisingly enough, in winter, negative efficiency occurs. The field of reeds exports phosphates then. Removal of phosphates (by about 10%) occurs only in summer. The yearly average efficiency of 2.7% should be regarded with caution and interpreted rather as the lack of phosphate removal.

BALANCE OF NITROGEN AND PHOSPHATES

The balances of both nitrogen and phosphorus are presented in Table 1. The loads entering the field were calculated as the product of average monthly concentrations and amounts of processed sewage. The loads leaving the field were calculated as the product of amounts of sewage leaving the field (Q_{out}) and concentrations. Q_{out} was calculated according to the following formula:

$$Q_{out} = \frac{S + P - E_T}{1000} F,$$

where: F = area of the field [m^2],
P = precipitation [mm/month],
S = hydraulic load to the field [mm/month], and
E_T = evapotranspiration [mm/month].

Water balance of the field was reported earlier (Obarska-Pempkowiak 1991). According to calculations, the field acquires 4221 kg of nitrogen and 225 kg of phosphorus per year.

EVALUATION OF EFFICIENCY

Overall efficiency of nutrient removal is composed of the removal caused by accumulation in biomass and sorption by soil. For nitrogen, denitrification contributes to the overall efficiency, too. In the studied field of reeds, production of biomass was estimated to be 1200 g/m^2 and fixation of nutrients was 65.2 g N/m^2 and 5.6 g P/m^2. The data are for the growing season. In winter both the biomass and contents of nutrients decreased as depicted in Figure 5. Therefore, reeds can be deemed a

TABLE 1
Balance of Nitrogen and Phosphorus in the Field Irrigated with Sewage

Parameter	Month												
	I	II	III	IV	V	VI	VII	VIII	IX	X	XI	XII	Year
Amount inflowing (m^3)	26,350	23,800	26,350	25,500	26,350	25,500	26,350	26,350	25,500	26,350	25,500	26,350	310,250
Amount inflowing (mm)	1,192	1,077	1,192	1,154	1,192	1,154	1,192	1,192	1,154	1,192	1,154	1,192	140,374
Inflowing nitrogen (kg)	1,491	1,471	1,539	1,158	1,009	1,046	1,204	1,073	1,308	1,370	1,461	1,462	15,592
Inflowing phosphorus (kg)	279	264	282	301	322	298	387	329	240	319	270	295	3,586
Amount outflowing (mm)	1,215	1,094	1,203	1,143	1,113	997	1,048	1,067	1,085	1,212	1,185	1,224	13,586
Amount outflowing (m^3)	26,852	24,177	26,586	25,260	24,597	22,034	23,161	23,581	23,979	26,785	26,189	27,050	300,251
Outflowing nitrogen (kg)	1,184	1,190	1,212	692	630	624	737	627	906	1,157	1,200	1,212	11,371
Outflowing phosphorus (kg)	301	264	269	263	273	231	315	274	221	338	301	311	3,361
Nitrogen retained (kg)	307	281	327	466	379	422	467	446	402	213	261	250	4,221
Phosphorus retained (kg)	–22	0	13	38	49	67	72	55	19	–19	9	–16	225

rather ineffective sink of nutrients. This conclusion is further supported by the observation that reeds in the nearby natural field (not irrigated with sewage) had the same density of stalks, although they were not as tall as those in the field irrigated with effluent.

The balance of nutrients in both fields is presented in Figure 5. If reeds were harvested, some 1440.9 kg N and 121.6 kg P would be removed from the system. Since no harvesting takes place, the nutrients are redirected to the system although at different times. The dry stalks form a layer of decaying plants which in time turn into peat containing trace amounts of nutrients, but acting as sorption matrix for sewage flowing through and above it.

CONCLUSIONS

The most important drawback of the assessed system is a very poor removal of nutrients. This is caused by insufficient area of the field. To achieve removal at the level of 50 to 60%, total area of the field should be increased some 8 to 10 times. This is quite possible considering local conditions and location of the system on the outskirts of the town.

Harvesting reeds in October would lead to additional removal of about 1000 kg N and 100 kg P per harvested hectare of the system.

Simple calculations swho that the layer of sewage covering field is too deep and surface flow predominates. The specific area of the field is 10% of that suggested for wetlands (10 m^2/PE vs. 100 m^2/PE) while the retention time is also a fraction (7 days, as compared to 14 days suggested for wetlands) but a much larger one. From this it follows that the relatively longer retention time is achieved by increasing the depth of sewage. This leads to insufficient, if any, removal of contaminants due to sorption on soil particles. The process of sorption is thought to play an important role in efficient wetland systems.

Another reason for poor efficiency is the way sewage is brought to the field. The point source of sewage means that it winds its way through the field using the area, small as it is, ineffectively. A system of dams enforcing serpentine flow of sewage would improve the efficiency of the system.

REFERENCES

Brix, H. 1987. "Treatment of wastewater in the rhizosphere of wetland plants by the root zone method." *Wat. Sci. Tech.* 19:107-118.

Bucksteeg, K. 1989. "Suitability of different biological sewage treatment systems." In: H. Odegaard (Ed.) *International Specialized Conference on Design and Operation of Small Wastewater Treatment Plants. Trondheim, Norway, 26-28 June 1989.* Pergamon Press, London, pp. 203-210.

Cooper, P.E., Boon A.G. 1987. "The use of *Phragmites* for wastewater by the root zone method: The U.K. approach." In: K.R. Reddy and W.H. Smith, (Eds.), *Aquatic plants for wastewater treatment and resource recovery.* Magnolia Publishing Inc., Orlando, FL, pp. 153-174.

Hermanowicz, W., Dozanska W., Dojlido W., Koziorowski B. 1976. *Fizyczno-chemiczne badania wody i sciekow.* Arkady, Warszawa, p. 817.

Kowalik, P. and Obarska-Pempkowiak H., in press. *Zasady oczyszczania sciekow przy udziale roslin w ekosystemie bagiennym*, IMUZ, Falenty.

Nichols, D.S. 1983. "Capacity of natural wetlands to remove nutrients from wastewater." *J. Water Pollut. Control. Fed.*, 55: 495-504.

Obarska-Pempkowiak H., Kowalik P. 1989. "The influence of small agrobusiness industry on the quality of waters in the Gdansk-region (northern Poland)." In: S. Tedeschi (ed.), *Research in hydraulic engineering, Proceedings of the 3rd Polish-Yugoslav Sympozium held in Gdansk, September 18-20, 1989.* Gradjevinski Institut. Zagreb 1990. pp. 217-231.

Obarska-Pempkowiak H. 1991. "Ocena mozliwosci wykorzystania stawow trzcinowych i plantacji wiklinowych do oczyszczania sciekow." Ph D. thesis, Politechnika Gdanska, Gdansk 1991.

Chapter 20

SAND-PLANT FILTER SYSTEMS

Johannes Heeb and Brigitta Züst

ABSTRACT

The Center for Applied Ecology Schattweid is situated at 730 m in the Swiss Prealps. This region, with its scattered settlements, is particularly suited for small wastewater treatment plants. For the last six years the settled domestic wastewater of the center has been treated in an aerobic sand filter and/or a predominantly anoxic planted soil filter. The effluent of each filter and thus the performance was analyzed. BOD/COD reduction is good in both filters, slightly better in the aerobic sand filter. Total P-content is predominantly reduced in the planted filter, the reduction being mostly above 80% (to a value in the order of 1 mg/l). The total reduction of N is not consistently measured, but certain steps of its transformations are monitored: ammonium, nitrite and nitrate. We work on computer modelling of nutrient flow patterns. Climatic values (precipitation, air, soil and water temperatures, potential evapotranspiration) are also recorded. Since the plant is located at our door-step, it is also suited for long-term observations (if a period of approximately ten years can be called "long term").

INTRODUCTION

The Center for Applied Ecology Schattweid is situated in the Swiss Prealps at 730 m above sea level. This region (Prealps) has dispersed settlements and remote communities; therefore small but nevertheless effective wastewater treatment plants are necessary. Sand-plant filters are particularly well suited for this area. Since 1986, the settled domestic wastewater of the center has been treated in an aerobic sand filter followed by a predominantly anaerobic planted soil filter.

The average annual temperature is about 8°C and the annual rainfall 1200 mm. Forty to fifty percent of this precipitation occurs between April and June. The climatological water balance is fairly high, at 700 mm (measured according to Hellmann; calculated according to Haude, 1955).

SAND-PLANT FILTER SYSTEMS: A NATURAL METHOD FOR CLEANING SEWAGE

In sewage treatment with a sand-plant filter system (SPF) natural processes such as the self-cleaning potential of running water and the filter effect of soil are put to use. All biodegradable domestic and industrial sewage can be purified. In Switzerland

0-87371-990-5/97/$0.00+$.50

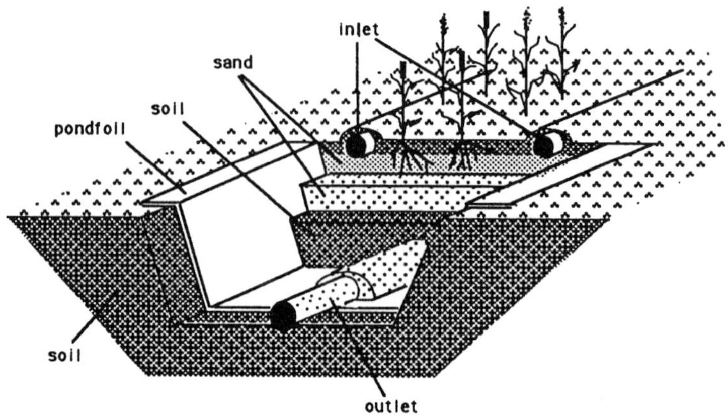

FIGURE 1. Construction of a sand-plant filter.

and Germany, over 30 SPFs are already in operation and other units are being planned in Switzerland, Germany, and Italy.

SPFs (developed at the *Zentrum für angewandte Ökologie Schattweid*, Switzerland) are constructed wetlands designed to compact naturally occurring processes into a shorter period of time and less space. Several other types of low technology soil filter systems are used (e.g., Cooper and Findlater, 1990). The SPF, however, and the phytofilt (Loeffler, Dresden) are the only ones with a double-layered filter bed (Figure 1).

The upper part consists of sand and gravel. The decomposition of sewage takes place in an aerobic environment.

The lower part consists of soil (usually from the construction site) which is permanently water logged; the decomposition is under anaerobic conditions.

The bed — about 1 m deep — is sealed with a polyethylene foil. It is planted for example with *Phragmites communis, Typha latifolia, Typha angustifolia, Iris pseudochorus, Acorus calmus, Juncus effusus, Juncus inflexus, Schoenoplectus lacustris.* The surface required is about 4 to 5 m^2/person.

The main advantage of SPFs is their suitability for remote areas. They are well able to deal with variations in quantity and quality of settled sewage. They are fully operational even during winter time: the very top layer may be frozen but below it the water circulates normally and cleaning occurs. Furthermore, SPFs can be used to polish treated effluent.

SPFs can be built for individual houses or entire communities.

THE CLEANING PROCESS

Mechanically treated sewage is pumped at intervals into the SPFs just below the surface, a measure which prevents stench and freezing in wintertime and assures a regular supply of fresh air in the aerobic layer. While trickling through the two layers of the filterbed (vertical infiltration), the water undergoes biological (e.g., nitrification, denitrification, or decomposition of organic carbon), chemical (e.g., fixation of phosphorus), and physical processes (filtration) which degrade and neutralize noxious

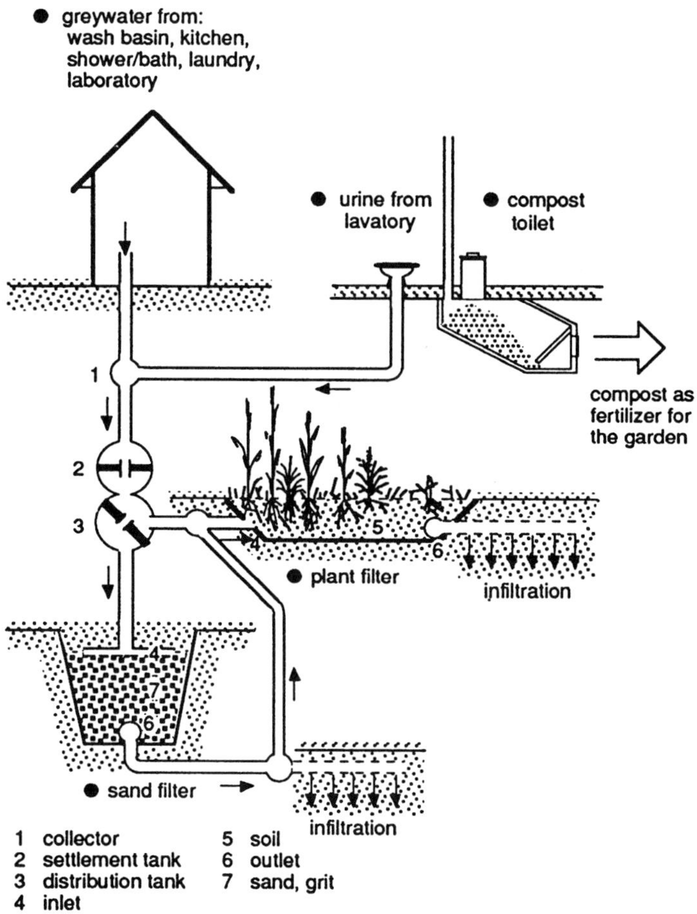

FIGURE 2. Schematic of the experimental sand-plant filter at Schattweid.

agents in the sewage. The treated water is collected in a drain at the bottom of the SPF and returned to local watercourse either directely or by underground infiltration or irrigation.

THE EFFICIENCY OF THE SAND-PLANT FILTER SYSTEM AT SCHATTWEID

Since the Schattweid treatment plant is a research plant, the aerobic sand filter (SF) and the anaerobic plant filter (PF) are separate (beside one another instead of one on top of the other), so as to allow individual efficiency evaluation of each filter. Normally, the water flows first through the SF and then through the PF (serial connection). Occasionally a parallel connection was used (see Figure 2).

At Schattweid the number of residents and visitors changes frequently, as shown in Figure 3. The sand-plant filter is particularly well-suited to cope efficiently with these variable amounts of water. The retention time of the water in the filter system

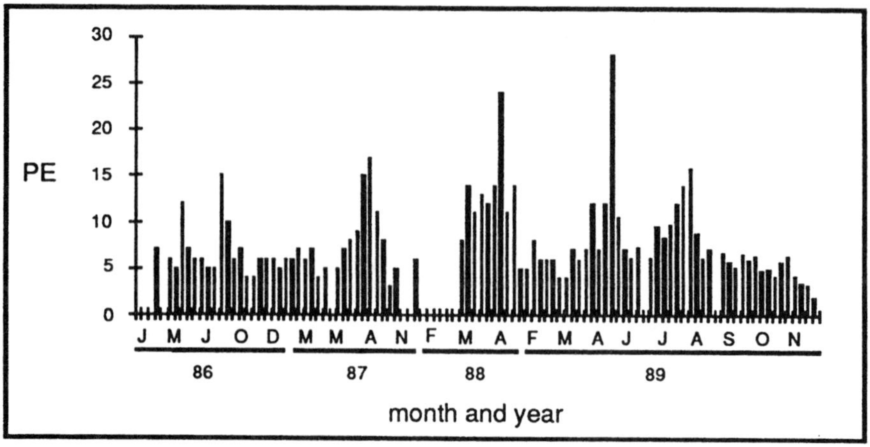

FIGURE 3. Population equivalents.

TABLE 1
Retention Time of the Water in the Filter System. Experiment 1: Uranin Tracer detected by Eye; Experiments 2 and 3: Uranin Tracer Detected by the Photometer

Retention time	Settlement and Distribution Tanks	Sand Filter	Plant Filter	Total
Theoretical value				
for 6 PE	3.3 d	5.8 d	12.7 d	22.4 d
for 12 PE	1.7 d	2.9 d	6.4 d	11.2 d
Empirical value				
Experiment 1		≈20 h	≈20 h	≈40 h
Experiment 2				8 h
Experiment 3		9 h	26 h	35 h

(Table 1) depends on the amount of water used (the average water consumption is low, 90 l/person/day), on the precipitation and the evaporation and on hydraulic and biological conditions in the filter substrate. The differences between the theoretical and the present empirical values of retention time are striking but it is clear that the empirical values need to be improved.

The efficiency of the sand-plant filter is — among other factors — dependent on the water temperature: it influences the rhythm of replication of the bacteria. The water temperature in the inlet and the outlet of the plant filter varies in a clearly seasonal way (Figure 4).

Results of the long-term monitoring of the sand-plant filter sytem at the Schattweid center are shown in the following figures. The sewage entering the Schattweid sand-plant filter system is highly concentrated (due to the low water consumption), as is illustrated in Figures 5 and 6.

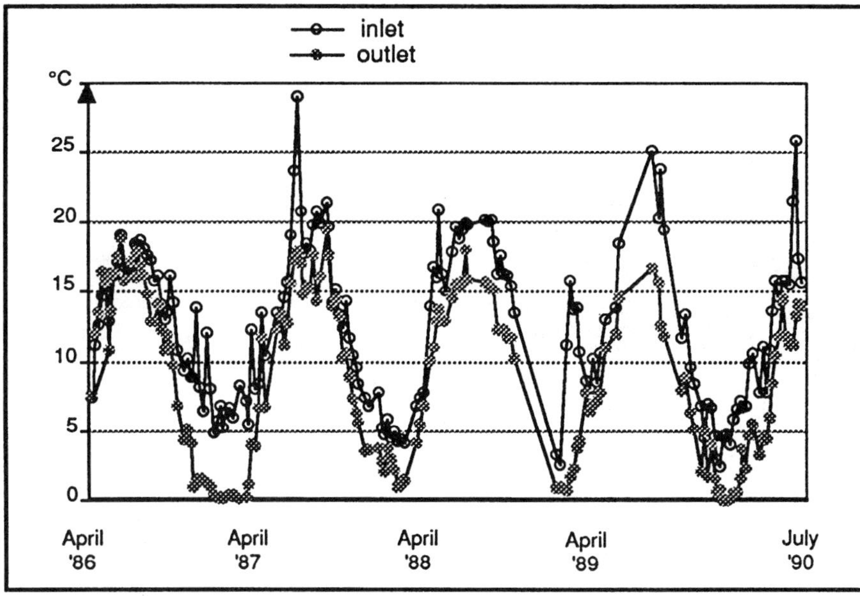

FIGURE 4. Water temperature at the inlet and outlet.

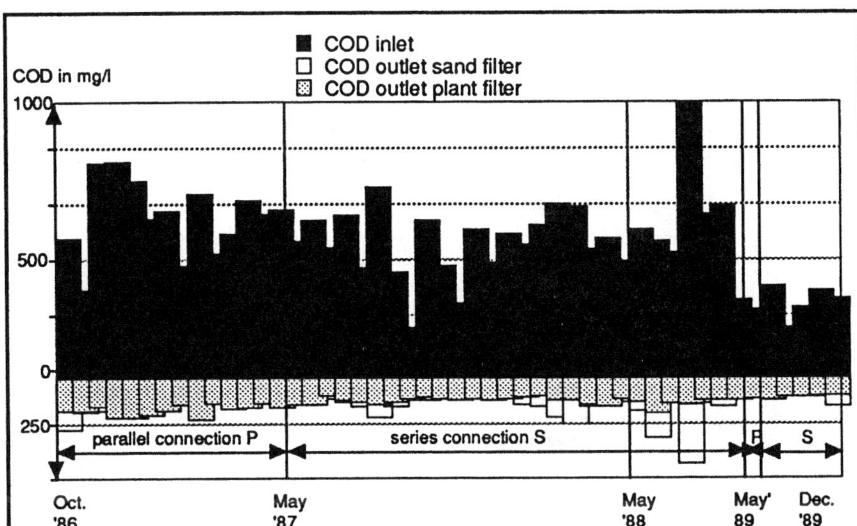

FIGURE 5. Chemical oxygen demand (COD) at the inlet and outlet of the sand filter and plant filter.

COD reduction is good in both filters, slightly better in the aerobic sand filter (Figure 5).

The total reduction of N has been systematically measured since October 1991 (Figure 9); before this date certain steps of nitrogen transformation were monitored. In a parallel connection, ammonium is degraded better in the aerobic sand filter; in

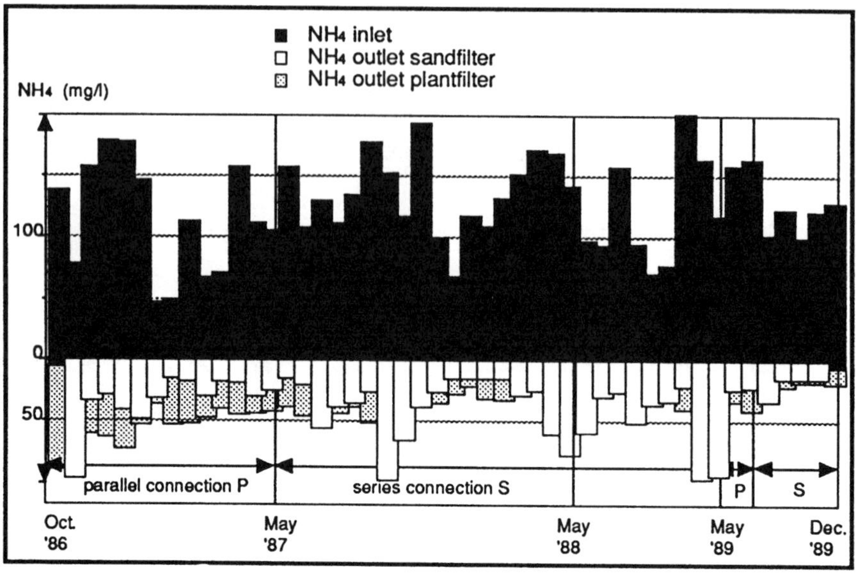

FIGURE 6. Ammonium concentrations at the inlet and outlet.

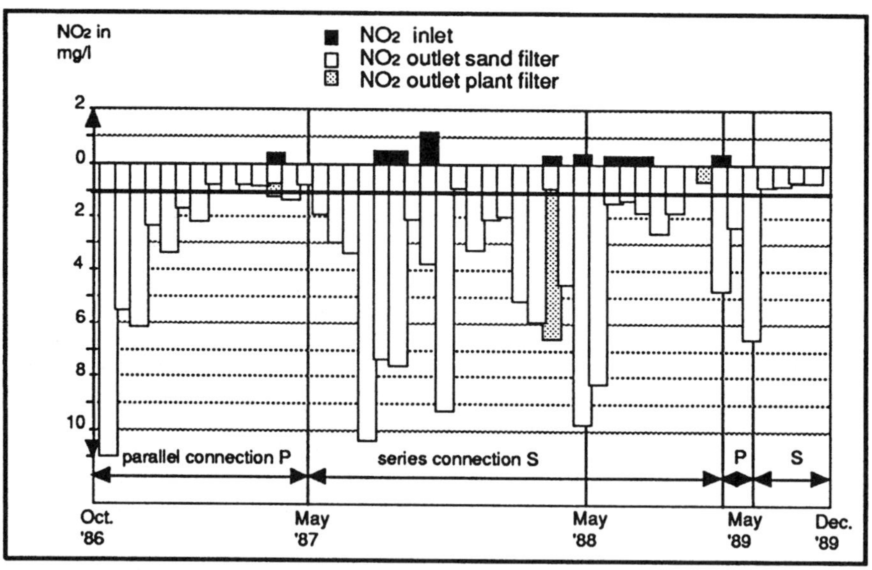

FIGURE 7. Nitrite concentrations at the inlet and outlet. The fat line at 1 mg/l NO_2 indicates the required effluent quality (Swiss standard).

series, the plant filter is in most cases able to reduce the ammonium concentration slightly better (Figure 6).

In the Schattweid plant there is often nitrite in the outlet of the sand filter (Figure 6), which indicates that nitrification is not complete; after the passage

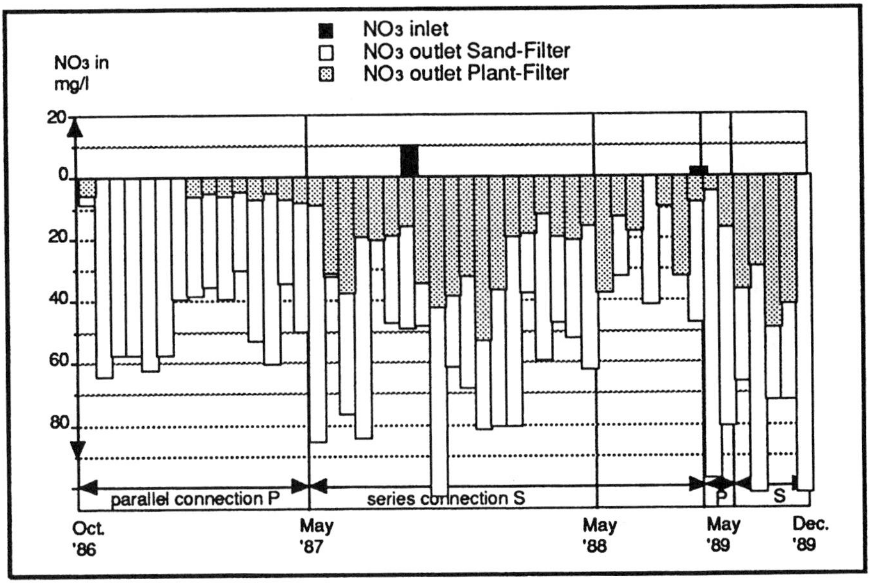

FIGURE 8. Nitrate concentrations at the inlet and outlet.

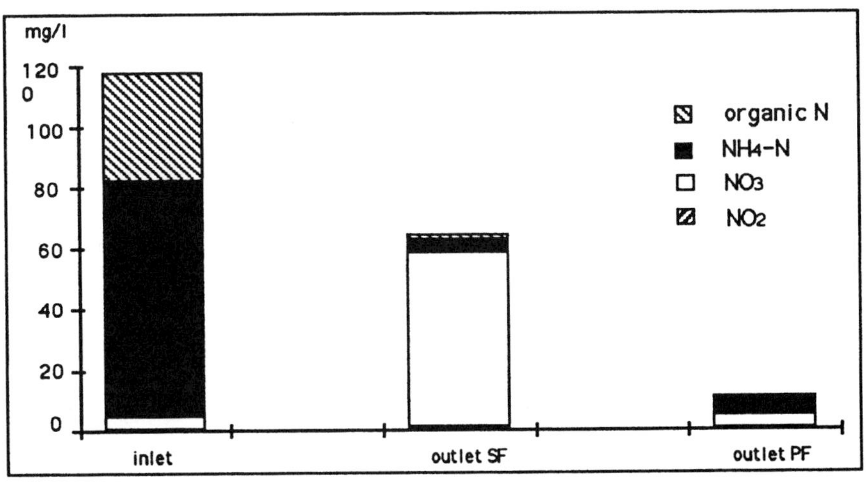

FIGURE 9. Total nitrogen in inlet, sand filter outlet and plant filter outlet, October 1991 to July 1992. The values are means (n = 12). The total height of the segmented columns corresponds to the total nitrogen.

through the plant filter, nitrite concentrations are in most cases reduced to below 1 mg/l.

Nitrate concentrations are reduced in the predominantly anaerobic plant filter (Figure 8). This suggests that at least part of the nitrate is denitrified and leaves the filter as nitrogen gas.

For a relatively short period (October 91 to July 92), a general view of the different nitrogen compounds in the filter system can be drawn up (Figure 9).

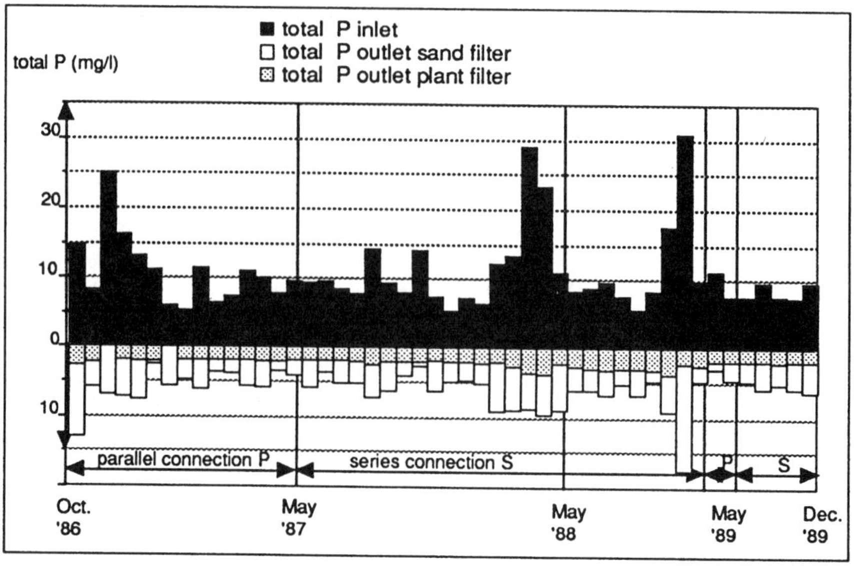

FIGURE 10. Phosphorus concentrations at the inlet and outlet. Concentration of total phosphorus in inlet and outlets of the sand filter and plant filter, October 1986 to December 1989. The intervals between measurements are not uniform. P = parallel connection, S = series connection (see text for details).

The total P-content is predominantly reduced in the anaerobic filter, the reduction being mostly above 80% (to a value in the order of 1 mg/l; Figure 10).

The elimination of bacteria (*E. coli*) was >99% in three different and independent analyses.

An interesting and important point is that the sand-plant filter system is, with few exceptions, able to meet the relevant — in this case the rigorous Swiss — effluent quality standards.

In a recent study on possible influences of microclimate on the purification efficiency of sand-plant filters, no clear dependence could be shown. Most likely this is not because there is none but because there are too many mutually influencing factors.

OPEN QUESTIONS

- Is it necessary/helpful to harvest the plants?
- What influence does the substrate have on the efficiency of sand-plant filter systems?
- When and where are bacteria (for example coliforms) eliminated in the sand-plant filter systems?
- What do nitrogen and phosphorus balances look like?
- What is the long-time behavior of sand-plant filter systems? Do they get clogged? Does mineralization occur?

FURTHER PROJECTS

- Nutrient flow patterns are studied with computer models.
- Long-term observations and documentation of the functioning of sand-plant filters (if, say, 10 years can be called "long term").
- A pilot project should give indications as to the sizing of SPFs treating cheese factory wastewaters instead of domestic waters.

PLANNING, CONSTRUCTION, OPERATION, COSTS

OEKAG (AG für angewandte Ökologie, Switzerland) or one of its licencees carry out the planning and supervision of the building. The parts are delivered as a set and can be assembled on the construction site. A separate element for composting the sediments of the settlement tank is available. For its operation, the SPF unit does not need electric power and requires only little maintenace. The costs per user vary from approximately SFr. 3000 (smallest units for 4 p.e.) to SFr. 1000 (largest units for 45 p.e.; for larger plants subunits will be built). SPFs units must be approved by local authorities.

REFERENCES

P.F. Cooper and B.C. Findlater (Eds.) 1990. *Use of constructed wetlands in water pollution control.* IAWPRC conference proceedings. Pergamon Press: Oxford.

Haude, W. 1955. "Zur Bestimmung der Verdunstung auf möglichst einfache Weise." Mitteil. Dt Wetterdienst

Note: A report on the 5th to 7th years (1990 to 1992) of the sand-plant filter at Schattweid was published in January 1994 by Andreas Schönborn and Brigitta Züst (in German). It can be ordered from the Center for Applied Ecology Schattweid.

Chapter 21

SUBMERSED MACROPHYTES FOR RECOVERY OF EUTROPHIED WATERS: NUTRIENT SPIRALING AND REMOVAL EFFICIENCY

Thomas Gumbricht

ABSTRACT

Non-point source (NPS) pollution is an effect of human destabilization of terrestrial ecosystems, and a cause of surface water contamination. Wetlands and riparian zones incorporate important resistance functions, accumulating NPS solute fluxes. The high turnover rates in such ecotones give rise to high resilience. Conserving, restoring and creating wetland environments can hence be a very cost-effective means of reducing watershed and watercourse degradation. In 1988 a full-scale experiment with *Elodea canadensis* and *Cladophora glomerata*, grown in eutrophied stream water in 16 parallel channels was started in southern Sweden. Between September 1988 and June 1991, the weekly mean reduction was 32% of nitrogen and 62% of phosphorus, equivalent to specific mean reductions of 0.5 g N $m^{-2}d^{-1}$ and 0.03 g P $m^{-2}d^{-1}$. Harvested biomass contained approximately 10% of removed nitrogen and 20% of removed phosphorus. Hence, processes other than direct uptake and subsequent harvest were important for nutrient removal.

Estimation of sediment nutrient content suggests that denitrification was the major nitrogen removal process, and burial in the sediment was the most important phosphorus removal mechanism. Statistical regression analysis showed that nitrogen removal was dependent on temperature and nitrogen load, again indicating that denitrification was the major nitrogen removal mechanism. Variations in phosphorus removal were largely explained by insolation, which is in agreement with the hypothesis that biomass uptake is important for the removal of phosphorus. Both nitrogen and phosphorus removal were correlated to detention time, the contact time between pollutants and active surfaces. Based on the nutrient spiraling concept, and a suggested nutrient phase-shifting concept, management of macrophyte systems for maximizing removal efficiency is discussed.

INTRODUCTION

Non-point source (NPS) pollution is the major cause of pollution of surface waters in many western countries (SNV 1988; Ryding and Rast 1989; Baker 1992). Human impact on the landscape has caused the outflow of substances from watersheds to increase several fold (Larsson et al. 1985; Peierls et al. 1991). This is a

0-87371-990-5/97/$0.00+$.50

strong sign of changes in nutrient cycling and decreased ecological stability in terrestrial ecosystems.

The recognized ability of wetlands to remove nutrients and other pollutants from non-point sources and wastewater has led to the idea of using constructed wetlands as cleaning media both for polluted wastewater and NPS eutrophied waters (Mitsch and Jørgensen 1989, Mitsch 1992), an example of ecological engineering (Mitsch 1996). Different types of wetlands have been tested and applied under different environmental and climatic condition (Brix and Schierup 1989; Hammer 1989; Cooper and Findlater 1990; Gumbricht 1993a). The removal mechanisms for nutrients and pollutants differ somewhat between different systems. However, the central part of most wetland treatment systems is macrophytes. Apart from biomass uptake, the macrophytes also play a crucial role by creating a favorable environment for a variety of complex chemical, biological, and physical processes that contribute to the removal and degradation of nutrients (Reed et al. 1988; Hofmann 1997).

Nitrogen removal is a consequence of assimilation, denitrification, and burial in the sediment (Gumbricht 1993a). In a harvested pond, denitrification and assimilation can be of the same magnitude. However, removal rates can be just as high in unharvested ponds because of higher denitrification rates in the denser macrophyte stand. Since both assimilation and denitrification are highly temperature dependent, nitrogen removal becomes temperature dependent. Denitrification rates also depend on nitrate level and availability of readily mineralizable carbon. Phosphorus removal is partly a result of assimilation; plant material is harvested or buried in the sediment (Gumbricht 1993a). The assimilation rate in macrophytes depend primarily on growth rates, and thus on insolation, temperature, and availability of nutrients. Sorption of phosphorus to Al, Fe, Ca, and clay minerals in the substrate is also an important removal process.

This study very briefly presents the results of a submersed macrophyte pond treating NPS polluted water (Gumbricht 1993b). Emphasis is on the conceptual understanding of retention mechanisms by applying the nutrient spiraling concept (Howard-Williams 1985).

THE NUTRIENT SPIRALING CONCEPT

The nutrient spiraling concept (Newbold et al. 1981; Pinay et al. 1990) emphasizes the interactive role between the river and its valley. It is originally based on the unidirectional flow of water through ecotones along the floodplain, and combines the concepts of nutrient cycling and transport. Spiraling length is defined as the expected downstream distance travelled by a nutrient atom as it completes a cycle between water and sediment (Figure 1). Water is the vector that constitute both the transport and the reaction medium for the downstream transfer of the nutrients.

Turning from the floodplain scale (Frissell et al. 1986) to a micro scale (Figure 1), it is clear that a nutrient atom, rather than spiraling, shifts phase between an unstructured (inorganic or soluble) and an structured (organic or crystallized) phase. This phase-shifting apparently occurs within the nutrient spirals (Figure 1) and is partly controlled by them. The water and sediment parts of the nutrient spiraling concept merely emphasize the difference in phase-shift frequency at the micro scale. Obviously,

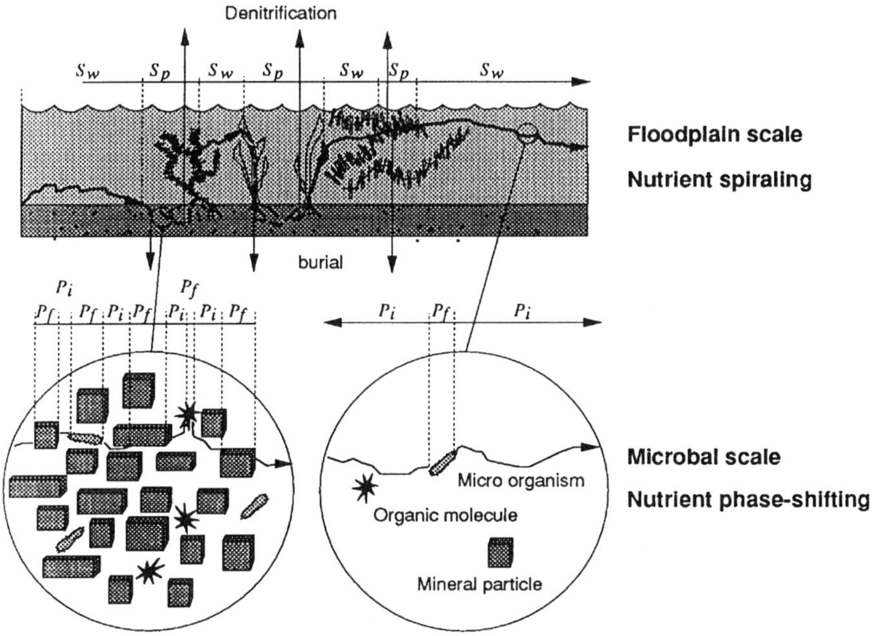

FIGURE 1. Nutrient spiraling and phase-shifting in a submersed macrophyte pond. Nutrient spiraling emphasises the floodplain scale of nutrient cycling and transport. A nutrient atom travels an average downstream distance S_w in the water before being taken up by particulates or assimilated in biomass in which form it travels distance S_p before returning to the water. Total spiraling length = S_w+S_p. The spiraling length decreases in patches with high retention capacity. At the microbal scale nutrients shift phase between a structured (P_s) and unstructured (P_s) phase, partly controlled by the spiraling scale. Permanent retention of nutrients can only be canalized through the P_s state of the phase-shifting scale.

it is only within the structured phase that permanent removal (i.e., denitrification, sorption, burial, or harvest) can occur.

The formation of organic matter and crystals utilizes entropy increases of an open thermodynamic system and they can be seen as non-equilibrium structures (dissipative structures; Prigogine, 1980). The nutrient phase-shifting concept emphasizes the laws of thermodynamics; the organic and crystallized phases are dissipative structures with lower entropy. If there are no phase-shifts (and no structuring processes) the energy dissipation will instead translate into equilibrium, and loss of nutrients and energy from the system. I suggest the term nutrient phase-shifting be used in this thermodynamic sense.

Nutrient spiraling and phase-shifting is closely connected with ecological stability, which is composed of two parts, resistance and resilience (Webster et al. 1975; Leibowitz et al. 1992). Resistance is the opposition to disturbance or displacement (i.e., inertia), and resilience the ability to return to a reference state following disturbance. The transversal ecotone (i.e., wetlands and riparian forests situated adjacent to the stream) is a system subject to continuous change. It seldom reaches its climax maturity. However, it has a high potential of resilience since it is a highly dynamic system with high turnover rates. The wetland or riparian zone is therefore

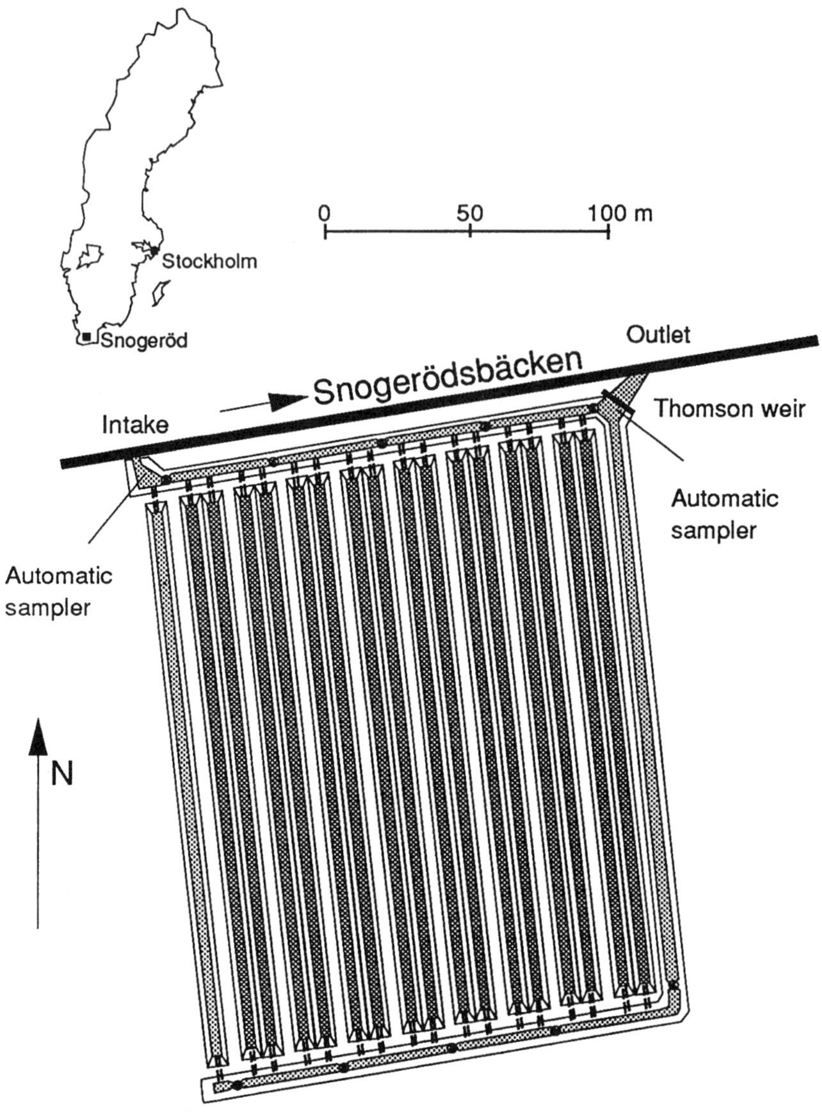

FIGURE 2. The submersed macrophyte pond at Snogeröd, Sweden.

also an important zone of resistance, accumulating NPS solute fluxes, especially in a changing environment. Conserving, restoring, and creating wetland environments can hence be a very cost-effective means to establish NPS pollution retention elements.

MATERIALS AND METHODS

In 1988, after some years of laboratory experiments, the firm K-Konsult started full scale experiments with *Elodea canadensis* and *Cladophora glomerata*, grown

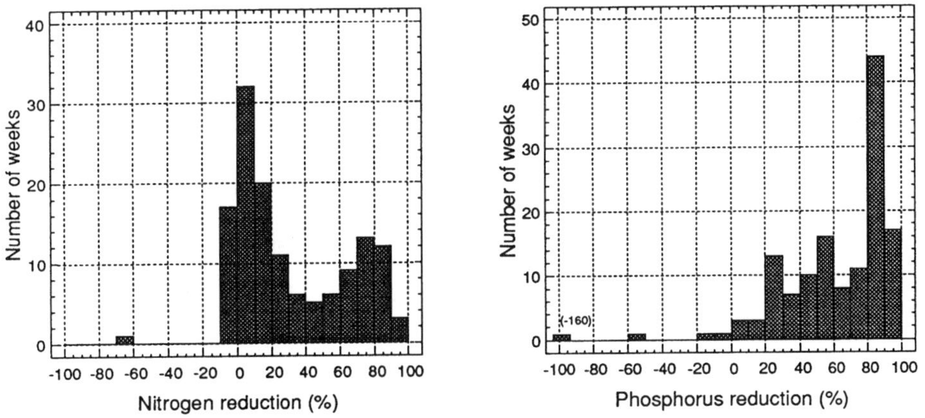

FIGURE 3. Frequency distribution of weekly results of removal efficiency for the submersed macrophyte pond at Snogeröd (September 1988 to June 1991).

in eutrophied stream water in 16 parallel channels in southern Sweden (Figure 2). Continuous samples from inlet and outlet water were analyzed weekly for total nitrogen (Koroleff 1970) and total phosphorus (Koroleff 1976) during the period September 1988 to June 1991. Water flow, temperature, and insolation were also measured. From May to September, the pond was harvested every other week. Materials and methods are more closely described in Gumbricht (1993b).

RESULTS AND DISCUSSION

For the 34-month period September 1988 to June 1991, weekly mean reduction was 32% for nitrogen and 62% for phosphorus (Figure 3), equivalent to specific annual mean reductions of 0.5 and 0.03 g m^{-2} d^{-1} respectively. On average, the incoming levels of nitrogen decreased from 9.8 to 7.5 mg l^{-1}, and phosphorus decreased from 0.26 to 0.07 mg l^{-1}; both changes were statistically significant ($p < 0.05$). The differences among %, mass, and mg l^{-1} removal are caused by seasonal fluctuations in both flow and reduction. Both nitrogen and phosphorus were dominated by inorganic (free) fractions (phases). Nitrate comprised 90% of the nitrogen fraction in both incoming and outgoing water. Phosphate made up 70% of the phosphorus fraction in both incoming and outgoing water.

The harvested biomass contained approximately 10% of removed nitrogen and 20% of removed phosphorus. Hence, processes other than direct uptake and subsequent harvest were of importance for nutrient removal. This is in agreement with results achieved under similar conditions (McNabb 1976; Eighmy and Bishop 1988).

Estimation of sediment nutrient content suggests that denitrification was the major nitrogen removal process, and burial in the sediment was the most important phosphorus removal mechanism. Statistical regression analysis showed that nitrogen removal was dependent on temperature ($r^2 = 55$) and specific nitrogen load ($r^2 = 32$), again indicating that denitrification was the major removal mechanism. Phosphorus removal was largely explained by insolation ($r^2 = 45$), which is in agreement with the hypothesis that biomass uptake is important for the removal of phosphorus. Both nitrogen and phosphorus removal were correlated with detention time ($r^2 = 47$ and $r^2 = 22$, respectively), increasing the contact time between pollutants and active surfaces.

Leaving one channel unharvested showed no significant difference in either nitrogen or phosphorus removal efficiency ($p > 0.05$). This once more points to denitrification and burial in the sediment being the major removal mechanisms.

CONCLUSION

The few published results of pond systems with submersed macrophytes are similar or slightly better than the results from this study (McNabb 1976; Nuttall 1985; Eighmy and Bishop 1988). A main conclusion from those and the present study is that higher removal efficiency can be reached by tuning the physico-chemical environment in the pond system (Tchobanoglous 1997; Gumbricht 1993a). It is especially vital to alter the composition of the cleaning media and pond management according to the kind of water that is to be treated. Harvesting was not promoting removal efficiency in this study. An alternative to harvesting could hence be to clear the sediments, for instance on an annual basis.

The results from this and other studies illustrate the importance of understanding the linkages in both nutrient cycles and their interactions with hydrological processes before management practices can be determined (e.g., Howard-Williams 1985). And, as emphasized by Hemond and Benoit (1988), cumulative effects cause wetland responses to external changes to be non-linear and unpredictable. Quantitative evaluation of wetland response to cumulative impacts is thus not possible (c.f. Leibowitz 1992). However, based on the concepts of nutrient spiraling and nutrient phase-shifting, some general management conclusions for improved removal efficiency in macrophyte pond systems can be drawn (c.f. Gumbricht 1993a).

A dissolved (inorganic) nutrient entering a submersed macrophyte pond must participate in at least one phase-shift if it is to have a chance of being removed from the water phase (Figure 1). At the Snogeröd pond, retention time has been too short for this during some winter periods (Gumbricht 1993b), as indicated by the net leakage occurring some weeks (Figure 3). Increasing the frequency of phase-shifting and decreasing the nutrient spiraling length optimizes the efficiency of the system.

One way of increasing the number of spiraling cycles is to recirculate water through the pond system (Tchobanoglous 1993). The results from the Snogeröd pond also show that when the water passes four channels connected in series, removal efficiency is higher than when the treated water only passes one channel (Gumbricht 1993b).

Phosphorus is sorbed to Fe, Al, and Ca in the substrate. Sorption to iron and aluminum predominate under neutral to acid conditions and sorption to calcium under alkaline conditions (Gumbricht 1993a). Hence, several reports indicate that phosphorus removal increases with higher content of iron, aluminum, and calcium both in water and sediments of different kinds of constructed wetlands (e.g., Cooke 1992; Gumbricht 1993a).

Of crucial importance is to bring the polluted water into contact with the active surfaces of the cleaning media (i.e., the macrophytes, floccules, microbes, etc.). Phase-shifting hence potentially increases with turbulent flow in comparison with laminar. Important sites for phase-shifting are oxygenated pores in anaerobic environments and the water-sediment interphase (Gumbricht 1993a). Enlargement of such domains increases the potential of nutrient removal. This can also be accomplished by temporal shifts (i.e., wetting and drying cycles), which has been reported to increase removal efficiency in several cases (Pinay and Décamps 1988; Zedler et al. 1992). In general, nitrification limits denitrification when reduced nitrogen is applied. Hence, several studies show that artificially oxygenated macrophyte systems have higher nitrogen removal capacity (Gumbricht 1993a).

A problem of many macrophyte treatment systems is that flow through soils and sediments probably is not uniform (e.g., van Genuchten and Wierenga 1976). Most of the research indicating preferential flow pathways in a matrix of immobile or "dead" zones has been done applying tracers in terrestrial soils. However, lately Cooper (1992) showed the presence of preferential flow pathways in wetland soils, thus raising the question of how large parts of wetlands participate in the renovation processes. When discharging point sources to wetlands, other studies also indicate that only parts of the wetlands are active in removing pollutants (Bowmer 1987; Brodrick et al. 1988).

ACKNOWLEDGMENTS

The Snogeröd project is supported by a consortium including the County of Malmhus, the communities of Eslöv, Hörby, and Höör, and K-Konsult. The project was financed by the consortium and the Swedish government. Ingvar Hörberg and Lars Kylefors, K-Konsult Sydost AB were project leaders. Monica Andersson (K-Konsult) has digitized and processed part of the data. The daily maintenance, sampling, etc. were done by Per-Axel Camper (K-Konsult).

REFERENCES

Baker, L.A., 1992. "Introduction to nonpoint source pollution in the United States and prospects for wetland use." *Ecol. Eng.* 1:1-26.

Bowmer, K.H., 1987. "Nutrient removal from effluents by an artificial wetland influence of rhizosphere aeration and preferential flow studied using bromide and dye tracers." *Water Res.* 21: 591-600.

Brix, H. and Schierup, H.H., 1989. "The use of aquatic macrophytes in water-pollution control." *Ambio* 18: 100-107.

Broderik, S.J., P. Cullen and W. Maher, 1988. "Denitrification in a natural wetland receiving secondary treated effluent." *Water Res.* 22: 431-439.

Cooke, J.G., 1992. Nutrient transformations in a natural wetland receiving sewage effluent and their implications to waste treatment. (Manuscript).

Cooper, A.B., 1992. Coupling wetland treatment to land treatment: an innovative method for nitrogen stripping? (Manuscript).

Cooper, P.F. and B.C. Findlater, (Eds), 1990. *Constructed wetlands in water pollution control.* Pergamon Press. Oxford. 605 pp.

Eighmy, T.T. and Bishop, P.L., 1988. "Autotrophic nitrification and its role in nitrogen removal in Elodea nuttalli-based aquatic treatment systems." *Water Supply* 6: 119-124.

Frissell, C.A., W.J. Liss, C.E. Warren and H.D. Hurley, 1986. "A hierarchical framework for stream habitat classification: viewing streams in a watershed context." *Environ. Manag.* 10: 199-214.

Genuchten, M.T. van, and P.J. Wierenga, 1976. "Mass transfer studies in sorbing porous media. I. Analytical solutions." *Soil Sci. Soc. Am. J.* 40: 473-480.

Gumbricht, T., 1993a. "Nutrient removal processes in freshwater submersed macrophyte systems." *Ecol. Eng.* 2:1-30

Gumbricht, T., 1993b. "Nutrient removal capacity in submersed macrophyte pond systems in a temperate climate." *Ecol. Eng.* 2:49-61.

Hammer, D.A. (ed.), 1989. *Constructed wetlands for wastewater treatment. Municipal, industrial and agricultural.* Boca Raton, FL: Lewis Publishers. 831 pp.

Hemond, H.F. and J. Benoit, 1988. "Cumulative impacts on water quality functions of wetlands." *Environ. Managem.* 12: 639-653.

Hofmann, K., 1997. "The role of plants in subsurface flow constructed wetlands." In: C. Etnier and B. Guterstam (eds.), *Ecological engineering for wastewater treatment.* Boca Raton, FL: Lewis Publishers.

Howard-Williams, C., 1985. "Cycling and retention of nitrogen and phosphorus in wetlands: a theoretical and applied perspective." *Freshwater Biology*, 15: 391-431.

Koroleff, F., 1970. "Determination of total nitrogen in natural waters with peroxodisulfate." International Council for the Exploration of the Sea, ICES. Council meeting 1969, paper C8 revised version 1970.

Koroleff, F., 1976. "Determination of phosphorus." In: K. Grasshoff (ed.), *Methods of Seawater Analysis.* Verlag Chemie. Weinheim/New York. pp. 117-126.

Larsson, U., R. Elmgren and F. Wulff, 1985. "Eutrophication and the Baltic Sea: Causes and consequences." *Ambio* 14: 9-14.

Leibowitz, S.G., B. Abbruzzese, P.R. Adamus, L.E. Hughes, J.T. Irish, 1992. "A synoptic approach to cumulative impact assessment. A proposed methodology." U.S. EPA/600/R-92/167. 129 pp.

McNabb, C.D., 1976. "The potential of submersed vascular plants for reclamation of wastewater in temperate zone ponds." In Tourbier, J. and Pierson, R.W. (eds.), *Biological control of water pollution.* Philadelphia, PA. University of Pennsylvania Press. pp. 123-132.

Mitsch, W.J. and S.E. Jørgensen (eds.), 1989. *Ecological engineering. An introduction to ecotechnology.* New York: John Wiley and Sons. 472 pp.

Mitsch, W.J., 1997. "Ecological engineering: the roots and rationale of a new ecological paradigm." In: C. Etnier and B. Guterstam (eds.), *Ecological engineering for wastewater treatment.* Boca Raton, FL: Lewis Publishers.

Mitsch, W.J., 1992. "Landscape design and the role of created, restored, and natural riparian wetlands in controlling nonpoint source pollution." *Ecol. Eng.* 1: 27-47.

Newbold, J.D., J.W. Elwood, R.V. O'Neill and W. van Winkle, 1981. "Measuring nutrient spiraling in streams." *Canadian Journal of Fisheries and Aquatic Sciences* 38: 860-863.

Nuttall, P.M., 1985. "Uptake of phosphorus and nitrogen by *Myriophyllum aquaticum* (*Velloza*) Verd. growing in wastewater treatment system. *Aust. J. Mar. Res.* 36: 493-507.

Peierls, B.L., N.F. Caraco, M.L. Pace and J.J. Cole, 1991. "Human influence on river nitrogen." *Nature* 350: 386-387.

Pinay, G. and H. Décamps, 1988. "The role of riparian woods in regulating nitrogen fluxes between the alluvial aquifer and surface water: A coneptual model." *Regulated rivers: Research and management* 2: 507-516.

Pinay, G., Décamps, H., Chauvet, E. and Fustec, E., 1990. "Functions of ecotones in fluvial systems." In Naiman, R.J. and Décamps, H. (eds.), *The ecology and management of aquatic-terrestrial ecotones.* Parthenon Press Publ., pp 141-169.

Prigogine, I. 1980. From being to becoming. W.H. Freeman, San Francisco. 272 pp.

Reed, S.C., Middlebrooks, E.J. and Crites, R.W., 1988. *Natural systems for waste management and treatment.* New York: McGraw Hill. 308 pp.

Ryding, S.O. and W. Rast, 1989. *The control of eutrophication of lakes and reservoirs.* Man and the biosphere series Volume 1. UNESCO, Paris. 312 pp.

SNV (National Swedish Environmental Protection Board), 1988. *Monitor 1988: Sweden's marine environment — ecosystems under pressure.* Solna, Sweden. 208 p.

Tchobanoglous, G., 1997. "Land-based systems, constructed wetlands, and aquatic plant system in the United States. An overview." In: C. Etnier and B. Guterstam (eds.), *Ecological engineering for wastewater treatment.* Boca Raton, FL: Lewis Publishers.

Webster, J.R., J.B. Waide and B.C. Patten, 1975. "Nutrient cycling and the stability of ecosystems." In F.G. Howell, J.B. Centry, and M.H. Smith (eds.), *Mineral cycling in Southarctic ecosystems.* ERDA symposium series. CONF 740513. Springfield, VA: Nat. Techn. Inform. Serv. pp 1-27.

Zedler, J.B., M. Busnardo, T. Sincrope, R. Langis, R. Gersberg, and S. Baczkowski, 1992. Pulsed-discharge wastewater wetlands: the potential for solving multiple problems by varying hydroperiod. Manuscript. Presented at INTECOL's IV International wetland conference. Columbus, OH, September 13-18, 1992.

Chapter 22

WASTEWATER TREATMENT BY AERATED PONDS AND CONSTRUCTED WETLANDS IN THE NORWEGIAN CLIMATE: RESULTS AND DESIGN CONSIDERATIONS

Petter D. Jenssen, Tore Krogstad, and Trond Mæhlum

ABSTRACT

Norway has committed itself to reduce the discharge of nitrogen (N) and phosphorus (P) to the North Sea by 50% before 1995. Constructed wetlands (CW) are low-cost, low-maintenance wastewater treatment systems that can be optimized for nitrogen removal. Their high efficiency in removing nitrogen makes these systems interesting as a polishing step after conventional treatment, and also an alternative to smaller conventional plants. The performance of CW systems in a cool temperate climate has been questioned and the experience with CW systems in Norway is limited. However, it is shown that CW systems can work well in climatic conditions close to those encountered in Norway and preliminary results from Norway are promising. We therefore believe that it is possible to optimize the design of CW systems to suit Norwegian conditions.

Rational design criteria for CW systems in cool temperate climate do not exist, but a considerable knowledge base is present from which design concepts can be deducted. Hydraulic considerations should always be the primary step in design of CW systems, and design based on surface area alone is inadequate. In order to gain sufficient hydraulic capacity in winter time, deeper systems might be necessary. Documentation of a high P removal is necessary in order to obtain widespread acceptance of CW systems in Norway. Optimizing design for N and P removal and recycling is discussed. Aerobic pretreatment might enhance both N and P removal in the following wetland system and is therefore recommended. Winter performance of wastewater stabilization ponds aerated by Flowforms is studied and shows that Flowforms are an efficient method of pond aeration at subfreezing temperatures. Sectioning and intermittent loading as well as multistage systems are suggested as means of enhancing nutrient removal. Preliminary results from a multistage CW system indicate that high removal of both N and P can be obtained in CW systems under Norwegian climatic conditions.

KEYWORDS

Constructed wetland systems, wastewater, oxidation ponds, Flowforms, cool temperate climate, phosphorus removal, nitrogen removal, hydraulic capacity, system design.

0-87371-990-5/97/$0.00+$.50

INTRODUCTION

The use of constructed wetlands (CW) and other aquatic ecosystems is widely accepted in reducing concentrations of many pollutants from a variety of sources (Reddy and Smith 1987, Hammer 1989, Cooper and Findlater 1990). Experience with CW and the general interest in ecologically sound wastewater treatment in Norway have been limited. Such systems are not part of the technologies generally accepted for wastewater treatment. Norway has committed itself to reduce the discharge of nitrogen (N) and phosphorus (P) to the North Sea by 50% before 1995. Constructed wetlands are low-cost, low-maintenance wastewater treatment systems that can be optimized for nitrogen removal. These systems are alternatives to conventional treatment of smaller waste flows, but are also interesting as a polishing step following conventional treatment.

There are several limitations to the use of CW in water pollution control in Norway. Norway has a cool temperate climate where the average monthly temperature varies from –15 to +15°C. Both biological and chemical treatment processes are temperature dependent, hence the winter performance of CW can be questioned. Norway is located between the 58° and 71° northern latitude, which is comparable with Alaska on the American continent but, due to warm ocean currents originating in the Gulf of Mexico, the climate, especially along the coast, is warmer than in Alaska. Along the coastline winter temperatures are seldom far below freezing. It is therefore in coastal regions that the potential for CW is most promising. The inland areas of the south-east of Norway and northern Norway have a climate with warm summers and long cold winters with temperatures often below minus 20°C. Another factor that needs to be considered is precipitation, which in certain coastal areas exceed 5000 mm/year. High precipitation combined with a high relief creates extreme hydrologic conditions that have to be accounted for in design of CW.

Norway has strict effluent limits, especially with respect to phosphorus, because of sensitive recipients. Documentation of high P removal is therefore necessary in order to obtain widespread acceptance of wetland systems in Norway. High P removal is reported by several authors (e.g., Bucksteeg 1990, Willadsen et al. 1990), but can high removal be achieved under Norwegian climatic conditions?

There have been contradictory results concerning the oxygen transport capabilities of plants in a CW system (Armstrong et al. 1990, Brix 1990, Brix and Schierup 1990). During the winter, the oxygen supplied by plants will definitely become limited. Without sufficient oxygen supply, the soil medium will become anaerobic and the redox level lower. This will limit nitrification and, hence, the subsequent denitrification. If the redox level is reduced there is also a chance of phosphate leaching due to reduction of iron(III) phosphate complexes (Holford and Patrick 1979). Design concepts with intermittent feeding of wetlands improves aeration (Burka and Lawrence 1990, Willadsen et al. 1990). Another possibility is aerobic pretreatment. This will improve the oxygen status prior to the wetland treatment as well as reduce the organic loading (Schierup and Brix 1990). There are several methods for aerobic treatment. The use of waste stabilization ponds aerated by Flowforms is an aesthetically attractive aerobic treatment process that can be used in combination with CW treatment (Mæhlum 1991). In this paper the winter performance of a pond and Flowform system is investigated and design criteria as well as possibilities of enhancing purification by a subsequent wetland system discussed. Preliminary results from a multistage CW system with aerobic pretreatment (sand filter) and two CW units are also presented.

WINTER PERFORMANCE OF A POND AND FLOWFORM SYSTEM

The investigated pond and Flowform system is located at Solborg Camphill, approximately 50 km north-west of Oslo. The sewage treatment plant consists of a three-chamber septic tank followed by three ponds in series. The ponds are 1.2 to 1.5 m deep, with a bottom layer of clay or a synthetic membrane. The first pond is 600 m^2 and the subsequent ponds are 350 m^2 respectively. The system has been in operation since 1978 and serves 75 people. The average flow is estimated at 15 m^3/day. Flowforms (Wilkes 1986) are used to aerate the water in the first pond. The first pond is continuously aerated by Flowforms throughout the year. The second and third ponds are not aerated. The water is continuously pumped from the bottom of the pond and lifted up 3 m above the pond surface and onto the Flowforms. The Flowforms at Solborg Camphill consist of a cascade made of 11 uniformly sized, specially designed, cup-shaped concrete forms. Flowforms (Figure 1) induce a figure-eight movement in running water, thus generating a vortical meander which pulsates rhythmically. The pump has a capacity of about 0.15 m^3/min.

The water quality in the pond system has been monitored since 1984, but this paper presents an intensive study during a winter season from December 1990 to March 1991. The mean air temperature during the period was -10°C, and the ice cover was between 30 and 50 cm thick.

The samples were analyzed for suspended solids (SS), chemical oxygen demand (COD), total-N, NO_3-N, NH_4-N, total-P, PO_4-P, chloride (Cl^-), conductivity, temperature, pH and oxygen. Oxygen measurements were performed for the septic tank effluent, Flowforms and in the three ponds. In the ponds 40 holes were drilled through the ice and the oxygen concentration measured at different depths. The oxygen measurements were repeated once a week for 1 month. Then the pump to the Flowforms was turned off, and oxygen concentrations measured again after 1 week without the Flowforms in operation. For further details of these investigations, refer to Jenssen et al. (1991b) and Mæhlum (1991).

The purification results from the Solborg Camphill sewage treatment plant are shown in Table 1 and the oxygen measurements in Figures 2 and 3.

During the investigation period (from 20.12.90 to 20.03.91) all the ponds, except pond 1, were completely covered with ice. In pond 1 there was a small area of approximately 4 m^2 located at the outlet from the Flowforms that was not frozen. The results of the oxygen measurements (Figures 2 and 3) indicate that aeration through the Flowforms is effective. At the end of the Flowforms the water is almost completely saturated with oxygen (Figure 2) and a high oxygen content is maintained almost down to the top of the sediment layer (Figure 3). When the pump is stopped, the oxygen concentration in pond 1 drops to the same level as in the septic tank effluent in a matter of a week. The total volume of septic tank effluent discharged in 1 week is only 1/10 of the total volume in pond 1. The decrease in oxygen content is therefore not just because of water exchange, but must also be due to oxygen-

FIGURE 1. Waste stabilization pond aerated by Flowforms at Järna, Sweden.

consuming processes in the pond. The average oxygen content of the water, with the Flowforms in operation, decreases from the outlet of the Flowforms to pond 3. This indicates that biological processes are occurring even at temperatures between 0 and 1.5°C, which were the temperatures measured in the ponds at Solborg. The temperature in the septic tank effluent was 9°C.

Table 1 shows that the removal of COD, TN, and TP is 80, 28, and 65%, respectively, in the pond system, but the main reduction occurs in pond 1, the aerated pond. Excessive P losses from pond systems due to anaerobic conditions under the ice and mobilization of phosphate have been reported in other systems (Økland 1982). This was not observed at Solborg.

TABLE 1
Average Concentrations and Purification Efficiency of the Solborg Camphill Wastewater Treatment System During the Period December 1990 to March 1991

Wastewater constituent	Effluent concentrations[a], mg/l			Removal efficiency	
	STE[b]	Pond 1	Pond 3	%	% Corrected[c]
COD	330	36	27	92	80
TSS	79	7	14	83	59
NH_4-N	49.4	13	12.6	75	39
NO_3-N	0.05	1.4	0.23		
TN	52	18	15	70	28
TP	9.5	2.0	1.4	86	65
Cl^-	33	15	14		

[a] Average values of 4 to 6 samples
[b] Septic tank effluent
[c] Cl^- concentrations indicate that the water is diluted 2.4 times by the time it reaches pond 3. The removal rates in this column are corrected for the dilution.

From Mæhlum 1991.

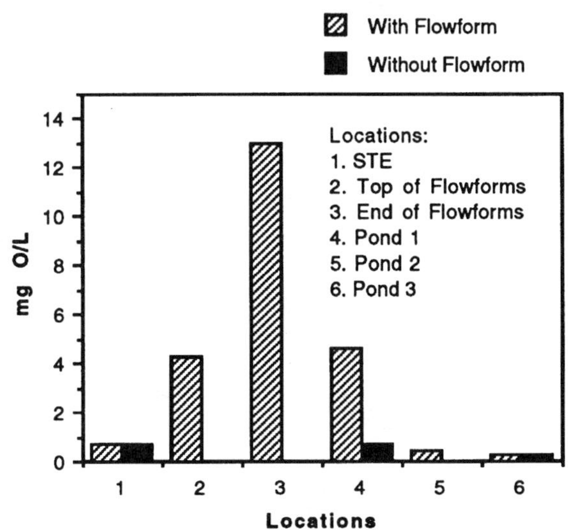

FIGURE 2. Oxygen concentrations with and without Flowforms in operation at different locations in the wastewater treatment system at Solborg Camphill.

Nitrification and denitrification are temperature-dependent processes that are very slow at low temperatures (Sawyer et al. 1973). However, Table 1 shows that some nitrification is occurring. Denitrification would also be possible in the upper anoxic part of the sediment layer of pond 1 and possibly also in pond 2 and 3.

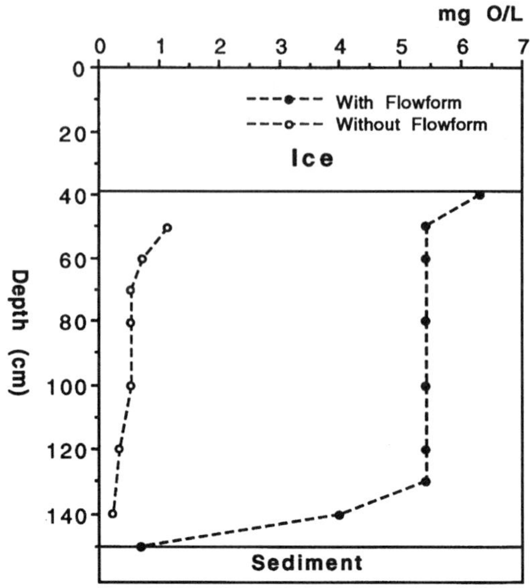

FIGURE 3. Oxygen concentrations as a function of depth in pond 1 at the wastewater treatment system at Solborg Camphill.

According to Sahrawat and Keeney (1986), nitrogen is also lost to the atmosphere as nitrogen gas during the nitrification step. This mechanism may also account for some of the N removal.

The investigations showed that the Flowforms effectively aerated pond 1 and that biological processes occurred at temperatures between 1.5 and 0°C. Flowforms are a sophisticated way to aerate water (Figure 1). Other means of supplying oxygen, as for example by compressed air devices, a rotating propeller, or a fountain, might be cheaper and more effective. However, Flowforms have the advantage of an aesthetic appearance, and will enhance the attraction of a sewage treatment plant based on ecological engineering.

There is not enough data to fully explain the high nutrient removal. The long retention time (theoretically 60 days in pond 1 and 120 days for all three ponds) does, however, allow slow incorporation in biological tissue and subsequent sedimentation. In addition, the oxidizing conditions created by the aeration of pond 1 greatly reduce the extent of anaerobic conditions and consequently reduce the possibilities for phosphate mobilization.

This indicates that an aerated waste stabilization pond can provide efficient treatment under winter conditions. It is therefore likely that reconstruction of pond 2 and 3 into a CW system would upgrade the effluent to very high, possibly tertiary standards.

OPTIMIZING DESIGN OF CONSTRUCTED WETLANDS IN A COOL CLIMATE

The following discussion concerns CW systems where horizontal subsurface flow conditions prevail (Figure 4). Design and operation guidelines for CW systems

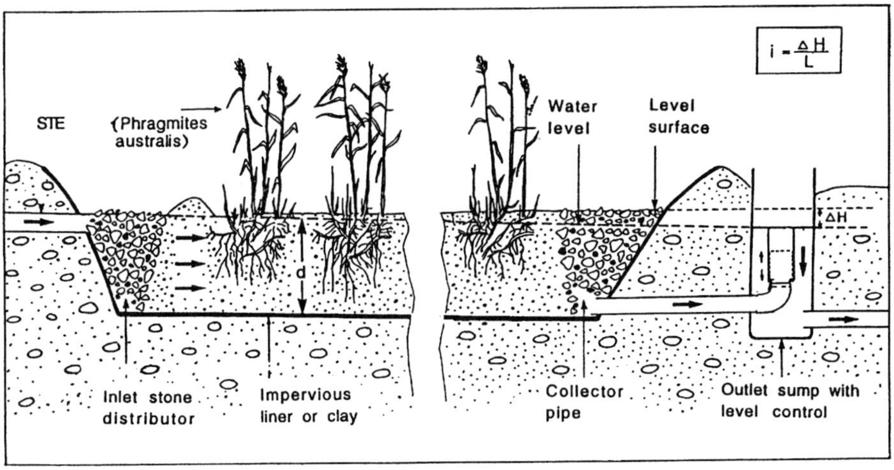

FIGURE 4. Typical arrangement of a CW system with subsurface horizontal flow.

exist (Bucksteeg 1990, Cooper 1990, Steiner et al. 1991), but the systems are in their infancy and many essential processes are not fully understood. Rational criteria for adoption of such systems to a cool temperate climate have not yet been developed. It is not documented how well CW systems will perform under Norwegian conditions, however, it is shown that CW systems can work well in climatic conditions close to those encountered in Norway. A considerable knowledge base is available (e.g., Reddy and Smith 1987, Hammer 1989, Cooper and Findlater 1990) and several possibilities for improved design exist. We therefore see many possibilities for optimizing the design to suit Norwegian conditions.

The cold winter climate in the eastern and northern parts of Norway may restrain the use of CW systems. If they are insulated, however, hydraulic failure due to freezing can be avoided. If the plant material is not harvested it will provide insulation, but whether this is sufficient, especially in the first years of operation, is questionable. Additional insulation may therefore be necessary unless a good snow cover arrives in the early autumn.

Surface area calculated on the basis of BOD loading is often used to assess the required size of a system, but using surface area alone is inadequate and disregards important hydraulic considerations. The design and sizing of a constructed wetland system should ideally depend upon:

- Hydraulics
- Purification goals

The hydraulics should always be considered first, because the hydraulic properties of the soil/porous media to be used will determine the minimum size of the system. If the purification goals also are met with this size no further design consideration, apart from aesthetics, needs consideration.

HYDRAULIC CONSIDERATIONS

The hydraulic capacity of a wetland system, Q (m^3/day), can be defined as the amount of water that can flow through the system per unit time without occurrence of surface flow. The influent volume, Q' (m^3/day), must never exceed Q. The hydraulic capacity is determined by the hydraulic conductivity, K (m/day), of the porous media, the cross sectional flow area, A (m^2), and the hydraulic gradient, i (Figure 4). This can be expressed by Darcy's law:

$$Q = K \times A \times i \tag{1}$$

A depends on the depth (d) and the width (w) of the system ($d \times w = A$). It is therefore possible to conduct the same amount of liquid provided

$$d \times w = C \text{ } (C \text{ is constant}) \tag{2}$$

In order to optimize the conditions for plant uptake and the micro-organisms in the root zone, a shallow system should be chosen ($d < 60$ cm)i however for a cool temperate climate we suggest increasing d to assure hydraulic capacity should the upper layer of the filter freeze. The average flow velocity and the retention time for water in the system can also be calculated based on Darcy's law expressed in one dimension (1). Equation 1 is not an exact model of the flow conditions, but is sufficient for most engineering purposes. The main uncertainty is in the estimation of the hydraulic conductivity. In a system consisting of more than one cell or bed we therefore suggest building one cell, testing this cell hydraulically, and reconsidering the sizing of the other cells based on this full-scale test.

The hydraulic conductivity of a wetland filter bed may decrease with time due to clogging of the interface between the inlet distribution trench and the soil/plant matrix. This possible development of clogging, which has not yet been thoroughly studied, may cause surface flow of liquid at the intake end of the wetland bed.

PURIFICATION PERFORMANCE

In a constructed wetland, impurities are removed by a complex array of interacting natural processes. A CW system can function satisfactorily hydraulically, but may not remove an adequate amount of the pollutants. This can be due to inadequate adsorption properties of the soil or insufficient uptake by the plants. If this is the case either the surface area must be enlarged to allow more plant uptake and/or the soil volume has to be increased to augment the adsorption capacity. This situation is most likely to occur when a coarse-grained porous medium, like gravel, is used and shows one case where purification considerations determine the size and geometry of a system.

It is not easy, however, to estimate the removal rates in a wetland system, because the processes occurring are not always well known. Designing a system so as to achieve a specific purification performance therefore becomes approximate. Chemical as well as biochemical reaction rates decrease with decreasing temperature and longer retention times might be necessary to achieve adequate purification. We therefore suggest that systems for Norwegian conditions be designed deeper and with larger surface area than systems used in a warmer climate. This will render the CW systems more expensive than similar systems in a warmer climate.

OPTIMIZING REMOVAL AND RECYCLING OF NITROGEN

In a CW system, nitrogen can be removed by denitrification, adsorption of ammonia by the soil, or uptake by plants. Incorporation into the soil biomass may also account for a small removal. In order to obtain an efficient recycling of nutrients, plant uptake should be optimized and the plants harvested. The experience with harvesting of reeds (*Phragmites communis*) suggests that harvesting may cause problems such as decay of the plant population (Geller et al. 1990). As a result, harvesting of plants is not common practice in CW.

According to authors referred to by Wrigley and Toerien (1988), the removal potential for *Phragmites* ranges from 330 to 880 kg N/ha. The N discharge from one person is estimated at 4.5 kg/year (Vråle 1987). This means a surface area between 50 to 136 m^2/person would be required in order to remove all the nitrogen by harvesting of *Phragmites*. The normal surface area of a CW system in Europe ranges between 3 to 10 m^2/p.e.

Unless other and more efficient plants than *Phragmites* are found or extremely large systems are constructed, removal of more than 10 to 15% of the nitrogen through harvesting of plants seems to be difficult. Denitrification is therefore the process that has to be optimized if substantial amounts of N are to be removed. Denitrification represents a loss of N and the end product of the denitrification process can be both N_2 and N_2O. N_2O is a contributor to global warming. However, in a CW system, conditions rendering N_2 as the major end product are likely to prevail (Sahrawat and Keeney 1986).

The conditions for denitrification, an anaerobic (anoxic) environment containing a carbon source, are easily met in a constructed wetland system, but the oxidizing conditions necessary for formation of nitrate are often limited (Brix and Schierup 1990), hence, nitrification restrains the rate of denitrification. Aerobic pretreatment is therefore suggested as a means of enhancing denitrification in a wetland system.

The results (Table 1) showed that 76% of the nitrogen in the effluent from the aerated pond was ammonium. Ammonium is easily adsorbed by soil under anaerobic conditions. It should therefore be possible to adsorb the ammonium discharged from the pond system in a subsequent CW system during the winter months. If a sandy soil with an adsorption capacity of 50 to 100 meq/kg is chosen, approximately 2 to 3 m^3 of soil is needed to adsorb the nitrogen from one person over a period of 6 months.

Provided the wetland system is constructed for intermittent feeding, the following approach can be suggested for enhancing N removal in a cool temperate climate:

1. Adsorb ammonium during the winter months.
2. Aerate the part of the system that has adsorbed ammonia during the winter by lowering the water level in the system. This will initiate nitrification.
3. Load the aerated part of the system with septic tank effluent in order to create anaerobic conditions and provide a carbon source for denitrification.

OPTIMIZING REMOVAL AND RECYCLING OF PHOSPHORUS

P removal in constructed wetland systems is variable. In general removal efficiencies are less than 40%, but there are also authors reporting removal in the range of 70 to 96% (e.g., Bucksteeg 1990, Willadsen et al. 1990). This shows that there is a potential for high P removal in CW systems, but good documentation of long term (> 10 years) high P removal is lacking.

The potential for P removal by plants is lower than the potential for N removal (Wrigley and Toerien 1988). The removal of P therefore has to rely on adsorption/precipitation reactions. Fe- and Al-rich mineral soil generally has a high capacity for phosphate adsorption and precipitation.

From Denmark, Willadsen et al. (1990) report removal efficiencies of more than 95% by adding quicklime (CaO) to the wastewater prior to discharge into the wetland system. Other investigators have found high removal efficiencies when using fine grained soils and in systems with low loading rates (Mark et al. 1989) or in soils with a high iron content (Bucksteeg 1990). Adding various iron and aluminum compounds has yielded higher removal efficiencies in traditional vertical flow sand filters (Brattebø 1983, Nilsson 1990). It is likely that such additives also will enhance P removal in wetland systems. Jenssen et al. (1991a) have found that light expanded clay aggregates (LECA®) have a high potential for P adsorption. Batch experiments indicate that the fraction 0 to 4 mm can adsorb more than 4 kg P/m^3. In addition the hydraulic conductivity is more than 1000 m/day. The LECA is often used as a growth medium for plants. This material therefore has very promising characteristics, and is now being tested in full-scale CW systems (see below).

Phosphorus adsorption sites can be blocked by a biofilm coating on the soil particles (Nilsson 1990). Application of low-BOD water may reduce the problem of biofilm production. Efficient aerobic pretreatment may therefore enhance P removal. The following multistage system can be suggested in order to improve P removal:

1. Aerobic pretreatment,
2. CW N removal unit, and
3. CW P removal unit,

where the P removal unit constitutes the final step.

P removal by chemical precipitation with lime in a pond can be suggested as an alternative to using a CW in stage 3. The P adsorption is finite and after some years (depending on system volume, P adsorption capacity, and loading rate) the P removal efficiency will decrease. However, it has been shown in studies of wastewater seepage beds (vertical infiltration) that the P adsorption capacity can be regenerated if the system is allowed to rest and dry (Miller and Wolf 1975, Sawhney and Hill 1975). In order to optimize P removal, CW systems should be constructed so that intermittent loading of different sections is possible. By intermittent loading it may be possible to extend the service life of the system considerably. When the system eventually ceases to remove P, the soil/porous media should be replaced. The excavated

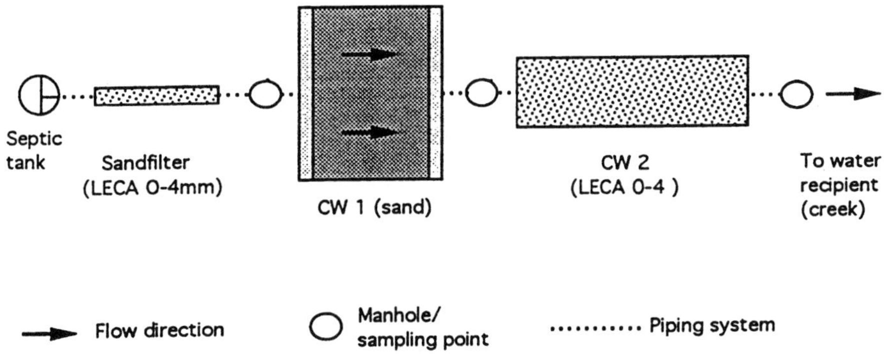

FIGURE 5. The layout of the multistage sewage treatment plant at Haugstein farm.

TABLE 2
Effluent Values (mg/l) and Overall Removal Efficiency (%) for the Haugstein Sewage Treatment Plant

	STE[a]	SFE[b]	CW1	CW2	% removal[c]
COD	250	100	127	55	78
BOD_7	210	90	45	20	90
TSS	56	27	26	10	82
TN	102	95	48	35	65
TP	11	8	1.1	0.3	97
Cl^-	52	52	44	38	

[a] Septic tank effluent.
[b] Sand filter effluent.
[c] Removal (%) is not adjusted for dilution due to precipitation based on the Cl^- concentrations.

soil will have a high nutrient content and may, if the level of other possible harmful constituents is low, be recycled into agricultural production.

PRELIMINARY RESULTS FROM A MULTISTAGE CONSTRUCTED WETLAND SYSTEM

In 1991 a multistage CW system (Figure 5) was built at Haugstein farm approximately 20 km east of Oslo in southern Norway.

The system serves two households and consists of a 3-chamber septic tank, a vertical flow single pass sand filter made of LECA (surface area: 6 m × 0.5 m, depth: 0.7 m), CW unit 1 with iron-rich sand (surface area: 6 m × 10 m, depth: 0.8 m, and CW unit 2 with LECA (surface area: 4 m × 10 m, depth: 0.9 m). The results from the first 18 months (Dec. 1991 to May 1993) of operation are shown in Table 2.

The overall removal is high for all the parameters. The COD values increase through the first CW unit (CW1). This may be due to organic matter (e.g., humic acids) leaching from the sand material and organic matter introduced when planting. The CW1 unit is designed to be the main N removal unit and the main removal of N occurs here. However, the nitrification in the sand filter has so far been limited due to high pH in the LECA material used in the sand filter. It is not known yet whether the N removal is mainly due to ammonia adsorption or if nitrification and subsequent denitrification has occurred in CW1.

Nearly all the P has been retained. The theoretical service life for P removal in CW2 is 15 years. In addition there is some P removal capacity in the sand filter and CW1. When the P removal drops, CW2 can be given a resting period or the LECA in CW2 can be replaced. The removed material will constitute a suitable soil conditioner and possibly fertilizer. However, the value of the used LECA as P fertilizer depends on the bioavailability of the associated P.

The results in Table 2 are preliminary and from a system where a good vegetative cover is not yet established. We expect the COD removal and maybe also the N removal to increase as the system matures. The P removal will decrease with time, but the rate at which this will happen remains to be seen. For more details on the Haugstein wastewater treatment system the reader is referred to Jenssen et al. (1992).

CONCLUSIONS

An aerated facultative waste stabilization pond can provide efficient treatment under winter conditions. Biological processes occur at temperatures between 0 and 1.5°C. Oxidizing conditions created by an aerated pond will reduce the extent of anaerobic conditions and consequently reduce the possibilities for phosphate mobilization. More knowledge is needed about purification processes at low temperatures in order to refine design criteria for natural wastewater treatment systems operating under cold conditions.

Preliminary results indicate that high nutrient removal can be obtained in constructed wetlands (subsurface horizontal flow systems) in a cool temperate climate. Multistage systems with aerobic pretreatment (e.g., aerated ponds or sand filter) and the use of reactive porous media are suggested as means of enhancing nutrient removal. Cold climates require larger systems with longer retention times. Hydraulic failure of constructed wetlands due to the cold climate can be avoided by insulation of the system.

REFERENCES

Armstrong, W., J. Armstrong and P.M. Beckett. 1990. "Measurement and modelling of oxygen release from roots of *Phragmites australis*." In: P.F. Cooper and B.C. Findlater (eds.), *Use of constructed wetlands in water pollution control*. Pergamon Press: Oxford. pp. 41-51.

Brattebø, H., 1983. Phosphorus removal from wastewater by fixed-bed adsorption on granular activated alumina. Ph.D. Thesis. Division of hydraulic and sanitary engineering, Norwegian Inst. of Technology, Trondheim.

Brix, H., 1990. "Gas exchange through the soil-atmosphere interphase and through dead culms of *Phragmites australis* in a constructed reed bed receiving domestic sewage." *Wat. Res.* 24: 259-266.

Brix, H. and H.-H. Schierup, 1990. "Soil oxygenation in constructed reed beds: The role of macrophyte and soil-atmosphere interface oxygen transport." In: P.F. Cooper and B.C. Findlater (Eds.), *Use of constructed wetlands in water pollution control*, Pergamon Press: Oxford. 41-51.

Bucksteeg, K., 1990. "Treatment of domestic sewage in emergent helophyte beds — German experiences and ATV-guidelines H 262." In: P.F. Cooper and B.C. Findlater (Eds.), *Use of constructed wetlands in water pollution control*, Pergamon Press: Oxford. pp. 505-515.

Burka, U. and C. Lawrence, 1990. "A new community approach to waste treatment with higher water plants." In: P.F. Cooper and B.C. Findlater (Eds.), *Use of constructed wetlands in water pollution control*, Pergamon Press: Oxford. pp. 359-371.

Cooper, P.F. (ed.). 1990. *European guidelines and operations — guidelines for reed bed treatment systems.* Report No: UI 17. WRc, Swindon, England.

Cooper, P.F. and B.C. Findlater (eds.), 1990. *Use of constructed wetlands in water pollution control*, Pergamon Press: Oxford. 605 pp.

Geller, G., K. Kleyn and A. Lenz, 1990. "Planted soil filters" for wastewater treatment: the complex system "planted soil filters", its components and their development. In: P.F. Cooper and B.C. Findlater (eds.), *Use of constructed wetlands in water pollution control*, Pergamon Press: Oxford. pp. 161-170.

Hammer, D.A. (ed.), 1989. *Constructed wetlands for wastewater treatment: Municipal, industrial and agricultural.* Boca Raton, FL: Lewis Publishers.

Holford, I.C.R. and W.H. Patrick, Jr., 1979. "Effects of reduction and pH changes on phosphate sorbtion and mobility in an acid soil." *Soil Sci. Soc. Am. J.* 43: 292-297.

Jenssen, P.D., 1990. "Methods for measuring the saturated hydraulic conductivity of tills." *Nordic Hydrology*, 21: 95-106.

Jenssen, P.D., T. Krogstad, T. Briseid and E. Norgaard, 1991a. "Testing of reactive filter media (LECA) for use in agricultural drainage systems." Proc. International seminar of the technical section of C.I.G.R. on environmental challenges and solutions in agricultural engineering, Agricultural Univ. of Norway, Ås, Norway, July 1-4, 1991.

Jenssen, P.D., T. Krogstad and T. Mæhlum, 1991b. "Wastewater treatment by constructed wetlands in the Norwegian climate: Pretreatment and optimal design." In: C. Etnier and B. Guterstam (eds.), *Ecological Engineering for Wastewater Treatment.* (first edition). Bokskogen: Gothenburg, Sweden. pp. 227-238

Jenssen, P.D., T. Mæhlum and T. Krogstad, 1992. "Adapting constructed wetlands for wastewater treatment to northern environments." Paper presented at INTECOL'S IV International Wetlands Conference: "Global Wetlands — Old World and New", 13-18 September 1992, Columbus, Ohio. Will appear in proceedings.

Mark, H.M., H. Madsen and B.S. Jebsen, 1989. Lavteknologisk spildevandsrensning for landbrugsejendomme. Report no. 41, Hedeselskabet, Denmark. 115 pp.

Miller, F.P. and D.C. Wolf, 1975. "Renovation of sewage effluents by soil." In: *Individual Onsite Systems.* Second National Conference, National Sanitation Foundation, Ann Arbor, MI, U.S.A., 87-102.

Mæhlum, T. 1991. Bruk av konstruerte våtmarker til rensing av avløpsvann i Norge. M.Sc. thesis, Agricultural Univ. of Norway, Ås.

Nilsson, P. 1990. Infiltration of wastewater — an applied study on treatment of wastewater by soil infiltration. Report no. 1002, Dept. Environ. Eng., Lund Univ., Sweden.

Økland, J. 1982. *Ferskvannets verden 1 — Miljö og prosesser i Innsje og elver.* Universitetsforlaget. 203 pp.

Reddy, K.R. and W.H. Smith (eds.), 1987. *Aquatic plants for water treatment and resource recovery.* Magnolia: Orlando, Florida.

Sahrawat, K.L. and D.R. Keeney, 1986. "Nitrous oxide emission from soils." In: B.A. Stewart (ed.), *Advances in soil science 4.* Springer-Verlag: New York. pp. 103-148.

Sawhney, B.L and D.E. Hill. 1975. "Phosphate sorbtion characteristics of soil treated with domestic waste water." *J. Environ. Qual.* 4: 342-346.

Sawyer, C.N, H.E. Wild and T.C. McMahon, 1973. Nitrification and denitrification facilities. U.S. Environmental Protection Agency, Technology Transfer Publ.

Schierup, H.-H. and H. Brix, 1990. "Danish experience with emergent hydrophyte treatment systems (EHTS) and prospects in the light of future requirements on outlet water quality." *Wat. Sci. Tech.* 22: (3/4, 65-72.

Steiner, G.R., J.T. Watson and K.D. Choate, 1991. *General design, construction, and operation guidelines: Constructed wetlands wastewater treatment systems for small users including individual residences.* TVA/WR/WQ — 91/2, Tennessee Valley Authority: Chattanooga, Tennessee.

Wilkes, J., 1986. *Flowforms: An innovative approach to the rhythmic properties and uses of water.* Planned Innovation, London.

Willadsen, C.T., O. Riger-Kusk and B. Qvist, 1990. "Removal of nutrient salts from two Danish root zone systems." In: P.F. Cooper and B.C. Findlater (Eds.), *Use of constructed wetlands in water pollution control*, Pergamon Press: Oxford. pp. 115-126.

Wrigley, T.J. and D.F. Toerien, 1988. "The ability of an artificially established wetland system to upgrade oxidation pond effluent to meet water quality criteria." *Water SA* 14 (4): 171-178.

Chapter 23

PERSPECTIVES ON VEGETATED SHOALS, PONDS, AND DITCHES AS EXTENSIVE OUTDOOR SYSTEMS OF WASTEWATER TREATMENT IN ESTONIA

Ülo Mander, Olev Matt, and Urmas Nugin

ABSTRACT

In recent years, different extensive outdoor systems for the treatment of wastewater from settlements and large farm complexes of Estonia have been designed and built up. They consist of various artificial wetland ecosystems where aquatic vascular plants (e.g., reeds, cattails, bulrushes, etc.) and also willows and alders are grown. Among them, vegetated bioponds and bioditches on natural soil, root-zone systems, high-moor filters, vegetated shoals in bodies of water, and vegetated riparian buffer strips/buffer zones have the greatest importance. The bioponds and ditches, root-zone systems and moor filters have been used mostly for secondary treatment of wastewater from point pollution sources, but vegetated shoals and riparian biotopes play an important role in control of non-point source pollution in rural areas. All of these systems can adsorb and/or transform organic matter, mineral nitrogen and phosphorus, heavy metals, pesticides, and other substances which in large amounts may exceed the corresponding capacity of conventional (indoor) systems.

INTRODUCTION

Wastewater treatment is one of the most important elements of environmental protection in Estonia. Wastewater causes a growing eutrophication of rivers, lakes, and coastal waters, and increasing level of pollutants (especially nitrates) in groundwater. Actually, only 30% of total amount of municipal and industrial wastewaters which are badly in need of treatment (i.e., 212 million m^3/yr, not including wastewater from oil-shale pits and aquacultures) has been purified both biologically and mechanically. All wastewater treatment is organized around about 1100 conventional systems with an average efficiency of 70 to 80% with respect to BOD_5 and 60 to 70% with respect to nitrogen and phosphorus (Maastik 1984). Among them are about 800 activated sludge plants (7 types for 25 to 50,000 population equivalents (p.e.), mostly combined with secondary treatment in aerobic and/or anaerobic bioponds), about 250 biopond complexes for primary treatment, 12 biofilters, and 45 mechanical filters (Estonian Board of Waters 1990). According to future plans, in 1995 up to 70% of wastewater must be treated biologically (Figure 1). This figure

0-87371-990-5/97/$0.00+$.50

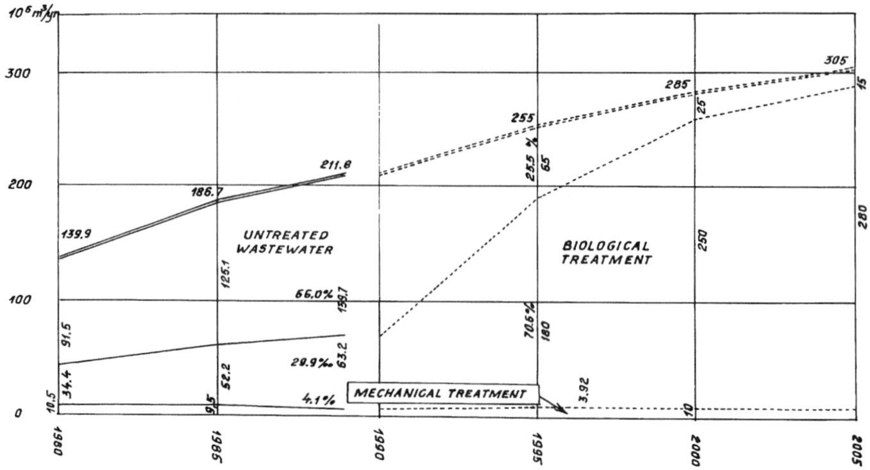

FIGURE 1. Perspectives on wastewater treatment in Estonia. Wastewater from oil-shale pits and fish-breeding beds not included. Data are until 1989; data from 1990 onward are projected.

can be reached primarily through the development of conventional activated sludge plants. The elimination of mineral nitrogen and phosphorus by chemical means is also planned.

But apart from conventional ways, some alternative and innovative processes based on the use of aquatic vascular plants in artificial wetlands have been planned for wastewater purification. It is already well known nowadays that such treatment systems are very attractive because of their high efficiency and relatively low cost (Haberl et al. 1984).

WASTEWATER PURIFICATION IN NATURAL ECOSYSTEMS

There are some practical examples of the purification effect of municipal and industrial wastewaters in natural/seminatural wetlands even without any previous treatment. For instance, the 300-m wide valley flat of River Valgejõgi by the small town Tapa in North Estonia has been used for 25 yr as a sink for concentrated municipal wastewater (with effluents from a distillery) with an average discharge of 60,000 to 80,000 m^3/d. Over an area of 18 ha, large stands of willows, reed, and cattail have formed. The retention of organic matter (BOD_5), mineral nitrogen, and total phosphorus during the vegetation period (May to October) is 96, 98, and 88%, respectively, and in winter 56, 62, and 27%. Each hectare of this seminatural wetland is able to accumulate up to 1100 kg phosphorus and 2100 kg N/yr (Figure 2). But owing to very high input concentrations (on an average: BOD_5 = 250 mg/l, N = 50 mg/l, P = 10 mg/l) the River Valgejõgi is extremely eutrophied (α-mesosaprobic) downstream.

This valley flat is planned as a secondary treatment system for a new activated sludge plant which is under construction.

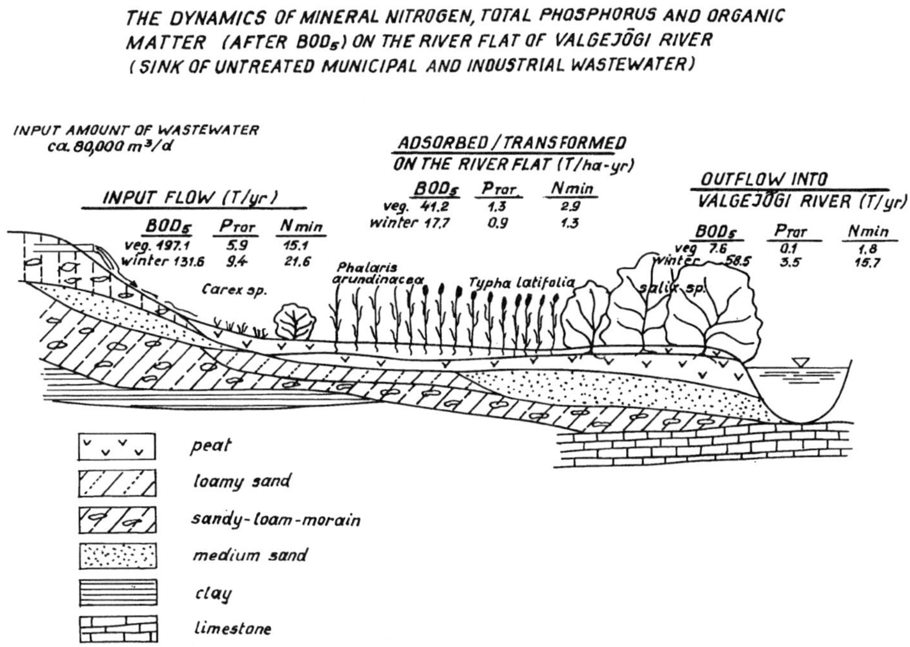

FIGURE 2. Purification of municipal wastewater on the valley flat of Valgejõgi River.

VEGETATED OUTDOOR SYSTEMS FOR WASTEWATER TREATMENT: PRINCIPLES, TECHNICAL PARAMETERS, AND EFFICIENCY

Principles of land treatment of wastewater were used in Estonia already some decades ago. One of these methods is the use of vegetated buffer strips along streams and lake shores to protect waters against non-point source pollution.

The design and construction of alternative aquatic treatment systems began in Estonia in 1988. The first ones were designed by the students of land reclamation of Estonian Agricultural Academy and engineers of the State Research and Design Institute "Estonian Land Reclamation Project" under the guidance of the authors of this article.

The following text gives a short overview of main types of alternative outdoor systems for wastewater treatment, based on the use of aquatic/terrestrial processes, which exist already in Estonia or have been designed and should be built.

VEGETATED SHALLOW BIOPONDS ON NATURAL SOIL

In these shallow bioponds, with an average area of 0.1 to 0.5 ha, one or more species of emergent and/or free-floating aquatic macrophytes are grown. Most have not been designed as bioponds, but are old aerobic bioponds, former storage ponds of water mills, and other previously existing bodies of water. In recent years, some

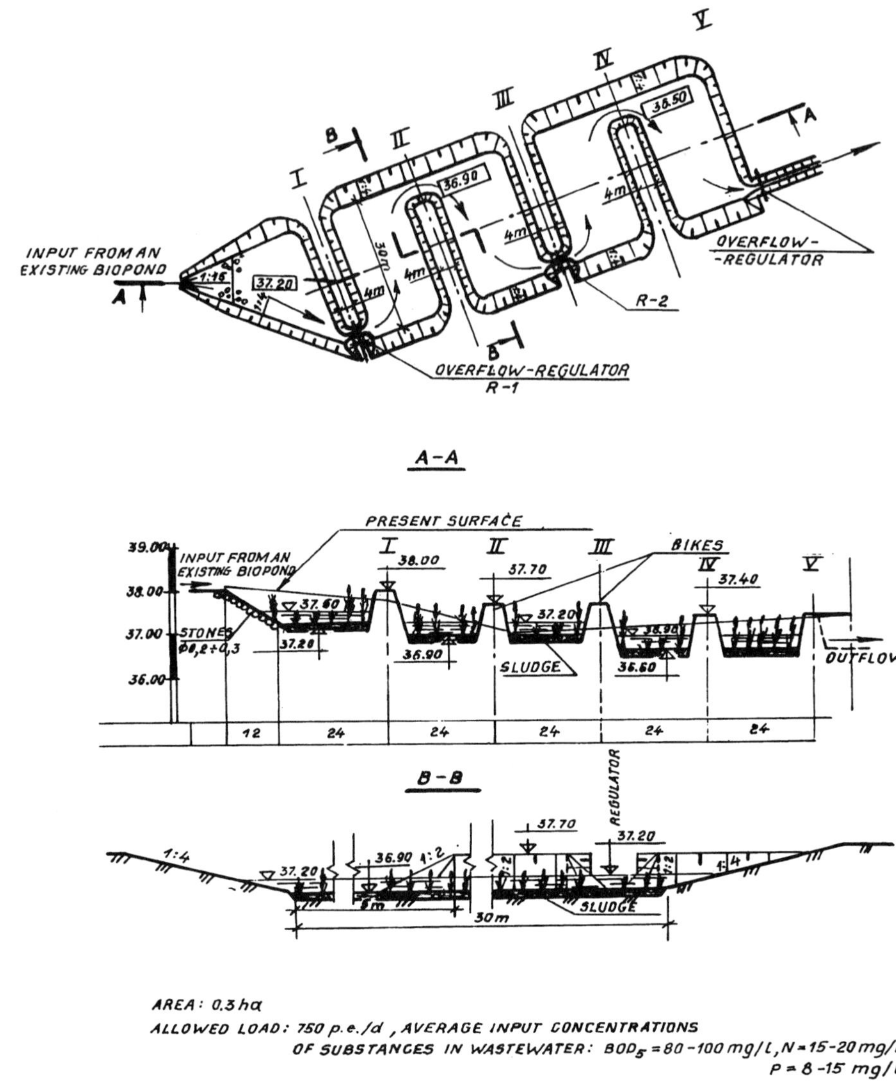

FIGURE 3. Vegetated biopond for secondary treatment of municipal and farm wastewater in Aaspere (West Viru region).

new vegetated biopоnds for secondary treatment of wastewater from villages (e.g., Reola in Tartu region; Aaspere in the West Viru region, see Figure 3), large farm complexes (e.g., Rahinge, see Figure 4; Sangla in Tartu region, and Haljala in Viru Region), and starch mills (e.g., Pajusi in Jõgeva region) have been both designed and constructed. They mostly consist of a cascade of three or four shallow ponds or have a serpentine-like form. Reed (*Phragmites communis* L.), cattail (*Typha latifolia* L.), bulrush (*Schoenoplectus lacustris* (L.) Palla), and other aquatic macrophytes are dominant in these ponds.

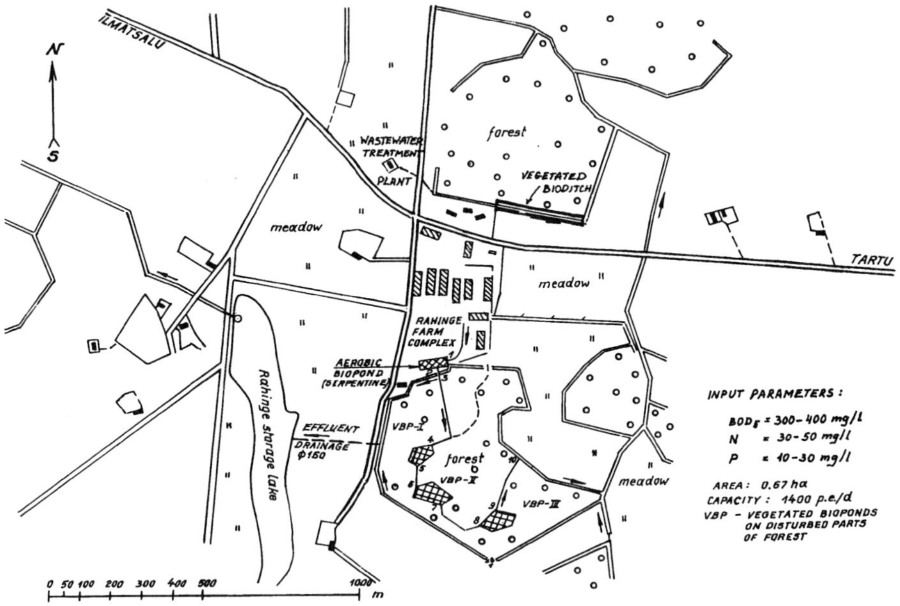

FIGURE 4. A cascade of vegetated biopods for farm wastewater treatment in Rahinge (Tartu region).

The installations of this type are for sources with approximately 10,000 p.e. The area required for secondary treatment is 5 to 6 m²/p.e. Retention of nitrogen in the system during the year is on the average 50%, and for phosphorus, 70%.

A cascade of bioponds/bioshoals ending with a standpipe overflow from which the water runs out over a stony area for aeration has been designed on the small Vända canal in the vicinity of Tartu (catchment area 419 ha), which is loaded through non-point source pollution from intensively used adjacent agricultural fields (Figure 7 and 10).

VEGETATED BIODITCHES

These are a subtype of vegetated bioponds. The main differences between ditches and bioponds are first, the faster flow of wastewater (shorter retention time), and second, the greater diversity of plant communities (in bioditches there are, in addition to the above-mentioned species, *Glyceria* spp., *Sparganum* spp., *Potamogeton* spp., *Phalaris arundinacea* L., *Elodea canadensis* Michx., and other water plants). These ditches are commonly parts of a drainage network and located typically downstream of conventional wastewater purification plants. During the vegetation period, the retention of phosphorus and nitrogen in bioditches is very effective: downstream of overloaded or irregularly functioning wastewater treatment facilities the retention of nitrogen can be up to 2 to 3% per running meter (see Table 1), with removal especially high in the first several meters.

TABLE 1
Efficiency of Some Alternative Outdoor Systems for Wastewater Treatment in the Tartu Region (Averages)

Name of facility	Period	Input (mg/l) BOD5	Input (mg/l) N	Input (mg/l) P	Output (mg/l) BOD5	Output (mg/l) N	Output (mg/l) P	Retention (%) BOD5	Retention (%) N	Retention (%) P
Vegetated biopond	Veget.	120	50	15	25	8	5	80	84	67
in Rannu[s] *Typha* sp.[1]	Winter	135	55	18	60	25	8	66	55	56
Root-zone system	Veget.	58	15	8	6	4	2	90	77	75
in Sangla[s] *Phalaris* sp.[2]	Winter	72	25	10	15	8	4	80	68	60
Bioditch in Rahinge[s,3]	Veget.	75	35	12	5	2	1	93	94	92
(1 = 200 m)	Winter	105	40	20	45	15	8	66	62	60

[s] Secondary treatment.
[1] Wastewater from swine farm (2000 pigs), after 3 bioponds; f = 0.15 ha.
[2] Wastewater from dairy farm (600 milk-cows), after 2 ponds; f = 0.4 ha.
[3] Wastewater from a settlement and dairy farm (2500 p.e.) f = 0.02 ha.

In some new drainage systems, such buffer sections of ditches have been built. They are to filter nutrients leached from soils during the construction of drainage or storage lakes.

ROOT-ZONE SYSTEMS

Root-zone systems are artificial wetland communities of different aquatic macrophytes where all above-mentioned species are dominant. This is the most common and most effective type of outdoor system for wastewater treatment. Unlike bioponds and bioditches, in these systems the main waterflow takes place not on the surface, but inside the root zone. Correspondingly, the main purification processes run inside the soil or filter substrate.

There are three subtypes of root-zone systems accepted on an international level which are present in Estonia as well (Haberl et al. 1984; Neurohr 1984):

- Marsh beds on sandy substrates with different macrophytes (after K. Seidel).
- Reed beds on natural (loamy) soils (after R. Kickuth).
- Vegetated gravel (sand) filters (after G. Geller and A. Lenz).

The sandy marsh beds covered with reed, cattails, and bulrushes have been designed in reconstruction projects of an old water-mill pond in Pajusi State Farm. The marsh bed system will be most attractive for wastewater treatment in new private farmsteads.

Professor R. Kickuth's reed beds on loamy soils are quite similar to vegetated bioponds. The main difference is that root-zone systems are normally not flooded by wastewater. These facilities are best for secondary treatment, e.g., the small systems for purification of wastewater from large farm complexes in Rannu (Tartu region; see Table 1) and Haljala (Viru region; Figure 5).

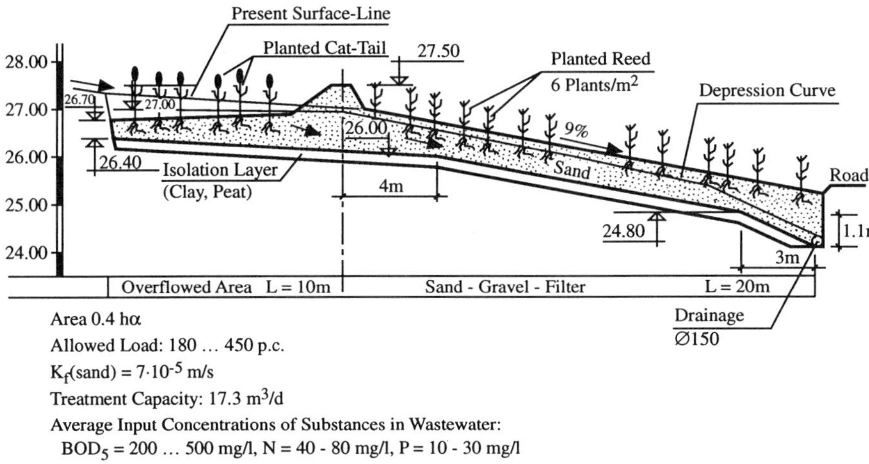

FIGURE 5. Combination of a vegetated shallow biopond and root zone system for wastewater treatment at Mullamäe farm (West Viru region).

The most promising root-zone systems are vegetated gravel or sand filters which guarantee an effective retention of nutrients (see Figure 5). The area requirements for root-zone systems are on an average 4 to 6 m^2/p.e. The most effective way to get maximal retention using these facilities is to combine different subtypes of root-zone systems in one complex, e.g., in the following order: 1) Geller-Lenz's system (aerobic, good for mineralization of organic compounds, for nitrification and phosphorus accumulation); 2) Kickuth's system (mostly anaerobic, good for nitrogen retention); 3) Geller-Lenz's system (removes the rest of the nitrogen and phosphorus).

It seems to be very attractive to use some root-zone systems for production of energy forests, e.g., with *Salix viminalis* L. Such a pilot project based on the Swedish and Danish experiences will be started this year.

A very effective root zone system was designed and built in 1988 in Rannu (Tartu region) for the secondary treatment of wastewater from a large dairy farm (about 600 milk cows). This system was built on a wet slope covered mostly with *Phalaris arundinacea* L. The retention of nutrients is very high even in winter (Table 1).

A root zone system combined with a coral-chalk filter (made of powdered *Lithothamnion* sp.) was designed for the purification of groundwater with very high concentration of nitrates (>150 mg NO_3-N/l). It will be useful for private farms located far from villages or other centers.

MOOR FILTERS

This type of extensive outdoor systems has been installed in some exhausted peat fields of Estonian high-moors. They are used both for primary and secondary treatment of wastewater. There are two subtypes of moor filters: with surface and underground input of wastewater (output of purified water flows through the underground drainage

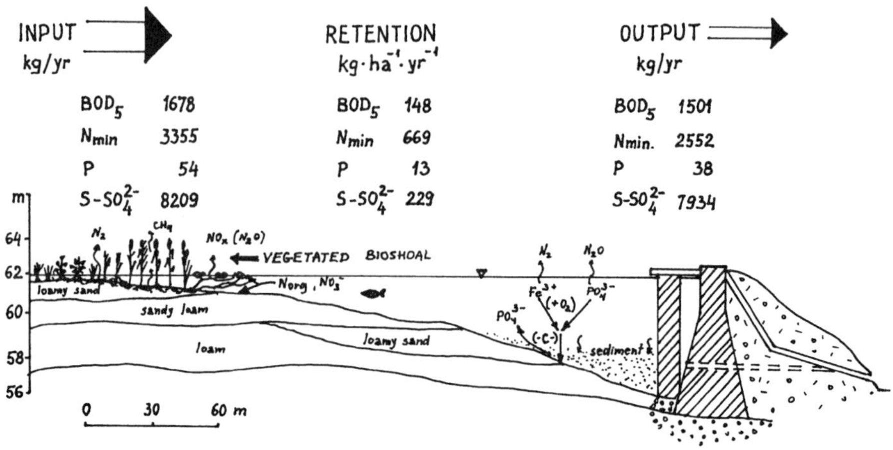

FIGURE 6. Nutrient balance in the storage pond (1.2 ha) on Vända canal (experimental watershed, Tartu region).

in both cases). But, owing to the very sterile character of peat (mostly from *Sphagnum* spp. and *Eriophorum* spp.) this system functions almost as a mechanical filter.

VEGETATED SHOALS IN BODIES OF WATER

These belong to natural or artificial buffer/filter systems which can essentially accumulate dissolved chemicals and suspended matter from water. Their role is greatest in storage lakes where they transform and adsorb mostly organic compounds and nitrogen (Figure 6).

A complex of vegetated shoals and different buffer ponds was designed for the new recreational storage lake on Konovere River. The reduction of nitrogen, phosphorus, and organic matter (BOD_5) that this reed-bulrush shoal will achieve is estimated to be 5, 8, and 11%, respectively. The area of the shoal takes less than 2% of the total lake area.

A similar system but combined with a standpipe overflow was designed for the experimental Vända canal (Figure 7).

VEGETATED BUFFER STRIPS AND ZONES ALONG RIVERS, DITCHES, AND LAKE SHORES

These buffer ecosystems have been used in water and landscape management already for some decades. There are two systems in this complex of measures in agricultural landscapes of Estonia:

- First, extensively managed perennial grasslands such as water protection zones (width 100 to 1300 m) between a body of water and intensively farmed adjacent fields.
- Second, forest and/or brush strips immediately on the banks of waterways, as water protection strips (5 to 50 m in width).

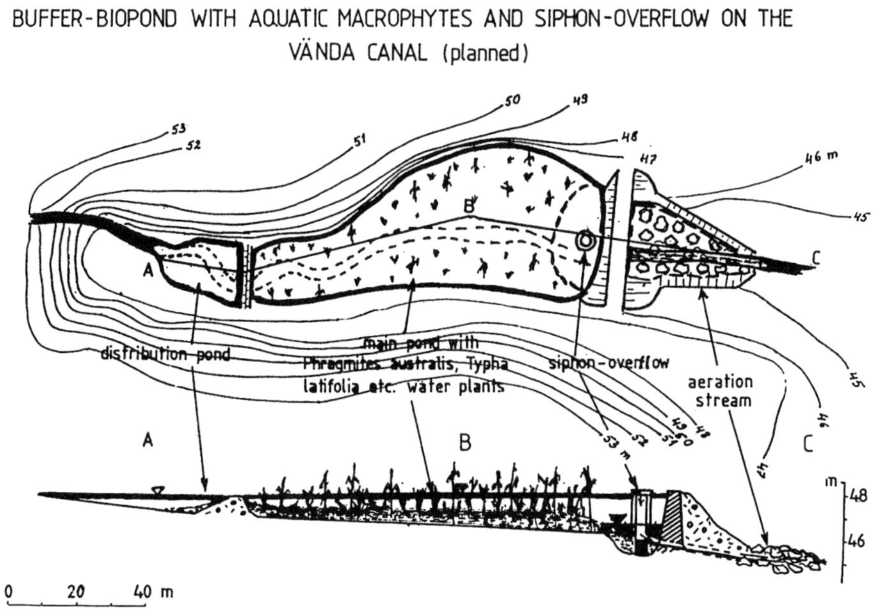

FIGURE 7. A designed biopond/shoal with a standpipe overflow on Vända canal (Tartu region).

These buffers belong to important land-water ecotones and to the ecological infrastructure of rural landscapes. The most effective biotopes for buffer strips are alder forests and willow brush. Ten-m wide strips of these biotopes can adsorb/transform nearly 100% of the phosphorus and organic matter and 50% of the nitrogen, lead and cadmium in the surface and subsurface flow. The 100-m wide zones combined with 10-m wide strips can adsorb almost all amounts of agrochemicals flowing through these filters (Mander 1989). During the vegetation period, one hectare of willow brush is able to retain up to 6000 kg BOD_5, 270 kg N and 70 kg P/day. These extreme values are achieved during intense rainfalls. The retention capability for alder woods and cultivated grasslands are two and three times lower, respectively. In the springtime, all these communities can retain only 5 to 10% of the amounts accumulated in summer (Figure 8). Data are from tests with sprinklers.

The filtration process in buffer biotopes has an exponential character: on the upper part of buffer strips (i.e., on the border strip between arable land and natural buffer community) the amounts of adsorbed and transformed substances are essentially larger than those in the middle and lower part of the buffer (Figure 9). That is why even narrow strips of forest or brush can play an important role in the control of material cycles in ecosystems.

DISCUSSION

All these alternative outdoor systems for wastewater management have a lot of advantages and also deficiencies. But not all aspects are known yet because of the innovative character of these facilities. The most important advantages are as follows:

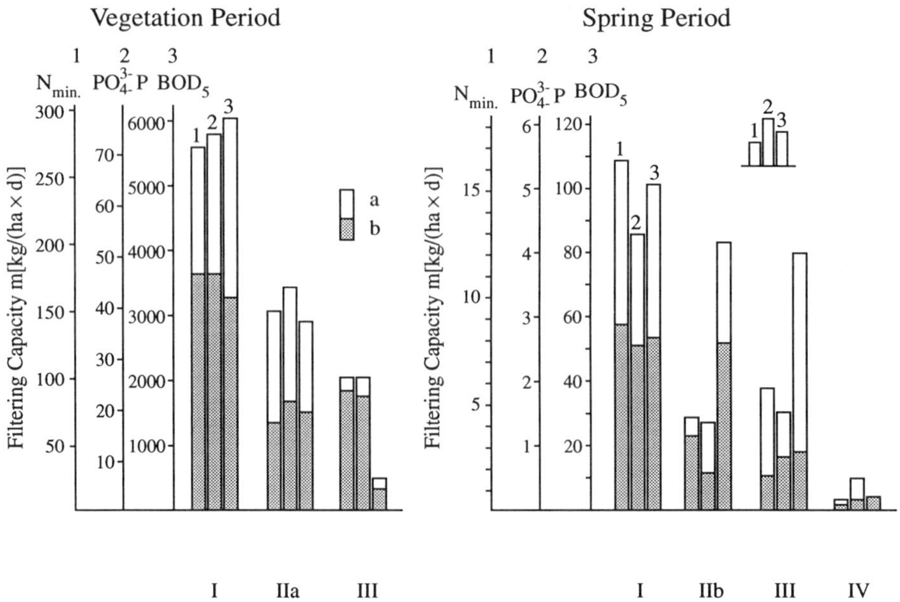

Filtering capacity of surface flow in different buffer strips:

I = Wet grassland with willow brush (Salix phylicifolia - Carex acuta),
II = Forest ecosystems:
IIa - alder wood (Alhnion incanae),
IIb - mixed forest (Picea abies, Betula pendula),
III = Intensively used grassland,
IV = Arable land;
a = The mass of substances adsorbed on the soil surface ($m_{surface}$),
b = In soil adsorbed substances (m_{soil}).

FIGURE 8. Nutrient retention in different riparian buffer biotopes during intense rainfalls and snow melts.

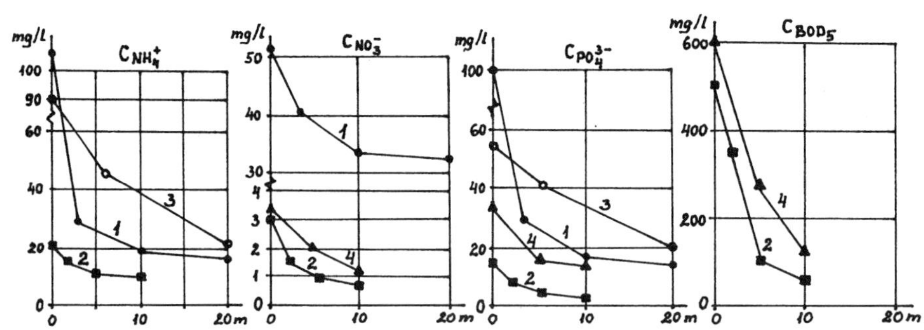

FIGURE 9. Exponential decrease of the content of different substances and the BOD_5 rate in contaminated surface flow from fields in vegetated buffer strips during the vegetation period: 1–3 — alder wood (*Alnus incana*), + — wet grassland (*Salix phylicifolia* — *Carex acuta*); x-axis indicates the distance from upper limit of buffer strips.

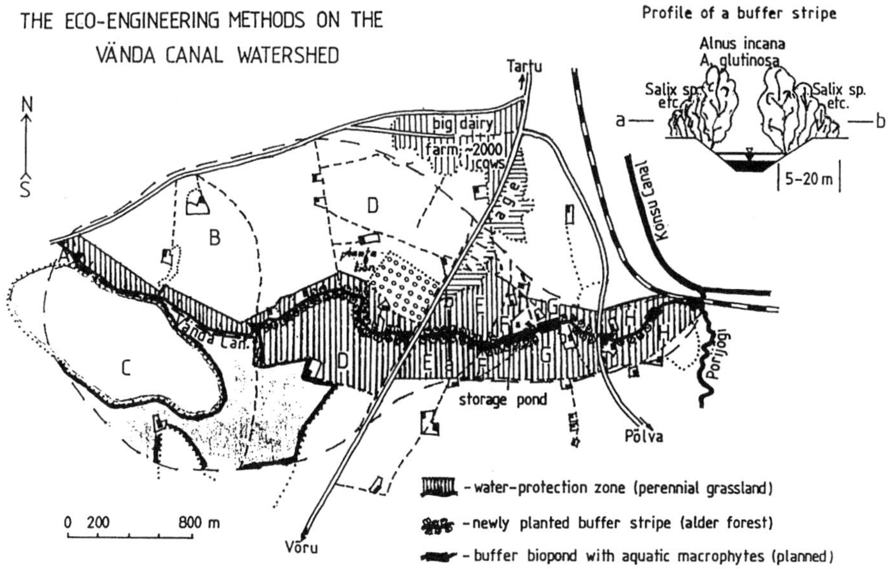

FIGURE 10. Scheme of experimental watershed of Vända canal with different eco-engineering measures to control nutrient balance.

- High efficiency in retention of all nutrients (especially mineral nitrogen and phosphorus).
- Low costs for construction and management (e.g., the root-zone systems do not need any special management after construction).
- Alternative outdoor systems are complementary habitats for biota, especially for water birds.

The most essential deficiencies are as follows:

- Loss of territory (not actually a very big problem in Estonia).
- Danger to groundwater (it is extremely important to recognize this problem in infiltration and transit areas of groundwater).
- Low efficiency in winter (it can be compensated for by combining conventional activated sludge systems or bioponds).
- Accumulation of heavy metals, pesticides, and other pollutants in the sludge and plants. The same problem occurs in all conventional systems; vegetated bioponds and root-zone systems cannot be used for agricultural purposes after exhaustion of filtration capacity.
- All outdoor purification plants are sources of "greenhouse gases" (NH_3, N_2O etc.).

The whole current situation in wastewater treatment and water management in Estonia needs the development and wide distribution of alternative outdoor purification systems.

ACKNOWLEDGMENTS

The authors are very grateful to Professor Aleksander Maastik from the Department of Water Management of Estonian Agricultural Academy and to Mr. Ülo Sults from the Tartu Department of the Estonian Nature Management Information Center for useful comments and for help in data collection.

REFERENCES

Estonian Board of Waters. 1990. *Perspectives of Water Management in Estonia*. (In Estonian).

Haberl, R., W. Diebold, R. Parsian 1984. "Darstellung verschiedener Systeme der Abwasserreinigung mit Makrophyten." *Österreichische Wasserwirtschaft* 35, 9/10.

Maastik, A. 1984. *Water Protection in Agriculture*. Tallinn: Valgus. (In Estonian)

Mander, Ü. 1989. *Kompensationsstreifen entlang der Ufer und Gewässerschutz*. Bericht am Landesamt f. Wasserhaushalt u. Küsten Schleswig/Holstein, Kiel.

Neurohr, G.A. 1984. "Abwasser — und Schlammbehandlung in Sumpfbeetfiltern: Stand und Entwicklung." *Monatsschrift für Bauwissenschaft* 11.

Chapter 24

ECOLOGICAL ENGINEERING FOR WASTEWATER CONTROL IN AGRICULTURAL CATCHMENT AREAS: THREE CASE STUDIES FROM ESTONIA

Ülo Mander, Valdo Kuusemets, Arvo Järvet, Katrin Häberli, Tiina Nõges, Arvo Tuvikene, and Tõnu Mauring

ABSTRACT

During the International Workshop on Ecological Engineering in Wastewater Treatment, Landscape and Settlement Planning organized by The Center for Ecological Engineering Tartu (CEET) in Tartu, Põlva and Otepää, Estonia from July 27 to August 7, 1992, three case studies for implementing eco-engineering measures in the agricultural catchment areas were carried out: the Vända ditch catchment area, the Kavilda River catchment area and the Lake Arbi with its drainage basin. In all areas, earlier long-term investigations served as a basis for finding solutions.

The first case study area of the Vända ditch catchment is characterized by non-point pollution problems and heavily contaminated shallow groundwater (mean annual NO_3-N content is 19 mg l^{-1}). In this catchment the following measures have been designed: water protection buffer zones (perennial grasslands) along the ditch and around wells, planted forest strips on the ditch banks, root-zone systems to clean up contaminated groundwater, and macrophyte ponds to retain nutrients in the ditch water.

For Kavilda River, which flows in a deep primeval valley covered with natural or semi-natural vegetation, the main problem is sewage from the town of Elva. Despite a new wastewater treatment plant, there are still high N and P concentrations in the outlet of the plant. Therefore, a semi-natural wetland with existing bioponds has been planned to use as an alternative purification system. Lake Arbi is a typical small eutrophic lake in the summer-resort town Elva. It receives much nitrogen and phosphorus from both sewage and fertilizers in the fields of the watershed. The restoration of the lake consisits of the manipulation of inflows (construction of a serpentine pond with aquatic macrophytes, establishment of a connecting ditch to lead polluted inflow directly to the outflow), biomanipulation of food webs inside the lake, and outflow management (establishment of a serpentine pond before entering the clean Elva River). Some of the planned measures will be implemented in 1994 to 1995.

0-87371-990-5/97/$0.00+$.50

INTRODUCTION

In the beginning of 1992, the Center for Ecological Engineering Tartu (CEET) was founded to spread innovative ideas, and accelerate the development of ecological engineering in Estonia. The idea originated from the days of the 1st International Workshop on Ecological Engineering for Wastewater Treatment, held at Stensund Folk College (Sweden, March 1991) where contacts between scientists and specialists from different countries of the whole world were established.

CEET is a non-profit organization uniting scientific and educational institutions and scientists who are working in the field of ecological engineering. CEET promotes information exchange among scientific, educational and producing institutions through organizing local and international courses and workshops on nature-conserving and energy-saving technologies (e.g., natural wastewater treatment systems, waste recycling technologies), landscape planning, and alternative energy sources. Other CEET activities include initiating and coordinating programs for common research and application, providing consulting or finding competent experts in ecological engineering, and editing publications in ecological engineering.

One of the first activities of CEET was the organization of the International Workshop on Ecological Engineering in Wastewater Treatment, Landscape and Settlement Planning in Tartu, Estonia, July 27 to August 7, 1992. In addition to CEET, the University of Tartu, Estonian Agricultural University, the Center for Applied Ecology Schattweid, Switzerland, and the Center for Ecological Education BIOS, Augsburg, Germany were the organizing institutions. This was practically the first meeting on ecological engineering organized in Eastern Europe. The main aim of the Workshop was to introduce the idea and practical possibilities of eco-engineering to specialists in wastewater treatment and regional planning working for the government and local authorities. Participants included over 60 government specialists plus scientists and engineers from Estonia, Switzerland, Germany, Sweden and Denmark.

During this workshop, a multi-layered planted soil filter (patented in Switzerland, realized at the Center for Applied Ecology Schattweid; Heeb and Züst 1994) was constructed for the treatment of raw municipal wastewater from a part of the small town Põlva in Southern Estonia. The maximal load of the system of 300 m^2 does not exceed 50 p.e. The first analyses of the work demonstrate a 90 to 95% efficiency for N and P elimination.

Three case studies on the use of eco-engineering measures in rural areas were also carried out. The study groups consisted of 10 to 15 people and were overseen by experienced specialists. Two catchment areas in the vicinity of Tartu represented intensively used agricultural areas with typical problems of water pollution. The complex of various eco-engineering measures to improve the ecological situation has been presented. One of the case study areas was a eutrophic lake Arbi in the small town Elva. Some possibilities to use planted soil filters, forest buffer strips and other buffering biotopes in the catchment area, as well as aeration and biomanipulation in the lake, have been discussed. The main results of all three cases studies are presented below.

STUDY AREA

The three case study areas are near Tartu, in the center of Estonia. They are located on the border of two landscape regions: the plain of southeast Estonia and the Otepää Heights (Varep 1964). The northern part of the study area lies within the southeast Estonian moraine plain in the vicinity of Tartu (58° 23' N; 26° 44' E). The absolute altitude of the moraine plain is from 30 to 60 m, relief is undulating (slopes normally up to 5 to 6%) and intersected by primeval valleys, 3 to 5 km wide and up to 40 m deep, formed by streams during the Pleistocene and remodeled by glaciers of the last glaciation. The southern part of the study area lies on the northern slope of Otepää Heights that is formed from moraine hills and kames with an extreme variety of glacial deposits. The altitude of this region is up to 120 m above sea level; the relative heights reach 30 to 35 m.

The bedrock of the whole area is formed by red Devonian sandstone (compact sandstone with clay and aleurolite layers of aruküla and burtnieki times) which is covered by loamy sand till of the Weichselian glaciation or fluvioglacial and glaciolacustrine sands and gravels. The Devonian sandstone lies at a depth of 2 m (in lower course) to 60 m (on the hills of Otepää Heights). The depth of the groundwater top layer varies significantly (0.5 to 20 m) depending on relief and geomorphologic conditions. The upland soils are mostly podzoluvisols, planosols, and podzols on loamy sands and fine sandy loams, with surface soil organic matter content and pH value in the cultivated fields of 1.6 to 1.9% and 5.6 to 6.5, respectively. On steeper gradients the soils have been eroded. About 50% of this area is used as arable land. In valleys and other lower areas the gleysols and peatlands, partly used as perennial grasslands, dominate. Fertilization rate of intensively managed fields reached 150 kg N, 70 kg P, and 100 kg K ha^{-1} yr^{-1}. At present, this rate has dropped significantly. In forested areas (about 25% of the territory), coniferous and mixed forests are most common. In riparian zones, alder forests and willow bushes occur. Bogs covered with pine or birch forests are the most common wetland communities on the watershed borders. The amount of rainfall is virtually identical in the whole area, but varies from one year to another from 402 to 885 mm yr^{-1}. The average annual rainfall during 1980 to 1992 was 675 mm. Forty to fifty percent of this precipitation occurs between June and September. Spring, lasting from March to May, is characterized by high water discharges due to thaw (35 to 45% of the annual total discharge). During the study period the winters have been relatively mild (mean air temperature for the coldest month, February, has been only approximately -5°C); snow cover lasted only a few days between November and April.

A more detailed description of landscape-ecological conditions is given for each case study area.

CASE STUDY 1: VÄNDA DITCH CATCHMENT AREA

Ülo Mander and Valdo Kuusemets

Since 1987 nutrient cycling in the Porijõgi River catchment area has been studied. This area (276 km^2) represents rural landscapes of south Estonia quite well. It has been divided into nine sub-watersheds with different land-use structure (Mander et al. 1989). From weirs installed in the mouth of each subwatershed, monthly water samples were taken to determine NH_4, NO_2, NO_3, N_{tot}, PO_4, P_{tot}, SO_4, Cl and Fe concentration and pH and BOD_5 values. Water discharge has been measured with propeller or triangle-shape overflows and compared with the daily measures in the lower course of the Porijõgi River. Also, precipitation and groundwater quality have been analyzed four times a year for N, P and S concentrations (Mander et al. 1994a). Since 1991 in six landscape profiles across different riparian ecotones (the most typical spectrum was arable land–cultivated grassland–natural grassland–white alder forest) soil water and groundwater, soil and biomass samples have been taken once a month to analyze nutrient transformation in ecotones. These transects are equipped with lysimeters, overland flow samplers, test plots for soil, plant biomass, and microbiological and decomposition investigations.

The intensively managed Vända ditch subwatershed (235 ha) has been divided into eight elementary watersheds. With the same interval as in the whole catchment area water samples from weirs were taken and water discharge measured. Additionally, fertilization, yield and crop quality in each field has been registered or estimated. Because of several complicated environmental problems and long-term investigations we decided to choose the Vända ditch catchment area as the first case study area of the Workshop. An overview of the Vända ditch catchment with groundwater quality data and planned eco-engineering measures is shown in Figure 1. The main environmental problems in this catchment area are:

- High concentration of NO_3-N in the shallow groundwater. The mean annual value is 19 mg l^{-1}, maximal values reach 50 mg l^{-1} (see Figure 1); about 70% of all wells have the NO_3-N concentration higher than EC standard (10 mg NO_3-N l^{-1}). Despite slightly decreasing fertilization intensity, the nitrate concentration in the groundwater is still very high. However, wells with buffer zones (non-fertilized grassland around wells) showed a significantly lower nitrate concentration in water (Mander et al. 1994a).
- Relatively high runoff values of total inorganic nitrogen (TIN), P_{tot} and BOD_5 from the watershed; however, the last 2 years these values have decreased slightly (Table 1, Figure 2, 5).
- Intensive erosion of the ditch banks due to very intensive crop-rotation and non-sustainable land use in the vicinity of the ditch; a large gully has formed in the middle course of the ditch (Figure 1).
- Poor management of slurry and manure composting sites; large areas in the vicinity of the ditch without any impermeable bottom layer have been used for depositing slurry (Figure 1).
- Unsatisfactory quality of wells in farmhouses. These last two factors are also reasons for bad drinking water quality of this region.

The nutrient budget in the Vända ditch catchment area shows quite low temporal changes both in inputs and outputs (the input in precipitation has been evaluated on the base of few bulk precipitation and snow samples; the N fixation by bacteria from the atmosphere has been estimated by the percentage of leguminous plants in crop-rotation).

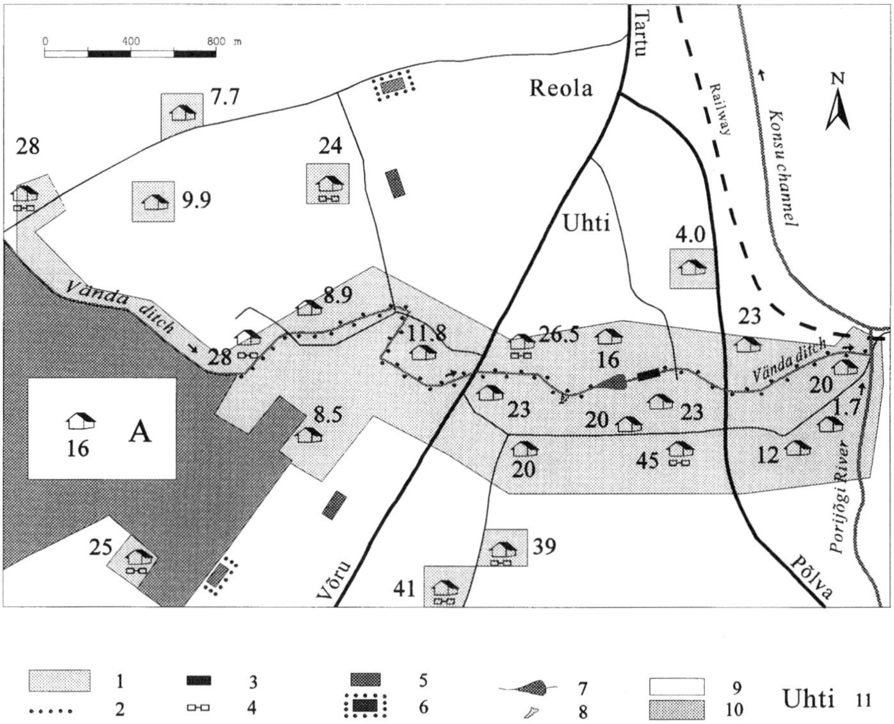

FIGURE 1. Complex of eco-engineering measures to control nutrient fluxes, and mean values (1991 to 1993) of nitrate nitrogen concentration in the top layer of groundwater in the Vända ditch catchment area. (A). Farms and mean NO_3-N concentration (mg l^{-1}) in wells. (1) Recommended buffer-zones (extensively used grasslands) around wells and along the ditch; (2) forest buffer-strips on the ditch bank; (3) constructed macrophyte pond with a stand-pipe (siphon) overflow (see Figure 5); (4) constructed root-zone systems with coral-chalk filters for cleaning the groundwater; (5) sites for manure composting to be reconstructed; (6) recommended sites for manure composting with an forest buffer strip and clay base; (7) a water reservoir (see Figure 3); (8) a gully to be stabilized; (9) arable land; (10) forest; (11) village.

Nevertheless, the effective use of nitrogen and phosphorus by crops is relatively low. So, losses into the groundwater and atmosphere are high. In the case of phosphorus, large amounts of this element accumulated every year in the soil could play a role of a future chemical time bomb.

A small reservoir built up on the Vända ditch in 1985 plays an important buffering role for water quality (Figure 3). Long-term investigations show that the organic material (measured as BOD_5) and TIN retention in this reservoir are quite high (0.6 and 4.1 t ha^{-1} yr^{-1}, respectively) and exceed the removal efficiency of similar water reservoirs within agricultural watersheds (Cooper and Knight 1990). The elimination of P_{tot} (5 kg ha^{-1} yr^{-1}) was very low and SO_4 was even released. This is probably a result of anaerobic conditions which dominate in the bottom of the reservoir. The temporal variation of nutrient input into the reservoir shows a significant decrease in the last 2 years (Figure 4). Seemingly, it is due to a spontaneously formed

TABLE 1
Nitrogen and Phosphorus Balance in the Vända Ditch Watershed

	1987		1988		1989		1990		1991		1992		1993	
Main fluxes kg/(ha yr)	**N**	**P**	**N**	**P**	**N**	**P**	**N**	**P**	**N**	**P**	**N**	**P**	**N**	**P**
Input														
Precipitation	10	0.13	17	0.13	16	0.13	18	0.13	21	0.13				
Fertilization	85	23	66	21	109	31	88	52	80	30				
N fixation[a]	50		50		75		60		60					
Output														
Moved with crops[b]			51	8	55	7	69	10	67	12	32	8		
Runoff in ditch water	16	1,03	15	0,26	14	0,15	24	0,22	13	0,22	15	0,18	19	0,33
Other[c]														
Accumulation in soil, leaching into groundwater, denitrification, etc.			79	14.9	74	14	107	20.9	86	39.9	114	22		

[a] Estimated on the base of age and percentage of leguminous plants in crop-rotation.
[b] Estimated after yield values and mean concentraton of N and P in crops.
[c] Other = Inputs–Outputs.

macrophyte shoal in the upper part of the reservoir that can retain nutrients more intensively than a non-vegetated shoal (Figure 3). Anyway, the buffering capacity of the reservoir is one of the most important stabilizing factors causing relatively stabile runoff parameters in the mouth of the ditch (Table 1). Only losses of P_{tot} were very high in the beginning of the investigations. That was probably a result of the newly built reservoir, which disturbed banks and caused phosphorus losses.

Considering the above-mentioned environmental problems, the following eco-technological measures are proposed (see Figure 1; Mander et al. 1994b):

- Planting forest buffer strips along the ditch (Mander 1989);
- Changing land-use structure of the watershed: implementation of a 5 to 6 year ecological crop-rotation with leguminous plants, conversion of arable lands along the ditch and around wells into perennial grasslands (water protection zones);
- Designing and establishing root-zone systems with a coral-chalk filter (produced by BioOrganics Svenska AB) to clean up the nitrate-contaminated drinking water;
- Establishing new, deeper pump wells instead of abandoned draw wells. However, even deeper pump wells are not the best managing practice for this area because of the lower quality of deeper groundwater layers (high sulfate concentration);
- Building up the constructed system of macrophyte ponds with siphon-overflow and stony rapids to remove nutrients from the ditch (Figure 5);
- Preserving an existing gully (2 m deep, 10 m wide and 20 m long) on the ditch bank by planting trees and stabilizing with stones;

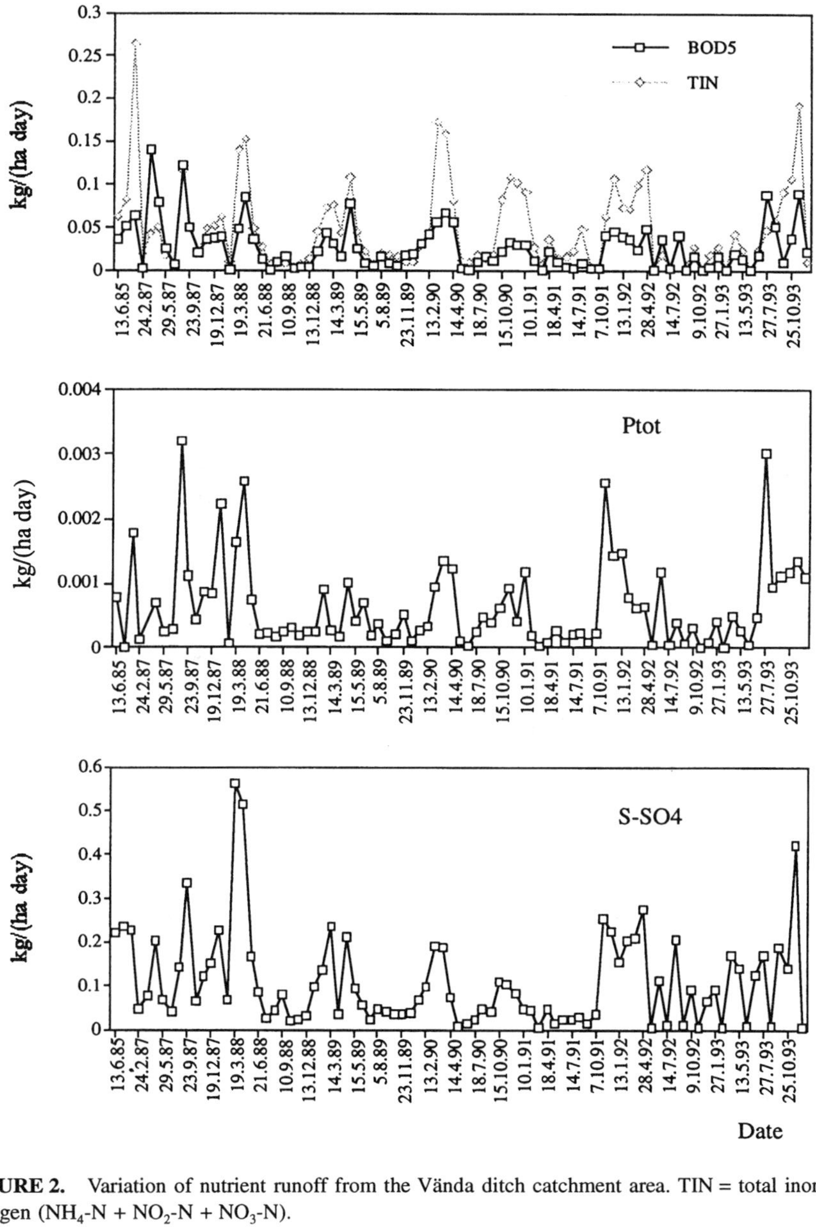

FIGURE 2. Variation of nutrient runoff from the Vända ditch catchment area. TIN = total inorganic nitrogen (NH_4-N + NO_2-N + NO_3-N).

- Restoring and removing manure composting sites with a clay base;
- Cooperating with farmers by organization of training courses and regulation of environmental taxes;
- Turning the catchment area into a demonstration and experimental watershed with long-term monitoring perspectives.

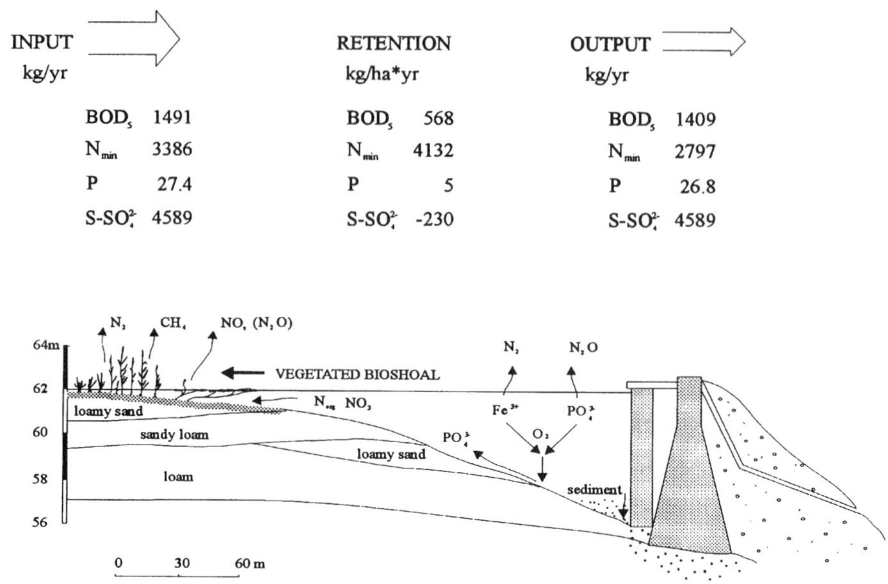

FIGURE 3. Nutrient budget of a small reservoir (0.12 ha) on Vända ditch (see Figure 2). Numbers indicate mean values of a 7-year study period (1987 to 1993). A vegetated shoal in the upper part of the reservoir has formed spontaneously.

This complex of measures has been planned to be implemented in 1994 to 1995 as a project of the HELCOM Baltic Agricultural Run-Off Action Programme (BAAP) in cooperation with the Swedish University of Agricultural Sciences and the University of Lund, Sweden. This project, called "Ecological Engineering and Landscape Planning as Means for Improving Water Quality in Agricultural Watersheds in Estonia," has been proposed for funding by the HELCOM Baltic Sea Environmental Action Programme.

CASE STUDY 2: KAVILDA RIVER CATCHMENT AREA

Arvo Järvet and Kathrin Häberli

The purpose of this case study is to give a concentrated presentation about the environmental problems of River Kavilda and to propose ecotechnological measures to reduce pollution load to the Kavilda River.

Kavilda is a small river in the south of Estonia. It starts close to the town of Elva and leads into Emajõgi River. The total catchment area of Kavilda River is 73 km^2. A map of the catchment area of Kavilda River and the sites of monitoring stations (sampling points) are given in Figure 6. The drainage basin varies in topography from flat moraine and glaciolacustrine plains to steep valley slopes with relative heights of 15 to 40 m. The absolute altitude lies between 34 to 91 m. The bottom of the valley is covered with fen peat and fluic-eutric-histosols. On the valley flats, fen forests (*Alnus* spp. and *Salix* spp.) are growing. The soils on the plateau are mainly podzoluvisols used in agriculture. Approximately 40% of the catchment area is used as arable land. In the valley clay soils and peatlands are partly used as

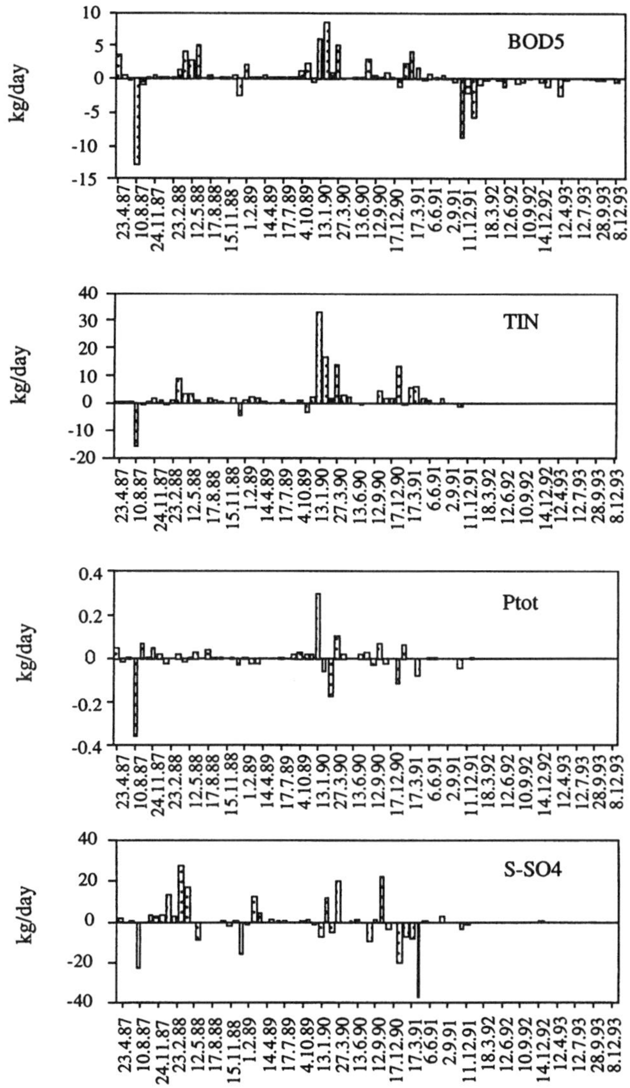

FIGURE 4. Variation in nutrient budget of the reservoir on Vända ditch.

grasslands. About one third of Kavilda watershed is covered with natural vegetation: swamps, moors, natural grasslands, brushwoods, and fen forests (Figure 7).

The pollution load of Kavilda River has been investigated by river water quality monitoring, which was started in 1979 by the Tartu Enviroment Protection Laboratory. From 1979 to 1985 water samples were taken once a month. The sampling strategy was changed after the construction of Elva wastewater treatment plant. Since 1986, only four samples per year have been taken.

The mean discharge during different periods was calculated on the basis of measurements made on Tarvastu River, 30 km west of Kavilda River catchment area.

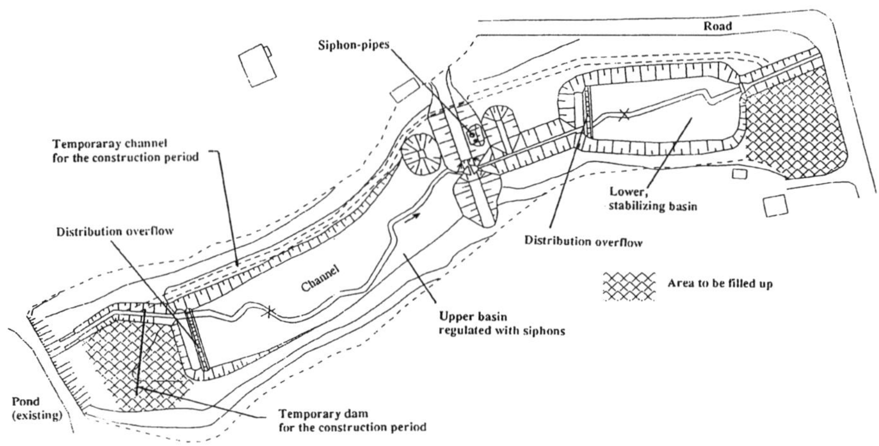

FIGURE 5. Scheme of the constructed system of macrophyte ponds with siphon overflow for nutrient retention on Vända ditch, to be established during 1994 to 95.

The river discharges are dependent on the hydrological factors in different years. The amount of discharged substances is greatest in years with the highest streamflow. From1980 to 1990, the annual mean specific runoff has been calculated as 9.5 l s^{-1} km^{-2}, with a minimum of 6.4 l s^{-1} km^{-2} in 1980 and a maximum of 13 l s^{-1} km^{-2} in 1990. The wastewater quantity and quality has been reported by the Tartu Environment Protection Agency (Järvet 1991).

River Kavilda is naturally poor in nutrients and has to be protected from nutrients coming out of agriculture and other non-point pollution sources by buffer zones. Along almost the whole Kavilda River there are buffer zones of usually 200 to 400 m width covered with swamp, natural grassland and fen forest. Buffer zones of less than 100 m width are situated around bridges, drainage inlets and along Uula polder, which is situated not far from the river mouth.

Elva, situated upstream of R. Kavilda, has about 6500 inhabitants. Of those, 3500 are connected with the existing wastewater treatment plant and 3000 inhabitants are living outside the wastewater network area. The dynamics of wastewater flow during the period 1980 to 1992 are presented in Figure 8. In 1993, the total wastewater amount coming into the treatment plant was 700 to 800 m^3 d^{-1}, of which domestic wastes were 600 m^3 d^{-1} (75%), dairy industry 80 m^3 d^{-1} (10%), other industries 120 m^3 d^{-1} (15%). During heavy rainfalls, a portion of the discharge (maximally 200 to 300 m^3 d^{-1}) is collected by the canalization system.

From 1978 to 1987, wastewater in Elva was discharged through biological ponds (with an area of 0.92 ha) into the R. Kavilda. The BOD removal efficiency of the overloaded ponds was only 25 to 30%. The wastewater treatment plant in Elva started to operate efficiently in 1988. The plant is equipped with an activated sludge treatment block and four bioponds with a total surface of 9000 m^2. Its capacity should be sufficient to treat 1800 m^3 d^{-1} of municipal wastewater, so at the moment the plant is only loaded up to 40% of its capacity. At the present time, chemical

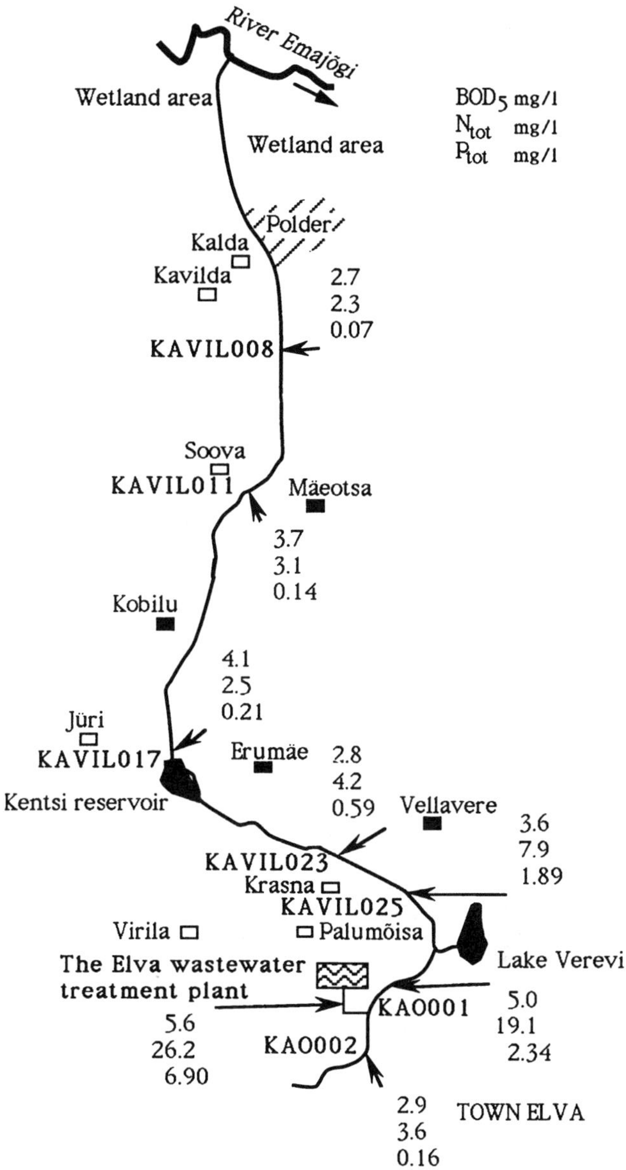

FIGURE 6. Location of water sampling sites in Kavilda River drainage basin with mean values of BOD_5, total-N and total-P concencentration in water (mg l^{-1}).

purification is not installed, however; phosphorus removal has been planned using flocculates in an aeration chamber. The electric energy consumption of the Elva wastewater treatment plant is relatively high — between 60,000 and 80,000 kWh per month, which is 17% of the total electric energy consumption of the whole town Elva.

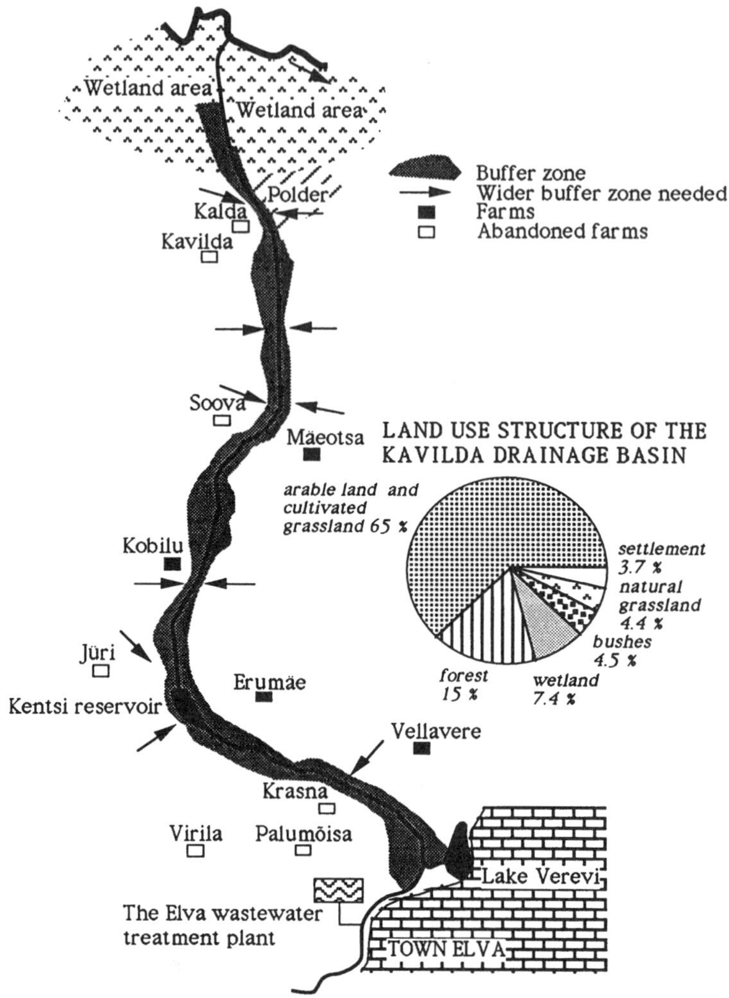

FIGURE 7. Natural buffer zones and land-use structure in the Kavilda River catchment area.

The purification efficiency of the Elva wastewater treatment plant by mean concentrations of different compounds during May 1992 to December 1993 is presented in Table 2 and in Figure 9.

The reduction of BOD_7 is very effective — 99%. The BOD concentration in the effluent of the plant is under the Estonian norm of 15 mg O_2 l^{-1}. Nevertheless, the nitrogen reduction is around 50%. The treatment system is still underloaded, because there are also nitrification processes taking place. The oxidation ponds do not seem to operate efficiently for removal of biogenic elements — only the concentration of N_{tot} is diminished by 34%.

Thus, the HELCOM (Helsinki Commission 1974) recommendations (for P_{tot} less than 1.5 mg l^{-1} and N_{tot} less than 15 mg l^{-1} in the effluent of wastewater treatment plants) are not achieved.

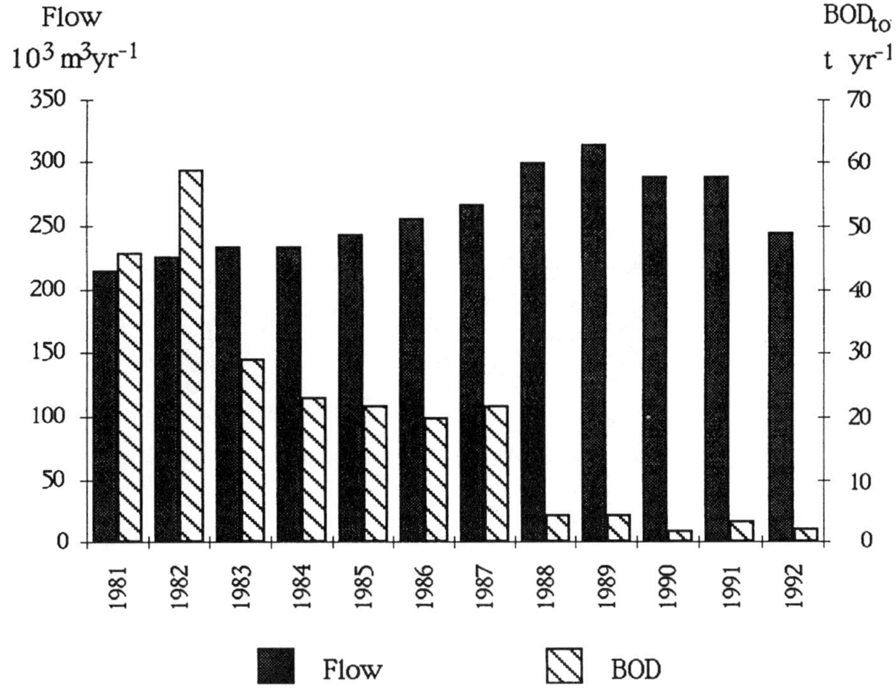

FIGURE 8. Variation of the flow and BOD_{tot} value of wastewater from the town Elva.

TABLE 2
Mean Concentration of Pollutants (mg l^{-1}) at the Elva Wastewater Treatment Plant (May 1992 to December 1993)

Sampling site and efficienty	BOD^7	TIN	N_{tot}	PO4-P	P_{tot}
Inflow of treatment plant	710		40.5	8.3	11.2
Outflow of activated sludge treatment block	15.1	25.6	31.2	6	7
Efficienty of the 1st step (%)	98		23	28	38
Outflow of 1st biopon	13.2	22.7	27.2	6.5	7.4
Outflow of 2nd biopond	9.7	22.3	26.1	5.9	7.1
Outflow of 3rd biopond	6.9	21.5	25.2	6.4	7
Outflow of 4th biopond	5.6	15.2	20.6	6.2	6.9
Efficiency of the 2nd step (%)	63	41	34	–3	1.5
Total efficiency	99		49	25	38

The concentration level of NO_3 even increased after the constructing of the Elva wastewater treatment plant due to ideal conditions for nitrification in the treatment plant. The actual concentration of NO_3 at points KAVILDA023 and KAVILDA017 is frequently very high, reaching a level of two or three times the value of the ecological concentration limit of 1.2 mg l^{-1}.

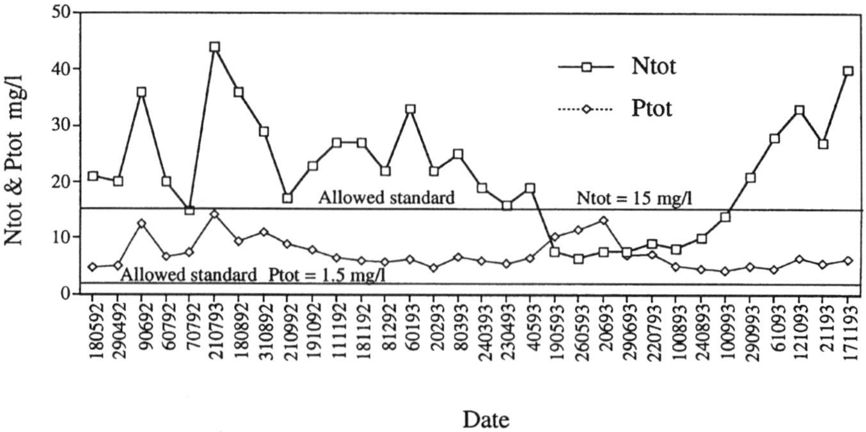

FIGURE 9. Variation of total-N and total-P concentrations at the outlet of the Elva wastewater treatment plant.

After the Elva wastewater treatment plant began operations, there were no more peaks of very high P_{tot} concentration, but the average concentration level stayed about the same. Actually, the recommended ecological limits for P_{tot} (0.1 mg l^{-1}) and PO_4 (0.03 mgP l^{-1}) are greatly exceeded in sampling point KAVILDA023 (Figure 6).

General principles for reducing the pollution load are:

1. Beginning-of-pipe-investments for reduction are usually much more efficient than end-of-pipe-investments.
2. The efficiency of different measures has to be compared — where does the smallest amount of investment most improve the situation?
3. All environmental aspects of reduction and elimination of anthropogenic impacts have to be considered. A wastewater treatment system eliminating a very high percentage of pollution is still not ecologically efficient if it also needs a very high amount of energy.

Considering the efficiency aspects, the following proposals can be made:

1. As much as possible, one third of the natural landscape in the watershed should be protected as water protection zones. The protection zones have to be declared during the running privatization of the land. It is much easier to protect existing natural ecosystems like wetlands, riparian buffer strips and forests than to reconstruct them later again. The principle of balanced fertilization combined with ecological farming is to be implemented
2. It should be checked very carefully where along Kavilda River and its natural and drainage inlets additional buffer strips are needed. So far, we suppose that larger buffer strips are needed especially close to the bridges, along Uula Polder and where farms are situated close to the Kavilda River, like in Soova and Kalda, where the buffer zone is less than 100 m wide.

3. Reservoirs and drainage systems represent strong impacts on the natural water regime of a river. Reservoirs and sedimentation ponds should rather be built in drainage ditches as bioponds before entering the main river. Further, drainage constructions should be avoided or built as natural as possible.
4. More water protection areas in Elva — especially around the lakes — should be declared. The adequacy of the existing laws (25 m from water no buildings are allowed and 100 m wide strips are declared as protection areas) should be investigated.
5. A comprehensive study of the sewer system in Elva is needed. The proposal for rehabilitation of the sewer network is a preliminary one, and the sewer system must be studied more carefully before starting the implementation. All developments for technologies decreasing the water consumption and for reduction of rain water infiltrating in the sewer pipes must be supported, e.g., by water-saving sanitary installations or by separating the stormwater from wastewater sewers.
6. An improved wastewater treatment system for Elva town. To increase the effectiveness of the treatment process, the following actions are recommended:
   a. Phosphorus elimination by flocculation as it is proposed in a recent study of Elva town authorities should be added to the Elva wastewater treatment plant as soon as possible. Phosphorus elimination by flocculation can only be done up to a removal efficiency of 80 to 90%. A tertiary treatment using root-zone systems to retain nutrients will also be needed.
   b. A natural reedbed system similar to the wastewater secondary treatment system (Geller et al. 1991, 1992) could be established in the wetlands from the Elva wastewater treatment plant down to Kavilda River. Due to partly anaerobic conditions in the natural reedbed, denitrification and losses of N_2 and N_2O into the atmosphere could occur. Denitrification needs nitrate, anaerobic conditions and a carbon source — all these three conditions are fulfilled in wetlands near the existing wastewater treatment plant of Elva.

As a basis for further discussions, we propose to establish an alternative outdoor system to clean up wastewater from the treatment plant. It will consist of a semi-natural reed bed of 20 ha and old oxidation ponds (Figure 10). The reed bed will be surrounded by dams on three sides. Great attention has to be paid to equal distribution of the water in and through the whole surface. A homogenous water flow through the soil guarantees good conditions for all retention reactions, especially for the microbial activity. During the dam construction, the sediments of the old oxidation ponds should be dug out and composted.

P- and N-removal in wetland systems is variable. Surface area derived from BOD loading is often used to assess the size of a system. The area required for a wetland has been estimated to be 5 m^2/population equivalent (p.e.; Geller et al. 1992). A daily wastewater load of 200 l/p.e. has been assumed. The estimated area required is 1.5 times higher than the actual infiltration area. The mean wastewater amount of Elva in 1993 was 580 m^3 day^{-1} and the BOD_5 load 360 kg day^{-1}. As the value of 1 p.e. is assumed 54 g BOD day^{-1}, the area needed for the proposed wetland is:

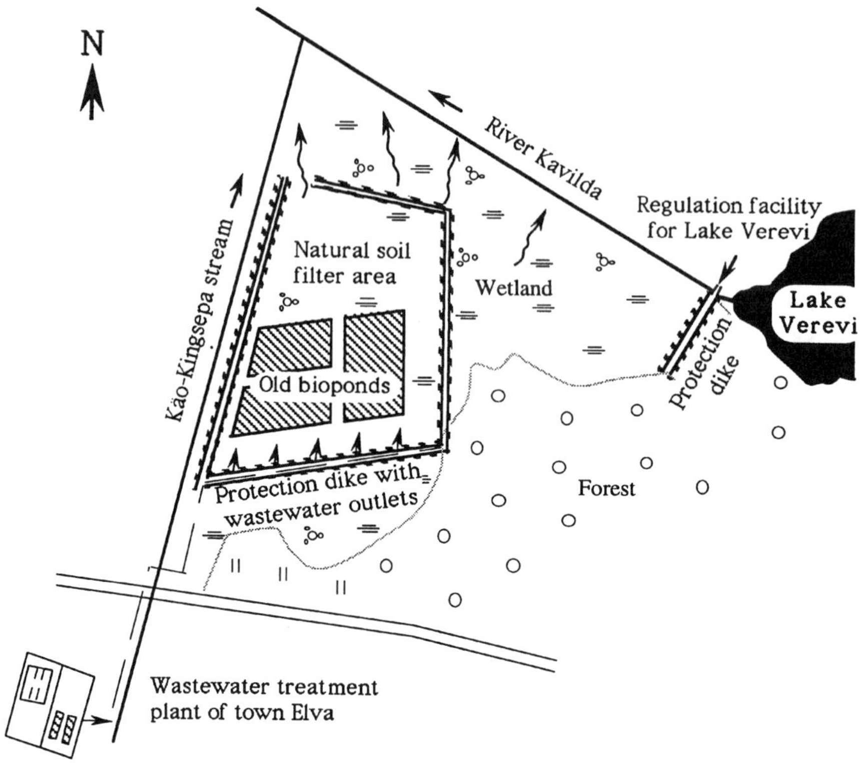

FIGURE 10. The design for an alternative system for secondary treatment of wastewater from Elva, consisting of a semi-natural wetland (soil filter area) and a biopond complex.

$$5 \cdot (360/54) \cdot 1.5 \cdot 0.1 = 5.0 \text{ ha}$$

The normal surface area of a constructed wetland system for removal of nutrients in Europe is 3 to 10 m^2 per p.e. (Cooper et al. 1989). The calculated area is sufficient also for removal of nitrogen (23.8 kg day^{-1}) and phosphorus (6.50 kg day^{-1}).

Based on regression equations for nutrient removal in wetlands that were published in the thesis of Tõnu Mauring (1992), the treating efficiency can be roughly estimated. A proposed wetland surface of 20 ha, a total phosphorus concentration of 9.7 mg l^{-1}, and total nitrogen concentration of 34 mg l^{-1} in the outlet of the existing plant are assumed. A total nitrogen removal of 60% and total phosphorus removal of even 73% can be calculated, based on the following equations:

$$N_{tot} \text{ elimination } (g\ m^{-2}\ day^{-1}) = 0.61 \times N_{tot} \text{ inlet } (g\ m^{-2}\ day^{-1}) - 0.005$$

and

$$P_{tot} \text{ elimination } (g\ m^{-2}\ day^{-1}) = 0.85 \times P_{tot} \text{ inlet } (g\ m^{-2}\ day^{-1}) - 0.013$$

As there is a large wetland surface available, this natural reed bed system in the natural wetland might also be used for one more purpose. Nutrients cannot only be eliminated by adsorption and denitrification but also by harvesting. For instance, these wetland can produce renewable resources like aquatic macrophytes (*Typha latifolia* or *Phragmites australis*) to produce biological insulation material for buildings.

CASE STUDY 3: THE ECOSYSTEM AND CATCHMENT AREA OF LAKE ARBI

Tiina Nõges, Arvo Tuvikene, and Tõnu Mauring

The main purpose of this case study was to restore the highly eutrophicated Lake Arbi ecosystem by reducing the nutrient load in its drainage basin and influents, and by reconstruction of the food chain in the lake via biomanipulation.

Lake Arbi is a small (4.5 ha, maximum depth 5.5 m, mean depth 3.6 m, water amount 162,000 m^3) water body, located inside the summer resort town of Elva. The lake is of great recreational importance. The water quality of the lake is poor. Secchi disc transparency in the summer is 0.4 to 0.9 m. Oxygen disappears rapidly at a depth of 2 to 3 m. The thermocline is not very distinct and appears at 3 to 4 m. The bottom is covered by an 18 m thick layer of mud. The total phosphorus (P_{tot}) and nitrogen (N_{tot}) concentrations (0.14 mg l^{-1} and 5 mg l^{-1}, respectively) and annual primary production (466 gC m^{-2}) make the lake a hypertrophic water body.

The spring phytoplankton maximum in April (biomass 44.5 g m^{-3}; chlorophyll a concentration 123 mg m^{-3}) was caused mainly by the diatom *Stephanodiscus hantzschii*. Summer maximum of primary production (4.2 gC m^{-2} day^{-1}) occurred in July and the highest phytoplankton biomass (60 g m^{-3}) and chlorophyll a concentration (200 mg m^{-3}) in August. Mainly blue-greens *Aphanizomenon gracile* and *Oscillatoria agardhii* were responsible for the latter.

Annual primary production (466 gC m^{-2}) exceeded annual destruction activity (300 gC m^{-2}) 1.55 times.

The biomass of summer zooplankton was 1 to 2 g m^{-3}. Dominant species: *Mesocyclops leucarti, M. oithenoides, Eudiaptomus graciloides, Bosmina longirostris, Daphnia cucullata, Keratella cochlearis.*

The total number of bacteria was 5.1×10^6 cells ml^{-1} and the production of heterotrophic bacteria 14.4 µgC l^{-1} h^{-1} (TdR, conversion factor 13×10^{18} cells mol^{-1}). Maximum was in July (2.9 gC m^{-2} day^{-1}). The total production of heterotrophic bacteria made up 70.5% of primary production.

Planktivorous fish [roach (*Rutilus rutilus*)], bream (*Abramis brama.*), tench (*Tinca tinca*), small perch (*Perca fluviatilis*) account for 72% of total fish biomass: perch (small + large) 29%; bream, 32%; roach, 26%; tench, 11%.

The greatest amount of nutrients comes from two main inflows (No. 2 and No. 4, see Figure 11, 14). The average (from March to November 1991) P_{tot} load into the lake was 1.9 kg day^{-1}, most of it from inflows (Figure 12). N_{tot} load was 49 kg day^{-1}, with more than a half of it coming from ground water (Figure 13).

Considering the average concentration of P_{tot} and N_{tot} in the lake water, we calculated the average total of phosphorus (22.4 kg) and nitrogen (811.3 kg) in the lake. Daily P-load constitutes 8.5% and N-load 6% of the total amount of P and N in the lake, respectively. During the investigated period (March to November, 1991) the water of Lake Arbi exchanged totally during 18 days. During this period the total P-load exceeded 1.5 times and N-load 1.07 times the amount of these elements

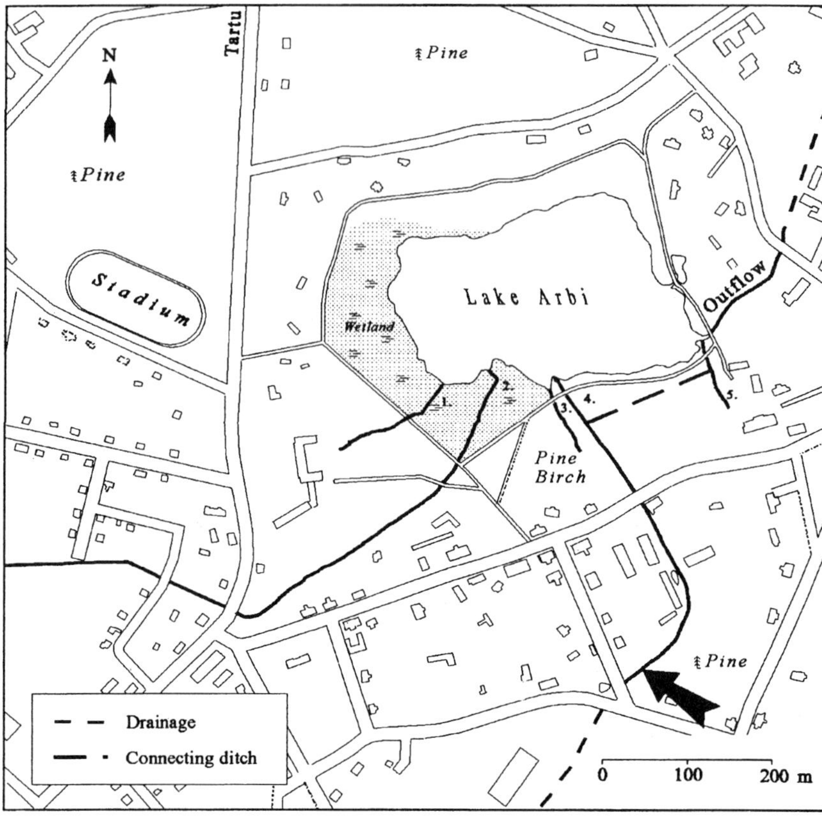

FIGURE 11. Lake Arbi and its catchment area in Elva. Arrow indicates location of a designed root-zone system (a serpentine pond) for nutrient retention (see Figure 15). 1 to 5 are inflows. A dashed line between inflows 4 and 5 indicates a connecting ditch to lead water from the most polluted inlet 4 directly to the outflow (established 1992 during the workshop).

in the lake. Consequently, the content of P and N in lake water is mainly formed by the load of these elements from inflows; recycling from sediments and N_2-fixation is less important. Considering the above-mentioned values and balances between inflows and outflow, the accumulation of P seems to be considerable (30% of the input, Figure 13). This assumption is supported by the annual ratio of primary production and destruction — 1.55. This value indicates that a considerable amount of produced organic matter accumulates in lake sediments.

The ratio of N_{tot}/P_{tot} in lake water varied mainly between 25 and 50, only occasionally (in May) reaching a very high value — 325. In any case, the N/P ratio higher than 16 indicates that phosphorus is the limiting factor for phytoplankton growth. This assumption is also supported by the fact that total assimilation of phosphorus by the phytoplankton in the lake (2.3 kg day^{-1}) even exceeds its total input (1.9 kg day^{-1}), but total assimilation of nitrogen (16.6 kg day^{-1}) constituted only 34% of its input (49 kg day^{-1}).

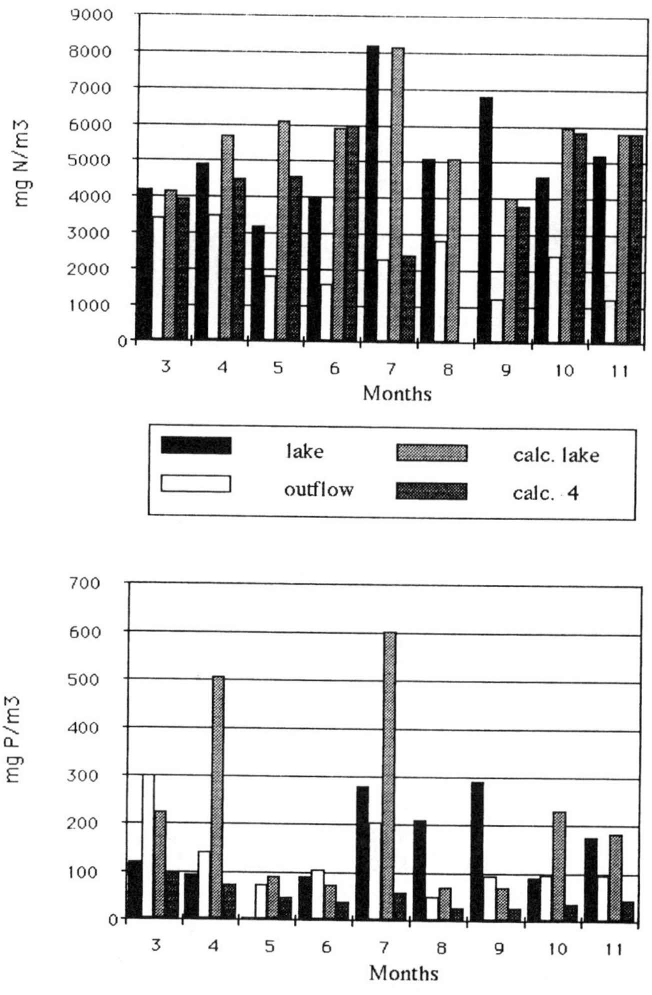

FIGURE 12. Variation of the total-N and total-P concentrations in Lake Arbi water in 1991.

Considering the key position of phosphorus in trophic state formation of Lake Arbi, the reduction of its load is needed for the restoration of the ecosystem. Inflow No. 4 (Figure 11) is responsible for 78% of total P-input. As the first step of restoration, this inflow will be excluded in 1992 (led to run by the lake). Our calculations revealed that this measure will decrease the P_{tot} in the lake by up to 48 mg m^{-3} and return the lake to being merely eutrophic. This value (calc. 4 on Figure 12) is 4.6 times lower than the calculated concentration in presence of inflow No. 4 (calc. lake on Figure 12) and 3 times lower than the actual concentration in the lake during the investigation period (Figure 12). The influence of elimination of the inflow No. 4 on nitrogen concentration is small — 1.2 and 1.1 times, respectively (Figure 12). Calculated concentration was figured out by dividing total load of P or

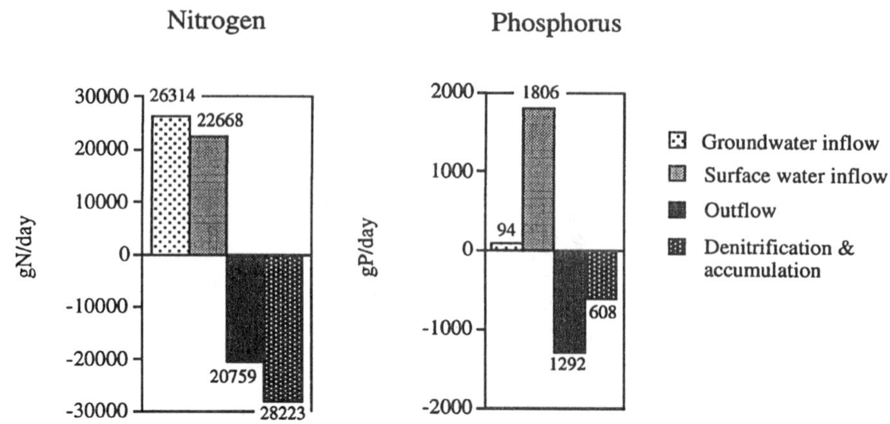

FIGURE 13. Nitrogen and phosphorus budget of Lake Arbi in 1991.

N from inflows (mg day^{-1}) by total hydrological load from the same inflows (m^{-3} day^{-1}). The investigation of the impact of this measure is planned for 1994.

The influence of the elimination of the main phosphorus input might not be as great as expected because of the second source of nutrient (phosphorus) load into the lake water — its release from sediments in anarerobic conditions. Our investigations in 1991 revealed the prevalence of P-accumulation on its release but after elimination of main source of P from outside, the release of that element from sediments might become of greater importance than we expect. The aeration of the bottom, removal or inactivation of sediments might to be needed to achieve the real recovering of the ecosystem.

Organic phosphorus is mineralized in lake water mainly by bacteria and zooplankton. If Fe^{3+} is present, then $FePO_4$ will precipitate. Under anaerobic conditions, bottom bacteria can use Fe^{3+} as an alternative electron acceptor, which causes phosphorus release. This process is called as "self-pollution" of a water body. It could be avoided by increasing the oxygen concentration at the bottom of the lake. Oxygen is produced in the process of photosynthesis of algae, higher plants and cyanobacteria. The indispensable factor is light. Photosynthesis can reach a depth roughly equal to three times the Secchi depth.

One possibility to aerate the bottom is increasing the extent of the euphotic zone down to the bottom by increasing of the water transparency. Seston, being responsible for decreasing of transparency, consists of plankton (mainly phytoplankton) and detritus (dead organic matter). To increase the water transparency, the accumulation of phytoplankton has to be avoided by stimulating its consumption by zooplankton. The organic matter produced by phytoplankton is consumed in the food chain:

$$\text{phytoplankton} \rightarrow \text{zooplankton} \rightarrow \text{planktivorous fish} \rightarrow \text{predatory fish}$$

Decreasing the amount of planktivorous fish and increasing the amount of predatory fish is favorable to zooplankton development. The latter will be able to control phytoplankton biomass and avoid its accumulation (see Gulati et al. 1990). The water

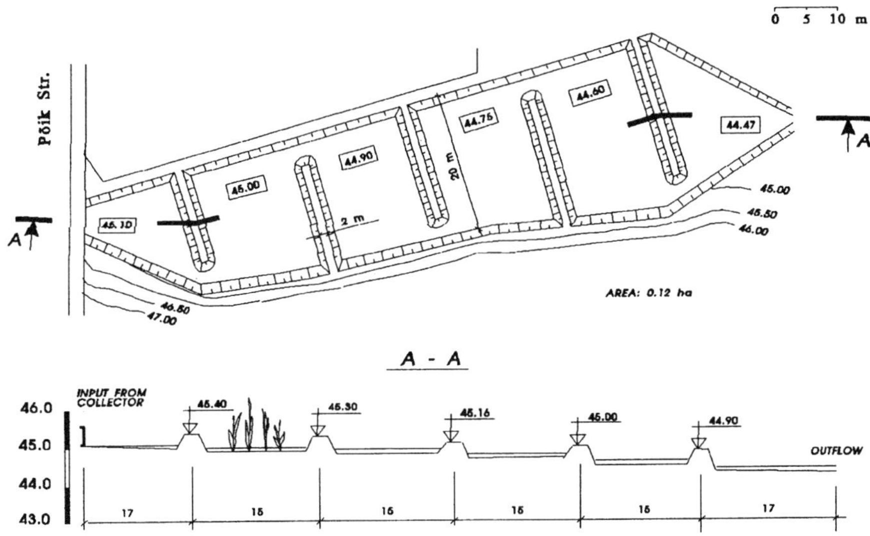

FIGURE 14. Design of the serpentine pond (root-zone system) on the main inflow of Lake Arbi for retaining nutrients from polluted ditch water (see also Figure 11).

transparency and therefore the extent of the euphotic zone will increase and the area of anoxic bottom will decrease.

The measures to improve the water quality by biomanipulation (planktivorous fish reduction and predatory fish introduction) will be undertaken in 1994 and their impact should be investigated in 1995. The main task of restoration is the increase of the water transparency up to 1.8 m for increasing the depth of euphotic zone down to the bottom of the lake. This enables natural oxygenation (photosynthesis) of the bottom water and prevents the release of phosphorus from sediments. For biomanipulation of the lake, the removal of ca 350 kg of roach and bream and the introduction of 150 to 200 kg of pike and large perch are needed.

For inflow and outflow treatment, the following measures have been planned. First, a connecting ditch has been built up to lead the water from inflow 4 as the main pollution source directly to the outflow (Figure 11). As a result, in 1993 the water quality parameters have been improved. For instance, in the former mouth of the inflow 4 the phytoplankton development has been less intensive than earlier.

Directing the polluted inflow 4 to the lake outlet, the potential pollution load downstream from the lake increased. To avoid additional loading of the Elva River, a serpentine pond planted with *Typha latifolia* was established (Figure 15). However, the sewage water pollution load in the former inlet 4 upstream of the lake has the same quality as before. Therefore, an additional root-zone system (a serpentine macrophyte pond) has been designed to eliminate nutrients from the channel water (see Figure 14). The active area of this serpentine pond (ca 1500 m^2) allows it to clean up wastewater with the maximum loading of 250 p.e. Normally, this rate will not be exceeded. Located in a small semi-natural forest, this outdoor system does not disturb urban environment.

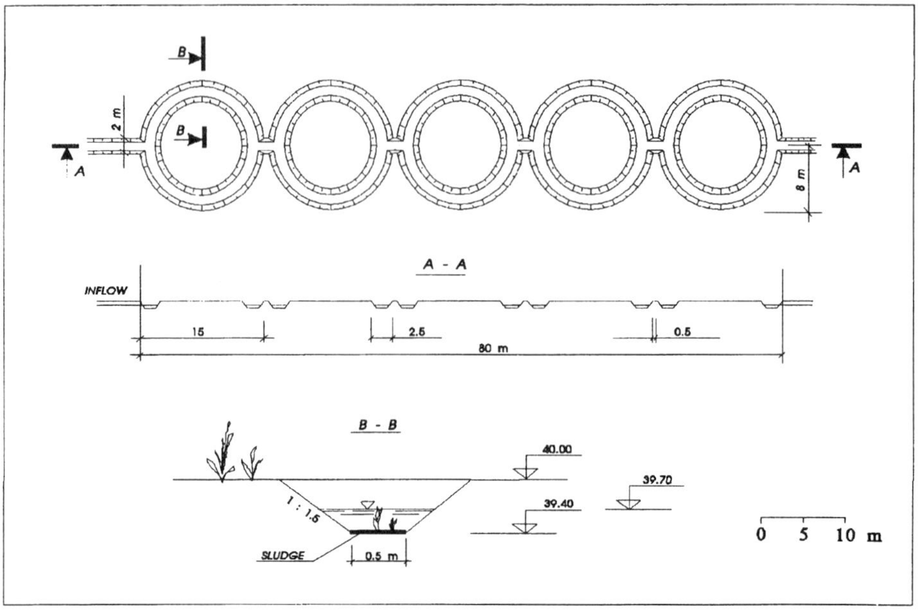

FIGURE 15. The established serpentine pond for treating water from Elva and Lake Arbi. Located on the outflow from Lake Arbi ca 2 km downstream of the lake and 0.5 km before entering the Elva River (see also Figure 11).

CONCLUSIONS

1. For the transition period of earlier socialist countries to a market economy, cost-effective eco-engineering measures for purification wastewater from both point and non-point pollution sources will be quite useful. This is valid especially for rural areas where the large investments in modern, high technology wastewater treatment plants are not available. In any case, such investments may remove much less phosphorus and nitrogen per dollar invested than eco-engineering measures like non-point pollution mitigation via buffer zones and buffer strips (Mitsch and Jørgensen 1989).
2. Ecotechnology can be implemented in complex ways for the whole territory of catchment areas. It allows comparison of different catchments and evaluation of the real efficiency.
3. We estimate the efficiency of established and constructed ecotechnological measures to be:

   Vända ditch catchment area: owing to a root-zone system, buffer zones and buffer strips, and changing crop-rotation, the nitrogen and phosphorus losses from the fields will decrease by up to 20 and 30%, respectively. This is in addition to drastically decreasing nutrient losses due to a lowered fertilization rate the last two years.

   Kavilda River catchment area: using semi-natural wetland as a root-zone system, the nitrogen and phosphorus load will decrease about 60%. This is because of the very low nutrient removing efficiency of existing wastewater treatment plant.

Lake Arbi and its catchment area: via more intensive nutrient removal in serpentine ponds, the summary pollution load of nitrogen and phosphorus will decrease up to 30 and 40%, respectively. Biomanipulation could change the share of organisms at different trophic levels up to 40%. However, to restore the lake and to improve its recreational value, other measures (bottom sediment removal or aeration, lake water aeration) are needed.

ACKNOWLEDGMENTS

Dr. Reet Laugaste and Dr. Aare Mäemets from The Limnology Station, Institute of Zoology and Botany, Estonian Academy of Sciences helped us in determining phytoplankton and zooplankton species and analyzing water quality in Lake Arbi. Also, we wish to acknowledge colleagues from the Tartu Environment Protection Laboratory and from the Institute of Environment Protection, Estonian Agricultural University for helping to analyze the water samples.

We also thank all organizers and participants of the International Workshop on Ecological Engineering in Wastewater Treatment, Landscape and Settlement Planning in Tartu, Estonia, July 27 to August 7, 1992 for active help in finding solutions for pollution problems. Special thanks we would like to extend to Dr. Johannes Heeb from the Center for Applied Ecology Schattweid, Switzerland and to Dr. Tõnu Oja from the Institute of Geography, University of Tartu for critically reading the manuscript.

REFERENCES

Cooper, C.M. and S.S. Knight 1990. "Nutrient trapping efficiency of a small sediment detention reservoir." *Agricultural Water Management* 18:2, 149-158.

Cooper, P. F., J. A. Hobson and S. Jones, 1989. "Sewage treatment by reed bed systems." *J. IWEM* 3, 60-74.

Geller, G., R. Netter, K. Kleyn, A. Lenz, 1992. "Bewachsene Bodenfilter zur Reinigung von Wässern-Ergebnisse und Empfehlungen aus einem 5-jährigen BMFT-Forschungsvorhaben." *Abwasserreinigung*, 6, 886-898.

Geller, G., E. Englmann, W. Haber, K. Kleyn, A. Lenz, R. Netter, 1991. "Bewachsene Bodenfilter zur Reinigung von Wässernein von Abwasser beeinflusstes System aus Boden und Pflanzen." *Wasser und Boden* 8, 477-495.

Gulati, R.D., E.H.R.R. Lammens, M.L. Meijer, and E. van Donk (eds). 1990. *Biomanipulation, Tool for Water Management.* Kluwer Academic Publishers. 628 pp.

Heeb, J and B. Züst, 1997. "Sand and plant filter systems." In: C. Etnier and B. Guterstam (eds.), *Ecological Engineering for Wastewater Treatment.* Boca Raton, FL: Lewis Publishers.

Järvet, A. 1991. Water quality of rivers in the drainage basin of the Lake Peipsi. In: *Amelerioration and environmental protection.* Estonian Ministry of the Environment: Tallinn, pp. 16-30. (In Estonian with a summary in English.)

Mander, Ü., 1989. *Kompensationsstreifen entlang der Ufer und Gewässerschutz.* Landesamt für Wasserhaushalt und Küsten Schleswig-Holstein, Bericht D10, 61 pp.

Mander, Ü., Matt, O., and Nugin, U. 1994. Perspectives on vegetated shoals, ponds, and ditches as extensive outdoor systems of wastewater treatment in Estonia. In: C. Etnier and B. Guterstam (eds.), *Ecological Engineering for Wastewater Treatment*. Boca Raton, FL: Lewis Publishers.

Mander, Ü., Kuusemets, V., and Treier, K. 1993. Variation in groundwater quality in South Estonian rural areas. In: IAHS Proceedings of GQM 93: *Groundwater Quality Management*, Tallinn, Estonia, 6–9 September 1993. (In press).

Mauring, T. 1992. Nitrogen and phosphorus retention in wetlands and riparian ecosystems. Diploma paper in Estonian Agricultural University, Tartu. 129 pp. (In Estonian, with a summary in English).

Mitsch, W. and Jørgensen, S.E. (Editors) 1989. *Ecological Engineering: an Introduction to Ecotechnology*. John Wiley and Sons. New York. 472 pp.

Stamm, C. 1992. *Überprüfung von Verfahren zur Ausscheidung von Pufferzonen gegen Stoffeintrag*. Zentrum für angewandte Ökologie Schattweid, Schweiz, 123 pp.

Chapter 25

RECOVERY OF SMALL RIVERS IN LATVIA

Andris.V. Urtans

Latvia, as a part of the Baltic Sea catchment area, has undergone drastic changes during the last 50 years, mainly due to abnormal industrialization and nutrient leakage from agricultural areas. According to our investigations, only 20 to 25% of 777 Latvian small streams and rivers have retained their normal, undisturbed state, and one third of them are partially or totally straightened. At the same time, the total length of ditches is 65,500 km, that is to say 1.7 times more than the total length of small rivers, thus increasing the possibility of nutrient leakage.

Under very limited technical and financial conditions in our country the only way to stabilize the situation is optimization of nutrient cycling within streams and their catchment areas. In the immediate future such optimization can be attained in three ways:

1. Construction of vegetated shoals. Unfortunately, during the past decades only a few attempts were made to construct these in the ditches. Most of them failed due mainly to a lack of basic knowledge and incorrect treatment.
2. Renovation and reconstruction of former millponds. Water mills, built on the streams and rivers, are a very characteristic feature for Latvia — nowadays there are about 740 ponds from such mills. Many of them are overgrown with macrophytes or even empty, but in recent years there has been a growing interest in them as potential sedimenters and accumulators of nutrients. Planting such ponds with macrophyte species with high accumulation capacity — *Typha latifolia, T. angustifolia, Schoenoplectus lacustris* (*Scirpus lacustris*), *Phragmites australis* — we can improve their capacity for self-purification and eliminate additional amounts of nutrients.
3. Recovery and rehabilitation of small rivers. This is the most likely strategy for protection and improvement of our damaged water resources and would involve active cooperation from local community associations, fishing authorities, and schools. The main aim would be to ensure conditions conducive to development and reproduction of fauna and flora while at the same time maintaining a high self-purification capacity. In each individual case the balance between conservation and self-purification must be decided by calculating the optimal structure of the biota for the given type of river.

So far these methods have been carried out in three rivers, of which the River Jaunupe is the most investigated. Located in the northwestern part of Latvia, its catchment area — including the River Svetupe — is 459 km^2.

The River Jaunupe, which flows through an area of intensive agriculture is the recipient of large amounts of phosphorus and nitrogen. This has resulted in mass development of vascular plants (macrophytes) covering up to 80% of the river's

0-87371-990-5/97/$0.00+$.50

surface. The river bed has become covered by a 30-cm thick layer of silt. In 1987, with the help of a tractor equipped with a shovel and cultivators, the macrophytes were cut out and, along with the uppermost layer of silt, pushed against the river bank. Lampreys (*Lampetra fluviatilis*) and other macrozoobenthic organisms trapped in this material were released back into the river and the macrophytes and silt were deposited on the river bank where it was left or used as compost. This mechanical removal of the macrophytes (about 70 to 80%), the removal of silt deposits and the additional cultivation of the river bed to destroy macrophyte root systems in the middle of the river led to a remarkable increase in fish and lamprey migration during the following years.

Since the recovery of the River Jaunupe in 1987, rehabilitation processes are carried out annually. It should be noted that immediately following the cleanup, the total number of dominating benthic organism groups sharply decreased. In the following 3 years the total number of organisms in the river stabilized and a new complex of organism groups, associated with riffle habitats was established.

Notwithstanding such promising results, we should stress that there is still an unsolved problem — nutrient leakage from the catchment area. This means that stability in the Jaunupe River depends exclusively on coordinated environmental action in both the actual river and the catchment area.

Summing up, we can draw some conclusions:

- The current state of small rivers in Latvia requires further development of alternative low cost methods for nutrient elimination.
- Recovery of small rivers is a complex task requiring parallel actions in river and catchment areas.
- To avoid a conflict between conservation of flora and fauna and a maximum self-purification capacity, the optimal structure of the biota must be calculated for each individual river.

Chapter 26

AN ENVIRONMENTAL RESTORATION PROJECT IN POLAND: THE APPLICATION OF NEW IDEAS

Janusz Niemczynowicz

ABSTRACT

The newly born freedom of Eastern European countries has brought frightening news about environmental neglect and degradation in these countries. The world community of researchers and technicians working with environmental questions should organize a cooperation leading to transfer of modern approach to pollution problems and up-to-date knowledge from western countries to Eastern Europe. A Polish-Swedish environmental restoration project, primarily aiming at transfer of modern approach and ecologically sound technology is being conducted in Poland. The paper gives details of the project and ideas behind it.

INTRODUCTION

Our world has recently experienced at least two major changes: (1) we have realized the global nature of environmental problems and the need for multidisciplinary approach; (2) new countries have joined the world community of free countries. The newly born freedom of Eastern European countries has brought frightening news about environmental neglect and degradation in these countries. To include this region of the world in our actions is a major responsibility and a challenge, from which we cannot escape because we are all citizens of the same world. Is there something that the world community of scientists and technicians working with environmental questions may do to help to break a vicious circle of economic, industrial, and agricultural problems leading to environmental damage and to stop and reverse the enormous degradation of the environment in these countries? One thing should and must be done: to organize a cooperation leading to a transfer of a modern approach to pollution problems and up-to-date knowledge.

MODERN APPROACH

The modern approach which I refer to is based on some simple key words:

- *Integrated system approach* in contrast to narrow-minded only-technical solutions. Multidisciplinary cooperation as well as an integrated system approach are the crucial conditions of success in solving complex problems. Relevant and environmentally sound technology can emerge as a result of such cooperation.

0-87371-990-5/97/$0.00+$.50

- *Small scale* in contrast to technological monumentalism. It is less expensive to construct small-scale treatment units than large ones. The distances of transport are shorter, the plants are less vulnerable, and it is possible to design plants adjusted to the local needs and conditions.
- *Source control* instead of the "end-of-pipe" approach (Ontario 1987, Keliar 1988).
- *Local disposal and reuse* instead of exploitation and wastefulness. Volumes of sewage waters and wastes may be reduced at the source by changing the routines at the production stage or/and by local disposal. Examples are infiltration of storm and wastewater, reduction of water use by structural and non-structural measures (Ontario 1987, Zhang and Qian 1991), infiltration of wastewater to the ground (Swedish EPA-SNV1985, 1985a), oxidation ponds, etc. Water may be reused for industry and for agriculture after local treatment, solid wastes may be reused after separation at the source, ashes may be used for road construction, sewage sludge may be composted (Ishii et al. 1991), heavy metals can be extracted from effluents using selective bacterial uptake, etc. (Wronowski 1990).
- *Pollution prevention* instead of reacting to damages.
- The use of *biological systems* in wastewater and solid waste management and *global, ecological thinking, ecological engineering.* Biological systems may substitute for traditional treatment facilities. Biological activity in the soil may serve as the best treatment plant. Bacteria, algae, plants, and animals may together constitute a system which selectively removes and concentrates all pollution components in any wastewater. Literature on this topic is extensive (U.S. EPA 1977, Grau et al.1991, Mara and Cairncross 1989, Middlebrooks 1982, Shuval et al.1986, Sundblad 1988, Widmer 1981, Reddy 1987, Ridderstolpe and Kindvall 1989). Treatment plants using biological systems may be much cheaper than traditional plants. Just to give some examples: according to Widmer (1981), a traditional plant with a trickling filter or activated sludge serving a population of 100,000 is five times more expensive than a waste stabilization pond system. According to Arthur (1983), a pond without bottom sealing costs between seven and ten times less than the same capacity conventional treatment plant. The land price may be a limiting factor. Work is going on to reduce land requirements (Yuki et al.1989).

NECESSITY OF A NEW APPROACH

Cities of western, highly industrialized countries still contribute — in spite of all efforts and costs for advanced sewerage systems and sophisticated treatment facilities — to continuing degradation of the global environment. The water treatment plants do not eliminate the most hazardous substances. Stormwater delivers heavy metals and other non-degradable toxic elements to the environment. Discharges from industrial and construction activities cause chemical contamination of surface waters and aquifers.

The result is that, increasingly, the traditional end-of-pipe solutions are losing credibility among scientist and technicians in the western world. Moreover, the traditional approach hardly provides the environmentally right solution, because it is based on the old, outdated philosophy of dilution. Obviously, diluted pollutants do not vanish — they just move to more remote places. Even from an economic point of view, it is unrealistic to believe that these methods and technologies can be applied in a full scale in the gigantic urban areas of the developing world. It can also be concluded that the Eastern European countries, whose infrastructure for water treatment and pollution abatement is insufficient or altogether lacking, should not blindly follow the traditional, western technology of developing sewage systems and water treatment plants. There are now other, alternative technologies based on the above characterized more up-to-date ways of thinking. These methods and technologies may substitute for or complement traditional solutions.

In industrialized countries, serious discussions among both water scientists and planners have already started. The goals of water management have been already reformulated from the traditional storm drainage and end-of-pipe approach to more ecologically sound solutions. The new goal is to avoid accumulation of non-degradable hazardous substances and to adapt the waste output to what nature can endure. There is a new realization that the struggle against global pollution can only be efficient if water management is integrated with management of other related urban subsystems such as solid wastes, industrial production systems, energy production and consumption, transportation systems, etc. This approach involves a deeper understanding that all material and energy are involved in large, long-term circulation systems of global scale.

This change of approach in water management has not occurred in developing countries and Eastern Europe yet. Western traditional technology of water treatment, protection against air pollution, and all environmental technologies in general, are still considered, especially by central authorities, as the only alternative. Thus, centralized plans of water resources management and environment protection actions still follow (with few exceptions) the traditional lines of mono-disciplinary, large-scale thinking, end-of-pipe approach instead of source control and local disposal, reaction instead of prevention, effluent dilution instead of selective concentration and reuse. The present economic situation in many countries stops or delays even these plans (Konstantynowicz 1986, Postawa et al. 1986). Scientists and technicians from both East and West should act together in order to accelerate a necessary change of technical paradigm; new technologies leading to conservation of resources instead of wastefulness should be promoted.

Money which many countries offer Eastern Europe for environmental restoration should not be spent on financing large constructions with yesterday's technology. It may do most good if spent on transfer of modern ecologically sound approach. This can be done by working together on small, well-defined areas where a really integrated approach to environmental questions is applied. Such a project is being jointly financed by Sweden and Poland in the Upper Silesian Region of southern Poland.

The project is an attempt to implement integrated environmental thinking to a well-defined catchment where present problems are complex. However, the real goal of the project goes far beyond its geographical area. The real goal of the project is the transfer of approaches based on a modern small-scale technology, source control and pollution prevention, demonstration of novel technologies, and thus, creation of a positive example of methodology that may be applied in other locations. A multidisciplinary team of researchers and practitioners from both countries was created in order to plan restoration activities and to find consensus between global environmental requirements and local needs.

Applied technologies are meant to demonstrate that several new options, far from traditional ones, exist. Constructed facilities are demonstration objects. The project is based on a modern small-scale technology, source control, and pollution prevention. Those solutions are often also the most economical ones. Economically weak countries should implement a new way of thinking and alternative technology since it is often the most economically effective on the long run. The project is conducted on one well-defined catchment situated in the most degraded region of Poland (Mikulczycki catchment in Zabrze region, see Figure 2).

BACKGROUND OF ENVIRONMENTAL PROBLEMS IN POLAND

For centuries Poland has been an agricultural country. Rich natural resources exist in Poland (see Figure 1). Deposits of hard coal, brown coal, iron, copper, lead, zinc, sulphur and rock salt could be a blessing for Poland and make it rich, but, instead, they have created the present ecological disaster. Between the First and the Second World Wars, the only period of political independence until the present, the process of fast industrialization began. After the war, the economic and industrial development was governed by political dependencies and ideas rather than by sound economic and social necessities. The development of heavy industry was, according to Stalin's ideas, the most important goal for the state policy. The industrial development and ruthless exploitation of natural resources laid the ground for the present environmental problems.

The effects of this policy are terrible. Poland delivers more than 30% of the nitrate and 50% of the phosphorous released to the Baltic Sea. Total emissions of SO_2 are 5.2 million tons per year (Lesniak 1989, Blaszczyk 1990). About 35,000 km^2, 11% of Poland's area, is classified as an area of ecological disaster. This area, inhabited by 35% of total population, is responsible for more than 60% of Poland's wastewater, 75% of SO_2, and 93% of its solid wastes (Postawa et al.1986). Let us look more closely at problems of most ecologically degraded region of Poland, the Upper Silesian region.

The Upper Silesian industrial region, situated in southern part of Poland, has been classified as one of the most degraded regions in the world. The city of Zabrze is situated in the heart of this region, with about 2,000,000 inhabitants in an area of 80 km^2. With its three coal mines, open-hearth furnaces, cokeries, chemical factories, five thermal-electric power stations and other industrial facilities situated within a densely populated area, it provides a classical example of environmental degradation at every level (OBKS 1985, Wilczek 1989, Wrona 1989). This example may illustrate the complexity of the existing problems; management of water resources is strictly connected to energy production, industry, the economy, and even the health and standard of living of inhabitants. The continuing environmental degradation of Poland and the Zabrze region in particular has begun to take its toll on the population; life expectancy is low and infant mortality is steadily increasing. The general decrease in health and quality of life is a result of environmental degradation. The average life expectancy in Poland is 67 years for men and 74 for woman, compared

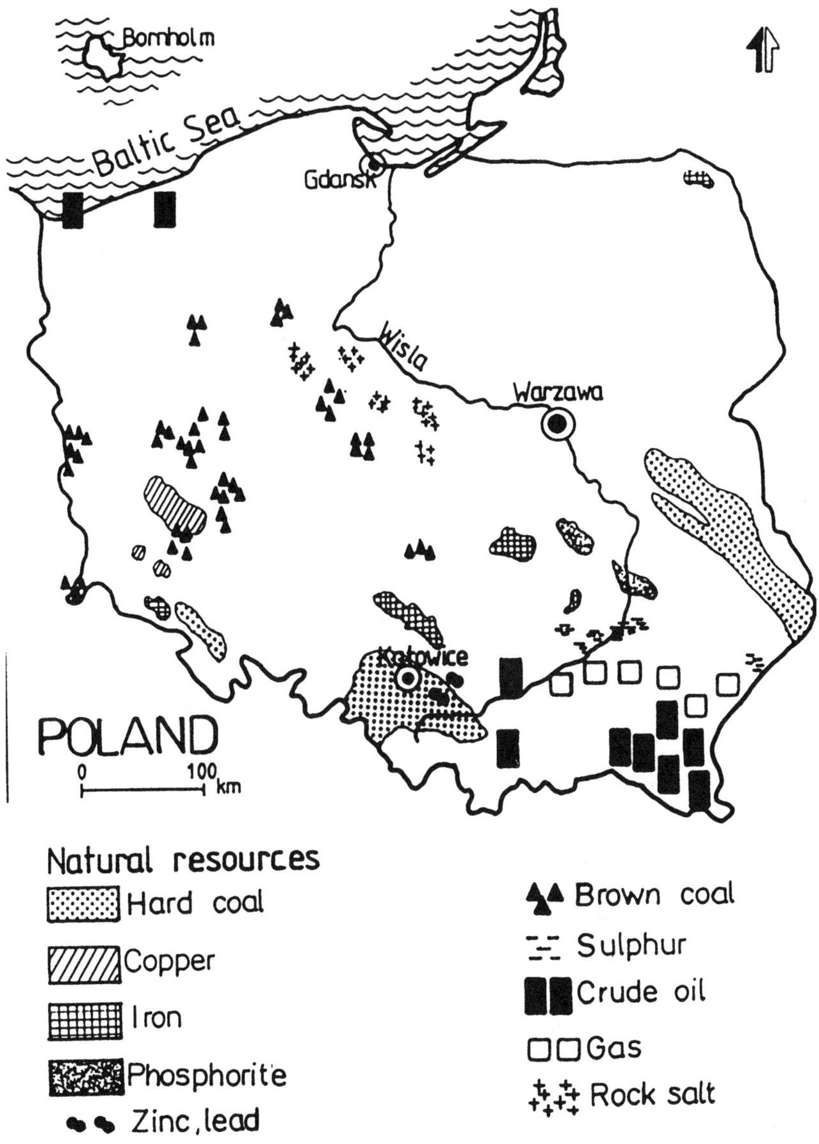

FIGURE 1. Natural resources in Poland.

with 74 and 81 years, respectively, in Sweden. In the Silesian region, life expectancy is further reduced by 5 years on the average. Death rate of children under 1 year is 19/1000 in Poland and 30/1000 in the Zabrze region, compared with 6/1000 in Sweden (Lithman et al. 1989).

Water resources are scarce in this region, and parts of the city periodically lack water. Heavy industry, which took top priority for historical and ideological reasons, consumes large quantities of water and releases large volumes of polluted effluents without any treatment. Most municipal wastewater is discharged to the rivers without

FIGURE 2. Location of the Upper Silesian Region, City of Zabrze and Mikulczycki catchment where the environmental restoration project is conducted with Polish-Swedish cooperation.

treatment or after mechanical treatment only. The only wastewater treatment plant in the city with biological treatment handles about one third of municipal wastewater from the northern part of the city.

About 70% of all households in the town are heated by coal stoves and local heating plants; the remaining 30% are connected to the central heating plant. The thermal plant and house stoves burn coal with 1.3% sulphur content (Swedish plants use coal with 0.3% sulphur). Neither the existing nor the newly-built thermal plants are equipped with any treatment of combustion gases. The new thermal plant, which is under construction, will be equipped with a soot separator. However, the new thermal plant requires cooling water of higher quality compared to the old one, which uses wastewater from a river situated nearby. The only solution which is available for the moment is to use communal drinking water as a coolant for the new thermal plant. This will reduce the already insufficient drinking water supply by about 40,000 m^3 daily.

A more than 300-year mining legacy has resulted in a labyrinth of underground extraction galleries extending more than 1000 m below the ground level. The whole land area is gradually subsiding. The land subsidence in the urban area is between 1.9 and 9.0 m; a record of 16 m was measured just outside the city. Huge cones of depression around deep mines split existing groundwater layers and make it difficult to tap groundwater resources. Buildings and pipes carrying drinking water, wastewater, or chemicals are often damaged (Niemczynowicz 1989). The existing plans for the environmental upgrading of the Zabrze-region contain only the most the most urgent measures that can possibly slow down the environmental degradation but would hardly reverse damages already done. Even these plans are being postponed or delayed, mostly for economic reasons (Postawa et al. 1986).

APPLICATION OF NEW IDEAS: AN ENVIRONMENTAL RESTORATION PROJECT IN POLAND

In the difficult situation in the city of Zabrze described above, it is tempting to apply multiobjective thinking and to implement really integrated water management, considering all pollution generating activities within one, well-defined catchment. Such an action should have a chance to break the present vicious circle and provide a complex solution. This is the aim of a Polish-Swedish multiobjective restoration project. The project proposal was formulated in cooperation within Polish-Swedish multidisciplinary working group and found acceptance and support on local, regional, and central administrative levels in Poland. The description below states what the project has in mind to do, i.e., what is believed can solve major environmental problems of the catchment. Let us look at the detailed goals and general lines of the proposed solution (Niemczynowicz et al. 1990).

The detailed goals of the project may be split into four parts:

1. To reduce pollution of the Mikulczycki brook to the degree necessary for industrial usage of water from the brook. This would release about 40,000 m^3 of water necessary for cooling of the new thermal plant. After the introduction of a new thermal plant, new residential areas will receive central heating. This will reduce air pollution because it will not be necessary to burn coal in individual houses and flats. The following measures are proposed:
   - The "key operation" which is necessary to reduce pollution level in the brook to the degree that other biological measures are effective is to construct a main sewer to collect municipal wastewater from the densely populated part of the area. Allocation of funds has been given first priority.
   - Upgrading the treatment capacity of the existing treatment plant by introducing an aeration stage and constructing a root-zone uptake.

- Improving incoming water quality by protective measures in upper reaches of the brook.
- Revising production methods and wastewater-generating processes in all industries within the area in order to reduce pollution release and the effluent volume at the source.
- Disconnecting most polluting industries from the river and sewage system. Local treatment of effluents from the three most polluting industries using infiltration to the ground and pre-aeration.
- Constructing small, local treatment units in order to selectively treat the effluents from the most polluting tributaries.
- Increasing the residence time of water and self-purification capacity in the river by 50% by changing straight stretchs to meandering courses. Flood risks will be decreased by attenuation of stormwater flow from the upper part of the catchment.
- Creating of vegetative filter strips and seasonally flooded riparian zones along the course of the brook and its tributaries.
- Creating small lakes designed as stormwater detention basins along the course of the brook.

2. To reduce water consumption in one housing area of the catchment (Helenka) using water-saving technology. This will reduce wastewater volumes and ease the drinking water shortage. A reduction of about 30% will be achieved by installation of water-saving toilets, using more efficient taps with better washers, restoration of water networks, and installation of water meters in each lodging.
3. To reduce energy consumption in the Helenka housing area by decrease of heat loses. Since 70% of houses are heated by coal-burning stoves, energy conservation is very important for environmental protection.
   - Installation of regulators in a local thermal plant in order to optimize the burning process.
   - Regulation of the heat flows in the heating network.
   - Installation of thermostats and instructing residents.
   - Introduction of electricity-saving schemes and energy-efficient fluorescent lamps.
4. To upgrade the aesthetic appearance of the Helenka housing area. The goal is to create a positive example of a case site where environmentally and economically sound management may also increase the aesthetic values of the area and, thus, indirectly increase the quality of life for inhabitants. The program includes upgrading green areas, improving the standard of sidewalks, painting houses, and introducing better routines for disposal and transport of solid wastes.

Since the project contains elements from several disciplines (physical planning, water resources planning, wastewater treatment by conventional and biological methods, management of energy, aesthetical aspects, etc.), a multidisciplinary team of researchers consisting of hydrologists, city planners, biologists, ecologists, and economists has been created. Local politicians are also actively involved in the project.

It is worth noticing that all elements of the project solution are interdependent, and therefore all elements must be performed in a logical sequence. For example, all biological measures in the river would not function without disconnecting municipal and industrial waste discharges. Stormwater flow must be attenuated before plant filters may be effective. Public acceptance and participation is essential for the success of the project. Calculated total costs of the proposed solution are about one third of the previously planned traditional solution containing the construction of a large river water treatment plant (Postawa et al. 1981).

The project has found understanding and active financial support by the local community and the municipality of Zabrze. In spite of constant financial problems, the project makes rapid progress. Disconnection of main pollution sources has significantly improved the water quality in the upper part of the brook. New species, characteristic of good water quality, are already abundant in this part of the brook. Upgrading of an existing water treatment plant is progressing rapidly. Construction of a wetland filter combined with a root-zone process for treatment of municipal wastewater from the northwestern part of the catchment is well advanced.

The main goal of the project — the transfer of ideas and approaches based on integration of measures, pollution prevention and the application of non-conventional, environmentally sound and economically effective technology — is already at least partially achieved. The Polish contractor, the Municipality of Zabrze, is determined to perform all actions proposed in the project, with or without Swedish money.

SUMMARY AND CONCLUSION

Our perception of the world and its problems has changed recently. We now realize the global nature of environmental problems, and we realize that environmental degradation is continuing in developing countries and in Eastern Europe.

In order to solve a fundamental material flow problem, i.e., the problem of accumulation of pollutants, a new approach to resource management must be applied in future societies. The goal of such an approach is to close the cycles of residuals that damage the environment. Highly developed countries have all the economic and human resources to implement such a new approach. However, existing traditional infrastructure and water-management related facilities limit the extent to which this change of approach is practically possible. The novel technologies may only complement and partially substitute for existing ones. Sustainable use of natural resources may only be applied if traditional sectorization in applied policies, actions, education systems and aid programs is broken. Not only scientific progress but also necessary major changes in the society depend on that.

The presently re-developing countries of Eastern-Central Europe that to a great extent lack such facilities, should not repeat the errors of the past committed in Western European countries due to lack of understanding of relationships between human activities and the response of Nature. What is to be done in developing and redeveloping countries should be done in accordance with the present knowledge and emerging novel technology.

Cities of western countries still contribute to continuing degradation of the global environment — in spite of all efforts and costs for advanced sewerage systems and sophisticated treatment facilities. The result is that, increasingly, the traditional end-of-pipe solutions are losing credibility. The goals of water management have already been reformulated from traditional sanitary engineering and storm drainage to more ecologically sound solutions. The new goal is to avoid accumulation of non-degradable hazardous substances and to adapt the waste output to what nature can endure.

There is a new realization that the struggle against global pollution can only be effective if water management is integrated with management of other related urban subsystems such as solid wastes, industrial production systems, energy production and consumption, transportation systems, etc. This approach involves a deeper understanding that all material and energy are involved in systems of a global scale with long-term circulation patterns

Present environmental problems in Poland are the result of ruthless exploitation of rich natural resources, unbalanced industrial development and environmental indifference during last four decades. Poland is only one example of many re-developing countries of Eastern-Central Europe with similar problems. The present economic situation in these countries — the necessity of preventing unemployment, the urgent need of restructurization of the economy and industry, the necessity of closing large industries, the urgent need of increasing water supply and improve sanitation and drainage — create the conditions under which the issues of environmental protection and restoration can hardly be given the first priority.

However, the price of environmental protection depends on the technology used. Pollution protection by source control is generally cheaper than treatment of effluents, especially if different type of effluents are mixed. Treatment of wastewater is very expensive if the traditional end-of-pipe solutions are applied. Fortunately, several novel, less expensive, environmentally sound technologies exist today and are ready for implementation in any country lacking treatment.

Performing demonstration projects, where the transfer of ideas and approaches is the major goal, is perhaps the most important action to which financial means should be allocated, instead of constructing facilities designed according to the previously governing technical paradigm. Aid programs conducted by the world's organizations should be revised accordingly.

REFERENCES

Arthur, J. P. 1983: "Notes on the design and operation of waste stabilization ponds in warm climates of developing countries." *World Bank Technical Paper No 7.* Washington.

Bartone, C.R. 1991: "International Perspective on Water Resources Management and Wastewater Reuse — Appropriate technologies." *Wat. Sci. Tech.* Vol. 23, Kyoto, Japan, pp 2039-2047, Japan.

Blaszczyk, B. (editor) 1990: "Informacja dotyczaca gospodarki odpadami przemyslowymi na terenie woj. Katowickiego." (Information about waste management in Katowice region) — in Polish; Unpublished.

Fleckseder, H. 1990. Abwasserableitung–Abwasserreinigung–Gewässerschutz. (Wastewater disposal–wastewater treatment–surface water protection. In German). Inst. fur Wassergute u. Landschaftswasserbau, Techn. Univ. Wien, A-1040 Wien, Karlsplatz 13.

Grau, P., Gujer, W., Ives, K.J., Jenkins, D., Krois, A.C., Di Pinto, M., Suzuki, M., Teorien, D.F., Milburn, A., Izod, E.J., Nagle, P.T. (Eds) 1990: *Water Pollution Research and Control — Proc. 5th Biennial Conf. IAWPRC, Kyoto, Japan, 29 July 3 August.* Pergamon Press.

Ishii, H., Tanaka, K., Aoki, M., Murakami, T., Yamada, M. 1991: "Sewage sludge composting by static pile method." *Wat. Sci Tech.* Vol 23, Kyoto, pp 1979-1989.

Keliar, D. 1988: "Storm Water Management Experience and Drainage Criteria of Markham." Ontario Proceed. Int. Symp. on Urban Hydrology and Municipal Engineering, Markham, Canada, June. Part A pp.A1-A5.

Konstantynowicz, E. (Ed). 1986: "Problems of protecting environment and natural resources in Katowice region." — In Polish. PTPNOZ, Sosnowiec, September.

Lesniak J. Personal communication, 1989.

Lithman, T., Dahn, I., Tell, I. 1989: "Hälso och sjukvård" /Health and medication/ In: "Raport från en ekologisk katastrof."/Report from an ecological catastrophe/, Miljödelegationen Västra Skåne, Info 3., October.

Mara, D., Cairncross, S. 1989: "Guidelines for the safe use of wastewater and excreta in agriculture and aquaculture: Measures for public health protection." World Health Organization, Geneva.

Middlebrooks, J.E. (Ed) 1982: *Water Reuse,* Ann Arbor Science, England.

Niemczynowicz J. 1989: "Vattenrelaterade miljöproblem"/Water-related environmental problems/in: "Raport från en ekologisk katastrof."/Report from an ecological catastrophe/, Miljödelegationen Västra Skåne, Info 3., October.

Niemczynowicz, J., Lazarek, S., Hammer, M., 1990: "Environmental restoration of the Mikulczycki catchment in Zabrze, Poland." Project proposal: Dep. of Water Res. Eng., University of Lund, Algapro, Lund, Swedish Univ. of Agicult. and Sc, Alnarp. August.

OBKS. 1985: "Opracowanie prognozy jakosci wody w pot. Mikulczyckim." (Prognosis of water quality in Mikulczycki brook). In Polish. BKS 05/93/83/BWI.

Postawa K., Kapala J., Kostecki M. 1986: "Koncepcja systemu wodno gospodarczego i ochrony wod w zlewni rzeki Bytomki." (Concept of restoration of water resources and environment in Bytomka river catchment). In Polish. Polska Akademia Nauk, Instytut Podstaw Inzynierii Srodowiska, Zabrze.

Reddy K.R., DeBusk, T.,A. 1987: State of the art utilization of aquatic plants in water pollution control. *Water Sci. Techn.,* Vol 19, No 10, 61-79.

Ridderstolpe, P., Kindvall, I. 1989: "Vattenbruk, vattenhantering och resursåtervinning."/Aquaculture, water treatment and reuse of resources — in Swedish/Royal Institute of Technology, Dept. of Land and Water Resources, Meddelande Trita-Kut 1050, Stockholm.

Shuval, H.,I., Adin, B., Fattal, E. Rawitz, E., Yekutiel, P. 1986: "Wastewater irrigation in Developing Countries: Health Effects and Technical Solutions" World Bank Technical Paper Number 51. Washington, D.C.: World Bank.

Sundblad, K. 1988: "Recycling of wastewater nutrients in a wetland filter." Ph.D. thesis, Linköping Studies in Arts and Sciences 27.

Swedish EPA-SNV. 1985: "Avloppsvatteninfiltration. Förutsättningar, Funktion, Miljökonsekvenser." (Infiltration of wastewater. Background, function and environmental impacts. In Swedish). Nordisk Samproduktion Naturvårdsverket, Nordiska Ministerrådet. ISBN 91-7590-221-4.

Swedish EPA-SNV. 1985a: "Infiltration av hushållsavlopsvatten." (Infiltration of municipal waste water. In Swedish). Statens naturvårdsverk, Liber förlag, Buhuslänningens Boktryckeri AB Uddevalla.

U.S. E.P.A. 1977. *Process Design Manual for Land Treatment of Municipal Wastewater.* U.S. E.P.A. 625/1-77-008, October.

Widmer, W.J. 1981: "Summary Previews of Waste Stabilization Ponds." In: *Waste Stabilization Ponds: Design and Operation.* WHO/EMPO Technical Publication No. 3:73-111.49.

Wilczek, A. 1989: "Badanie zmian fizjografichnych. Czesc 1. (Study of physiographic changes — in Polish). Part 1" in Urban-conservatory studies in Zabrze.

Wrona, A. 1989: "Katowice district a major region of ecological threat in Poland." In: Tylko jedna ziemia (only one Earth) Biuletyn informacyjny No 4(178) pp. 4-12.

Wronowskai, A.U. 1990. "Selection of bacterial and fungal strains for bioacumulation of heavy metals from aquenos solutions." *Wat. Sci. Tech.,* Kyoto, Japan, Vol. 23:309-318.

Yuki, Y., Takayanagi, E., Abe, T. 1991. "Design of multi-storey sewage treatment facilities." *Wat. Sci. Tech.,* Kyoto, Japan, Vol. 23, pp 1733-1742.

Chapter 27

CONTAMINATION OF GROUNDWATER NEAR LANDFILLS IN POLAND AND POSSIBILITIES FOR PROTECTING IT USING PLANTS

Marek Agopsowicz

INTRODUCTION

In some Western countries, macrophytes are used for landfill leachate treatment (Ettala 1988, Hasselgren 1989, Robinson 1990).

Under Finnish, Swedish, and English conditions, this is done with success. Transfer of this new technology to Poland should be made after careful recognition of Polish conditions. Landfills for municipal solid wastes in Poland generally do not meet the requirements of environmental protection. Both municipal and industrial wastes are delivered to landfills without any segregation. We can very often find hazardous and liquid wastes there as well.

Generally, the wastes are dumped on an unsealed bottom. No other procedures of water protection are carried out there very often. Therefore, the water which flows through the waste leachate contaminates the ground water. In the northeast part of Poland, investigations of a few landfills' effects on water in their area have been carried out (Agopsowicz 1990a, 1990b). This chapter presents two landfills.

LANDFILL IN LINOWO NEAR SZCZYTNO

Landfill use began in 1986. Solid and liquid wastes from Szczytno (about 30,000 inhabitants) were dumped without any site preparation and without application of environmental protection procedures. The area of waste dumping was about 1.5 ha. In 1987, hydrogeological investigations were carried out. In 1990, investigations of the effect of landfills on ground water were undertaken. It was found that liquid wastes brought to the landfill and surface water on the landfill area were as highly polluted as municipal sewage. Intensive processes of self-purification of water took place in a 7-m aeration layer. Ground water close to the landfill was also polluted, although less than the surface water on the landfill. Within 130 to 240 m of the landfill, ground water was quite pure and did not differ from ground water far from the landfill. In the future, if pollution is found to be spreading in ground water, the heap of wastes will be covered by an impermeable layer and dumping will be continued. Leachate collected in the drainage will be directed to a reed bed, as is done in England (Robinson 1990).

The above procedures help us develop our knowledge in this field.

0-87371-990-5/97/$0.00+$.50

LANDFILL IN LIDZBARK WARMINSKI

The landfill has been used for about 20 years. Solid and liquid wastes were brought to this landfill without any earlier construction and application of environmental protection procedures. Nowadays, its area is about 3 ha. Depth of wastes is about 10 to 20 m. In 1989, hydrogeological investigations were carried out. The effects of the landfill on water environment were studied. It was found that leachate from the landfill and liquid wastes flowed in two directions. To the northeast, the wastes are collected in a pond and flow into a drainage ditch (partly on the surface and partly underground). To the southwest, they flow underground about 80 m to a pond situated behind a nearby road.

The investigations proved that liquid wastes were highly polluted and they violated the standards for municipal sewage. Similar qualities were valid also for wastes in the pond in the northern part of the landfill. Ground water flowing northeast was also highly contaminated, much more than is permitted for municipal sewage. Ground water to the southwest was less polluted than the water flowing northeast, but was similar to municipal sewage. Self-purification of wastes was intensive in the ground. Water collected in drainage ditches was less polluted and corresponded to the standards for pure water or exceeded them a little.

In the future, dumping will be continued. Leachate collected in ground water and surface water from the pond in the southwest part will be spread on willow fields located on the west side of the landfill. The northeast side between the landfill and drainage ditch will be planted with willow and reed for increasing the self-purification of waters in this area.

In laboratory experiments at the University of Agriculture and Technology in Olsztyn, species of willows were found that could grow on the undiluted and very heavily polluted leachate of Lidzbark Warminski landfill.

CONCLUSIONS

Macrophytes (willows, reeds) are used for the treatment of landfill leachate in Western countries. There are no similar solutions in Poland. Leachate of Polish landfills is much more polluted than in Western countries, so transfer of this new technology to Poland should be made very carefully and developed in accordance with local conditions.

REFERENCES

Agopsowicz, M. (1990a), "Wplyw wysypiska-wylewiska odpadów komunalnych w Linowie k/Szczytna na srodowisko wodne" (Effect of solid and liquid waste landfill in Linowo on water environment), *Report Nr 06.03.25. of University of Agriculture and Technology in Olsztyn.*

Agopsowicz, M. (1990b), "Badania wplywu wysypiska-wylewiska odpadów komunalnych w Lidzbarku Warminskim na srodowisko wodne" (Investigation of effect of solid and liquid waste landfill in Lidzbark Warminski on water environment), *Report Nr 743/Wm/89 of Polish Society of Friends of Earth Sciences in Olsztyn.*

Ettala, M. (1988), "Short-rotation tree plantations at sanitary landfills," *Waste Management and Research* 6, 291-302.

Hasselgren, K. (1989), "Landfill leachate treatment and reuse in soil-plant system," *SARDINIA '89-2nd International Landfill Symposium.*

Robinson, H. (1990), "Leachate Treatment to Surface Water Standards Using Reed Bed Polishing," (Report) *The Use of Macrophytes in Water Pollution Control*, 3, 32.

Chapter 28

ECOLOGICALLY SOUND WASTEWATER TREATMENT: CONCEPTS AND IMPLEMENTATION

Petter D. Jenssen and Arild Vatn

ABSTRACT

Ecologically sound wastewater treatment can be defined as treatment concepts that promote a high degree of recycling as well as minimize environmental stress. However, most sewage systems currently in use discharge large amounts of plant nutrients into watercourses or the sea. Ecologically sound technologies that allow recycling of nutrients into plant or animal production and ultimately return water to pristine quality are available, yet their application is limited. This paper describes how it is possible, by the use of decentralized technologies and separate treatment of blackwater and graywater, to achieve a goal of nearly complete nutrient recycling. If the total amount of nitrogen (N) and phosphorus (P) in domestic sewage in Norway was reclaimed and reused in agriculture, the consumption of artificial fertilizer could be reduced by 15%. Ecologically sound wastewater treatment concepts for both urban and rural areas are presented. The concepts are all based on well-known technologies, hence it should be possible to apply these concepts with a minimum of testing and optimization. The presented concepts will, however, require a change in sewage system infrastructure. All the presented concepts exceed the purification performance of conventional systems and almost completely eliminate pollutant discharge to water recipients.

Ecological wastewater engineering is a young discipline and problems encountered when implementing ecological systems are system maturity, public acceptance, and legislative obstacles. Lack of professional competence as well as professional conservatism are major hindrances to more widespread use of ecologically sound wastewater treatment.

It is anticipated that the suggested systems will be more cost-effective than conventional sewage systems, especially if the benefits of reduced water pollution and the value of the reclaimed plant nutrients are accounted for.

INTRODUCTION

Domestic sewage contains substantial amounts of plant nutrients and organic matter. Even though nutrients, especially phosphorus, are removed in conventional sewage treatment plants, only a small amount is recycled, hence the nutrients ultimately are discharged into water recipients. Ecologically sound systems that promote recycling of nutrients into plant or animal production and reduce environmental stress are available, yet their application is limited. This paper briefly describes how

0-87371-990-5/97/$0.00+$.50

TABLE 1
Plant Nutrients and Organic Matter in Domestic Wastewater

Substance	kg/person/year
Nitrogen (N)[a]	4.4
Phosphorus (P)[a]	0.6
Potassium (K)[b]	1.0
Organic matter (COD)[a]	34

Note: Values are given in kg/person and year.

[a] Holtan and Åstebøl 1990
[b] Huledal et al. 1988

it is possible, with new technologies and changes of sewage system infrastructure, to achieve a goal of nearly complete nutrient recycling. The problems of implementing more ecologically sound sewage handling are considered. Finally, economic aspects (costs and benefits) of the systems promoted are briefly discussed.

WASTEWATER COMPOSITION AND COST

At present only a small amount of the nutrients in wastewater is recycled and the rest is discharged to water recipients where it may cause severe pollution. If these nutrients were recycled we could cut water pollution significantly and thereby also obtain environmental benefits. The amount of plant nutrients and organic matter in domestic wastewater is shown in Table 1.

An essential question is what costs the nutrients in Table 1 represent when they are pollutants. It is not easy to assess the environmental costs that may arise of out of improper wastewater management. Adequate tools for their assessment are not readily available. An indirect way to assess these costs is to look at the costs of N and P removal in sewage treatment plants. These costs can be regarded as the costs society is willing to pay to avoid adverse environmental effects in the recipients. These "purification costs" ought not to be higher than the potential costs to the environment. The nutrient removal costs are dependent on the size of the treatment plant and varies for P in the range of 75 to 500 U.S.$/person/year when Norwegian figures are used (Johnsen 1989). If nitrogen also is to be removed, it can be calculated that the costs per person will be nearly doubled (Aall 1990). This means high costs are involved in avoiding potential damage to recipients.

The wastewater constituents that create problems in water systems become valuable nutrients when recycled into silviculture or agriculture. Recycling will reduce the total costs of the wastewater treatment system because the environmental costs will be lower and because the nutrients have commercial value. However, the commercial value is probably small compared to the reduction in environmental damages obtained by increased recycling.

Because its composition of N, P, and K is close to an optimum ratio for plant growth (Foth 1984), domestic wastewater constitutes a fairly good plant fertilizer.

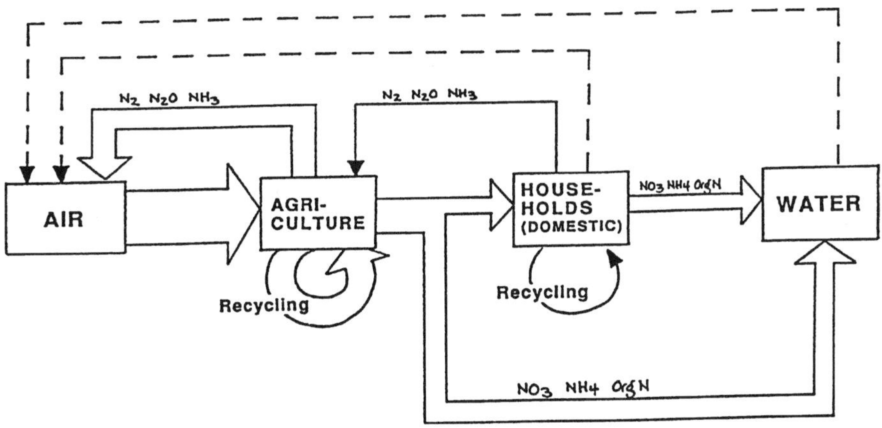

FIGURE 1. Some major pathways for nitrogen compounds.

The commercial value of the nutrients given in Table 1 is estimated at U.S.$9, based on current prices for artificial fertilizer in Norway. The quality of the remaining P ores is variable (Bøckman et al. 1991) and some have a higher cadmium content. Unless new technologies are developed, it is therefore likely that the prices of mineral P fertilizer will increase over the coming decades. In addition, environmental taxes are being applied to artificial N fertilizer in Norway in order to cut nitrogen pollution. The value of alternative fertilizer increases accordingly.

In many areas, the soil's organic matter is depleted. Wastewater contains considerable amounts of organic matter, hence recycling of wastewater constituents would also affect soil condition and increase the possibility of future high yields. It is therefore likely that the value of the wastewater constituents as an alternative source of fertilizer will increase in the years to come. This means that technologies that promote recycling of wastewater constituents back into plant production will become increasingly more competitive. An interesting question in this context is how much of the artificial fertilizer can be replaced if all the nutrients present in wastewater were recycled. Figure 1 shows some major pathways for nitrogen compounds.

The figure disregards contribution from industry and transportation and the arrows are not scaled to give a completely correct picture of the mass flow. The main input of N in modern agriculture comes from atmospheric N_2 gas that is made available to farmers through industrial processes. It is estimated that 15% of the N input in artificial fertilizer and imported feedstuff to Norwegian agriculture flows from the agricultural domain to the domestic domain in the form of food (Dahl 1990). This equals the estimated amount of nitrogen in domestic wastewater (SFT 1990). Thus, with present technology, which removes very little N, there is an almost total loss of the N input to the domestic domain.

Phosphorus figures on the input in terms of food are not available, but the amount discharged to the sewage systems equals about 15% of the mineral P fertilizer input. Hence, the input of artificial N and P fertilizer could be reduced by 15% if technology that allowed complete recycling of the plant nutrients was available. We do not have figures showing these percentages on a world scale, but in countries with relatively more plant production than Norway it can be assumed that a higher percentage of the nutrient input to agriculture follows the food chain to the domestic domain. Export and import of food will therefore influence the ratio between plant nutrients present in domestic wastewater and nutrients present in artificial fertilizer. In countries or regions with high imports, the ratio will be higher than in countries or regions that are large exporters of agricultural products.

ECOLOGICAL TREATMENT CONCEPTS

Ecologically sound wastewater treatment can be defined as treatment concepts that promote a high degree of recycling as well as minimize environmental stress. Environmental stress is caused by activities that may result in damage to or undesired change in the ecosystems and can in economic terms be expressed as an environmental costs. Two important principles that facilitate design of ecologically sound technologies are:

1. Recycling
2. Decentralized/on-site treatment

The ecological wastewater treatment concepts presented in this text are based on the above principles.

Conventional sewage treatment systems are mostly an end-of-pipe technique. That is, wastewater from a large area and many sources is collected and treated in a central plant. In centralized systems, domestic and industrial sources often discharge wastewater to the same system, hence there is a risk of contamination of the sludge from industrial wastewater due to heavy metals and other environmentally suspicious constituents. In Norway this has significantly reduced the motivation for recycling of sewage sludge into agriculture.

Conventional sewage systems can be optimized for removal and recycling of P and removal of N. However, recycling of N into plant production is more difficult. In addition, conventional treatment processes are often energy consuming and thereby cause environmental stresses of their own.

Though the design of wastewater treatment systems based on ecological principles is a young discipline, technologies and components are available so that a more ecologically sound wastewater treatment system can already be implemented. This will, however, as the concepts outlined below show, require a change in the sewage system infrastructure.

THE TOILET-BLACKWATER TREATMENT

All the concepts outlined below are based on separate blackwater (toilet wastewater) treatment, since main source of pollutants in domestic sewage is the water toilet (WC). The values of Otis et al. (1991) shown in Table 2 represent U.S. figures and include garbage disposal and phosphate-containing detergents. The values from Holtan and Åstebøl (1990) are the most recent Norwegian data. Garbage disposals and detergents containing more than 0.2% phosphorus are not used in Norway (Holtan and Åstebøl 1990). In an investigation at an ecological housing development,

TABLE 2
Pollutant Load Originating from the Toilet in % of Total Content in Domestic Sewage

Component	%[a]	%[b]
Nitrogen	82	90
Phosphorus	30	77
BOD_5	37	
BOD_7	39	

[a] Otis et al. 1991
[b] Holtan and Åstebøl 1990

Kristiansen and Skaarer (1979) reported that the contributions from the toilet were 73% and 83% for BOD_7 and phosphorus, respectively. This probably shows values that are possible to obtain when people are motivated. In addition to the constituents in Table 2, the majority of the pathogenic organisms have their origin in toilet waste and up to 1/3 of the water consumption is due to the toilet. If, therefore, the toilet waste could be treated separately and recycled, the majority of the nutrients in wastewater would be removed.

Several alternatives to the traditional WC exist (Engen 1991). In our opinion the most interesting options for use in ecological wastewater engineering are biological toilets and certain kinds of low-flush toilets.

Through the use of biological toilets (also called dry or composting toilets), treatment and recycling on site can be achieved and water consumption cut. However, biological toilets require more maintenance and care than a WC. From a user's standpoint, a biological toilet is therefore not considered to be an equally acceptable alternative to the WC. Biological toilets exist that are approved for full-time use in single dwellings (Engen 1991). A properly sized and functioning biological toilet produces no — or very little — effluent and the composted waste can be used as garden fertilizer and soil conditioner. In a biological toilet, P is totally retained, but some N is lost due to evaporation of ammonia. The amount of N lost has not been estimated.

Urban Areas

In urban areas, the use of currently available versions of biological toilets is difficult, but may be possible in the future (Engen 1991). The concept outlined below (Figure 2) is based on the use of toilets with extremely low water consumption. Several "extreme" low-flush toilets are on the market. The more sophisticated types use a vacuum to enhance transport. Vacuum toilets offer the same comfort as the traditional WC, but use only 1 to 1.5 l/flush (Haddal 1991). Gravity type low flush toilets using only 0.8 l are also available (Dynoplast 1991).

Low-flush toilets will reduce the water consumption substantially, thus yielding a more concentrated blackwater. Assuming a water consumption of 1 l per flush, 20 families will produce approximately 10 m^3 blackwater per month. Such an amount

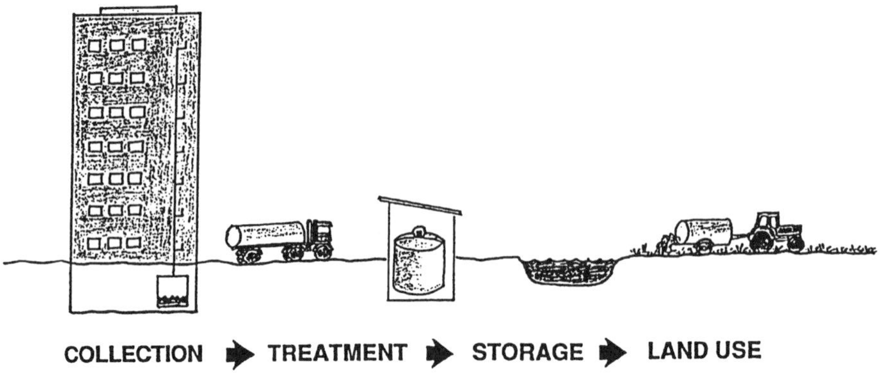

FIGURE 2. Blackwater treatment concept for urban areas.

can easily be trucked or transported by a low diameter pressure or vacuum sewer to a nearby treatment facility.

Several possibilities exist and can be suggested for blackwater treatment (Figure 2). Anaerobic treatment in biogas reactors (Tjernshaugen 1978) or liquid composting (Skjelhaugen 1991) are both well-known techniques that can be applied. Both these techniques involve processes that can kill pathogens and render the end product hygienically safer (Hansen and Tjernshaugen 1981). In addition, mineralization will occur. Mineralization is the transfer of N and P from an organic to an ionic state, in which the nutrients are easily adsorbed by plants (Myhr 1990).

The concentration of organic matter in blackwater is dependent on the toilet used, and even with very efficient low-flush toilets the concentration of organic matter is generally lower than in sewage sludge or animal manure. Consequently, blackwater alone may be difficult to process by liquid composting or in a biogas reactor. Co-composting of blackwater and other types of organic waste is therefore suggested. Tchobanoglous (1991) reports successful co-composting of sewage sludge and wastepaper yielding biogas and a soil-like end product. Energy production is another interesting aspect of both the aerobic and anaerobic treatment suggested in Figure 2. Heat can be retrieved from the liquid composting, and high quality energy in the form of gas is produced in biogas reactors. This energy can be used for a variety of purposes such as heating, generating electricity, or in the process of waste treatment (Figure 2).

The end products of both the aerobic and anaerobic digestion (Figure 2) are high in water content and storage dams will be necessary. This storage represents a cost, especially if capacity to hold all production outside the growing season is needed. Drying of the end product from the biogas reactor or liquid composting should therefore be evaluated. Drying will reduce the volume and greatly enhance the handling and sales potential. However, valuable components, especially N compounds, might be lost in the process of drying. Several systems for sludge drying exist (e.g., Utvik 1991). Drying requires energy and may therefore represent an environmental cost as well as additional treatment cost. Use of the produced biogas and solar energy can be suggested in order to reduce costs of drying.

In both biogas reactors and liquid composting, some N is lost to the atmosphere. However, Skjelhaugen (1991) reports that by tuning the process of liquid composting, the N loss can be reduced to less than 1%.

The blackwater treatment options (Figure 2) offer the possibility to convert different organic waste types to a valuable fertilizer or soil conditioner. The organic waste created by human activity can therefore be regarded as valuable raw material for production of fertilizer and energy. Most of the technologies are well known. We therefore anticipate that a limited amount of research is needed in order to optimize the techniques to co-process blackwater and other organic waste types. There is a possible odor and pathogen problem connected to the blackwater treatment; this must be solved. The success of blackwater treatment will also depend on costs and environmental benefits. This is discussed below.

In Figure 2, the graywater (wastewater from kitchen sinks, showers etc.) is not accounted for. The graywater constitutes a considerable volume of liquid that needs to be handled. Recycling of graywater on site (Figure 3) is one option that could greatly reduce the need for disposal. It is possible to construct facilities for graywater treatment and recycling indoors. However, efficient treatment units for local use then need to be developed.

If soil infiltration is possible, the graywater can simply be infiltrated and the need for piping is thereby reduced. However, in most urban areas infiltration into natural soil will probably be difficult. At present it is therefore most likely that some kind of collector system will be required.

The transport of only graywater will reduce the requirements on a collector system substantially. Because of reduced pollution loads the recipient requirements are less. Local discharge may therefore be possible. Local discharges into rivers and waterways will reduce the need for pumping stations and large collecting tunnels. We therefore anticipate that a graywater collector system can be constructed at a much lower price than a system for transport of combined (graywater and blackwater) wastewater.

In existing urban areas, a pipeline system is already available. However, this system is often overloaded or in poor condition, but if only graywater is to be transported the service life of the existing pipeline system can probably be extended. Existing pipelines often have leakages, and sometimes pollution is transferred to the drinking water mains (SIFF 1988) or to the groundwater. This pollution problem is reduced if only graywater is transported.

Up to 90% of the N and 80% of the P (Table 2) can be removed if the blackwater is handled separately. The need for graywater treatment is therefore questionable. However, graywater does contain organic matter, particles, and significant amounts of indicator bacteria (U.S. E.P.A. 1978, Kristiansen and Skaarer 1979). Unless very good recipients are available, some treatment will be necessary.

Graywater treatment can be performed by simple mechanical or biological methods. The use of constructed wetlands and wastewater stabilization ponds in combination with Flowforms (Jenssen et al. 1991) and/or hydroponic treatment in greenhouses (Guterstam 1991) offers aesthetically and pedagogically interesting decentralized options for graywater treatment that can be integrated in urban areas. However, before any free surface flow of treated graywater is permitted in urban areas, sufficient hygienic quality must be assured.

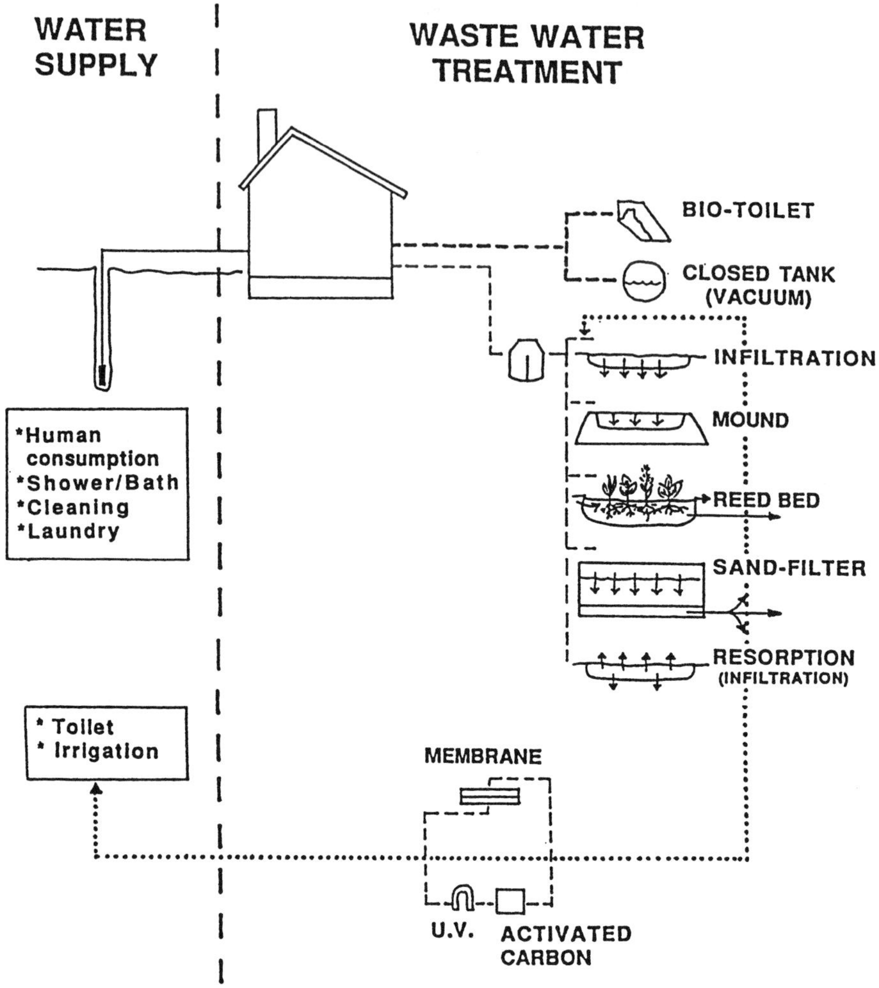

FIGURE 3. On-site water and wastewater treatment concepts based on separate black- and graywater treatment.

Rural and Suburban Housing Areas

In rural areas, conventional wastewater treatment becomes expensive because of piping costs. On-site treatment of wastewater predominates, but existing solutions commonly incorporate a traditional WC. Through the use of separate blackwater treatment, ecologically sound solutions are more easily obtained. Several concepts based on separate treatment of blackwater and graywater exist (Figure 3). A few of these concepts are commented on below.

In rural areas, we recommend the use of biological toilets, because in our opinion this is the simplest way to obtain more ecologically sound wastewater treatment. If the blackwater is treated separately, several options for graywater treatment exist. All the graywater treatment options (Figure 3), except some of the options suggested

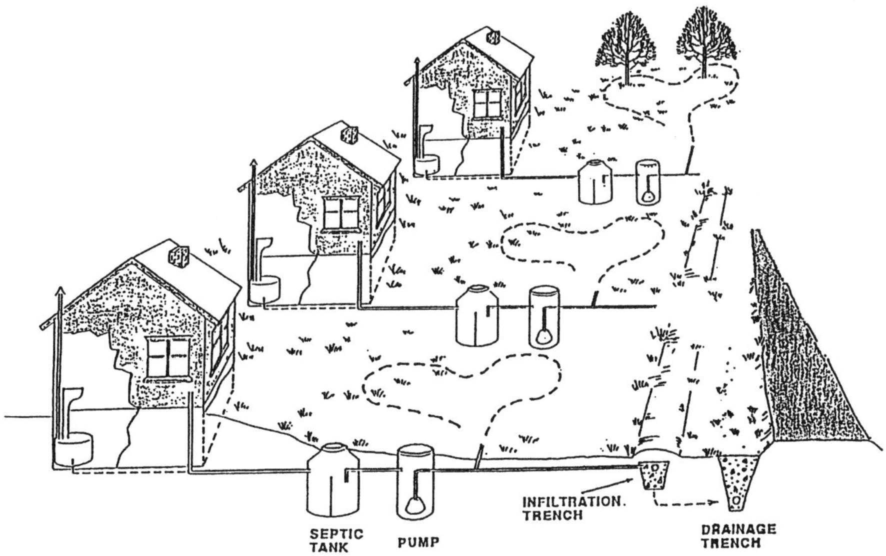

FIGURE 4. Graywater treatment in soils with a low hydraulic capacity. Summer treatment in shallow trenches in the root zone and winter treatment in a drained leach field.

for recycling of graywater into the household water supply, are based on the use of simple low-maintenance soil or soil/plant filters. Graywater may be difficult to treat by conventional technology due to its low concentration of pollutants, but soil has been proven to treat graywater efficiently (U.S. E.P.A. 1978).

If the soil fulfills the requirements for wastewater infiltration (U.S. E.P.A. 1980, MD 1985), cheap and simple disposal of graywater is achieved by the use of a traditional septic tank/soil infiltration system. However, the soil does not always fulfill the requirements for building a traditional septic tank system. This is the case in heavy clay soils or in dense till where the hydraulic capacity often is limited (Jenssen 1988). In these soils, a system as outlined in Figure 4 is suggested.

In summer, the plants will be able to absorb water and nutrients present in the graywater. Under such conditions, a shallow trench system will function even in areas with very low hydraulic capacity (Skaarer 1980). In the Norwegian climate, a drained leach field must be used in order to avoid water saturation of the garden soil in the fall, winter and spring. Suburban housing areas could easily be served by such a solution. The community sewer could then be replaced by a "community drainage trench."

Purification of combined wastewater by soil has been extensively studied and near complete mineralization and removal of pathogens and phosphorus can be obtained within one meter of travel through soil (McCoy and Ziebell 1975, Lewis et al. 1982, Jenssen and Siegrist 1990). When graywater is filtered through some distance of fine-grained soil (soils of low hydraulic capacity are usually fine-grained), a very high effluent quality can be expected. It should therefore be possible to discharge the effluent from the "community drainage trench" into the nearest stream. By the use of separate blackwater treatment (Figure 2), closing of the streams to human access can be avoided and the streams can be preserved as a valuable landscape element in housing areas.

Figure 3 also suggests other concepts for graywater treatment such as sand filtration and the use of reed beds. If the graywater is collected during the winter months, it can be treated by a shallow infiltration/resorption in the root zone in the summer. This concept has been tried with success in the Norwegian climate (Skaarer 1980). The system was tried in an area with 30 to 50 cm of clay soil over bedrock, and produced no effluent from the lot.

IMPLEMENTATION

Ecologically sounder wastewater treatment systems are available, but why is their use not more widespread? One possibility is that the solutions are immature. It is obvious that not every link in the systems outlined above has been thoroughly tested and approved. Vacuum toilets work well on boats and trains, but still require too much maintenance for widespread use in public housing. According to our information, work is now being carried out that will enhance operational stability of vacuum toilets. Testing and improvement of other components of the outlined concepts is also under way (Kåre Haddal, pers. comm.). We are therefore convinced that it should be possible to build systems based on the above concepts with a limited amount of testing and research, and if the right decisions are taken, widespread use of these concepts could commence in the very near future.

But there are other obstacles. The public might not be ready for these solutions. However, the public opposition is likely to depend upon how great a change a more ecologically sound system represents, both economically and in terms of comfort. The WC is the only component of the sewage system that most people ever are familiar with. The use of low-flush (e.g., vacuum) toilets is very similar to the WC and is probably easily accepted. On the other hand, conversion to a biological toilet is likely to meet certain psychological barriers and acceptance will require carefully planned promotion, perhaps with the aid of economic incentives. Changes occurring outside the bathroom will probably cause very little public opposition as long as these changes do not increase apparent pollution and personal taxes/payments. We therefore think it should be possible to make the political decisions necessary for a gradual conversion to a more ecologically sound sewage treatment system. In Norway, better political decisions are necessary, because present legislation and regulations promote the use of centralized sewage treatment systems.

An ecologically based sewage system cannot gain recognition without the necessary competence and capacity among professionals. Present education at engineering schools in Norway does not emphasize ecological wastewater engineering. Even if political decisions are made that encourage an evolution towards a more ecologically sound wastewater treatment system, there will still be an enormous lack of competence in the society. On the other hand, political decisions in favor of ecological systems are unlikely as long as the traditional professional society is conservative. Conservatism among professionals will also influence public opinion and is therefore probably one of the main obstacles restricting implementation of more ecologically sound wastewater treatment systems.

ECONOMY

More ecologically sound wastewater treatment systems mean applying systems with a higher degree of recycling and lower pressure on the environment. But this does not mean that these systems are to be chosen at any cost. It is possible that they will require high investment and maintenance costs and it is also likely that recycling systems will not themselves be totally free of environmental stress. All these costs have to be accounted for when making a complete economic evaluation of different wastewater treatment systems.

The following is a brief discussion of the principles and factors involved in an economic evaluation. The total costs for conventional systems can be divided into three groups.

- Investment, operation, and maintenance of WC and pipeline systems/collector system
- Investment, operation, and maintenance of treatment plants
- Environmental costs

In 1987 the yearly investment, operation, and maintenance costs in water and wastewater treatment systems in Norway was estimated at 200 U.S.$ per capita (Mosevoll and Ræstad 1988). About 80% of these costs were due to the piping systems. At this annual cost, it was estimated that Norwegian sewer lines would have to serve 300 to 400 years. The service life of sewer lines is 30 to 50 years (Aall 1990). Even though some lines may have a longer service life, the general state of the sewer lines is very poor (Mosevoll and Ræstad 1988, Sirum 1990). Connecting storm water to the sewage water system and crushing of pipes when filling the trench is common even in newer systems (MD 1978). Hence, pipelines are the weakest and yet most expensive link of conventional sewer systems. The result is extremely high flow, diluted water and hydraulic overloading of the treatment plants in periods of high precipitation. In periods of dry weather, large amounts of wastewater are lost through leakages and never treated in a sewage treatment plant. This greatly reduces the environmental efficiency and increases the environmental costs of the system due to adverse effects in the recipients.

In addition, many Norwegian cities lack sewage treatment or have only primary treatment (Aall 1990). In order to obtain a conventional sewage system that is environmentally safe and meets the present requirements, the costs must increase and the State Pollution Control Authority (Norwegian EPA) therefore proposes large investments in conventional sewer systems in the years to come (SFT 1990). The above figures give reason to anticipate that a doubling or more of the yearly per capita costs is necessary in order to achieve a fully functioning conventional sewage system.

The environmental costs of the existing conventional sewage system include damage in recipients due to discharges that are controlled (from treatment plants) or uncontrolled (e.g., leakages, overflow at pumping stations, seepage from sludge deposits).

The total costs for an ecologically oriented system can be grouped in a similar way to the costs of conventional systems. In the following we focus on the concept for urban areas outlined in Figure 2:

- Investment, operation, and maintenance costs for the blackwater collection and storage system (e.g., toilets, tanks, dams)
- Transportation costs (transport of blackwater to the treatment facility and farms)
- Graywater transport and treatment costs
- Environmental costs

The cost of blackwater treatment includes installation, transport vehicles, and the blackwater treatment and storage facilities. If a low-flush toilet system and a collecting tank is installed in an existing building, the costs will be much higher than if this system is installed when constructing the building. This will affect the comparison of conventional and ecological systems as outlined below.

The main environmental costs considered when evaluating the ecological concept outlined in Figure 2 are damage to natural recipients due to emissions from transportation vehicles and runoff from arable land. They are assumed to be much lower than those occurring for a conventional system.

Since we have benefits from recycling of nutrients and organic matter, the costs must be reduced according to their value as soil conditioner and fertilizer in agriculture or silviculture. In addition, the value of the energy produced when processing the blackwater (ecological system, Figure 2) or sludge treatment (conventional systems) also must be subtracted from the total costs. The above factors are assumed to reduce the costs of the ecological concept more than the costs of the conventional system. This reduction will be considerably higher for the ecological system than the present conventional system.

The main economic question is, then: Which system has the lowest total costs? So far there have been no complete calculations on the full investment, operation, and maintenance costs as well as environmental costs for each system. In Figure 5, the authors have tried to express an economic comparison of the costs of an ecological (C_E) and conventional (C_C) sewage treatment system in an area where no sewage system existed previously. Where conventional systems already exist the situation is somewhat different. This is discussed below.

The figure illustrates the costs for the two systems as a function of the distance to biorecipient, because the competitiveness of the ecological system, in our evaluation, will depend much upon distance to the areas for recycling of the nutrients.

For simplicity, we assume that the C_C is independent of the distance to the biorecipient, whereas C_E, as Figure 5 shows, will greatly depend on distance due to ordinary transportation costs. If we disregard environmental costs, the ecological system is competitive up to a distance A1, but if environmental costs are included, then the distance is increased to A2. Environmental costs are not easily assessed in terms of economic figures. Due to recycling, the environmental costs connected to the ecological system (C_{E2}–C_{E1}) are much smaller than the environmental costs (C_{C2}–C_{C1}) connected to the conventional system.

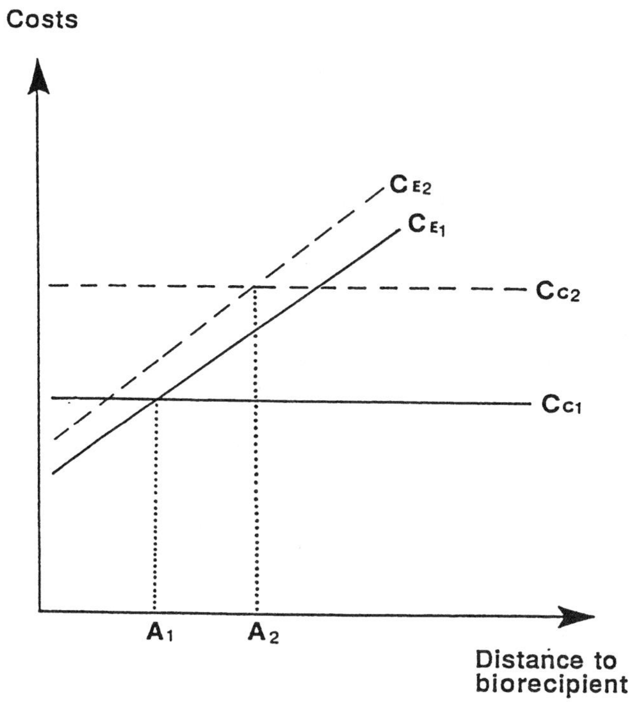

FIGURE 5. Costs of an ecological (C_E) and conventional (C_C) sewage treatment system vs. distance to biorecipient (areas for recycling/land use in silviculture or agriculture). The ecological system refers to the urban concept described above and in Figure 2.

The above presentation is simplified and does not account for time dependence of the costs which will vary in the two systems. But this is probably not of great importance in situations where no sewage system exists initially. However, this is different when conventional collection and treatment facilities already exist. Calculation of the present value is then an adequate tool for handling the comparison of the two systems:

$$Q_{out} = \frac{S + P - E_T}{1000} F \qquad (1)$$

C_{Et} and C_{Ct} represent the costs at year t for the ecological and conventional systems, respectively, environmental costs included. The real interest rate is r. The costs are added for all years. The ecological system will be competitive if the net present value is negative, i.e., the costs for the ecological system is lower than the conventional system when all future costs are capitalized to present time. When comparing systems according to Equation 1, the result will depend on at what time the major costs come and the real interest rate. Where a conventional system already exists, the costs related to continued use of this system will be small in the first years. The situation is reversed for the ecological system where all the initial investments are lacking. The relative gains of implementing ecological engineering accrue in the future. In a situation with high real interest rates, as today, an ecological system will probably not be competitive according to Equation 1. High interest rates mean little emphasis on future consumption and standard of living. This is hardly consistent with the demand for a more sustainable development. However, for the municipalities that have to finance their investments in future systems with loans at the present interest rate, future benefits of an ecologically sound system are often disregarded. High interest rates are therefore probably another factor that helps explain the current limited use of ecological systems.

In Norway the future costs of conventional systems will increase considerably. This is evident in publications from the national pollution authorities (SFT 1990) and from a practical evaluation of the state of the present conventional sewer system (see above). We are therefore standing at the crossroads and have to decide whether we want to continue to invest in a conventional system that will become increasingly expensive over the coming decade or take higher initial costs and convert to an ecological system with potentially lower future costs and higher future life standard. A thorough evaluation of ecological soundness of the concepts herein discussed is therefore urgent if we are to avoid committing increasing economic resources to conventional technology.

REFERENCES

Aall, C., 1990. *Miljøårboka.* Det Norske Samlaget, Oslo.

Bøckman, O.C., Kaarstad, O., Lie, O.H. and Richards, I., 1991. *Landbruk og gjødsling — mineralgjødsel i perspektiv,* Landbruksdivisjonen, Norsk Hydro, Oslo.

Dahl, E. 1990. *Økologi, landbruk og forurensninger.* LØF 4/90. Department of Economy and Social Sciences, Agr. Univ. Norway, Ås.

Engen, Ø. 1991. "Can enhanced biological toilets solve the problem?" Paper presented at the International Conference "Ecological Engineering for Wastewater Treatment," 24–28 March, Stensund Folk College, Sweden.

Dynoplast, 1991. Sales information brochure. DYNOPLAST A/S, Cipax, N-1940 Bjørkelangen, Norway.

Foth, H.D. 1984. *Fundamentals of Soil Science.* New York: John Wiley & Sons.

Guterstam, B. 1997. Ecological engineering for wastewater treatment: Theoretical foundations and practical realities. In: C. Etnier and B. Gutersam (eds.), Ecological engineering for wastewater treatment. Boca Raton, FL: Lewis Publishers.

Haddal, K., 1991. Personal communication. JETS Inc., N-6060 Hareid, Norway.

Hansen, J.F. and Tjernshaugen, O., 1981. *Hygienisering ved aerob omsetning av kloakkslam og husdyrgjødsel.* IBT internserie 2/75, Dept. of Agr. Engineering, Agr. Univ. Norway, 1432 Ås-NLH.

Holtan, H. and Åstebøl, S.O., 1990. Håndbok i innsamling av data om forurensningstilførsler til vassdrag og fjorder. NIVA/JORDFORSK rapport. N-1432 Ås, Norway. 52 p.

Huledal, P., Nordberg, V. and Rydberg, A., 1988. *Ekologiskt avlopp i Övertorneå.* Kursrapport SB 320, Techn. Univ. Luleå, Sweden.

Jenssen, P.D., 1988. "Design criteria for wastewater infiltration systems." In: R. Bahmidimarri (ed.). *Alternative wastewater treatment systems.* London: Elsevier, pp. 93-107.

Jenssen, P.D., and Åsgård, T., 1979. "On-site water and wastewater treatment in the Lian-Vådanområdet in Trondheim commune." M. Sc. Theiss, Dept. of Biology and Nature Conservation, Agr. Univ. Norway, Ås (in Norwegian).

Jenssen, P.D. and Siegrist, R.L., 1990. "Technology assessment of wastewater treatment by soil infiltration systems." *Wat. Sci. Tech.* 22 (3/4):83-92.

Jenssen, P.D., Krogstad, T., Briseid, T. and Norgaard, E., 1991. "Testing of reactive filter media (LECA) for use in agricultural drainage systems." To be presented at: International Seminar of the Technical Section of C.I.G.R. on Environmental Challenges and Solutions in Agricultural Engineering, Agricultural Univ. of Norway, Ås, Norway, July 1–4, 1991.

Johnsen, F.H., 1989. *Økonomiske analyser av tiltak mot fosforavrenning.* Department of Economy and Social Sciences, The Agricultural University of Norway, Ås.

Kristiansen, R. and Skaarer, N., 1979. "BOV vannets mengde og sammensetning." *Vann* 2:151 — 156.

Lewis, W.J., Foster, S.S.D. and Drasar, B.S., 1982. *The risk of groundwater pollution by onsite sanitation in developing countries. A literature review.* IRCWD Report no. 01/82, Dubendorf Switzerland.

McCoy, E. and Ziebell, W.A., 1975. "The effects of effluents on groundwater, bacteriological aspects." In: *Second National Conference on Individual Onsite Wastewater Systems*, National Sanitation Foundation, Ann Arbor Michigan, pp. 67-78.

MD 1978. *Aksjon Mjøsa — Utkast til situasjonsrapport.* Ministry of Environment, Oslo, Norway.

MD 1985. *Forskrift om utslipp fra separate avløpsanlegg.* Ministry of Environment, Oslo, Norway.

Mosevoll, G. and Ræstad, C., 1988. *Planlegging av drift, vedlikehold og fornyelse av ledningsnettet for vann og avløp.* Brukerrapport 5A/87, Inst. for Vassbygging, Norw. Inst. of Technology, Trondheim, Norway.

Myhr, K., 1990. "Verknad av våtkompostert og ubehandla gjødsel på jord avling og miljø." *Landbrukstidende* no. 8, pp. 252-255

SFT 1990. *The future is now. A summary of long term plans.* State Pollution Control Authority, Oslo.

SIFF 1988. *Ledningers innbyrdes plassering i grøft — hygieniske forhold.* Statens Institutt for Folkehelse. Oslo.

Sirum, Ø., 1990. "Lekkasjer på avløpsnettet er ikke problemet." *Vann* 2:149-158.

Skaarer, N., 1980. *Rensing av avløpsvann ved infiltrasjon i rotsonen (resorpsjon) kombinert med lager. Ressursvennlige boligformer.* Rapport nr. 3., Agricultural Univ. of Norway, Ås. Skjelhaugen, O.J. 1991. "Liquid composting unit." To be presented at: International Seminar of the Technical Section of C.I.G.R. on Environmental Challenges and Solutions in Agricultural Engineering, Agricultural Univ. of Norway, Ås, Norway, July 1–4, 1991.

Tchobanoglous, G., 1991. "Land-based systems, constructed wetlands, and aquatic plant systems in the United States: An overview." In: C. Etnier and B. Guterstam (eds.). *Ecological Engineering for Wastewater Treatment, Proceedings of the International Conference, 24–28 March 1991, Stensund Folk College.* Bokskogen. Gothenburg, Sweden.

Tjernshaugen, O., 1976. *Våtkompostring av husdyrgjødsel ved kontinuerlig behandling.* NLH melding no. 88, Agr. Univ. Norway, 1432 Ås-NLH.

Tjernshaugen, O., 1978. *Naturgjødsel og vegetabilsk avfall for fremstilling av biogass.* IBT rapport no.151. Dept. of Agr. Engineering, Agr. Univ. Norway, 1432 Ås-NLH.

U.S. E.P.A., 1978. *Management of small waste flows. U. costs.* Environmental Protection Agency, Municipal Environ. Research Lab., Cincinnati OH.

U.S. E.P.A., 1980. *Design manual: Onsite wastewater treatment and disposal systems. U. costs.* Environmental Protection Agency, Municipal Environ. Research Lab., Cincinnati, OH.

Chapter 29

ECOLOGICAL APPROACHES TO WASTEWATER MANAGEMENT IN URBAN REGIONS IN THE NETHERLANDS

Tjeerd Deelstra

ABSTRACT

Our presentation deals with the following questions:

- What is the state of the art with respect to the technical and design aspects of wastewater purification by means of vegetated wetlands in the Netherlands?
- What has the Dutch government done to encourage ecological engineering in waste water management?
- How do ecological and conventional engineering compare financially?
- How does ecological engineering fit into Dutch city planning?

We conclude our presentation with an outlook on future developments in water and wastewater management in the Netherlands.

TYPOLOGY AND TECHNOLOGY OF ECOLOGICAL WASTEWATER SYSTEMS IN THE NETHERLANDS

INTRODUCTION

Water purification by means of vegetated wetlands is gaining recognition in the Netherlands and other countries. It has been proved that these wetlands can make an important contribution to the improvement of the quality of water. Even so, some are skeptical of the use of this technique for the purification of waste and surface water. The most important reasons for this are the relatively large areas it requires, the uncertain results, and the possible problems of smell and insects.

In the Netherlands at the moment, raw wastewater from camping sites is purified in this way on a small scale. Another use for this system is the purification of sewage overflows in order to prevent pollution of surface water. Due to the high "input" of polluting agents, the wetland system of filtering in this instance is referred to as "highly burdened."

Thought is also given to the use of this system in surface waters, in situations where water rich in nutrients is let into nutrient-poor natural areas. The expectations are that in such situations the pre-purification of the surface water via a wetlands filter should lead to an improvement in quality of the water in general. If such "low burdened" systems show good results in water purification, then perhaps it will be possible to use the wetland filter system to pre-purify the water used for the potable

0-87371-990-5/97/$0.00+$.50

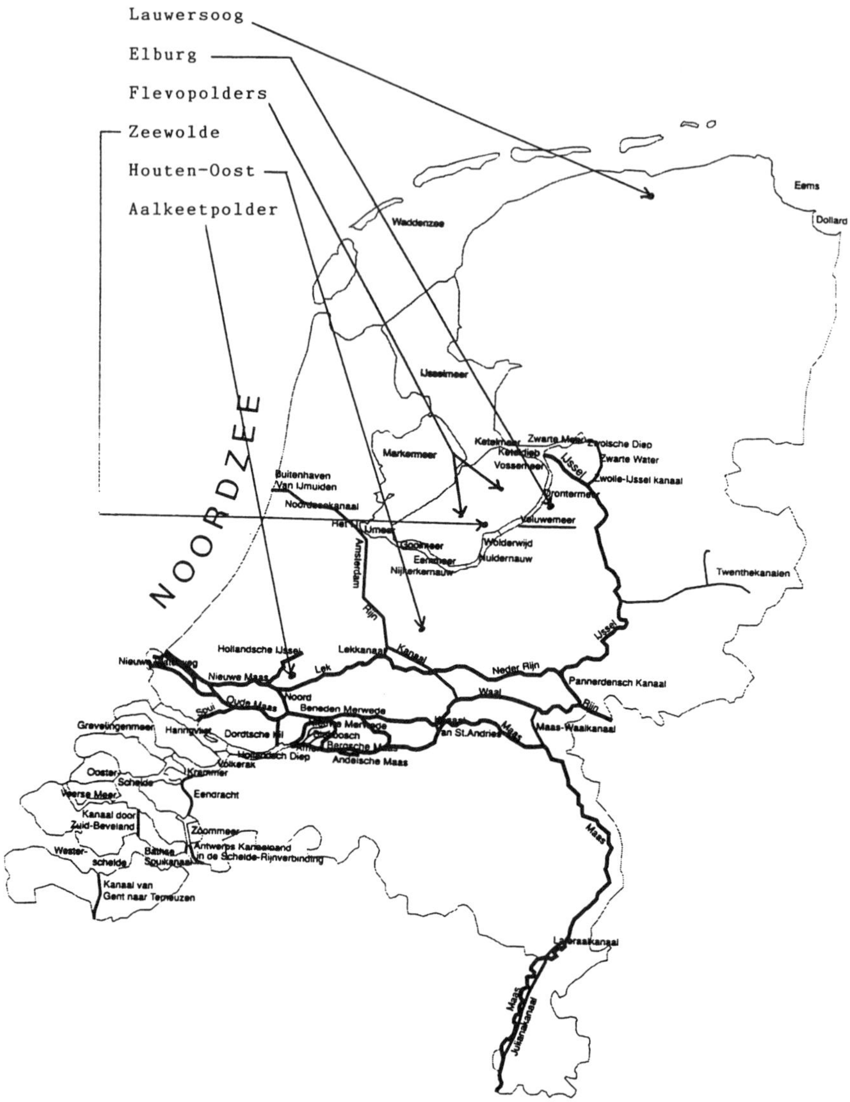

FIGURE 1. Locations of the projects in the Netherlands mentioned in this paper.

water process. In addition to use in water purification, reed (*Phragmites australis*) can also be used for dehydration of sludge.

There are some 30 wetland projects in the Netherlands. The first projects were completed about 20 years ago. The projects were initiated by different organizations. Some of the wetlands were designed by water boards in order to treat the effluent of a purification plant, others were laid out by municipalities to treat sewage overflows. Some projects are natural wetlands used by the water boards to protect sensitive wildlife areas. Some are owned by factories wanting to purify their wastewater. It is

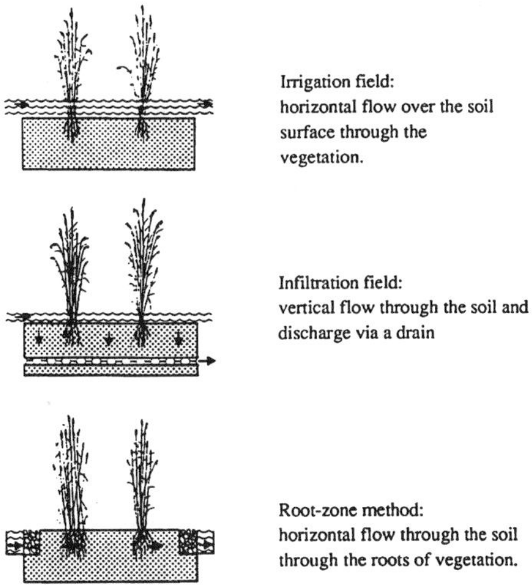

FIGURE 2. Types of artifical reed bed systems.

not possible to look closely at all the projects. To give some idea about the specific features of the several wetlands, examples will be given together with the different types of use of the vegetated wetlands. Figure 1 gives a map of the Netherlands with the location of the projects dealt with in this chapter.

TYPES OF VEGETATED WETLANDS

In general, the following systems can be used for wastewater purification (Figure 2).

Vegetated wetlands that are used for the purification of surface water do not have to serve only technical purposes. A natural system can also provide secondary functions such as improvement of the natural qualities of the area and recreation. The system in this case is multifunctional and conforms with the landscape.

POSSIBLE USES FOR VEGETATED WETLANDS

The use of vegetated wetland filters in the purification of wastewater has definite limitations in the winter months as far as quantity and cleansing capacity are concerned. Realistic possibilities for application include the following.

Wastewater Purification

At present more than 95% of households, businesses, and recreational facilities are connected to the municipal sewerage system. The remaining 5% are seasonal facilities such as bungalow parks, camping areas, etc., or houses outside the urban network. These locations make collection and purification of wastewater through the municipal system too expensive per person equivalent (p.e.). Local treatment of wastewater in conventional purification systems is not possible due to the often temporary nature of the location and the supervision necessary. Drainage from recreation sites usually takes place in the summer. This is the time of year when

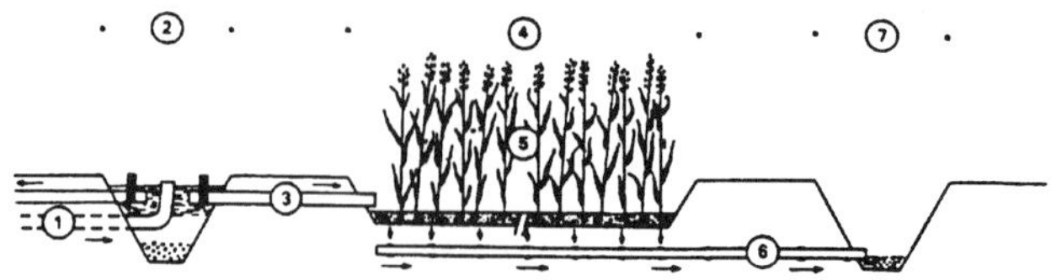

1: influent pipe
2: preliminary settling/distribution ditch
3: lockable pipe
4: reed infiltration field
5: reed plants
6: drain
7: effluent ditch.

FIGURE 3. Schematic cross section of the vegetated infiltration system near Lauwersoog.

vegetated wetlands are most efficient and productive, so that supply of wastewater and purification process are coherent. From the beginning of the 1970s, vegetated flow and filter systems have been used for the purification of wastewater from recreation sites outside the municipal network in the Flevopolders and near Lauwersoog.

This field near Lauwersoog was constructed in 1974 and is still in use. It is fed with wastewater, especially from recreation sites. The consequence is that there is a great difference in pollution load during summer (1700 p.e.) and winter (300 p.e.). The original design was based on a hydraulic load of 20 to 30 mm per 24-hour period and a capacity of 3000 p.e. The wet surface area is 1.3 ha, the gross surface area is 2.1 ha.

This system is comprised of a preliminary settling/distribution ditch, four infiltration compartments and an effluent ditch (Figure 3). The infiltration ponds have been planted with reed and provided with drainage 0.55 m beneath ground level.

The pond works as follows: raw domestic wastewater is discharged into the preliminary settling/distribution ditch. After settling, the wastewater is mixed with surface water from the effluent ditch and led into one of the infiltration compartments every three to four days.

The other three compartments are dry for a period of 10 to 11 days. The water filters through the sand layer and is discharged by the drainage systems into the effluent ditch.

Sewerage Outlet and Overflow

The possibilities for reducing pollution of surface water caused by sewerage overflows are minimum. The reasons for this are the number of overflows on the one hand and the difficulty of the task on the other hand. Due to the unpredictable flow of wastewater, most conventional purification systems are not suitable. The use of vegetated wetlands provides an alternative for collecting overflow water and purifying it. In 1986 a reed field outside Houten-Oost was put into use as an additional facility to the sewerage system. This field was designed as a treatment for the overflow from the eastern part of the sewage system.

The main characteristics of the field are the following: Overflow to the reed fields will occur a few times a year. The treatment in the fields will take place by means of sedimentation and absorption of the pollution. The fields also reduce the amount of polluting compounds by mixing sewage overflow with clear water present in the fields. The purified water will be drained off to the receiving water. Occasionally the storage capacity of the fields will be insufficient, causing a direct overflow to the surface water. Nevertheless, during these extreme conditions the water will still be treated but for sedimentation only.

Because the reed fields hold back the larger part of the pollution, the oxygen content might drop strongly. Special measures (like the installation of a cascade aerator) have been taken to prevent any possible nuisance due to odor which might occur during the extreme oxygen drop.

Cleansing of Surface Water

Cleansing of in-flow water. The inflow of nutrient-rich water into marsh areas low in nutrients can be of essential importance in dry periods when incoming water from rain and seepage is not enough. The in-flow water could possibly be precleansed by leading it through a vegetated wetland filter system. One such wetland is situated near Rotterdam. Its main objective is to purify water which is pumped during the summer from another area into the nature reserve of the Aalkeetpolder. The wetland is 5 ha and is divided in three main ditches with a total length of 2000 m. One part of the ditches is planted with reed, the other part is not. The water level is higher in one part of the system than the other in order to observe the effect of such measures on the purification rate.

A buffer between agricultural and natural areas. Many natural marsh areas occur as leftover areas from agricultural industry and are threatened by phosphate and nitrogen residues. Vegetated wetlands could form a buffer zone around the natural areas, offering some protection against the threat of pollution.

Purification of streams that flow into waters sensitive to eutrophication. Some projects are operating in the Netherlands to learn about the specific characteristics of man-made estuary systems, for instance the Schuitebeek estuary at the Veluwe Randmeren. The water from the stream is led, together with surplus water from the nearest polder, through a reed field of 180 ha. The main purification effect will come from sedimentation. This results in an increase in water clarity, which gives plants the opportunity to grow and thus further increasing the water clarity.

Environmentally friendly river banks. In National Water Board's project, "Environmentally friendly river banks" (PMO), the creation of vegetated banks is encouraged as an alternative to "hard" constructions of timber, steel, and concrete.

Vegetation on the river banks is important to the ecology of rivers and as a connection between natural areas along the banks. One possibility not yet considered is the role of the river bank vegetation in cleaning surface water. This is perhaps not as interesting for the vegetation along the larger canals as it is for thousands of kilometers of waterways along roads and in agricultural areas. Up to the present, supervision of dikes has concentrated on the control of water quantity. Control of water quality through planning of maintenance and choice of maintenance methods has not yet been considered.

THE POLITICAL CONSIDERATIONS

There are three national policy plans in which the use of helophyte filters is proposed. All of the three plans use these filters for their own specific means of use.

In the fourth National Report of Physical Planning (1988, updated 1990), it is suggested that vegetated wetlands be incorporated in city and land-use planning. The report states that water can be purified in large wastewater treatment plants, which are expensive and energy consuming. When more space is available, there are possibilities for decentralized and smaller purification locations which are less costly and can serve many purposes. They can be used for the pre-purification of drinking water, pre-purification of water which flows into sensitive wildlife areas or the "after-purification" of sewage. Recent studies prove that the combination of water purification, nature development, recreation, and agriculture is possible.

In the National Policy Document on Water Management (1989), helophyte filters are proposed in order to protect sensitive areas from "foreign water." This "foreign water" is brought to agricultural or natural areas in the summer to keep a steady water level. The quality of this water is low, because it is heavily burdened with nutrients. This causes a decrease in the species variety of plants and animals. The helophyte filters absorbs these nutrients before the water enters the area.

In the Nature Policy Plan of the Netherlands (1990), the helophyte filters are referred to in almost the same way: to pre-purify water before it flows into areas of great natural value. Also the quality of nature and landscape values are meant to be reinforced by the development of vegetated wetlands.

The three national plans concentrate on policy sectors of land-use planning, water management, and nature management. The National Environment Policy Plans (1989 and 1990) are meant to integrate these and other sectoral policies into a wider, cohesive concept. This concept is based on the idea of sustainable development. This is a development which provides for the needs of this generation, without endangering the possibility for future generations to provide for their needs. In order to further this development, a combination of policies is implemented:

- Recycling of goods: managing the various phases in the complete cycle of the extraction of raw material, production and processing, the use of products, and managing the waste stage.
- Saving energy, together with an increase of efficiency and the use of sustainable energy sources.
- Quality improvement of products to extend the use of goods in the economic cycle.

Among the many implementations of the environmental policy is a program of the National Physical Planning Service to support and monitor projects that serve as examples of good practice in the area of flow management (including water). These projects aim to improve the quality of the living and working environment in towns and cities.

TABLE 1
Phosphate and *E. coli* Removal at the Reed Field at Elburg

Phosphate Removal

Year	1978	1979	1980	1984	1985	Average
Influent	3.1	1.6	4.3	1.9	2.4	
Effluent	1.8	1.8	2.8	1.9	1.8	
Removal %	42	–13	35	0	25	23

***E. coli* Removal**

Year	1978	1979	1980	1984	1985	Average
Influent	>6350	3452	3844	4084	5864	
Effluent	116	448	>75	1	695	
Removal %	>98	87	<98	99.9	88	94

ECONOMIC CONSIDERATIONS

INTRODUCTION

It is very important that the financial aspects of wastewater purification with helophytes is understood. When the financial balance is in favor of such a system, it will get the attention it deserves.

Because relatively little is known about the economics of these types of purification systems, several assumptions concerning purification rate and building costs have to be made. In order to show how the financial balance is determined in the Netherlands, two examples will be given. One of the projects is profitable, the other is not. Finally, a model is presented to determine the costs for a new reed percolation field with a capacity of 10,000 p.e.

REED FIELD IN ELBURG

In 1977, the Department of Public Works initiated the construction of a reed field to polish the effluent of a sewage treatment plant. It is specifically aimed at the removal of phosphate and *E. coli* bacteria. The results are given in Table 1.

Phosphate Removal

The data show that the results are insufficient to meet the standards. To learn more about reed fields, the governmental body decided not to close the fields but to maintain them for research purposes.

The flow through the fields was decreased (after 1985) from 10,000 m^3/d to 1750 m^3/d. The research results from the period after 1985 will be important to determine the financial possibilities of these fields.

Disinfection

The disinfection of the effluent is good. It is not good enough to meet national standards for swimming water but a substantial number of bacteria are removed.

TABLE 2
Estimated Building Costs of the Reed Field at Elburg

			·1000 Dfl
Sand	42.000 m³ · Dfl 5,12		215
Sheet piling	250 · Dfl 85,=		21
Rubble	520 · Dfl 15,=		8
Construction temporary culvert			10
Temporary use of a pump			10
Supply and transportation of equipment			155
Pumping equipment 20,000 m³h			60
Inlet and outlet construction (for water)			60
Energy supply and placement of cables			50
Buying and planting of reed	391.500 m³ * Dfl 0.70		274
Raising sand			400
			1263
		Rounded off	1300
Incidental expenditure	15%		195
			1495
		Rounded off	1500
V.A.T. and engineering cost	50%		750
		TOTAL COST	Dfl 2.250.000

Costs

The building costs of such a reed field would be Dfl 2.25 million in 1990 (see Table 2).

Because of the location and the shallowness of the field, the costs were rather low. For instance, the dike could be made of sand, which is probably not possible anywhere else because of the danger of being washed away by the water.

Comparison with Other Disinfection Methods

The building costs of the field can be compared with a conventional chlorination installation. To chlorinate the effluent of a sewage treatment plant of 100,000 p.e., the building costs would be Dfl 820,000 and the yearly costs Dfl 361,000. The disinfection by chlorination has some disadvantages, for instance the side-effects which can occur. The costs of disinfection by means of UV-radiation would be higher: Dfl 607,000/year.

When these two methods are compared with the natural system, the reed field would cost less: Dfl 20,000 a year to maintain the field and Dfl 260,000/year capital costs. This gives a total of Dfl 280,000/year (price level 1990). From a financial point of view, it is useful to use this way of disinfecting the effluent.

REED FIELD ZEEWOLDE

The effluent of sewage treatment plant Zeewolde is led through a reed field to remove phosphate and several other pollutants. The data from the research being done indicates a removal of 700 p.e. and about 700 kg P/year.

TABLE 3
Annual Costs of Vegetated Wetlands at Zeewolde

Management and maintenance			
Mowing grass between ditches	2.8 · 3 · Dfl 950,=		Dfl 8000,=
Maintenance reed			Dfl 0,=
Supervision			Dfl 1000,=
Capital cost			Dfl 42400,=
Depreciation			Dfl 8333,=
		TOTAL	Dfl 59733,=
		Rounded off	Dfl 60.000,=

Mowing grass	2.8 ha, three times a year and Dfl 950,= inclusive V.A.T./ha.
Maintenance reed	A reed trader is managing the reed fields. The yield of the fields is his income, in return for the maintenance of the fields.
Supervision	30 hours a year and Dfl 70.000,= a year equals Dfl 1000,=
Capital cost	Build cost were Dfl 166.000 50% V.A.T. and engineering makes Dfl 250.000 leads to a total cost of Dfl 400.000. Capital cost with 10% interest are estimated at Dfl 42.400,=
Depreciation	Of the investments without the land (Dfl 250.000) over 30 years leads to Dfl 8.333,= a year.

Table 3 shows the costs to maintain the field, Dfl 60,000/year. The savings for the retention of 700 p.e. is Dfl 42,700. This is a negative result. The removal of 700 kg P/year is difficult to express in money because there are no governmental restrictions on this subject. When the P is removed chemically, it costs Dfl 14,000. The total profit of this natural wetland will still be negative.

MODEL OF A NEW REED FIELD

The model is based on the comparison of a conventional system (oxidation ditch) and a reed percolation field, both with a capacity of 10,000 p.e. The starting points are: a gross area of 7 ha, which gives a gross area of 7 and a wet area of 5 m^2/p.e. The two most expensive objects are the foil to protect the ground water from being polluted and the buying and planting of the young reed plants. Table 4 gives more information on this.

When the two systems are compared financially, the natural system is cheaper: Dfl 245/p.e. and Dfl 440/p.e. for the conventional system (investment costs). The yearly costs are Dfl 40/p.e. for the reed field and Dfl 55/p.e. for the oxidation ditch.

One of the assumptions is an acceptable removal of pollutants during the winter. This might be a bit optimistic.

When the costs for smaller installations are compared, the financial advantages will be greater for the natural system.

ECOLOGICAL ENGINEERING IN CITIES

To concentrate on wastewater is in fact to concentrate on one element of the system only. The system is a city, with all kinds of incoming and outgoing flows (see Figure 4).

TABLE 4
Estimated Construction Costs for a Reed Infiltration Field

Subject		Construction	Equipment	Total (· 1000 Dfl)
Influent pump		50	40	90
Pre-purification		50	50	100
Ground removal		110		110
Drainage		17		17
Foil		650		650
Roads, paths		75		75
Planting reed		175		175
SUB TOTAL		1127	90	1217
Incidental expenditure			15%	183
				1400
Engineering and V.A.T.			50%	700
				2100
Buying the land				350
			TOTAL	Dfl 2,450,000,=
Yearly cost				
Personel:	0.5 person/year a Dfl 70.000,=			35
Maintenance:	0.5% of the civil part			6
	2.0% of the electro part			2
Energy:	100,000 kWh a Dfl 0,25 per kWh			25
Annuity:	10% interest over 30 years			
	2.45 million · 0.106			260
Depreciation over 30 years without the land cost				70
	TOTAL YEARLY COST			Dfl 400,000,=

One of the flows is water. Looking closer at the flow of water, six different kinds of subflows can be discovered: potable water, sewage, precipitation, evapotranspiration, surface water, and ground water (see Figure 5).

The handling of these subflows inside the city, together with the two different types of sewerage, are shown in Figures 6 and 7.

The main difference between the mixed and the separated system is the fact that the separated system contains two sewerages, one for the precipitation and one for the sewage. By using this system it is possible to prevent the flow of sewage from coming in contact with the surface water during heavy showers, a common danger in cities.

Another danger to the city is the ground water level rising and falling. At present the difference between summer and winter can be 3 m. Wooden piles below the houses and buildings rot, which causes collapse. Sensitive flora and fauna will disappear.

A third danger is the lack of awareness of the people in the city regarding the importance of water and the difficulty of preparing potable water.

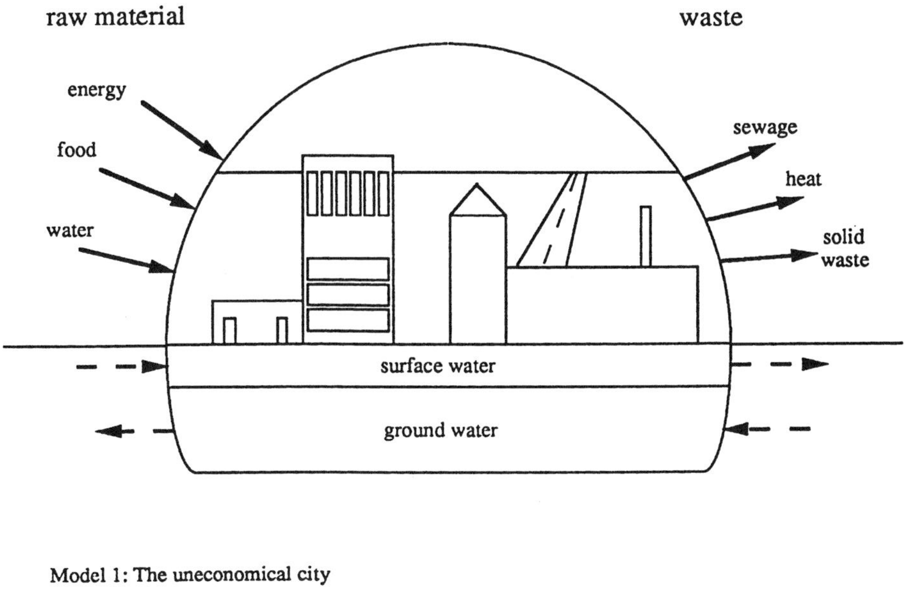

FIGURE 4. The uneconomical city.

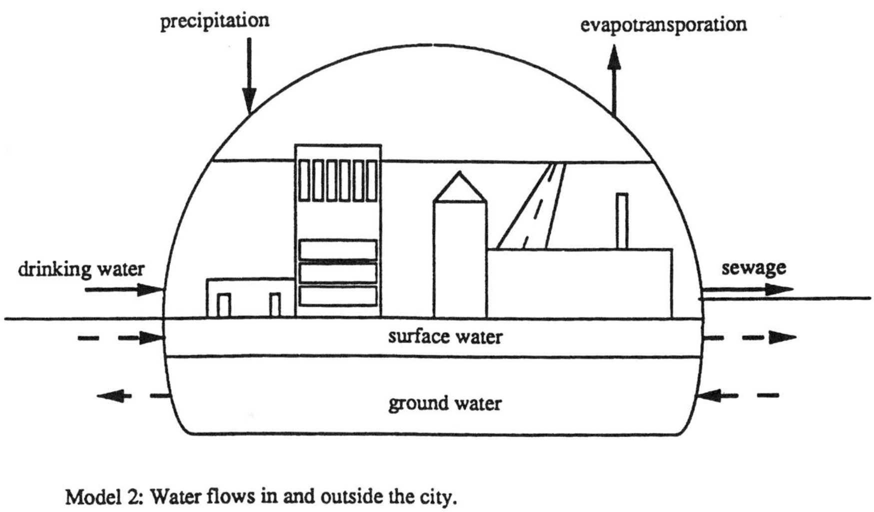

FIGURE 5. Water flows in and out of the city.

The handling of these subflows inside the city, together with the two different types of sewerage are shown in the two following pictures.

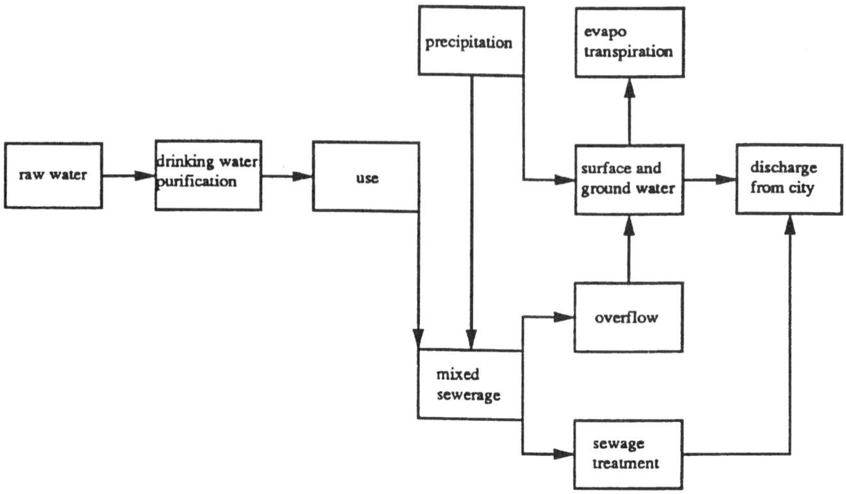

Model 3: the mixed sewerage system

FIGURE 6. The mixed sewerage system.

In order to avoid these dangers, five targets and eight means can be formulated:

Target 1: The city should put less pressure on its surroundings through pollution.

Means 1: Increase the efficiency of the sewage treatment plant by forcing an equal flow to the plant.

Means 2: Prevent contact between sewage and surface water through the use of better sewerages or the treatment of the diluted sewage in natural wetlands.

Target 2: The city should put less pressure on its surroundings through the reduction of consumption of potable water.

Means 3: The water-saving equipment in houses and buildings.

Means 4: Install a graywater circuit together with a natural purification system at the local level.

Target 3: The city should put less pressure on its own environment by decreasing ground water extraction.

Means 5: Install infiltration ponds in which precipitation is collected and lead this water to the ground water.

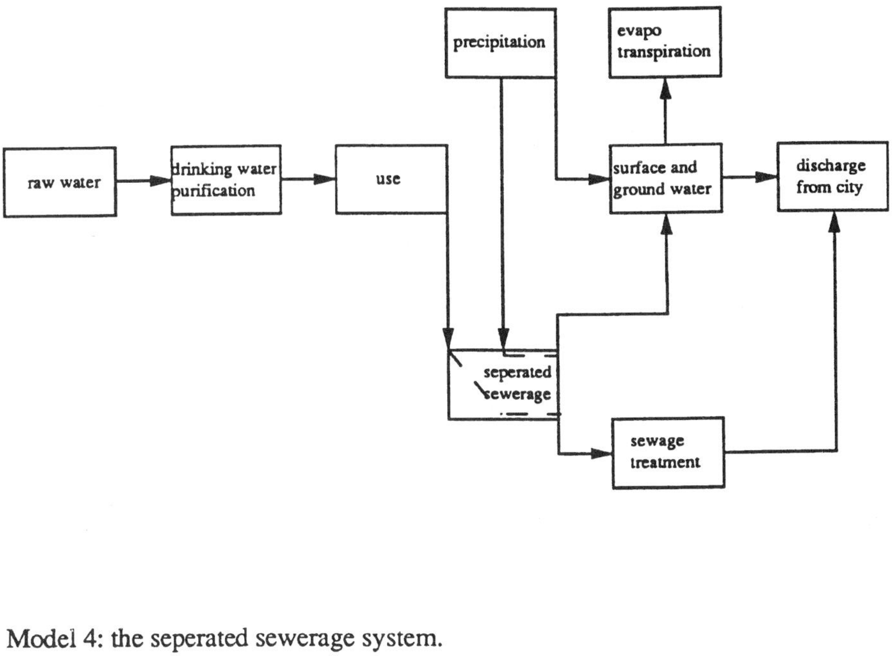

FIGURE 7. The separated sewerage system.

Target 4: The city should put less pressure on its own environment through factors which influence the livability.

Means 6: Improve the atmosphere in the city by enhancing the evapotranspiration during the summer, causing a cooler and less dusty sub-climate.

Means 7: Enhance the attractiveness of the city by creating parks with ponds and fountains.

Target 5: The people should be able to feel themselves involved with water management.

Means 8: Visualizing the water cycle through introducing decentralized water management systems.

The possible results of these measures can be demonstrated in a simple model (Figure 8).

There will be less potable water use (which puts less pressure on raw water use outside the city), the sewage flow will decrease (thus increasing removal efficiency of the purification plant), and there will be more evapotranspiration and more water will percolate in the soil (when needed in the summer).

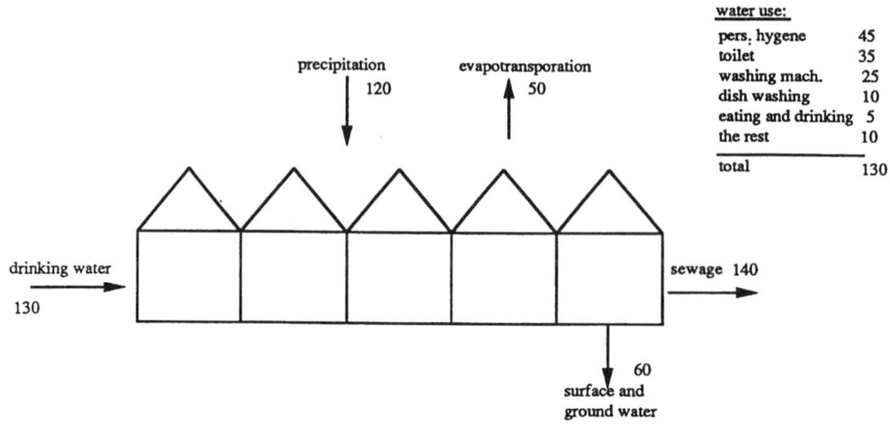

FIGURE 8A. The existing situation, before the recommended actions (means 1 to 8) are implemented. This and Figure 8b are based on a neighborhood with a separated sewerage. The numbers are l/person/day. The area per house is 150 m^2, including roof, garden, and part of the road and square. There are an average of 2.5 persons in each house.

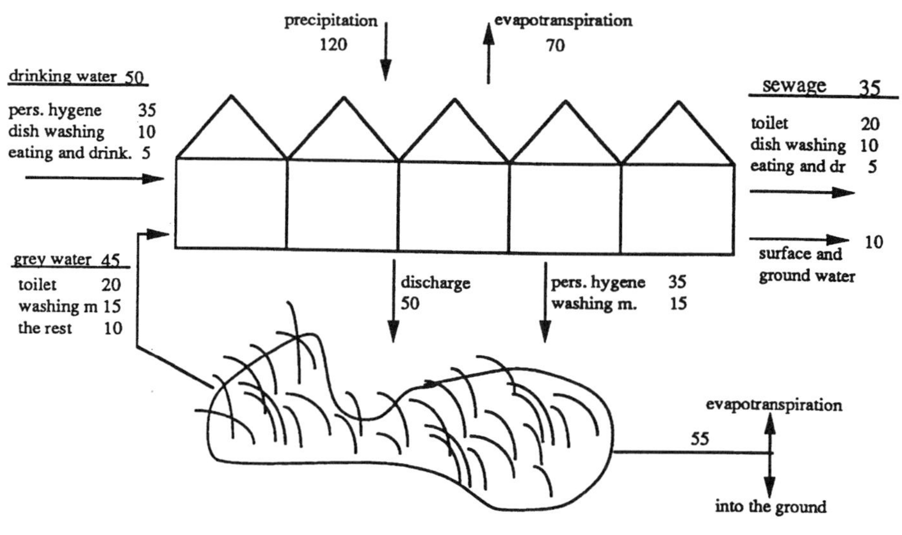

FIGURE 8B. The situation after the recommendations have been implemented.

CONCLUSIONS

The implementation of new concepts for water management in cities will lead to three types of measures:

- The building of graywater systems combined with infiltration ponds, natural purification systems, and wet places for children to play and enjoy themselves.
- The building of improved separated sewerages and natural purification reed fields to treat rainwater sewage before it flows into the surface water.
- The building of fewer paved areas by using vegetation roofs, open sidewalks, and efficiently designed green areas.

At present our Institute is discussing these policy guidelines with the government of the Netherlands.

ACKNOWLEDGMENTS

The authors would like to thank the following persons for their information and help in preparing this report: G.B.J. Rijs, DBW/Riza, Lelystad; J.L. Fiselier DHV-water, Amersfoort; L. Lijklema and H. Duel, Agricultural University Wageningen, SCMO/TNO, Delft; L.J. Kappe, RWS Directie Flevoland, Lelystad; N. Nelissen, Katholieke Universiteit Nijmegen.

Chapter 30

GRAYWATER RECYCLING IN URBAN AREAS

Peter Thomas and Joachim Zeisel, Translation: Mike Hughes

ABSTRACT

Clean drinking water in urban areas is often subject to shortages, at the same time that it is used for purposes that do not require water of such high purity, e.g., flushing toilets and watering gardens. A system for recycling graywater is described and the potential savings of drinking water calculated.

WASTEWATER STRATEGIES

"The reserves of good water are not inexhaustible. It is therefore becoming more and more important to preserve them, to use them sparingly and, where ever possible to increase them."

European Water Charter, May 6th, 1968

The realization of these guiding ecological ideas requires a logical strategy of water saving that can be integrated in the existing centralized structure of the urban water supply (Figure 1).

1. Direct conservation of drinking water. Water conservation hardware and education which, conserve drinking water without reducing comfort for the user.
2. Conservation of drinking water through substitution. Living space and community-related measures that substitute "lower" quality water for precious drinking water when that is not necessary.
   - Use of rainwater, surface water, groundwater.
   - Use of wastewater, e.g., graywater recycling

Graywater recycling. The linear structure of the urban water system can, through the use of decentralized recycling plants, be enhanced with an open circulatory system, using the less polluted part of the household wastewater (graywater). The effect of using this water once more can lead to a saving of 60%. On the waste side, the concentration of household wastewater leads to a significant improvement of the wastewater cleaning performance and a reduction of pressure on the outlet channels.

0-87371-990-5/97/$0.00+$.50

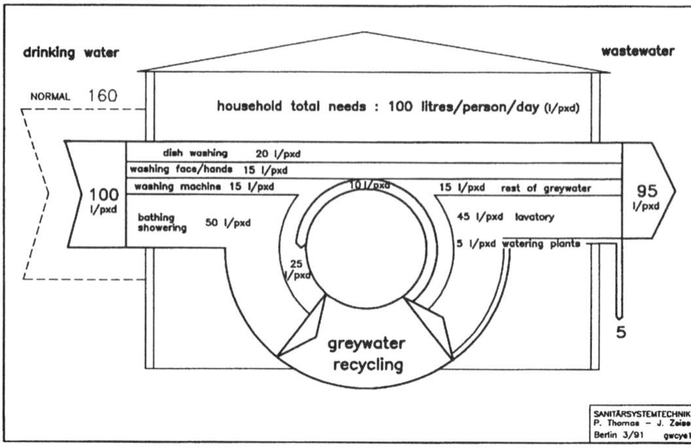

FIGURE 1. Potential for drinking water conservation with a graywater recycling system. Water from washing machines and washing face and hands is not included in the recycling system.

GRAYWATER RECYCLING SYSTEMATICS

STRUCTURE OF HOUSEHOLD WATER NEEDS

The centralized water supply imposes on the user drinking water quality for all of his/her needs because no other kind of water is available in the city.

Water that meets drinking water standards is actually only needed in food preparation (5 l/person/day; Figure 2); high hygienic quality is also needed for dishwashing, cleaning, and washing/bathing/showering (80 l/person/day). All other uses could be covered by a lower quality water, which we call "system water," for example for toilet flushing, watering of plants, and washing clothes (75 l/person/day).

Centralized disposal of wastewater is anonymous and the degree of pollution depends on the behavior of the user. It is possible in principle, however, to differentiate between different qualities of wastewater:

- Wastewater from the toilet (blackwater), carrying feces and urine, and kitchen wastewater with solid organic waste (65 l/person/day).
- Slightly polluted wastewater from the bath or shower and washing machine (what we refer to as graywater in this paper; 90 l/person/day).

A decentralized inner-city graywater recycling plant can use the centralized water system (Figure 3). Accumulating sludge on site, for example, is not necessary because the sewer connection can be used. Apart from a user-related, natural component and an extra disinfection plant, a recycling plant uses principally the same plant mechanics as classical wastewater cleaning techniques, only in miniature format and with certain modifications of details.

To demonstrate these principles, the authors have constructed a biological graywater cleaning plant with a dip and trickle wheel and "vertical swamp," located on Berlin's Manteuffelstrasse 40/41, in the region Kreuzberg. This is also called

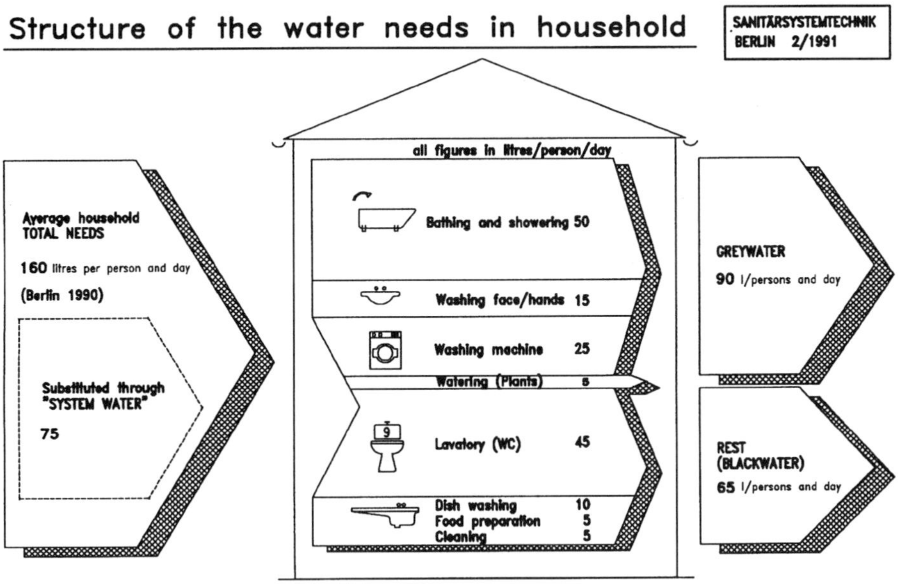

FIGURE 2. A breakdown of the water needs in a typical household.

Town/Ecological Model Experiment Block 103, Berlin Kreuzberg. The owner is STATTBAU GmbH.

PLANT DESCRIPTION

Collecting and initial cleaning. The graywater ([1]; all [bracketed] numbers refer to Figure 4) comes from bath, shower, basin, and washing machines from four partially collectivized, four-story houses and is collected together in a settling tank [2]. The mechanical, initial cleaning is followed by a second, buffer tank [3] which evens out peak loads.

Biological cleaning. This sedimented water reserve is continuously scooped with a scoop wheel [4] into the biological cleaning phase of the dip and trickle wheel ([5/6]; see also Figures 5 and 6). The dip and trickle wheel is made up of polyethylene filter pipes with a large surface area that are welded together to form a cylinder that is 0.80 m wide with a diameter of 1.10 m. This produces a total growth area of approximately 110 m^2. The dip and trickle wheel rotates with 0.5 rpm and is 60% under water.

Final cleaning phase. The water which leaves the dip and trickle phase carrying a certain amount of biomass is led into a two-chamber tank [7] where it settles out.

The "vertical swamp." The user-oriented and natural component consists of a collection of cascades which have been built onto a windowless house wall ([10]; see also Figure 7), with a good deal of sunshine. A vertical series of containers has

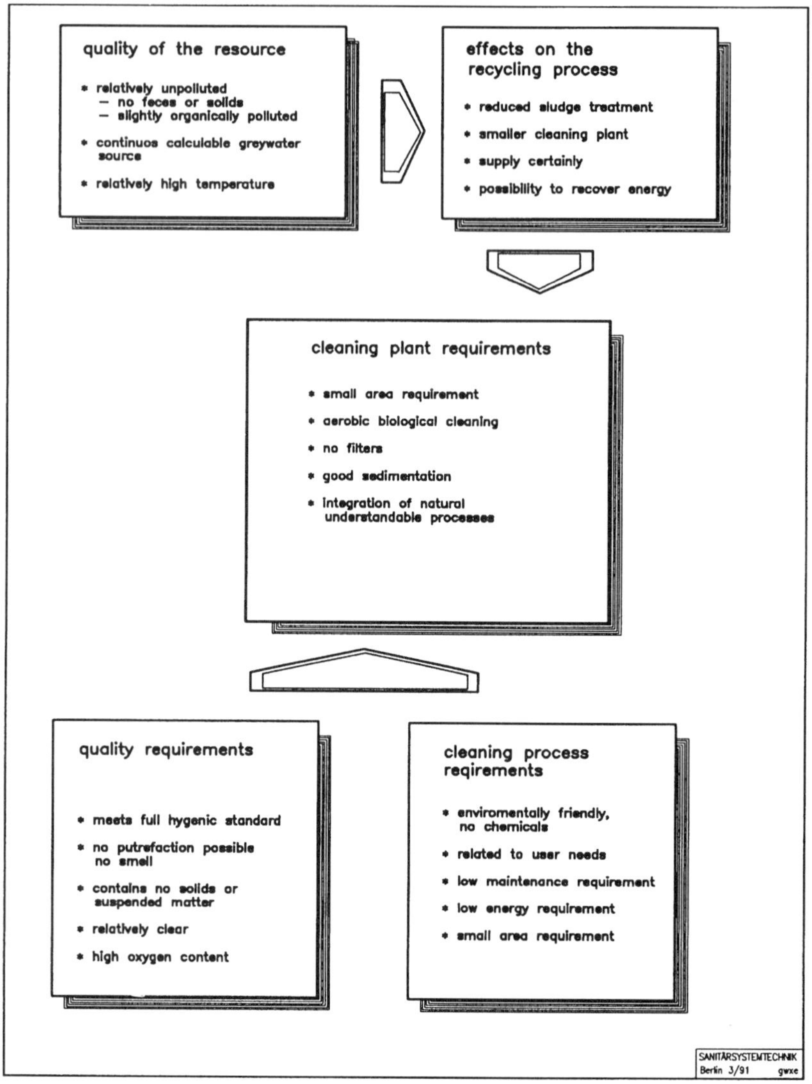

FIGURE 3. An overview of goals and requirements for a graywater cleaning plant.

been filled with pottery balls and sand and planted with different swamp plants. The graywater is pumped over the top containers and drips down into the lower ones. Besides additionally cleaning the water, the swamp improves the microclimate through the plants' evaporation of water. The vertical-swamp can be employed as the main and final cleaning phase.

Disinfection. Disinfection of the biologically cleaned graywater is provided by using a small pump to carry the water from the last (sedimentation) phase of the

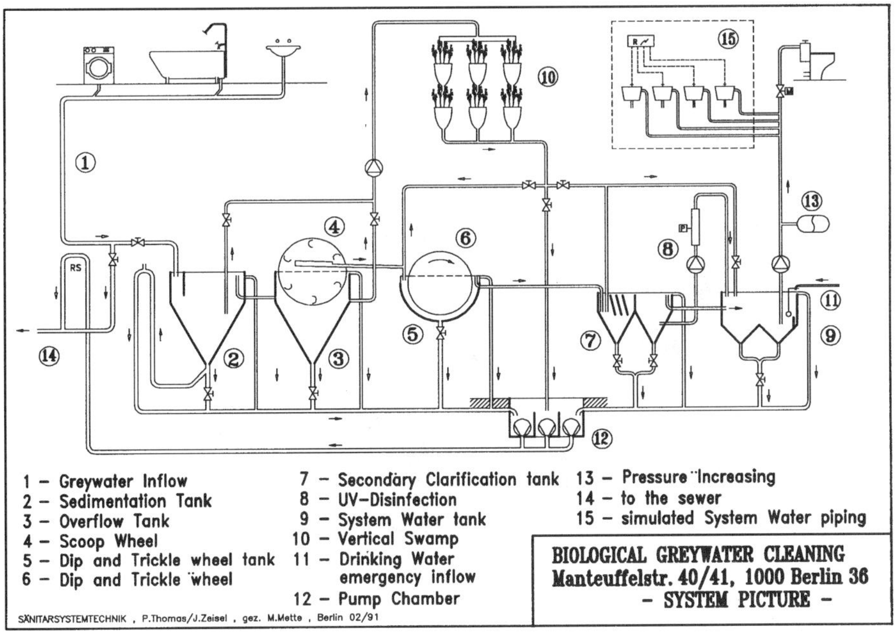

FIGURE 4. The biological graywater cleaning syster at Manteuffelstr. 40/41, Berlin.

FIGURE 5. The graywater cleaning plant in the basement of one of the apartment buildings served.

FIGURE 6. Counterclockwise from lower left: The dip and trickle wheel with the scoop wheel, the sedimentation tank, and the overflow tank.

process to an ultraviolet disinfection plant [8] and into the system water tank [9] through which an additional amount of oxygen enters the water.

System water supply. The cleaned system water is pumped to WC flush cisterns in three houses through a machine used to increase the water pressure. To ensure a continuous supply of flush water, an emergency drinking water pipe [11] is attached to the system water tank and operates with a float valve and open channel. It can supply flush water if the system water is not enough for all flushing needs.

A summary of the system's dimensions and capacity can be found in Table 1.

RESEARCH RESULTS TO DATE

Wastewater technology. The graywater plant was first used in September 1989 and has been researched in four different operating situations. The supply of the system water will begin in summer 1991. We have seen the following results:

- The formation of a biological film in graywater that is poor in nutrients and only slightly polluted is heavily dependent on twist forces (the speed of the dip/trickle wheel).
- The cleaned graywater is organically only marginally polluted, with under 5 mg BOD_7/l, and cannot putrefy. It is useable as system water.
- Despite the minimal rotation, the oxygen supply of the biological mass is supplied during the time it is above water

FIGURE 7. "Vertical swamp" mounted on an outside wall in the courtyard.

- The reduction of the organic pollution of the graywater can be achieved at a loading of 3 g $BOD_7/m^2/d$ with very good results.
- The addition of nutrients (nitrogen, phosphorus) can be avoided, although their use can very much help the decomposition.

HYGIENIC STANDARDS

The results of the hygienic research can be summarized as follows:

- The hygienic standards of the EEC guidelines for bathing water were always met.
- The persistence of potentially pathological bacteria in graywater is slight (Exception: *Pseudomonas aeroginosa*)
- There have been no cases of the reappearance of relevant pathological bacteria. This lead to an optimization of the disinfection plant: The UV-disinfection phase was reduced and placed before the system water tank.

Additional data on the plant's hygienic standards and performance can be found in Figures 8 and 9 and Table 2)

TABLE 1
Dimensions and Capacity of the Graywater Cleaning System at Manteuffelstr. 40/41, Berlin

Base Data	
Graywater produced by 55–60 persons (70 l/person/day; Manteuffelstr. 39/40/41/42)	3800 l/d
System water requirement for 45–50 persons (toilet flushing: 30 l/person/day; Manteuffelstr. 40/41/42)	1500/l/d
Initial Clarification	
Sediment tank volume	1100 l
Residence time (fill volume Q_{10} 380 l)	3 h
Residence time with a hydraulic peak load of approx. 960 l/h	Approx. 2 h
Sediment funnel angle	55°
Biological Cleaning	
Cleaning performance	2000 l/d
Organic load	110 mg BOD_7/l
Daily sediment load	0.22 kg BOD_7
Specific surface land (chosen)	2 g $BOD_7/m^2/d$
Specific growth area	110 m^2
Dip and trickle wheel volume	0.75 m^3
Drive performance motor	90 W
Secondary Clarification	
Continous inflow	0.083 m^3/h
Surface feed	0.09 $m^3/m^2/h$
Flow-through time (sediment funnel angle: 2 × 60°; 1 × 40°)	6.5 h
System Water Tank	
Volume (50 persons × 2 toilet flushes of 6 l each)	600 l
Disinfection	
UV-disinfection plant, power requirement	96 W
Vertical Swamp	
5 cascades with 47 containers	
Container volume, each	50–65 l
Total surface area of containers	7 m^2
Treatment capacity	700–1000 l/d
Planted with:	*Phragmites communis* *Scirpus lacustris* *Iris versicolor*

EXPERIENCE IN USE

- The energy requirement (drive, biological cleaning, disinfection, pumps, light, etc.) is approx. 2.2 kWh/m^3 system water.
- The maintenance requirement can be fulfilled with 1 to 2 hours weekly by persons with rudimentary mechanical know-how.

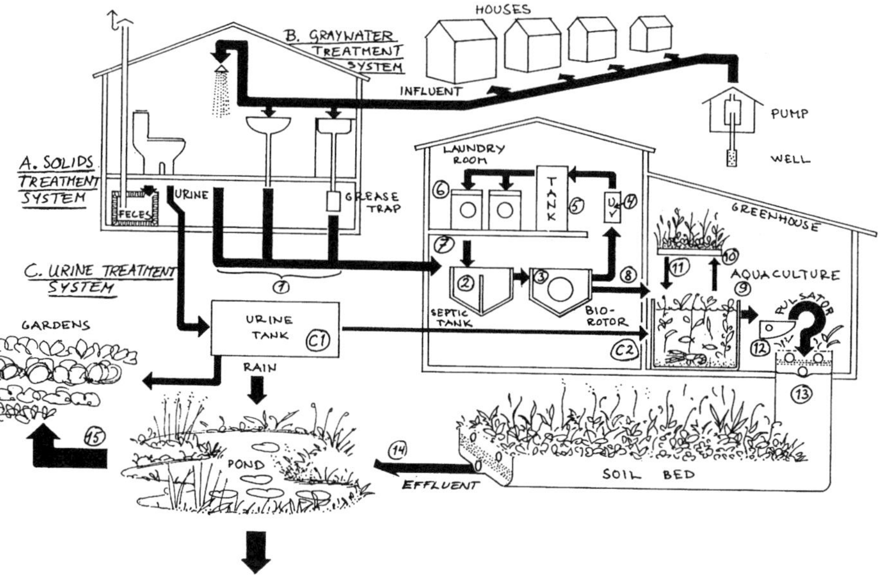

FIGURE 8. BOD_7 of effluent water, residence time, and specific surface loading on the dip and trickle wheel for four sets of conditions at the graywater cleaning system, Manteuffelstr. 40/41, Berlin.

Phase 1: Graywater from the shower and bath
Intake load: 100 mg BOD7/l; 850 l/d
Speed: 1.7 rpm

Phase 2: Graywater with washing machine
Intake load: 150 mg BOD7/l; 1500 l/d
Speed: 1.7 rpm

Phase 3: Graywater with washing machine
Intake load: 130 mg BOD7/l; 2500 l/d
Speed: 1.7 rpm
Addition of nutrients (P,N)

Phase 4: Graywater with washing machine
Intake load: 120 mg BOD7/l; 2500 l/d
Speed: 0.5 rpm

- The motivation of the user is high. Up until now there have been no accidents (poison in the water).
- The regeneration time of the biological film after a plant stoppage is very short (5 to 7 days).

PERSPECTIVES

APARTMENTS, NEW HOUSING

Graywater recycling should be integrated into a comprehensive water saving system with the following components:

- Communication, encouraging less wasteful attitudes to water on the part of general user
- Drinking water conservation measures: installation of water saving fittings
- Replacement of drinking water with system water (lower quality water), e.g., clarified graywater, for those uses that do not require drinking water quality.

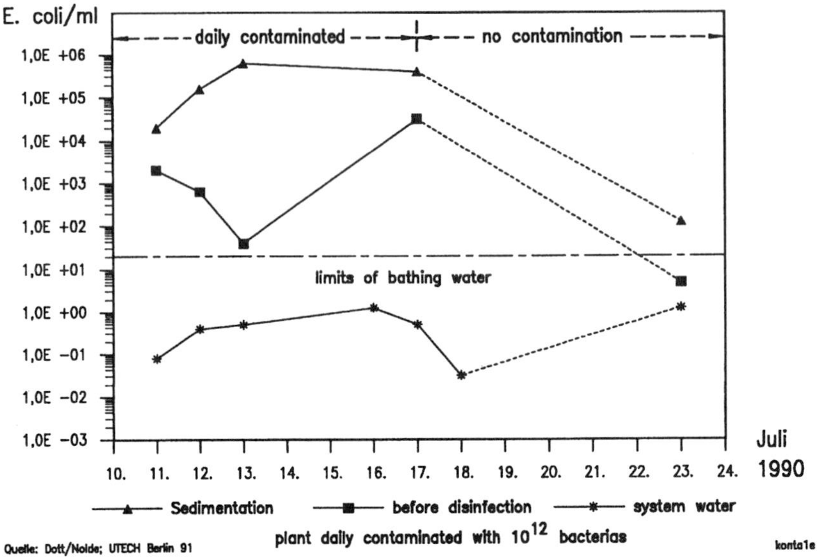

FIGURE 9. Results of contamination trials with *Eschirichia coli.* The effluent (system water) maintained a quality significantly better than bathing water during the entire set of trials.

TABLE 2
Hygienic Guidelines*

Parameter	Decree for drinking water (Germany)	EEC guidelines bathing water	System water of graywater
Total coliforms	0/100 ml	G: 500/100 ml I: 10000/100 ml	0/0,01 ml
Fecal coliforms *Escherichia coli*	0/100 ml	G: 100/100 ml I: 2000/100 ml	0/0,01 ml
Fecal streptococcus	0/100 ml	G: (500/100 ml)	0/0,1 ml
Salmonella sp	*	I: (0/1000 ml)	0/100 ml
P. aeruginosa	*		0/ ml
Legionella sp			0/10 ml
S. aureus	*		0/ ml
C. albicans			0/0,1 ml
cfu 20°C	<100/ ml		
36°C	<100/ ml		
Intestinal virus		I: (0/10 l)	

Note: * = only if necessary; (. . .) = research by suspect; G: = guide; I: = imperativ.

System water guidelnes were developed based on the EEC guidelines for bathing water August 12, 1975, together with the German department of health and Prof. W. Dott, and E. Nolde of the Institute for Hygiene of the Technical University of Berlin.

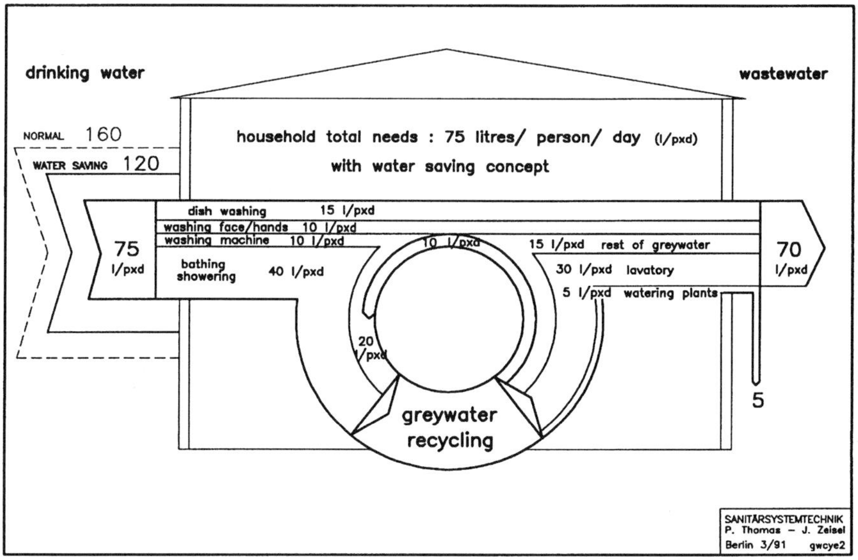

FIGURE 10. Potential for drinking water conservation with a graywater recycling system. Water from washing machines and dish washing is not included in the recycling system.

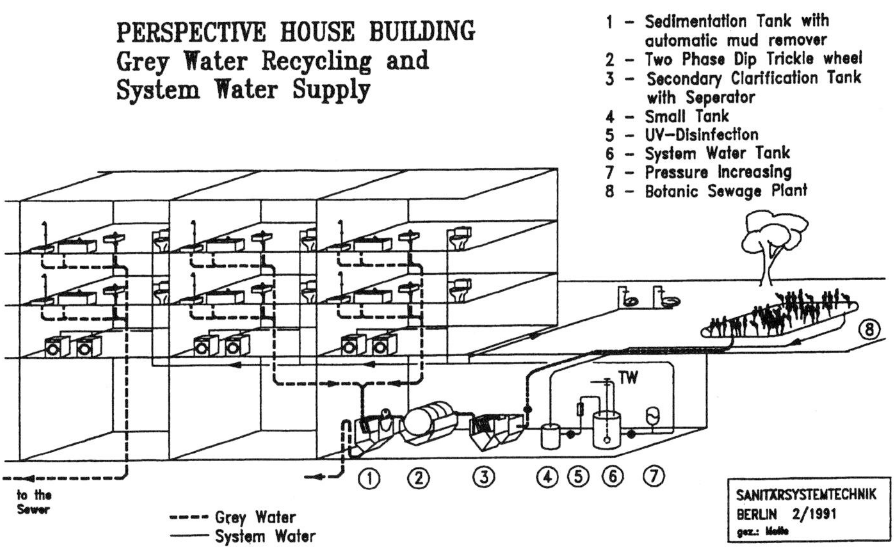

FIGURE 11. Graywater recycling in a multi-family building.

If water conservation were carried out on all possible levels and combined with graywater recycling, it would be possible to reduce the drinking water needs of an average building by approximately 50% (Figures 10 and 11).

TABLE 3
Calculations for the Smallest Graywater Plant Economically Feasible in Berlin*

Size: 130 persons

(1) Yearly system water volume (WC-flush = 30 l; washing machine without rinsing = 10 l, watering (120 d/year) = 5 l)	1,976 m^3
Complete costs for graywater system (based on a four story house with three people per flat)	138,000 DM
(2) Yearly capital costs (including 30% government subsidy)	9,660 DM
(3) Yearly repair costs at 1% investment on 50,000.00 DM	500 DM
(4) Energy costs for system water	1.00 DM/m^3

Maintenance costs:
Not explicitly estimated, carried out by landlord

Price of the system water = [(2) + (3) + (4)$_*$ (1)]/(1)	6.14 DM/m^3
By way of comparison: drinking/wastewater fee in Kassel 1991	6.00 DM/m^3

With a concentration of about 130 persons, the cost of a graywater recycling system becomes about the same as that of using the city drinking water it replaces.

An optimized graywater recycling system is constructed according to the necessary system water volume requirement. It is then possible to reduce the capacity of the graywater outflow pipes by 20 to 50%.

Graywater recycling makes economic sense when the savings brought in by reduced drinking water and wastewater costs equal the yearly additional costs of the plant and its functions. An estimate (Table 3) shows that this is reached when around 130 to 140 people use the system.

HOTEL USE

The wastewater/graywater recycling and the use of system water has a clear future in the hotel trade.

The tourist industry is developing criteria for an environmentally conscious tourist. The theme water is most important. Careful use of restricted water resources in many tourist areas through the recycling of wastewater is of enormous ecological importance. These recycling plants are usually a useful extension of already planned wastewater and sewage systems.

Inner-city hotels of an average size, with their compact dimensions and a high potential usage, should have already reached a stage where the installation of such a plant makes economic sense.

Chapter 31

COMMERCIAL PROSPECTS FOR ECOLOGICAL ENGINEERING

Susan Peterson

ABSTRACT

Commercial prospects for ecologically engineered wastewater systems are excellent. Not only is there world-wide understanding of the pressing need to protect the environment, but there is also ample evidence that we need to repair environments damaged by past practices. Traditional technology is expensive and often unsatisfactory in its results. This paper discusses prospects for and limitations to ecologically engineered wastewater systems, including capital and operating costs, performance, reliability, and acceptability. The general points covered include: intensive vs. extensive use of space, siting, seasonal performance, byproducts including water reuse, acceptability to environmental or wastewater engineers, and regulatory constraints. The discussion relies heavily upon data from the United States, for which the author apologizes.

INTRODUCTION

Most practicing ecological engineers who work in the field of wastewater treatment are self-taught because few universities offered, until very recently, formal programs in ecological engineering. The "natural treatment" methods for dealing with wastewater evolved from the general field of sanitary engineering, now commonly referred to as environmental engineering. At the same time, the discipline of ecology has evolved and matured. Ecological engineering combines the best features of environmental engineering with ecology. While most professional ecologists are natural scientists, many other academic disciplines now incorporate ecological perspectives in their curricula, including economics, anthropology, geography, and engineering (see Hammer, 1989 and Reed et al., 1989).

Ecologically engineered solutions to wastewater treatment problems typically incorporate as many features of the existing natural environment as possible, relying upon local vegetation, soils, and microbes to break down the polluting compounds in wastewater. Using decades of observations and intense analyses of the purifying abilities of natural wetland systems, ecological engineers use constructed wetlands as an important technology for wastewater treatment. These systems are intended to be sensitive to existing land use patterns, the local economy, and the opportunity to enhance wetland habitat. The 1988 publication, "Constructed Wetlands and Aquatic Plant Systems for Municipal Wastewater Treatment" (US EPA), helped to

0-87371-990-5/97/$0.00+$.50

establish ecologically engineered systems as viable options which environmental engineers could offer to clients.

Experience with constructed wetlands, general public disdain for reliance upon artificial/chemical processes for wastewater treatment, and the growing acceptability of natural systems for wastewater treatment have set the stage for broader applications and refined technology.

WHAT DO ECOLOGICAL ENGINEERS HAVE TO SELL?

Equipment and/or specifications have been developed by ecological engineers, including bioreactors, rotating biological contactors, activated sludge systems, composting systems, constructed wetlands, Solar Aquatics®, and a wide range of environmental monitoring equipment. Most of the technological innovations recognize the important role of microorganisms in degrading harmful compounds. Sanitary engineering focused upon collection systems, hydraulic loading, mechanics for managing large flows, aeration, settling, solids management, and disinfection. Ecological engineering relies upon and builds from this earlier work of sanitary engineers.

Knowledge and/or experience in using natural systems for wastewater treatment is growing. The published literature on natural wastewater treatment systems for a range of climates, waste types, and flows grows daily. The documentation shows a broadening knowledge base for system design, as well as operating experience. The techniques needed in ecological engineering include information on construction of artificial ecosystems and the assembly of useful plants, animals, and microbes. Also needed are techniques for ecosystem repair or enhancement, methods to improve effluent quality from existing conventional facilities using natural technologies, and knowledge of ecological systems and function, including a basic understanding of biogeochemistry, microbial ecology, hydraulics, and natural variability.

WHAT IS THE COST OF AN ECOLOGICALLY ENGINEERED SYSTEM?

The cost to construct a system using one of the natural technologies should be compared to the cost of building conventional tertiary systems, since most natural systems are designed to meet very high effluent quality. Natural systems are reported to have construction costs ranging from \$1000 U.S. to \$3000 U.S./m^3 of treatment. Conventional systems in the U.S. range from \$2000 to \$3000/m^3 (EPA, 1986, 1987, 1989; Massachusetts, 1988).

Operating costs for natural systems can be very low since many of them are extensive systems, relying upon gravity, a good collection system, and infrequent operator attention. Others are intensely managed and have higher operating costs. Published information in the U.S. indicates these costs can range from \$0.20 U.S. to \$1.90 U.S./m^3/day. Conventional secondary systems in the U.S. have operating costs which range from \$0.80 U.S. to \$1.70 U.S./m^3/day; these costs are inversely correlated with the volume of flow. Small systems, less than 400 m^3/day, are expensive to build and operate.

The cost of services of ecological engineers ranges enormously. Many ecological engineers are academics who provide advice to clients at no cost through extension services. Ecological engineers hired by major engineering consulting firms may charge fees in the range of $1500 U.S./day or more.

DO ECOLOGICALLY ENGINEERED SYSTEMS ALWAYS WORK?

Reliability is an important issue for wastewater treatment technologies since public health concerns are involved. For the most part, public health has been protected by requiring disinfection of wastewater effluent; that is a common requirement for natural systems as well. As public awareness of environmental quality has broadened beyond the narrow focus of public health to environmental health, demand has grown for wastewater treatment technology to remove contaminants (VOCs, hydrocarbons, nutrients, herbicides and pesticides) in addition to pathogens. Natural systems are ideal for breaking down harmful compounds with a variety of bacteria, binding up phosphorus on sediment, and eliminating a large proportion of nitrogen from the effluent stream. The constructed wetlands systems, many built since 1980, have proven effective and reliable over the relatively short time they have been operating.

Ecologically engineered systems are as resilient as the part of nature they mimic. More complex natural systems (water hyacinth/fish, Solar Aquatics®) are developing operating histories which indicate that they function reliably under different loading rates and in a range of climates.

In order to improve the reliability of natural systems, we need to develop better understanding of the operating parameters of ecologically engineered systems. There is an underlying presumption that natural systems will take care of themselves; observations to date indicate that natural systems, carefully managed, continue to provide excellent quality effluent over time, but that they do need to be managed. A good operating and maintenance manual is as important to a constructed wetland wastewater treatment plant as it is to a conventional trickling filter system.

WHO WANTS TO BUY NATURAL SYSTEMS?

There is a strong demand for natural, chemical-free, affordable technology to treat wastewater. Society has become suspicious of systems that smell bad, use chemicals, chlorinate their discharge, contribute to lake, river and ocean pollution by adding unwanted nutrients, and fail to meet discharge standards. In general, natural systems can be less expensive to build than traditional systems, and thus are or will be in demand by a wide range of customers seeking new or upgraded systems. Some types of customers include:

- Municipalities which need high quality discharge because they are in environmentally fragile areas such as sole-source aquifers.
- Small towns that are unwilling to incur unsupportable debt for conventional systems.
- Industries that need to clean up their reputations along with their discharges.
- Home owners faced with rising sewer fees for their conventional municipal systems who want better quality discharge.

- Developers wanting to make a statement about environmental awareness, provide additional wildlife habitat, and show an attractive system.
- Resort owners who need on-site systems that do not detract from the physical beauty of the location.

WHY DO CUSTOMERS WANT TO BUY NATURAL SYSTEMS FOR WASTEWATER TREATMENT?

- Green technology is vital to the future of society.
- Natural systems are scaled appropriately for current needs and, coupled together with water-saving devices, public education, and recycling, they are logical.
- Natural systems can be beautiful and an asset to a property.
- The cost can be less than or equal to conventional systems.
- Using natural systems solves part of the problem of siting an often-controversial structure.
- Natural systems perform multiple functions to clean water not only of BOD_5 and Total Suspended Solids, but also of Volatile Organic Compounds, nitrogen, and fats, oils, and greases.
- There are the byproducts of ornamental vegetation, habitat, reusable water, and potential food crops.

WHY AREN'T THERE MORE NATURAL SYSTEMS TREATING WASTEWATER?

As ecological engineering matures as a discipline, there will be rapid growth in the numbers of natural systems treating wastewater. Most consulting engineers prefer to advocate "proven technology" to their clients, and since few natural systems have 20-year operating histories, many qualified engineers refrain from recommending them. This is gradually changing as more data become available. Natural systems also face high hurdles when being reviewed by regulatory agencies, since few agency staff have been trained as ecological engineers.

There continue to be levels of uncertainty regarding operations, maintenance, repair, improvement and staffing. These problems will subside as more and more Operating & Maintenance manuals are prepared by practicing ecological engineers.

Most constructed wetlands and many aquatic systems require more space than conventional technology. Natural systems perform better when the wastewater has adequate residence time in the microenvironments. Conventional technology is typically designed with less than 12 hours residence time; a natural system may need 2 to 30 or more days. Solar Aquatics has addressed this problem by compacting the natural processes within a greenhouse to ensure rapid and effective treatment. However, space remains a limiting factor for large, urban wastewater treatment systems. One way to use natural system technology in urban areas is to decentralize the wastewater treatment into units which are manageable and use the natural systems for landscaping or to enhance wildlife potential in urban areas.

The most commonly asked question is: "But does it work in winter?" Most municipal natural systems in the U.S. are in the sub-tropics and marginal temperate zones and most are outdoors. Performance levels in those systems decline somewhat in winter, but effluent quality can be maintained within permit limits by designing a system which recognizes seasonal variations. In the last decade, a number of natural system designers have begun to require better environmental controls during winter; most rely upon commercial greenhouses to cover the system and maintain performance. In some very cold and dark climates, an option to be considered is storing wastewater in winter and treating it in summer.

CONCLUSIONS

Natural systems for wastewater treatment rely upon system design which recognizes resilience, variability and function which is driven by microorganisms. The microbial populations in the soils, the air, and the water need to be nurtured, protected, and enhanced. A well-designed natural system considers local conditions, cost, land use patterns, the nature of the waste, and the potential for byproducts. Clean water is needed to maintain health, grow crops, and run industry. Ecological engineering is the way to restore the water.

REFERENCES

Hammer, Donald A., 1989. *Constructed Wetlands for Wastewater Treatment.* Boca Raton, FL: Lewis Publishers, Inc.

Massachusetts, Commonwealth of, 1988. *Municipal Sewerage User Fee Survey.* Bureau of Water Quality Management: 10525.

Reed, S.C., E.J. Middlebrooks and R.W. Crites, 1989. *Natural Systems for Waste Management and Treatment.* New York: McGraw-Hill Book Co.

U.S. EPA, 1988 *Constructed Wetlands and Aquatic Plant Systems for Municipal Wastewater Treatment.* Office of Research and Development: EPA/625/1-88/022.

U.S. EPA, 1989 *Assessment of Needed Publicly Owned Wastewater Treatment Facilities in the United States.* Office of Municipal Pollution Control: EPA/430/09-89-001.

U.S. EPA, 1987 *A Statistical Abstract of the Unsewered U.S. Population.* Water Engineering Research Laboratory, Office of Research and Development: EPA/600/5-87/004.

U.S. EPA, 1986 *U.S. EPA Region 1 User Charge Survey,* September, 1986. Boston: JFK Federal Building.

Chapter 32

BIOLOGICAL SURFACES, METABOLIC CAPACITANCE, GROWTH, AND DIFFERENTIATION: A THEORETICAL EXPLORATION OF THERMODYNAMIC, ECONOMIC, AND MATERIAL EFFICIENCIES IN FLUID PURIFICATION SYSTEMS

Paul S. Mankiewicz

ABSTRACT

Because biological systems partition their surroundings through growth and development, ecological engineering circumvents a group of solution chemistry limits inherent to conventional wastewater treatment. Specifically, conventional vat-based wastewater technologies cannot develop the steep concentration and velocity gradients and the concomitant catalytic functional efficiencies which characterize well designed ecologically engineered systems. Without spatially and temporally differentiated gradients, these standard wastewater facilities incur increasingly large economic and energetic costs in the purchase of worthwhile, but essentially limited, health and purificatory benefits.

This discussion outlines in general terms how ecologically engineered water purification systems order energy and material differentials across more than five orders of magnitude, a dimensional range unavailable to traditional water purification technologies for a given unit of energy, material, and capital input.

INTRODUCTION

This brief review explores fluid dynamic and interfacial bounds on the efficiencies of fluid purification. While some of these mechanical and scalar parameters have occupied the central focus of western science and engineering from its Galilean beginnings (Galileo 1638), these principles have only rarely been applied to biologically and ecologically based technologies for water and air purification. Beginning with interfacial fundaments of physical-chemical relations, we explore the extensive surface catalytic properties of organisms, and their inherent capacity to grow and develop interstitially within the flow of matter and energy, structuring this flow in the process. This paper points to the largely unrealized capacities for ecologically engineered filtration, chemical transformation and fluid purification in terms of the fundamental perspectives of fluid dynamics, thermodynamics and statistical mechanics.

There is one primary family of differences between any configuration of flow channels and connectors vs. any ecological system:

0-87371-990-5/97/$0.00+$.50

- Conventional systems dissipate energy in turbulence and frictional interaction between fluid and containment vessel, augmenting biochemical purification mainly by increasing rates of diffusion through mixing;
- Ecological systems dissipate energy by growing into the flow and thus entraining the movement of the surrounding fluids in the functional work of exchange through processes of development.

In essence, the drag and resistance of plant bodies accomplishes dissipative work because increasing organismic surface increases rates of exchange (see Figure 1, 2, and 3). Examples in much of the discussion that follows largely feature plants, but all multicellular organisms and films may be seen in similar light.

At present, the largest organic 'body' on the earth is apparently the macromolecular matrix of land plant cell walls (U.S. Dept of Energy 1988). Taking 7.2×10^{13} kg carbon as the amount fixed by land plants each year (see Lieth and Whittaker 1975, Nobel 1983, *inter alia*), this would provide enough one micron sheets of fixed carbon to cover the earth more than 500 times. Described in other terms, the land plants elaborate the equivalent each year of a 500 μm or half millimeter filter of "activated carbon" over the whole globe, from rain forests to deserts to open ocean. A productive wetland has much more concentrated power still, producing the equivalent of a 10 μm to 50 μm filter of activated carbon each day! We offer the hypothesis that this living and then decaying material plays the primary regulatory role in the biogeochemical processes that filter and purify water in the terrestrial and limnological ecosystems. It is here proposed that this flow of biomass acts directly as enzymatic surface, and as attachment substratum for additional enzymatic surfaces of bacteria, fungi, ciliates, and other single and multicellular organisms, recognizing that it also plays roles beyond the scope of this paper as ligand and colloidal binder to metals, nutrients, and pollutants.

Global patterns of biology are in general grounded in regionally integrated systems, and these more local phenomena, such as wastewater treatment systems, will be explored here in terms of the biophysics, fluid dynamics, and the thermodynamics of surfaces.

THE STRUCTURE OF EXCHANGE: LAMINAR FLOW, TURBULENCE, AND SURFACE BOUNDARY LAYERS

Fluids move slowly near surfaces, such as smooth concrete walls or the glassy plane of a pond. In such circumstances, fluids flow in lamina, layers which do not mix with one another. A low flow rate has two primary effects: first, material interchange between lamina is largely through diffusion; and second, fluid viscosity builds thick layers around bodies, thereby minimizing exchange. As fluids move more rapidly, eddies shear off from the flow, mixing materials in three dimensions. Laminar flow thus creates thick boundary layers where exchange occurs only through diffusion between layers, while rapid turbulent flow homogenizes concentration and temperature differences by breaking down the layering of the fluid strata.

Multicellular plants structure the flow of matter by growing towards zones of turbulence, creating dramatic potentials for mass and momentum exchange, which further power growth and development. At the same time, surfaces of plants, from

TABLE 1
Radius and Flow Rates in Selected Plant Parts

Structure	Radius	Flow rate, Q, in ml/sec
Tracheid of an herb	50 μm	4×10^{-7}
Vessel of vine	500 μm (.5 mm)	4×10^{-3}
Capillary space in moss or peat	5 mm	40

roots to stems and leaves, structure the roughness of the surface boundary layer, integrating relations between velocity, shear forces, and energy dissipation. These relationships are implicit in the Hagen-Poiseuille equation, which gives the essential geometry of flow rate and scale:

$$Q = \frac{\pi \Delta P r^4}{8 \mu l}$$

where Q is flow rate, ΔP is the pressure difference between two ends of a pipe or channel, r is the channel radius, *l* its length, and μ the viscosity of water. Following this relationship, for each ten-fold increase in the radius of capillary conducting spaces, the conductive capacities of volume of flow increase ten thousand times (as $10^4 = 10{,}000$). For example, along a pressure drop of about a fifth of an atmosphere, plant structures conduct water at substantial rates, as indicated in Table 1.

Resistance follows the inverse of this relationship. As Vogel (1981, p. 168) points out,

$$\text{resistance, } R = \frac{\Delta P}{Q} = \frac{8 \mu l}{\pi r^4}$$

Vogel further indicates that pipe resistance comes in unusual dimensions: force times time divided by length to the fifth power. In any constructed system, ecologically engineered or otherwise, resistance to flow sets power requirements, in terms of the necessary head or pressure over the total length to maintain flow.

Thus the length of the system, all else being equal, determines overall resistance, but when making a comparison between ecologically engineered and standard wastewater treatment systems, this is somewhat misleading. Length, as a variable of construction, is a cost: the greater the length of pipe or channel, the more space and capital required. In ecologically based wastewater treatment, length of roots, shoots, fungal hyphae, and microbe colonies *is* the working capital of the purification system. These systems are generated by the waste stream itself after inoculation. Through growth and development, they minimize concentrations of pollutants and pathogens. Since resistance is one measure of frictional or turbulent interaction and exchange, it may be stated that this variable is a passive result of configuration and construction in standard wastewater treatment systems, but an actively regulated component in any ecologically based systems because growth and development structure exchange and dissipation. While resistance to flow may play the predominant role in sand and

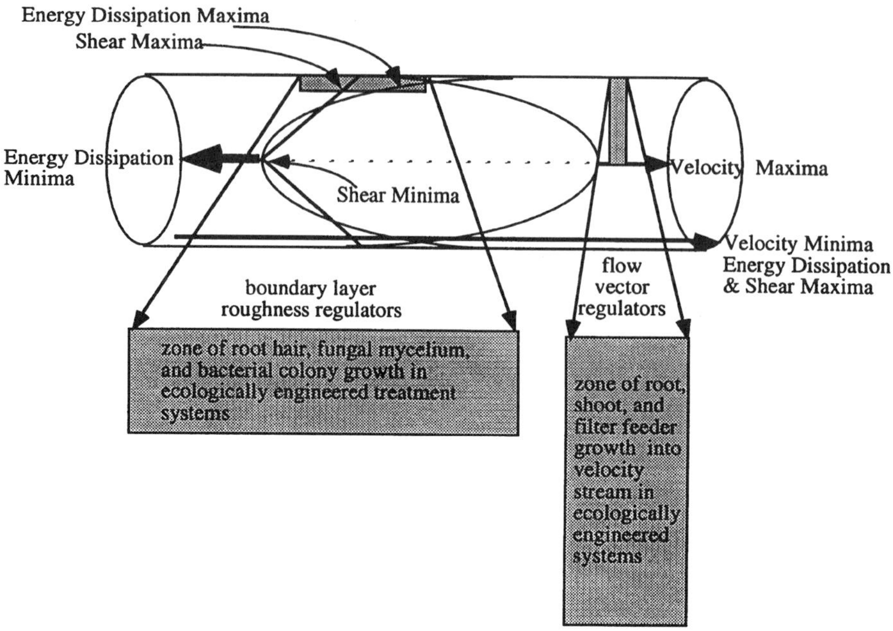

FIGURE 1. Adopted from the classical literature in fluid dynamics (Rouse 1946, p 161). The above diagram shows the schematic distribution of velocity, shear, and energy dissipation maxima and minima, with fluid moving from right to left through an idealized channel. Conventional systems cannot functionally entrain shear, dissipation, or velocity, since they cannot directly respond or 'adapt' to these environmental properties. On the other hand, the organismic units of ecologically engineered systems order the dissipation of chemical energy by positioning themselves to structure surrounding flow, thus favoring the exchange of matter and momentum. Design innovations should be able to further enhance this organismic entrainment of the fluid medium.

other physical filters, the dimensions of resistance are not organized by these systems to maximize or optimize the catalytic chemistry on which the solids reduction of water treatment depends.

In the worst case scenario of laminar flow, exchange is entirely limited by diffusion, and most of the flowing material moves down the center of a pipe or channel in a three dimensional parabolic velocity distribution, often called a paraboloid of rotation, with minimal interaction with hard edges or surfaces. The stippled areas in Figure 1 show where growth and development can exploit surface and energetic relations of the flow of fluids in ways inaccessible to any standard wastewater facility. The horizontal box indicates how the "skin friction" of the channel is ordered by its container or, all the more so, by organismic growth to steepen gradients parallel to flow. The vertically stippled rectangle, on the other hand, indicates the dimension of organismic growth perpendicular to the mean flow path. Organisms thus modify the roughness of edges parallel to flow, thereby, in fluid dynamic terms, positioning exchange systems in regions of maximum energy dissipation and shear stress.

Perpendicular to flow, organisms partition the free mean path of the parabolic velocity distribution by growing into the flow path itself. This leads to specific differentials which are unattainable in conventional wastewater treatment facilities. This is because, although velocity gradients and interfacial phenomena have been studied for half a century or more in the wastewater literature, early characterizations of systems properties have been in terms of the relationship of power requirements for mixing and the volume and viscosity of the system (Camp and Stein 1942). While these are relevant variables and parameters, the integrated set of surface and physical chemical interactions that regulate virtually all ecological processes are dependant on exchange surfaces and metabolic volumes. Thus, while virtually all wastewater treatment approaches address flow characterizations such as the Reynolds number, a dimensionless number expressing the relationship between velocity, density, dimension, and viscosity, thus far, no general approach to wastewater purification utilizes surface area as the fundamental variable in material transformation. This is all the more surprising, since another dimensionless number out of fluid dynamics, the Sherwood number, relates mass transfer, the concentration difference between surface and free stream, diffusion constants, and surface area of the objects exchanging mass.

$$\mathrm{Sh} = \frac{\mathrm{qml}}{\mathrm{D_c S \Delta [C]}}$$

where: qml = the rate of mass transfer by convection
D_c = the diffusion constant or diffusion coefficient
S = the surface area of the object exchanging mass
$\Delta[C]$ = the concentration difference between the free stream and the solid sink or source of diffusing material.

The major point of this paper then, is that to be efficient, wastewater systems must be scaled such that flow is proportional to exchange surfaces. The Reynolds number which characterizes flow is itself inadequate to evaluate a treatment plant without some measure of exchange surface area, such as the Sherwood number. Vat-based conventional technologies cannot structure themselves to increase exchange capacity or augment diffusion, as do plant and other macroscopic organismic surfaces (see Figure 2 and 3), by maximizing or optimizing surface area in response to flow and concentration. They thus will never achieve the efficiencies of ecologically based systems which exploit much higher Sherwood numbers, thereby facilitating exchange of nutrients and waste matter between sources and sinks through surfaces.

There is no clear physical understanding of turbulent velocity gradients on which to base designs of conventional or ecologically based wastewater treatment systems (Nelkin 1992). Nevertheless, this much is known, that friction dissipates energy throughout a turbulent fluid. Unlike the laminar flow conditions cited above, in turbulent environments surfaces of containers or structures in the fluid do not necessarily increase friction, although exchange can be maximized at these interfaces. As the classical literature in fluid dynamics indicates, relations between turbulence and coefficients of drag become decoupled when surface roughness, the size of grains or fibers on a smooth body (indicated by the scale of the horizontal stippled patch in Figure 1) exceeds rather small critical values (Goldstein 1938, esp. Vol. II; Rouse 1946, p. 201 ff.; Shapiro 1961). This means that fibrous, rootlike, or leaf blade type structures can maximize concentration gradients at their surfaces without maximizing the power requirement of pumping to overcome friction.

Wastewater treatment efficiencies depend on a group of coupled chemical reactions constrained by concentration, temperature, and chemical potential. This is necessarily true of both standard *and* ecologically engineered treatment facilities (true, in fact, of any system). These two approaches differ, however, in that vat-based wastewater treatment systems are essentially limited by processes of diffusion since they contain no macroscopic self-structuring components to maximize exchange without maximizing frictive interactions through turbulent mixing. In essence, because conventional vat-based systems cannot decouple mixing and exchange, the flux of molecules into the work units, i.e., bacterial and other cells, is a function of both the volume of the system, and the power loaded into that volume to overcome inherent forces operative in the fluid wastewater. This was characterized by Camp and Stein in 1942:

$$G = \sqrt{P/\mu V}$$

where: G = mean velocity gradient, 1/s
P = power requirement, ft-lbs (Watts)
μ = dynamic viscosity, lbs/ft^2 (N s/m^2)
V = floculator volume ft^3 (m^3)

Note that exchange surface does not come into play in this framework since the power loadings cannot address specific patterns of distribution of matter, only a specific level of turbulence of fluid matter.

In the scaling of all biological systems, the ratio of exchange or surface area to metabolic volume is inherent in the integration of developmental processes. Mass transfer has, in fact, been identified as an explanation of metabolic scaling in aquatic invertebrates and algae (Patterson 1992).* Specifically, the log body size plotted against the log of metabolic rate has been known to give a slope characteristic of biological systems (Schmidt-Nielsen 1979 and references therein). Patterson, however, has shown that the mass and metabolism of specific taxa of algae and invertebrates follow an allometric relation as do the Reynolds and Sherwood numbers characteristic of flow in the habitats of these organisms and the mass transfer relations of their organismic surfaces. While these slopes are not identical, the graph in Figure 2 indicates in the stippled region how mass vs. respiration, and flow vs. mass transfer map into two dimensional space.

* Patterson's work appeared after this paper was presented at the International Conference on Wastewater Treatment in Trosa in 1991. While I had alluded to a relationship between Reynolds and Sherwood numbers, Patterson worked this relation out allometrically at organismic scale. Nonetheless, the Patterson paper implies the point I am making here, namely that the relationship between flow and mass transfer holds across the many orders of magnitude of biological systems.

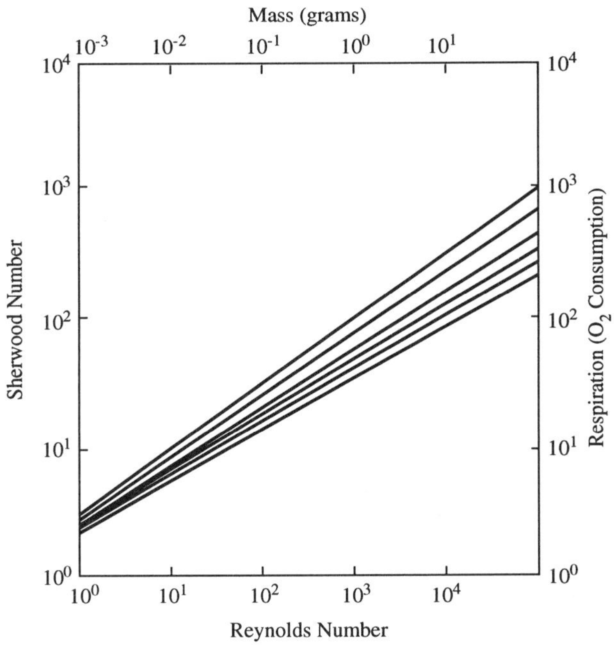

FIGURE 2. Mass scales to metabolic rate similarly to the way flow scales to mass transfer and exchange surface (adapted from Patterson 1992). Flow is characterized by the Reynolds number, a dimensionless number relating velocity, density, and dimension of flow to the viscosity of the fluid. Mass transfer per surface area is described by the Sherwood number, a similarly dimensionless ratio of mass transfer to the surface area times the concentration difference times the diffusion constant.

In the final section below, the generality of these results for metabolically based fluid purification systems will be discussed. Here, let us turn to the question of the flow of material at smaller scale, as a function of the change in concentration of a given material with distance:

$$J_j = D_j \partial c/\partial x$$

where J_j is the flux of material j, $\partial c/\partial x$ is the concentration gradient, or driving force of diffusion, and D_j is the diffusion coefficient of species j.

While local equilibria and inertia can be overcome with mixing and various forms of imposed flow, these only homogenize the concentration of materials without and direct steepening of gradients or organizing flow in relation to exchange surfaces. What is termed increased efficiency in traditional water treatment facilities requires expenditures of energy, measured in dollar costs as well as in increased waste and pollution production. It costs roughly a kilowatt to induce a flow of a cubic meter of water to a height of one meter each minute, or to completely mix a cubic meter of water by pumping the water in a cycle, as in Figure 3A, below. This should be compared to the average solar wattage over a cubic meter of ecologically engineered facility. Sunlight input is on the order of 200 watts, averaged over all latitudes and

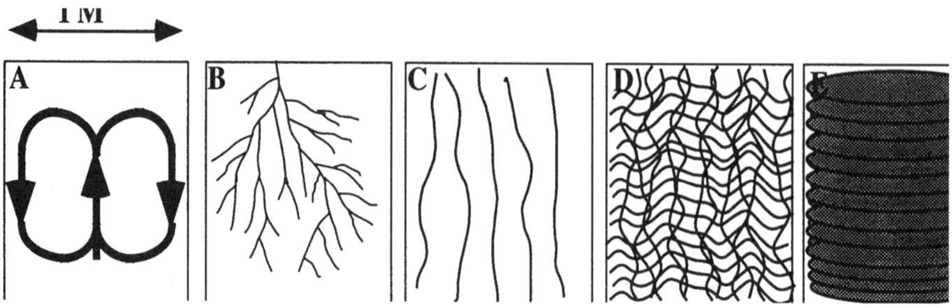

FIGURE 3. To move a cubic meter of water to a height of a meter each minute requires about a kilowatt of energy, as indicated schematically in A. Plants fill root space through branching strategies as in B. On average, about a fifth of a kilowatt, or 200 watts, falls on a square meter of the earth's surface. The material equivalent of one days productivity would be a one millimeter root more than 5 meters long, indicated in C, or a lattice of 500,000 root hairs, each one meter long, proportionally, about twenty times the density of the lines drawn in D. Were plant productivity turned into 5 grams of humic acids, its total surface area would be some 10,000 m^2, illustrated by the wafer-like structures in E (actual area would be some 200 times that pictured).

If about 1% of solar energy is built daily into plant, animal, and microbe structure, then, by compound interest rates, in a period of days to weeks this would effectively 'mix' a fluid by growing through it, making contact between phases of matter and thus augmenting diffusion to an increasingly greater degree than a kilowatt of fossil fuel-driven pumping.

weather conditions, day and night. The amount of fixed energy in plants and organismic structure is, in gross terms, 1 to 5 watts or 5 to 50 grams of carbon per m^2 per day.

It should be noted here that under comparison is the energetic expenditure of pumping water vs. the compound interest investment of biomass produced in and partly by virtue of the filtration process. A gain of one solar watt, over the expenditure of a thousand to move a cubic meter of water to a height of a meter comes to a difference of five orders of magnitude. The last section of this paper on network thermodynamics outlines how, in general, such exponential differences between functional capacities of ecologically designed vs. vat-based systems are a general consequence of the ratio of biomass to system mass or volume.

In general, rates of enzymatic activity increase with steeper concentration gradients which concomitantly increase material flow or flux, Jj. The maintenance of this flux depends on the surface area of the catalytic system as well as its mass or capacity to utilize the diffusing material. In other words, while the output of the system is diffusion limited, diffusion itself can only be maintained by mixing materials in the fluid and moving them through biological surfaces and transforming them into the working mass of the system, effectively removing them and thus maintaining the gradient. Mixing can only reduce this limitation in proportion to the biological surfaces and biomass in the system. Figures 3B through E indicate how ecologically engineered systems can increase biological surfaces much beyond the microbial interchange surfaces in conventional vat-based systems. The section on network thermodynamics below indicates how the systems biological volume acts to steepen concentration gradients and increase the dissipative capacity of self-organized subsystems in the process.

Plant growth and development thus structures the concentration gradient so that it becomes more favorable to exchange, thereby increasing overall efficiencies for ecologically engineered systems. Mixing and induced flow would have similar costs here as in conventional facilities, except that multicellular plant components can orient themselves to the flow, both along the boundary layer and across the main velocity gradient axis (as in Figure 1), thereby transforming the exchange parameters. In simple terms, the integrated interaction of biological systems enhance exchange capacity by using energy inputs to create instabilities in the system which lead to new structure and increased dissipation of free energy. The primary design aim of wastewater treatment can be restated in congruence with this framework: water purification through oriented surface waste catalysis.

PARTITIONING, MASS TRANSFER, AND THE DISSIPATION OF ENERGY

The work done in these systems, or equivalently, the free energy dissipated, can be described in terms of mass transfer between the solution and bacterial and other cells, and also in terms of the overall kinetics of the system. For simplicity, we first describe the mass transfer in terms of the equilibrium between free energy change on the left and the mass transfer of moles of material between two different phases on the right:

$$dG = dn\ (\mu_1 - \mu_2)$$

where dG is free energy change, dn is mass transfer in moles, and μ_1 and μ_2 are phase states in contact.

This can be seen as a thermodynamic partition function in that free energy is partitioned between phases of matter present — solid, liquid, gas, and colloid. Since each phase combination allows the system greater capacity for lowering its free energy, the addition of a missing or limiting phase increases overall efficiencies of the system.

In wastewater treatment, the waste solution, atmosphere, and bodies of organisms constitute the phase states of the system. As indicated in Figure 4, bacteria entrained in the waste solution arrive at local equilibria defined by the system influx and efflux (following Morowitz 1970). Energy added to the system as mixing can overcome viscous inertia and local equilibria through turbulence, as in Figure 4, but the scale of this behavior is limited by the instantaneous energy inputs, i.e., further development of the system is limited by energy added, diffusion, and the configuration of the components. In systems terms, initial conditions largely set the outcome.

Plants and their cobionts, however, do not permit the establishment of local equilibria. The growth of plants is a non-equilibrium process which structures the environment to enhance exchange. Elaboration of biological surfaces maximizes free energy dissipation by developing steepening environmental gradients. In systems terms, the outcome is not set by *initial* conditions, but by *boundary* conditions which continually push the system far from equilibrium (see Lamprecht and Zotin 1985 and Webber et al. 1988 for general discussions of this framework).

Growth itself may be seen to cause instabilities or disequilibria, and these may be described in terms of two different, but complementary perspectives (Figure 5):

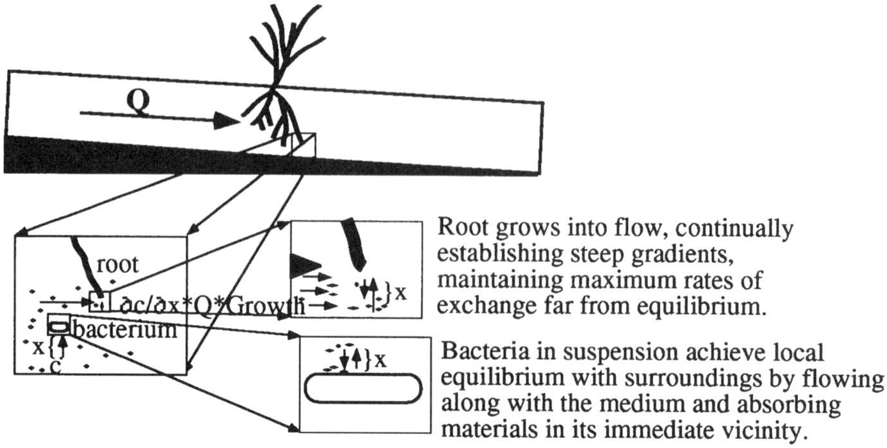

FIGURE 4. Efficiencies of vat based systems are limited by the concentration difference between microbes and the surrounding medium and the regulation of microbe populations by ciliates and other organisms. Ecologically engineered systems are able to entrain the flow of materials by growing into the flux, thus increasing the effective concentration differences between medium and metabolic machinery, and thereby, the efficiencies of the system. $\partial c/\partial x$ indicates concentration difference. Length of arrows within brackets indicate relative steepness of gradients. $\partial c/\partial x * Q * \text{Growth}$ is a complex non-linear function of the concentration gradient times flow rate times the mass specific growth rate.

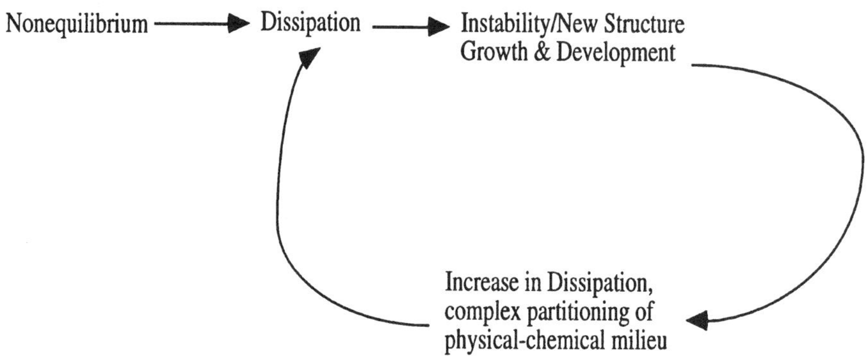

FIGURE 5. Non-equilibrium leads to the dissipation of free energy. Far from equilibrium, dissipation leads to specific instabilities or fluxes which are structured into the growth and development of biological and ecological systems. This structure increases dissipation capacity by partitioning physical and chemical systems, closing the loop of selforganization by moving further from equilibrium and creating further dissipation. Modified from Prigogine and Lefever's work "Stability and Self-organization in Open Systems" (Prigogine and Lefever 1975).

1. In the general terms of irreversible thermodynamics, non-equilibrium leads to instabilities which become structured to increase the dissipation of free energy.
2. In terms of partition functions, disequilibria may be described in terms of the elaboration of macromolecular and structural complexity of a given system. Or in specific chemical terms, "the sum of all macromolecular species determines the partition function of the system . . . " (Wyman and Gill 1990, p. 83).

The theoretical overview of Wyman and Gill indicates that macromolecular diversity sets the bounds on the binding capacity in ligand chemistry generally. From here, we must take a long but necessary step from macromolecular to organismic and community scale (O'Neill et al. 1986). The major groups and specific taxa of organisms all have unique macromolecular signatures. The diverse biochemical individuality of species and higher taxa can be said to set the bounds on the chemical partitioning capacities of ecological systems, since macromolecular diversity is constrained by biological diversity at organismic scale as well as interactions at community scale (see Chiariello et al. 1982; Hadley 1980; Safir et al. 1971; Walter Stadelmann 1968 for specific examples of physical and chemical partitioning by organisms; the classic work is that of Gibbs 1906).

Partition functions have still more general formulations. Chemisorption surfaces of the Langmuir model (see Zangwill 1988, p. 206 ff.) are described at the scale of atoms and their binding sites by the grand partition function, X, which relates chemical potential and binding energy to vibrational parameters of atomic and molecular systems.

$$X = 1 + Z_{vib}e^{b(\mu - eb)}$$

where: Z_{vib} = the vibrational canonical partition function
e = the root of all natural logarithms
b = 1/kT
μ = chemical potential
e_b = binding energy

At the two-dimensional quantum scale, this relationship indicates how the vibrational component functionally partitions interfaces in terms of chemical potential and binding energy differentials. However, burdensome this formula may be for calculations, it tells us that surfaces partition materials according to their binding energies and chemical potentials. The latter is especially significant, since biological and chemical oxygen demand are important but indirect measures of this fundamental property of wastewater.

Macroscopic surfaces interact with macroscopic fluids, and are not fully describable by methods that work one molecule at a time. It becomes necessary to put the behavior of one surface site in the context of the many surfacial interactions, and this is formally accomplished with a great deal of rigor in the framework of statistical mechanics. Here, a large number of chemical potentials and binding energies can be summed as energy states, e_n, of the system. This makes it possible to deal with a vibrational partition function, $Q_{vib,}$ at the much greater scale of very large groups of molecules and binding sites (see Adamson 1976 for a fine general discussion of this matter).

$$Q_{vib} = \Sigma_n e^{-e_n/kT}$$

Since energy states, e_n, may be defined in terms of μ, chemical potential, and e_b, binding energies, the two equations are analogous: since X is functionally related to $e^{b(\mu - e_b)}$, and since b is defined a 1/kT, $e^{b(\mu - e_b)} = e^{(\mu - e_b)/kT}$, and is an analogue of $e^{-e_n/kT}$.

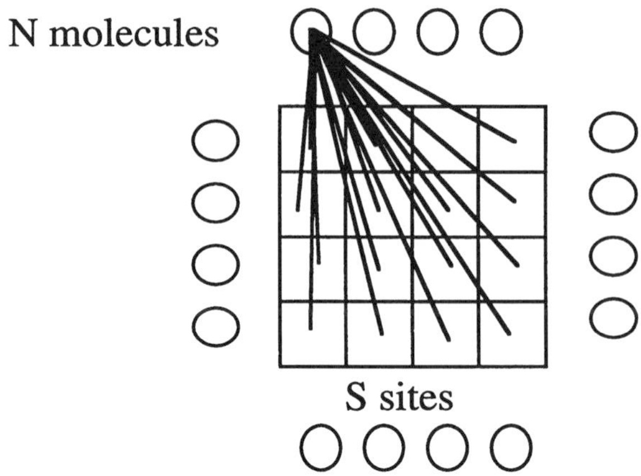

FIGURE 6. The molecules indicated by the circles can occupy any of the sites, indicated by the boxes. The first molecule can occupy any of 16 sites in this example, the second any of 15, and so on, until all the sites are occupied.

What do partition functions partition? Given a fixed number of molecules, N, and a finite number of sites, S, then molecules will be distributed on the sites according to how many are available. The first molecule has a choice of all S sites. The second molecule has a choice of all but one site, or S-1 sites, and so on. The partition function for a surface, Q^s_{tot}, on which N molecules are distributed over S sites (Figure 6) is (Adamson 1976; p. 555):

$$Q^S_{tot} = \frac{S!}{(S-N)!*N!}\left(Q^S\right)^N$$

In the simplified case of Figure 6, the first molecule has 16 possible adsorption sites, the second has 15 or S-1, the third 14 or S-2, etc. The total is S!, in this case, 2.09×10^{13} combinations or partitions of molecules by sites. In this case, the position of the last molecule would be determined by the positions of the previous, or S-15, leaving only one free energy lowering surface position.

Molecular systems thus distribute or partition themselves on surfaces, and partition functions indicate that this partitioning is accomplished according to their energy states. This is probably the governing step in the removal and breakdown of wastes, and again, it is most often not couched in regard to the functional surfaces involved, but rather in terms of the properties of the volume of fluids treated, evaluated by measures such as BOD and COD.

The reason for putting these only partly adequate measures into the framework of statistical thermodynamics is to make it clear that wastewater treatment has been functioning without a theoretical framework properly based on the catalytic surfaces that organize and facilitate the processes involved. As pointed out by Adamson (1976, p. 59), "The function of thermodynamics is to provide phenomenological relationships whose validity has the authority of the laws of thermodynamics themselves." The lawful relations of physical and surface chemistry are used here to provide a fundamental approach to wastewater treatment in terms of partitioning and adsorption, the integrated means by which BOD, COD, ammonia, nitrates, phosphates, and toxins are removed from the flow stream.

Taking specific entropy production, i.e., entropy production per mole of solute, as the systems measure, again following Prigogine and Lefever, we arrive at a very interesting conclusion:

"... when the polymerization reaction is switched from a regime in which it occurs mainly by simple condensation of the monomers to a regime in which they are assembled on a template, a discrete and important jump of the specific entropy production is observed. In other words, it is found that an instability by which the system manifestly reaches a more complex stage of functional organization, at the same time creates conditions which favor a subsequent evolution of this state through non-equilibrium instabilities." (Prigogine and Lefever 1975).

Cell wall fractions, templates produced by plants, fungi, and bacteria, are probably the most important regulators of terrestrial, aquatic, and estuarine ecological systems. These templates serve as nodes in material and energy flow. In general terms, the surface energies of these cell wall templates create the instabilities which increase specific entropy production (Ruis and Goodwin 1985). Hypotheses based on such template behavior would be most interesting to test in the wastewater treatment framework. As outlined, it should be possible for these experiments to integrate the powerful frameworks of statistical mechanics and ligand chemistry with functional measures of biological diversity.

Other fundamental interfacial relations such as physisorption can be discussed in similar terms since they also describe the behavior of surfaces which lower free energy, including those of organisms. Again, the main difference between 'inanimate' and biological surfaces is that the latter are continually elaborated, utilizing conditions of surface entropy and Gibbs Potential, but establishing a non-equilibrium framework by continually elaborating membrane and cell wall with their own characteristic surface energies:

$$\Delta G_s = \Delta H_s - T\Delta S_s$$

where: G_s = surface Gibbs potential
H_s = surface enthalpy
S_s = surface entropy.

Organismic surfaces move these relations into an irreversible field where membranes and biological polymers are continually elaborated in response to concentration gradients, flow rate, and temperature (see especially Ruis and Goodwin 1985), as well as mechanical stresses (Preston 1974; Raven 1977; Wainwright et al. 1976).

PARTITIONING CHEMICAL ENERGY

Tests and measures of systems properties of chemical potential are facilitated by one of the most inclusive and elegant sets of relations in all of biology (Nobel 1983). The chemical potential, μ, of component j, may be partitioned into its components:

$$\mu_j = \mu_j^* + RT \ln a_j^* + V_jP + z_jFE + m_jgh$$

where: μ_j^* is a reference point or standard measure for chemical potential.

RT ln a_j is the concentration or 'activity' term of chemical potential in energy/mole:

R is the gas constant, 8.3 Joules per mole per°K;

T is absolute temperature in°K;

a_j is the activity or effective concentration of species j.

V_jP represents the contribution of pressure to chemical potential:

V_j is increase in volume per quantity of species j added;

P is pressure, usually that in excess of atmospheric.

z_jFE is the electrochemical contribution to the chemical potential of species j:

z_j is the charge number of species j;

F is the farad, in coulomb/V;

E is the electrical potential, millivolts.

m_jgh is the gravitational term of chemical potential,

m_j is mass of species j;

g is the gravitational constant;

h is height.

This equation functionally relates concentration, pressure, electrochemical, and gravitational components inherent in chemical potential. Even better, this formula is not only not burdensome, but its application has been well worked out with biological examples in one of the finest texts written in biology, P.S. Nobel's *Biophysical Plant Physiology and Ecology* (1983 p. 61 ff.).

The concentration gradient or activity term of this equation describes the contribution of the forces that govern diffusion to overall chemical potential and set, as described above, general limits on any water treatment system. By focusing on the chemical potential of water itself, the water potential of a waste treatment system may be considered.

By way of example, an ecologically engineered wastewater treatment system such as a series of constructed marshes could be designed, utilizing the chemical potential framework as a set of guidelines. In this context, evaporative potential of an area of the marsh could be scaled to lower the water concentration in order to maintain a favorable solute concentration gradient for water treatment. In other words, the very well-corroborated framework of chemical potential in physiology can be utilized in water treatment facilities to maintain favorable rates of breakdown or absorption of specific nutrients or pollutants by maintaining a specific activity or concentration gradient in the system.

Chemical potential also describes the oxidation/reduction or redox work of ecological communities. At several different chemical potential poles, aerobic, anaerobic, and photosynthetic systems function to develop complex gradients of values which extend from those of molecular oxygen to reduced sulphur and carbon. Such potential differences can exist in the space of millimeters, at a potential difference of 1200 mv (–400 mv to +800 mv; see Margulis 1982 for a discussion of these communities). While such potential differences can be established by bubbling, mixing, and induced flow in vat-based systems, the growth of single-celled organisms in suspension cannot structurally couple thermal gradients with electron and proton cascades as do the chemical potentials of microbial films plant and peat surfaces and films.

PARTITIONING RADIANT ENERGY

Vat-based technologies are often partially covered systems with a minimum footprint. While this definitely lowers certain costs, it limits the use of the greatest energy flux on the earth's surface, solar radiation. Most of the energy that passes through the environment, and through all organisms, is radiant energy, heat flow or flux, moving from source to sink, from one body to another, on its way into and out of the fluid envelope of the earth.

Partitioning of radient and thermal energy is quite well understood, with major work dating back more than three decades, and extensive, well corroborated applications to plants and other organisms (see Gates 1980, 1976; Lee and Gates 1964).

$$Q_a = e\partial T^4 + H(T_p\text{-}T_f) + LE + P$$

where:
- Q_a = amount of radiation absorbed
- e = emissivity of plant body (tendency to radiate heat)
- ∂ = Boltzmann's constant for blackbody radiation
- H = the convection coefficient
- T_p = plant body surface temperature
- T_f = fluid temperature
- L = heat of vaporization of water (≈580 cal/g H_2O)
- E = amount of water evaporated by leaf
- P = photosynthesis

Thermal partitioning is structured by four major regulatory relationships: reradiation $e\partial T^4$; convection $H(T_p\text{-}T_f)$; evaporation LE; and photosynthesis P. These thermal relations partially map onto other partitioning systems developed by organisms and communities. For example, LE, heat of vaporization times amount of water evaporated by a leaf, is functionally related to the chemical potential of water and the evaporative rates cited in the section above. Photosynthesis is the productive engine which creates the biomass which structures virtually all the partition functions discussed (for one design application, see Kinsinger, Mankiewicz, and Mankiewicz 1996).

SURFACE TENSION, CAPILLARITY, AND WATER PURIFICATION

Soil, peat, and bryophyte (mosses and their allies) structures entrain many multipliers of functional capacity which would be most useful in ecologically engineered wastewater treatment systems. Specifically, water adsorption, absorption and resistance to water loss show hysteresis, i.e., dependance on previous state (Haines 1925, 1927, 1930; Fisher 1926 and 1928). However, since water is a differential regulator of diffusion and uptake, soil, peat, and communities of moss-like plants achieve periodic fluctuations of oxidation-reduction potential and hydration levels. *More importantly, the properties of surface tension and capillary architecture scale the thickness of water films in contact with filtrative and purificatory surfaces, regulating*

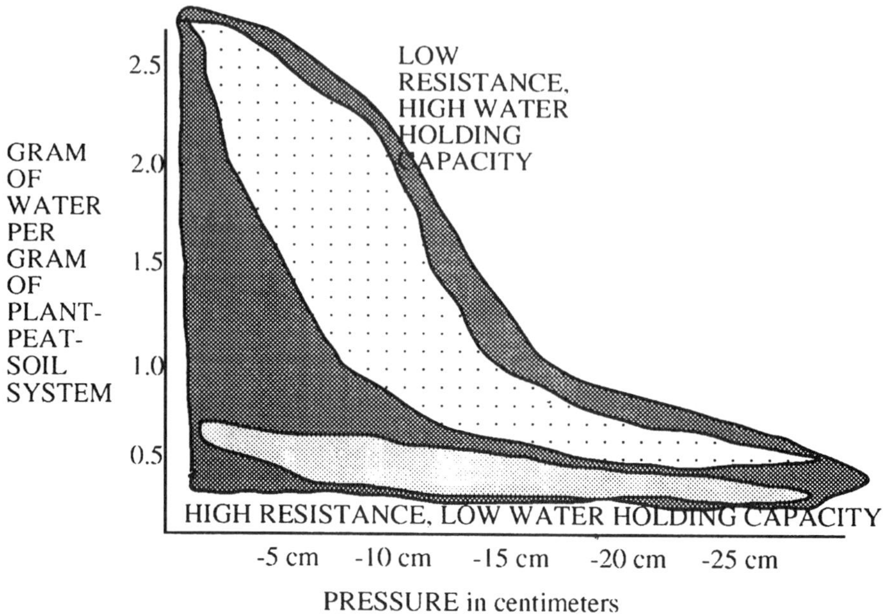

FIGURE 7. The lightly stippled domain indicates a plant/peat system with high water holding capacity but low resistance to water loss or intrinsic capacity to pull water from the surroundings by capillary forces. The darkly stippled domain more parallel to the x axis indicates a plant/peat/soil capillary system with a flat water holding capacity per gram dry weight but high resistance to water loss and capacity to move water into the capillary matrix. Systems have evolved which cover virtually this whole range of behaviors, or, in other terms, fill virtually this whole phase space, indicated by the grey envelope in the background.

in the process the length of flow in these contact films and the temporal domain of contact.

As Figure 7 indicates, soil/plant/peat domains exhibit differential capacities to hold water and resist water loss (Proctor 1979; Clymo & Hayward 1982; Hayward & Clymo 1982; Mankiewicz 1987). Even more powerful is the regulation of filtration and purification which occurs as a function of hysteresis and the entrained flow of capillarity. As the system moves from 300% to 150% of its dry weight*, the thickness of the water film on the plant decreases from a maximum radius of some 5 mm to a maximum radius of less than 0.05 mm, with diffusion distances decreasing more than a hundred times in the process. Where water loss is driven by evaporation, the concentration of nutrients in the water volume will also become a hundred times as steep.

Flux density in diffusion, i.e., the movement of materials from points of higher to points of lower concentration, depends on the amount of material moving through a given area over time. This is formally expressed:

* Dry weight here means after centrifugation at approximately 300 Gs. See Mankiewicz 1987b for a description.

$$\text{flux density} = \frac{\text{(amount of matter)}}{\text{(area)(time)}}$$

However, because diffusion depends on the random walks of molecules through the thermal noise of a system, this equation does not fully express the slowness of the process at macroscopic scale. As Schmidt-Nielsen (1979) has pointed out, transfer of oxygen from capillary to cell may take 0.01 seconds, while diffusion across a millimeter would take 100 seconds. *From lung to hand or foot, the random paths of molecular diffusion would take some 3 years.* This is why so much energy must be put into wastewater treatment systems to overcome viscous inertia and augment diffusion.

Ecological engineered systems, including various kinds of constructed wetlands, may be described as utilizing the surface tension of water to overcome its viscosity. Plant and peat surfaces pull water along their surfaces into capillary films, which both increase evaporative rates and steepen concentration gradients. Plant surfaces thus multiply diffusion rates by two to four orders of magnitude by capillarity and evaporation, but the scale of metabolic behavior as sink for nitrate, phosphate, and other materials further multiplies these differentials.

Water is both architect and architecture in natural systems. Intriguingly, the area of a body of water, a surface which by definition minimizes free energy, is much less than the surface of the macromolecular matrix of any plant coverage of soil, from dry grassland to rain forest. In a word, the area of plant surfaces compared to free liquid water surfaces is immense, and while the volume of water on the earth is much greater, atmosphere, water and solid substratum may interact much more intensely in conjunction with life (Hutchinson 1970; Mankiewicz 1991). This brings us to our last topic: the thermodynamic capacity of biological mass in purification systems.

NETWORK THERMODYNAMICS

While process control in wastewater treatment has been conceptualized in terms of flow and mixing rates, surface area to volume ratios of various tank types, and even rates of accumulation of microorganisms (see especially Chapter 8 in Metcalf and Eddy Inc. 1991), no mention is generally made in the engineering literature of the *volume* of organisms and the *volume* of their byproducts as a measure and indicator of system performance. Basically, wastes are treated without a knowledge of the scale of the chemiosmotic machinery that is doing the work, and therefore, the catalytic biology of wastewater treatment cannot be rationalized in terms of the metabolic capacity or capacitance of the treatment system. This problem, with suggestions for its solution, will be discussed in the context of network thermodynamics. What follows is based on the work of Mikulecky (1985). For an introduction to this field the reader is directed to this paper and references therein.

While ecological engineering has not yet articulated a rationale of wastewater treatment configurations in terms of biomass and the volume of metabolic machinery, flagrant imitation and appropriation of natural systems has allowed many ecologically engineered projects thousands to millions of times more biological mass per system volume than conventional vat-based systems. The treatment equation, by including macrophytes (and in the case of the Chinese experiment of the last three millennia, fish; see Yan and Ma 1997; Li 1997), has achieved much greater efficiencies because it contains a much higher biological diversity which itself increases the complexity of surface chemistry. They also contain an immensely greater volume of energy stored in biomass within the system. Most importantly from a thermodynamic perspective, entropy production per mole of solute is maximized where the greatest number of moles of solute are in contact with free energy lowering interfacial surfaces which surround the biological volumes of catabolic machinery which utilize and break down wastes.

In Figure 8B, the nodes marked 1 and 2 exist in standard wastewater treatment facilities as well as in ecologically engineered projects. The node marked 3, however, appears only where larger photosynthetic organisms are structured into wastewater flow. Where this occurs, efficiencies are multiplied, since plant surfaces provide habit for microbes (Corpe and Rheem 1989), and because plants move water and absorb solutes, thereby increasing uptake and regulating exchange (Blackwell 1963; Dacey and Howes 1984).

The network diagram in Figure 8 indicates that the interactive complexity of a system of capacitors and dissipators increases greatly as components are added. Processes or rates are not, however, additive. Where concentration and diffusion limit most conventional water purification systems, increases in either of these dimensions will contribute exponentially to the increase in scale of a biological system, and with such increases, dissipative capacity of the system also increases in a non-linear fashion.

Energy dissipation follows from the second law of thermodynamics: where entropy is maximized, or, alternatively, free energy, the ability to do work, minimized. In simple terms, energy must be dissipated in any process. The chemiosmotic work accomplished by living systems is both structural and chemical. Chemical potential in molecular bonds is broken down and turned into organic form. These forms insinuate themselves into the flow of energy and materials at scales ranging from the molecular to the organismic and larger. These living forms then produce refractory organic matter, a form of biomass flow through soils and lakes, rivers and estuaries worldwide in the immense volume and surface area of humic acids. This flow of carbon compounds may be substantially larger than the "living biomass" at any given moment.

The capacity to dissipate energy depends on the entrainment of flow, shear, and velocity (Figure 1 to 3) by structures and surfaces capable of transforming this entrainment into favorable absorptive conditions and concentration gradients. The only 'machines' that accomplish these ends on this planet at this point in the history of engineering are organisms. The surface area and volume of plants and their cobionts increase proportionally the effects of the four dimensions of flow, increasing potential exchange efficiencies more than five orders of magnitude by a complex multiple of surface area and volume. Even beyond this increase in sink capacity for nutrients and pollutants provided by vegetation is the much greater catalytic chemistry of microbes of more than ten different phyla and an aggregate surface area many times greater than all of the higher plants combined.

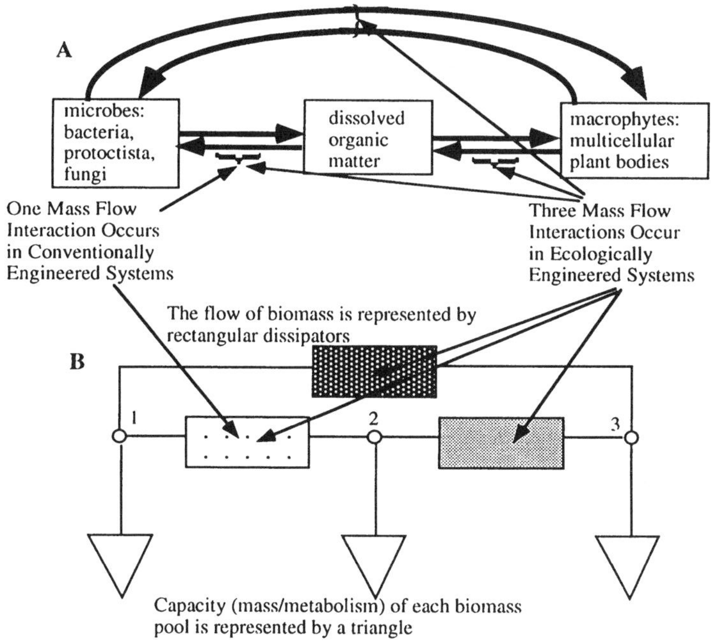

FIGURE 8. Biomass 'pools' characterize all water treatment systems: pictured above is a three component system of microbes, dissolved organic matter, and macrophytes or large plants. Biomass or metabolic volume of the pools is represented by the triangular volume capacitors. These are connected by rectangular dissipators. Biomass flows through the dissipators, measured in mass/time, yielding rates of chemical transformation.

The capacity of the system to dissipate free energy is a function of concentration, or biomass/volume. This includes a comparison of the biomass differences between the pools which is a complex non-linear function of concentration of nutrients and inhibitors, flow rate, and mass specific growth rates (dNxQx-Growth in Figure 4, above). See Mikulecky 1985 for discussion of biomass volume in this thermodynamic framework.

Ecological engineering adds the two top and two right hand arrows in 7 A. These connect the four dimensions of flow, a volume moving through space, to the two dimensions of organismic surface and the three dimensions of their metabolic volume.

The volume of biomass per system volume, and the volume flux of biomass from one pool to another within a treatment facility, have yet to be addressed, much less quantified, in the context of ecological engineering. While the work of Patterson (1992) cited above relates mass exchange to biological surface, the ratios of Reynolds numbers to Sherwood numbers have not yet been used to characterize intrinsic efficiencies of wastewater treatment. In the preceding argument, the flow of mass through biological dissipators has been shown to entrain the four dimensional geometry of flow with the catalytic surfaces that increase diffusion rates by several orders of magnitude. However, as wastewater treatment systems are analyzed in terms of the network thermodynamic parameters of biomass flow, the ratio of the Reynolds number characterizing flow to the Sherwood number characterizing mass transfer through biological surfaces into metabolic dissipators may serve as the basis for design of the next generation of ecologically based facilities.

CONCLUSIONS

Taking the source-to-sink diffusion rate, $Jj = Dj\partial c/\partial x$, as a major limit on the behavior of fluid treatment systems, what kind of dimensional regulation do organisms exert over this process? The exponential powers of ecologically based systems can perhaps best be described in terms of dimensional analysis, since more than three dimensions must be considered simultaneously. Functionally, two dimensional adsorptive surfaces constrain the behavior of the one dimensional scalar of diffusion, $\partial c/\partial x$. What is meant here is simply that organisms entrain three of the four dimensions of flow (Q is proportional to r^4) along the two dimensional interface of their surfaces. At the same time, these surfaces expand, and grow in a whole range of functional orientations, from perpendicular to parallel to the flow itself, as indicated in Figure 1. At the same time, the absorptive volume of plants increases, filling three dimensional space, adding another behavioral dimension to their surfaces by modifying adsorption constants through absorption and active transport. Electrochemical charges sustained on organismic surfaces constitute a kind of field behavior which changes the dimensionality of diffusion of many molecules since the diffusion constant, Dj, can be strongly modified by electromotive forces around growing cells and roots.

Finally, the four dimensional structure of flow itself is functionally partitioned by four dimensional plant growth strategies over time while concentrations of suspended and dissolved matter in the water column is regulated through byproducts of plants and their cobionts, humic acids, and many other ligands.

The temporal, developmental dimension of ecologically engineered systems using multicellular plants may be conceptualized in terms of the 1% of incident solar energy input which acts as a compound interest investment in the working surface and volume of the system. A diversity of organisms and habitats should permit the chemical partition function of an ecologically engineered system to be maximized, since each phylum, taxon, and individual is characterized by unique biochemical signatures (following Wyman and Gill 1990). Hysteresis of soils, muds, and plant communities in ecological systems must add dimensional degrees of freedom and increase the diversity of outputs, since this phenomenon regulates water and solute concentration by incorporating solar energy and wind shear into the water holding and resistance to water loss equation.

Thermodynamic reasoning provides a broad and well-grounded foundation for the applied sciences generally, since it embodies fundamental and empirical results of physics, chemistry, and biology. While the field is too complex to cover fully in a short review, this paper demonstrates that only organisms partition energy and material flow to maximize energy use, and, therefore, specific entropy production. By eliminating systems which entrain flow to dissipate chemical energy, at its best, wastewater engineering is confined to search for infrastructure configurations which minimize footprint and fossil-fuel investment.

Ecological engineering will doubtless have its own nutrient and pollutant loading limits, but it is predicted that these will be much less constrained by the inherent inertia of diffusion rates than by the relations between total system volume, biomass flow, and the scaling of biological mass, surface, and diversity. In the process of discovering these constraints, it is hoped that the tools of biology and thermodynamics may forge in applied ecology and ecological engineering an interdisciplinary field large enough to solve these pressing problems of human economy.

REFERENCES

Adamson, A.W. 1976. *Physical Chemistry of Surfaces.* New York, Wiley Interscience.

Blackwell, M.J. 1963. The role of evaporation in the surface energy balance. In *The Water Relations of Plants*, ed. by A. J. Ritter and F. H. Whitehead. New York, Wiley.

Camp, T.R. & P.C. Stein. 1942. "Velocity Gradients and Internal Work in Fluid Motion." *Journal of the Boston Society of Civil Engineering*. 30: 209.

Chiariello, N, J.C. Hickman and H.A. Mooney. 1982. "Endomycorrhizal role for interspecific transfer of phosphorus in a community of annual plants." *Science* 217: 941-943.

Clymo, R.A. & P.M. Hayward. 1982. "The ecology of *Sphagnum*," pp. 229-290. In A.J.E. Smith (ed.), *Bryophyte Ecology*. Chapman and Hall, New York.

Corpe, W. A. and S. Rheem. 1989. "Ecology of the methylotrophic bacteria on living leaf surfaces." *Federation of European Microbiological Societies. Microbiology Ecology* 62(1989): 243-250.

Dacey, J. W.H. and B.L. Howes. 1984. "Water uptake by roots controls water table movement and sediment oxidation in short *Spartina* marsh." *Science* 224: 487-489.

Fisher, R.A. 1926. "On the capillary forces in an ideal soil; correction of formulae given by W.B. Haines." *Journal of Agricultural Science 16: 492-505.*

Fisher, R.A. 1928. "Further note on the capillary forces in an ideal soil." Journal of Agricultural Science 18: 406-410.

Galileo, G. 1638. (1974). Two New Sciences, Including Centers of Gravity & Force of Percussion. Translated by S. Drake. Madison, WI: University of Wisconsin Press.

Gates, D.M. 1980. *Biophysical Ecology.* New York: Springer-Verlag.

Gates, D.M. 1976. "Energy exchange and transpiration." In *Water and Plant Life: Ecological Studies 19*: 137-147. Lang, O.L., Kappen,L. and Schulze, E.D., ed. New York: Springer-Verlag.

Gibbs, J. Willard. 1961 (1906). *The Scientific Papers of J. Willard Gibbs.* New York: Dover Publications.

Goldstein, S. 1938. *Modern Developments in Fluid Dynamics: An Account of Theory and Experiment Relating to Boundary Layers, Turbulent Motion and Wakes. Vol. I and II.* New York: Dover Publications.

Hadley, N.F. 1980. "Surface waxes and integumentary permeability." *American Scientist* 68: 546-553.

Haines, W.B. 1930. "Studies in the physical properties of soil. V. The hysteresis effect in capillary properties, and the modes of moisture distribution associated therewith." *Journal of Agricultural Science* 20: 97-116

Haines, W.B. 1927. Studies in the physical properties of soil. IV. A further contribution to the theory of capillary phenomena in soil. *Journal of Agricultural Science* 17: 264-290.

Haines, W.B. 1925. "Studies in the physical properties of soil. II. A note on the cohesion developed by capillary forces in an ideal soil." *Journal of Agricultural Science* 15: 529-35.

Hayward, P.M. & R.S. Clymo. 1982. "Profiles of water content and pore size in *Sphagnum* and peat, and their relation to peat bog ecology." *Proceedings of the Royal Society of London*, Series B, 215: 299-325.

Hutchinson, G.E. 1970. The Biosphere. *Scientific American* 223 (3): 44-53.

Kinsinger, W.C., P.S. Mankiewicz & J.A. Mankiewicz. 1997. Wastewater, ground water and air purification schematics for the Réne Dubos Bioshelter. In: C. Etnier and B. Guterstam (eds.), *Ecological engineering for wastewater treatment.* Boca Raton, FL: Lewis Publishers.

Lamprecht, I. and A. I. Zotin, eds. 1985. *Thermodynamics and Regulation of Biological Processes*. Berlin: Walter de Gruyter.

Lee, R., and D.M. Gates. 1964. "Diffusion resistance in leaves as related to their stomatal anatomy and micro-structure." *American Journal of Botany* 51: 963-975.

Li S. 1996. "Agriculture and its role in ecological wastewater treatment." In: C. Etnier and B. Guterstam (eds.), *Ecological engineering for wastewater treatment: Proceedings of the international conference, 24-28 March 1991, Stensund Folk College, Sweden.* Boca Raton, FL: Lewis.

Lieth, H. and R.H. Whittaker, eds. 1975. *The primary production of the biosphere*. New York: Springer-Verlag.

Mankiewicz, P. S. 1991. The Macromolecular Matrix of Plant Cell Walls as a Major Gaian Interfacial Regulator in Terrestrial Environments, pp 309-319. In *Scientists on Gaia*, ed. by P. Boston and S. Schneider. MIT Press, Cambridge, Mass.

Mankiewicz, P. S. 1987. "The low pressure field porometer: A new, low cost technique for characterizing external capillary water conduction in whole colonies of bryophytes and other small plants." *The Bryologist* 90(3), pp 253-262.

Margulis, L. 1982. *Early Life*. Science Books International. Boston, MA.

Metcalf & Eddy Inc. 1991. *Wastewater Engineering: Treatment, Disposal, and Reuse*. 3rd ed. revised by Tchobanoglous, G and F.L. Burton. eds. New York: McGraw Hill.

Mikulecky, D.C. 1985. "Network Thermodynamics in Biology and Ecology: An Introduction." In *Ecosystem theory for biological oceanography*, ed. by Ulanowicz, R.E. and T. Platfield. Canadian Bulletin of Fisheries and Aquatic Sciences, 260 pp.

Morowitz, H.J. 1970. *Entropy for Biologists*. Academic Press, New York.

Nelkin, M. 1992. "In What Sense is Turbulence an Unsolved Problem?" Science 255: 566-569.

Nobel, P.S. 1983. *Biophysical Plant Physiology and Ecology*. San Francisco: W.H. Freeman and Co.

O'Neill, R.V., D.L. DeAngelis, J.B. Wade and T.F.H. Allen. 1986, *A Hierarchical Concept of Ecosystems*. Princeton University Press.

Patterson, M.R. 1992. "A Mass Transfer Explanation of Metabolic Scaling Relations in Some Aquatic Invertebrates and Algae." Science 255: 1421-1423.

Preston, R.D. 1974. *The Physical Biology of Plant Cell Walls*. London: Chapman and Hall.

Proctor, M.C.F. 1979. "Structure and eco-physiological adaptation in bryophytes." In *Bryophyte Systematics*, Clark, G.C.S. and Duckett, J.G., eds., pp. 479-509. New York: Academic Press.

Prigogine, I. and R. Lefever. 1975. "Stability and Self-Organization in Open Systems." In *Membranes, Dissipative Structures, and Evolution*, edited by G. Nicolis and R. Lefever. New York: Wiley.

Raven, J.A. 1977, "The Evolution of Vascular Land Plants in Relation to Supracellular Transport Processes." In *Advances in Botanical Research*, ed. by H.W. Woolhouse. New York: Academic Press.

Rouse, H. 1946. *Elementary Mechanics of Fluids*. New York: Dover Publications.

Ruis, J. L. and B. C. Goodwin. 1985. "Continuous Phase Transitions and Morphogenesis." In *Thermodynamics and Regulation of Biological Processes*. Lamprecht, I. and A. I. Zotin, eds. Berlin: Walter de Gruyter.

Safir, G.R., J.S. Boyer, and J.W. Gerdemann. 1971. "Mycorrhizal enhancement of water transport in soybean." *Science* 172: 581-583

Schmidt-Nielsen, K. 1979. *Animal Physiology: Adaptation and Environment*. 2nd ed. London: Cambridge University Press.

Shapiro, A.H. 1961. *Shape and Flow: The Fluid Dynamics of Drag*. New York: Doubleday and Company.

U.S. Dept of Energy. 1988. "Summary Report of a Workshop on Lignin Research: Needs and Opportunities" held at the University of Illinois, April 18-19, 1988. CONF-880423 — SUMM., U.S. Dept of Energy, Office of Energy Research, Division of Energy Biosciences, Washington, D.C. 20545.

Ulanowicz, R.E. 1986. *Growth and Development: Ecosystems Phenomenology*. New York: Springer-Verlag.

Vogel, S. 1981. *Life in moving fluids: The physical biology of flow*. Boston: Willard Grant Press.

Wainwright, S.A., W.D. Biggs, J.D. Currey, and J.M. Gosline. 1976. *Mechanical design in organisms*. New York: John Wiley and Sons

Walter, H. and E. Stadelmann. 1968. "The physiological prerequisites for the transition of autotrophic plants from water to terrestrial life." *BioScience* 18: 694-701.

Webber, B.H., D.J. Depew, and James D. Smith, eds. 1988. *Entropy, Information, and Evolution*. Cambridge, MA: MIT Press.

Wyman, J. and S.J. Gill. 1990. *Binding and Linkage: Functional Chemistry of Biological Macromolecules*. Mill Valley, CA: University Science Books.

Yan J. & Ma S. 1996. "The function of ecological engineering in environment conservation with some case studies from China." In: C. Etnier and B. Guterstam (eds.), *Ecological engineering for wastewater treatment*. Boca Raton, FL: Lewis Publishers.

Zangwill,A. 1988. *Physics at Surfaces*. New York: Cambridge University Press.

Chapter 33

ECOLOGICAL APPROACHES TO TREATMENT OF WASTEWATER FROM POTATO PROCESSING

Maret Merisaar

When comparing the branches of food processing industry in Estonia, one can say that the biggest pollution load for the water bodies has originated from potato starch manufacturing — up to 40% of the raw material is turned into the waste products potato pulp and potato juice (Figure 1).

In conventional technology, 0.75 m^3 concentrated potato juice with a BOD_5 of 60,000 mg O_2/1 is separated from every ton of mashed potatoes. This is diluted with processing waters and directed into sewage systems, thus giving 10 to 14 m^3 of sewage water. The nutrient resource of these waters (BOD_5:N:P) is 100:4.5:0.6. COD is 2700 mg O_2/1, pH = 5.4.

Potato pulp has been used as a livestock feed since the first days of potato starch manufacture. As the product is difficult to transport and does not keep fresh for long, it is often pressed and/or silaged. Potato juice contains almost the entire amount of potato protein, known to be of high biological value. It is especially rich in lysine, but the other amino acids are also present in ratios that are appropriate for use as feed. Figure 2 depicts the results of the analyses of the contents of the amino acids in the potato pulp and potato juice from our pilot plant. Columns 1 to 10 depict essential and 11 to 18 nonessential amino acids. A technical problem prevented our completing the tryptophan analyses, but it is known (Talley et al. 1984) that 1 g potato tuber dry matter contains 0.63 to 1.18 mg tryptophan. In addition to protein, the potato juice contains sugars, soluble minerals, organic acids (citric, malic, oxalic, etc.) and the vitamins C, B_1, and B_2 (Figure 2).

Even when half of the wastes — potato pulp — is used for cattle feed, the main polluter, potato juice, is not used in spite of its valuable nutrients. This is for the following reasons:

- The juice is too diluted (less than 1% dry matter (DM)).
- It foams very strongly.
- It does not keep fresh for long and may give off toxic substances when the nitrogenous compounds break down.
- Although many expensive technologies are known for recycling potato protein, they have not been introduced into our economy because of various obstacles, the most important of which is the absence of private property.

So, in the water protection laboratory of the Tallinn Technical University, the problem was solved by modernizing the starch manufacturing technology. In our low-waste technology, the juice is separated containing 4% of DM, mixed with dewatered pulp (10% DM) and heated with hot steam. The liquid feed that is obtained as the byproduct is still very diluted (6 to 8% DM), but contains all the nutrients of

0-87371-990-5/97/$0.00+$.50

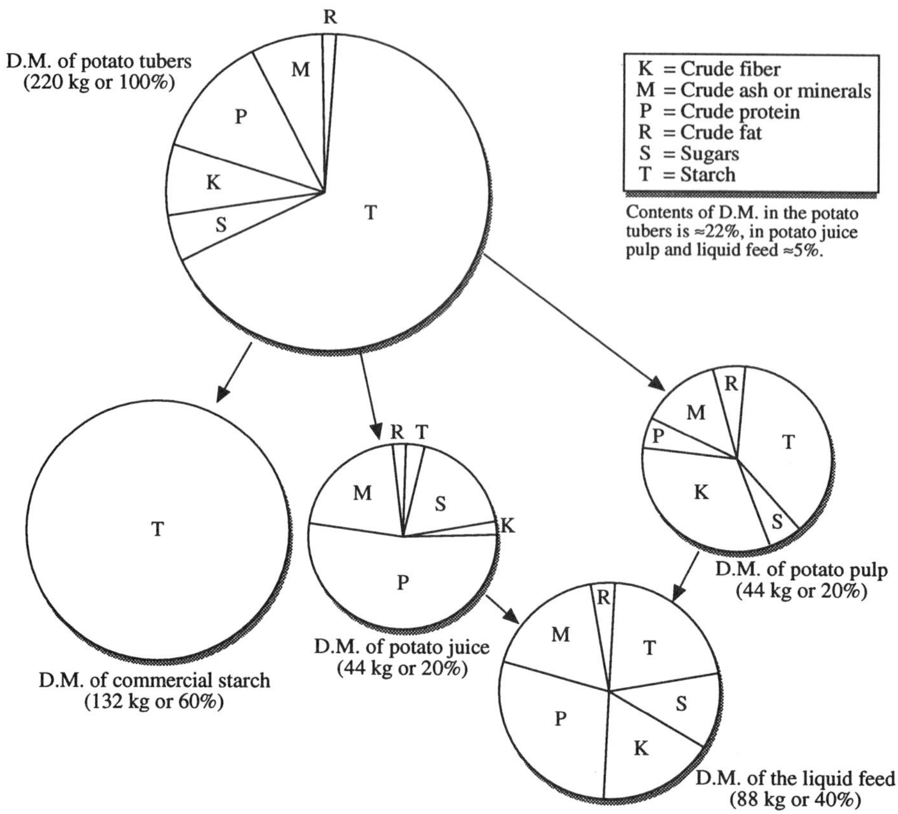

FIGURE 1. Distribution of dry matter (D.M.) of 1 ton of potato tubers that are processed into commercial starch and liquid feed.

the pulp and the juice. One feed unit (our F.U. = 1414 kcal) is equivalent to 20 l of liquid feed and contains about 142 g digestible protein. A feed unit is also approximately equivalent to 1 kg of the concentrate that is often fed the animals.

Controlled feeding experiments were carried out with two groups of 10 beef calves (for 74 days), two groups of 8 dairy cows (for 71 days), and two groups of 180 pigs (for 87 days). The experiments showed that:

- The daily maximum amount of liquid feed per capita for ruminants was 30 l (in place of 1.5 kg concentrate) and for pigs, depending on their age, 1 to 3 l (in place of 50 to 150 g concentrate). Depending on the feeding value, 50 kg concentrate can be saved with 1 ton of liquid feed.
- It is economically profitable to transport liquid feed up to 30 km.
- There were no health troubles noticed and the daily live weight gains and milk production were sufficient.
- The differences in the yields of meat and milk production between the groups were insignificant, but the costs per 100 kg produced were lower in the groups that were given liquid feed: 9.55% for beef, 2.86% for milk, and 4.49% for pork.

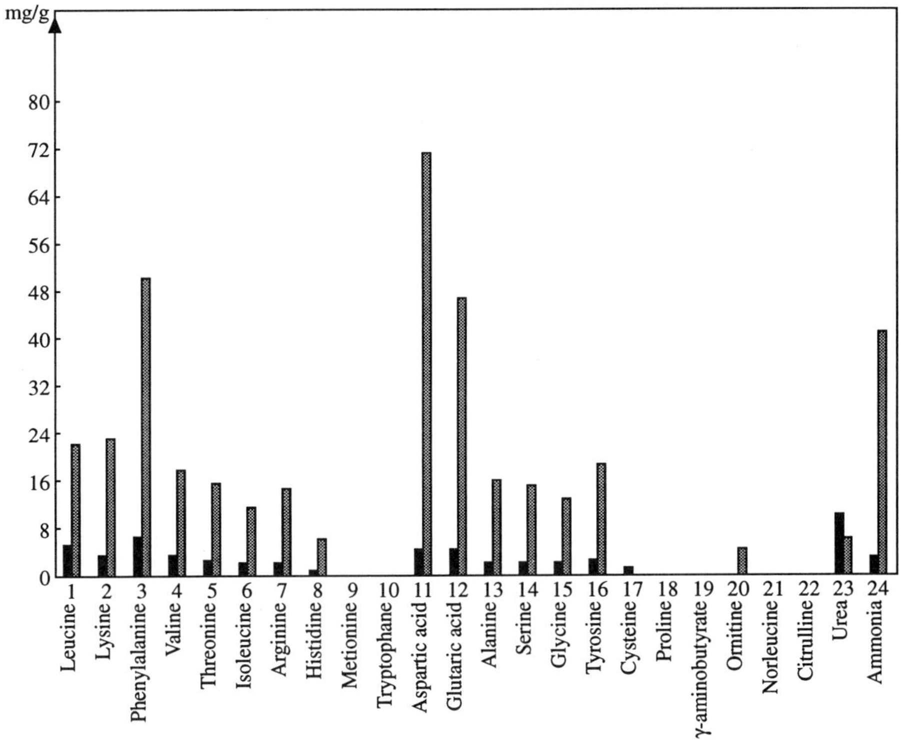

FIGURE 2. Contents of nitrogenous compounds (crude protein) in the dry matter of potato pulp (dark column) and potato juice (light column) from our pilot plant.

In Estonia, 80,000 tons of potatoes are processed into starch yearly. This could yield 120,000 tons of liquid feed as a byproduct, saving 6000 tons of concentrate. As only 10% of the potato juice and pulp of the whole republic are utilized in our pilot plant, 900 tons of beef or 5400 tons of milk could have been produced annually with the feed wasted up to now.

CONCLUSIONS

With the traditional starch manufacture technology, 13% of the pollution load originates from washing and transportation of the tubers and 87% from potato juice. Potato juice used for liquid feed need not be a waste product. That means that the pollution load for every ton of processed potatoes can be reduced from 800 population equivalents to 125 p.e.

The most suitable purification plant is an aerated biopond.

The building cost for the purification systems is significantly reduced also, from 20,000 rubles to 3100 rubles/ton of processed raw material.

In our pilot plant, the yearly economic benefit from production of the new byproduct is 35,420 rubles and for lowering the costs for building the purification systems is 168,000 rubles. In the whole republic the yearly benefit could be 9 million rubles (Kirt and Merisaar 1988).

REFERENCES

Kirt, E. and M. Merisaar. 1988. "New technological process in food industry." *Toiduainetetööstus*, No. 26, pp. 44-58. (in Estonian).

Talley, E.A., Toma, R.B., and P.H. Orr. 1984. "Amino acid composition of freshly harvested and stored potatoes." *American Potato Journal*, Vol. 61, No. 5, pp. 267-280.

Chapter 34

THE ECOLOGICAL DEVELOPMENT OF THE PÄRNU SEWERAGE AND TREATMENT SYSTEMS

Udo Asmus and R. Vaba

INTRODUCTION

Pärnu is one of the best-known health resorts of our republic and of the former Soviet Union. That is the reason why much attention has been paid to the improvement of the conditions of the local environment. Since the importance of the town as a center of fishing industry is also significant, the specialists who have planned the environmental protection measures have had to take into account all the specific factors that influence the conditions of the Pärnu Bay and the Pärnu River (see Figure 1).

THE ESTONIAN PROJECT FOR INTEGRATED WASTEWATER TREATMENT

At the beginning of the 1980s, the Eesti Projekt made a unique proposal for directing the remaining slurry of the anaerobically treated manure from a pig farm to an extended biological treatment system of the town. As the interval in the phases of the construction of the treatment plant has been 2 years, we have to be content with the situation where the liquid manure is conducted to the fields, already over-fertilized with nitrogen (more than the standard 300 kg N/ha/year). At the same time, we have to apply for more land (as the plant wishes) and hope that the peat bogs will be able to take care of more nutrients during a shorter period of time. The flora promoted by the slurry will probably change back into the ordinary one in 2 to 3 years. (See also Mander et al. 1997 on moor-filters.)

The other purpose of the co-treatment is to use up all the sludge so that the circulation of all substances would be closed and that this would be done at reduced costs and with minimal nutrient losses.

RESULTS

According to the latest information, we can state that since September 1990 the nitrification-denitrification process has been established in the biological stage of the Pärnu sewage treatment plant. The process was started at half load in July and worked at full capacity of 25,000 m^3/d in September and 30,000 m^3/d in November 1990. The reduction efficiencies for September 1990, registered by the engineer in charge, R. Vaba, were 90% for suspended solids, 98% for BOD_7, 70% for P_{tot}, and 100% for NH_4-N (17% was nitrified and analyzed as NO_3-N in the effluent).

0-87371-990-5/97/$0.00+$.50

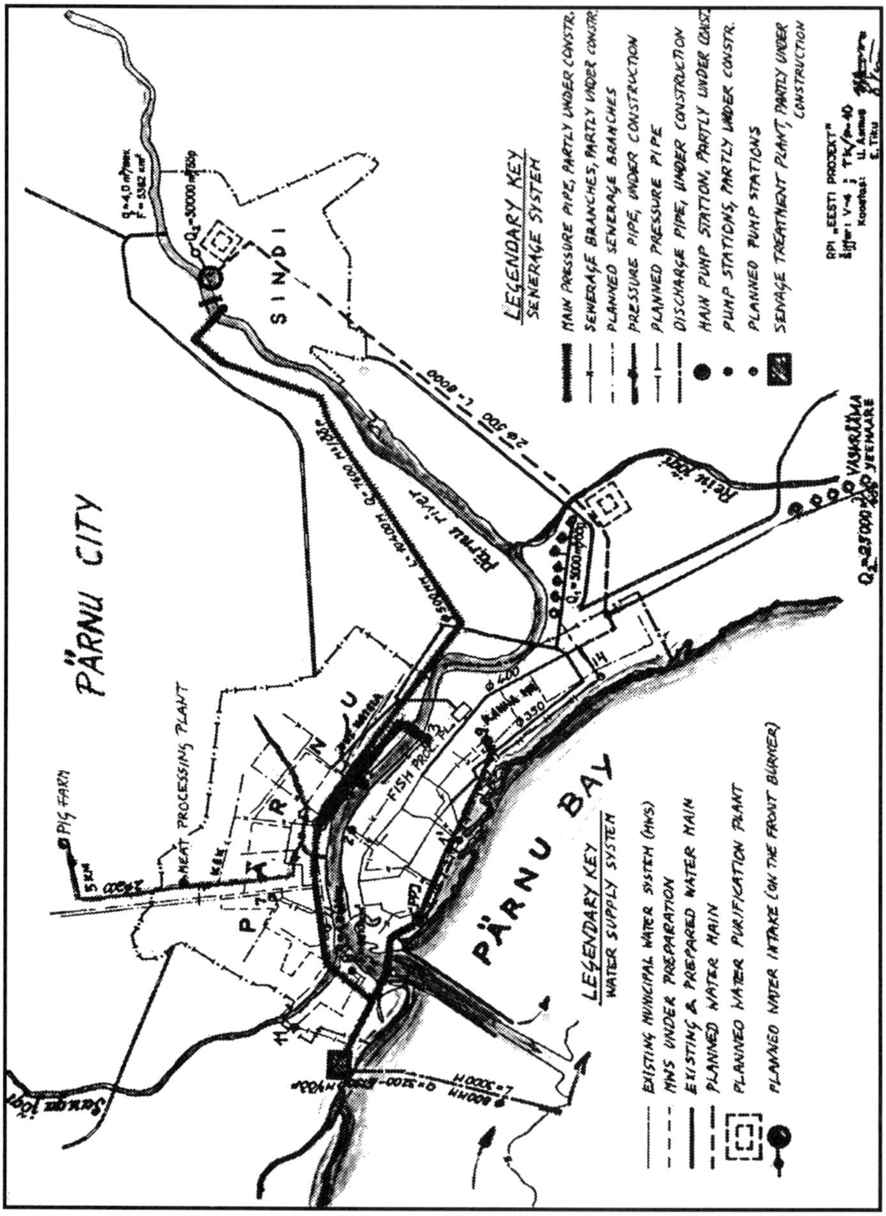
PÄRNU CITY
PIG FARM
MEAT PROCESSING PLANT
FISH PROC. PL.
PÄRNU BAY
Pärnu river
SINDI
PÄRNU
LEGENDARY KEY
SEWERAGE SYSTEM
MAIN PRESSURE PIPE, PARTLY UNDER CONSTR.
SEWERAGE BRANCHES, PARTLY UNDER CONSTR.
PLANNED SEWERAGE BRANCHES
PRESSURE PIPE, UNDER CONSTRUCTION
PLANNED PRESSURE PIPE
DISCHARGE PIPE, UNDER CONSTRUCTION
MAIN PUMP STATION, PARTLY UNDER CONSTR.
PUMP STATIONS, PARTLY UNDER CONSTR.
PLANNED PUMP STATIONS
SEWAGE TREATMENT PLANT, PARTLY UNDER CONSTRUCTION
LEGENDARY KEY
WATER SUPPLY SYSTEM
EXISTING MUNICIPAL WATER SYSTEM (MWS)
MWS UNDER PREPARATION
EXISTING & PREPARED WATER MAIN
PLANNED WATER MAIN
PLANNED WATER PURIFICATION PLANT
PLANNED WATER INTAKE (ON THE FRONT BURNER)
RPI „EESTI PROJEKT"

FIGURE 1

DISCUSSION

We must not forget that the load will increase up to 150 to 170% because of additional milk and fish processing wastewater in 1 to 2 years and owing to pre-treated, digested pig manure a little later. The design for this high load was tested in November 1990 when one of the four sections was emptied for reconstruction of the aeration system and changing to disc aerators. The process went on with less nitrification caused by switching off one of the two aerators. The retention time was 14 hours. As a result, no reagents were used and the efficiency remained still high.

The development of the above methods for sewage treatment and sludge processing will give enough hope for normalization of the sanitary conditions in the whole region, if only the following circumstances are taken into account:

1. It is necessary to remove the substances hindering the treatment process, such as fat, fibers, and textile scraps, already at the source. For that purpose probably some imported devices will be needed.
2. The separation rate of the pig farm slurry should meet the designed values so that after irrigation (limited at 300 kg N/ha/year) the remaining matter could be delivered to the municipal sewage system and not to the bogs anymore. The sludge cake will be composted and recycled again to the fields as fertilizer.
3. In times of need, the storm water can go from the town districts through the partially operational sewerage to the treatment plant, which in the future will have some hydraulic reserves in each element of the process and at the same time a very high purification level. Some of the rain water will overflow with sewage to a reed bed area where self-purification will take place. The loads which can be allowed must be tested.

REFERENCES

Mander, Ülo, M. Olev, and U. Nugin 1996. "Perspectives on vegatated shoals, ponds, and ditches as extensive outdoor systems of wastewater treatment in Estonia." In: C. Etnier and B. Guterstam (eds.), *Ecological engineering for wastewater treatment: Proceedings of the international conference, 24-28 March 1991, Stensund Folk College, Sweden.* Boca Raton, FL: Lewis Publishers.

Chapter 35

ECOLOGICAL EFFECTS OF HUMAN ACTIVITY ON THE COURIC LAGOON AND THE COASTAL ZONE OF LITHUANIA

Kirlys Vaizgantas and Laima-Lilija Lazauskiene

On the basis of the analysis of material collected during complex scientific research work, we have made an evaluation of the effect of building and using of the international railway ferry from the former USSR to the former German Democratic Republic on the ecosystem and fish reserves in the Couric Lagoon (Kurshiu Marios) of the Baltic Sea. We have determined the effects of ferry traffic on the changes in bottom relief, as well as the hydrodynamics of the Klaipeda Channel. The changes in the dynamics of suspended matter in the strait were monitored. Hydrochemical and geochemical parameters have shown that, as a result of ferry traffic in the Klaipeda Strait, the area of distribution of sludge sediments changes and the transition of a number of pollutants into suspension intensifies. Bottom sediments in many cases exceed the maximum permissible concentration. Among organic pollutants, 50 to 60 toxic substances were identified, undoubtedly exerting a negative impact on the ecosystem of the Couric Lagoon.

During the railway ferry's construction and use, there occurred marked changes destabilizing, to a certain extent, the equilibrium of natural processes in the northern part of the Couric Lagoon and particularly in the Klaipeda Channel.

Interrelation of changes of physical processes with those of biota makes it possible to determine the intensity and trend of these processes, to estimate ecological effects and consequences under the impact of ferry traffic. Calculations were made on the changes in the dynamics of aquatic populations and the structure of biocenoses. The ferry traffic was shown to affect the structure of populations and communities of the ichthyofauna (the fish), their migration, spawning, vital activity, and productivity.

We have evaluated the ecological and economic damage of ferry traffic on the ecosystem of the Couric Lagoon. Recommendations and nature conservation measures for the regeneration of ecological equilibrium in the Klaipeda Channel and the coastal zone of the Baltic Sea have been established.

0-87371-990-5/97/$0.00+$.50

Chapter 36

ECOLOGICAL ENGINEERING FOR INCREASED PRODUCTION AND ENVIRONMENTAL IMPROVEMENT IN OPEN SEA AQUACULTURE

Nils Kautsky, Max Troell, and Carl Folke

ABSTRACT

Aquaculture is assuming an increasingly important role in the world's food production, but its continued development is now becoming restricted by lack of cheap feeds and suitable environments and increasing pollution from aquaculture itself.

It is generally not percieved that the cultivation of seaweeds, mussels, and fish are ecologically completely different, putting entirely different demands on the environment, and affecting the ecosystem in completely different and even opposite ways.

We compare the ecological resource requirements, the wastes generated, and the environmental impacts of seaweed culturing, long-line mussel rearing, and salmon cage-farming in coastal waters. Basic ecological engineering principles are discussed through a comparison of conventional throughput-based monocultures with an integrated recycling-based culturing system of seaweeds, mussels, and salmon.

INTRODUCTION

Commercial aquaculture is usually practised as large monocultures. Like all types of monoculture, large-scale, and especially intensive aquaculture, will result in negative environmental impacts, such as eutrophication, oxygen depletion, and pollution of the surrounding waters, and make these less suitable or even unusable for other purposes including further culturing and harvesting of natural stocks (reviews in Gowen and Bradbury 1987, Braaten et al. 1988, Folke and Kautsky 1989, Troell et al. 1994). This is a common feature of all economic activities which concentrate resources that have been collected over vast areas and which use them in a linear fashion without managing the feedbacks with the environment. If this concentration of resources is not recognized, there will be waste problems, which only in the short term and only to some extent can be reduced by the selection of sites with a good water exchange, by changing feeding practices and so on (Folke 1988, Folke and Kautsky 1989, 1991, 1992).

We know from China, for example, that a sustainable food production via aquaculture can be achieved by carefully combining different types of aquatic and terrestrial cultivation systems. In this way it is possible to optimize food production and minimize negative impacts in the area (Li 1987, Ma and Yan 1989, Liu 1987).

0-87371-990-5/97/$0.00+$.50

Our idea is to diversify and integrate aquaculture activities in coastal areas, analogously to what the Chinese have been doing for a long time in integrated pond culturing. As aquaculture may take place on all trophic levels, it is possible that wastes from one type of cultivation can be used as a resource in others, if properly managed. A challenge for aquaculture is to develop cultivation systems that are not only resource efficient and waste minimizing, but also contribute to enhancing the productivity and health of surrounding ecosystems.

COMPARISON OF RESOURCE UTILIZATION AND WASTE GENERATION IN DIFFERENT TYPES OF COASTAL SEA CULTIVATION

It is generally not understood that the cultivation of seaweeds, mussels and fish are ecologically completely different, putting entirely different demands on the environment, and affecting the ecosystem in completely different and even opposite ways (Kautsky and Folke 1989, 1990). Below we briefly discuss the effects of solar energy, dissolved nutrients, plankton, and particulate matter on the three different monoculture systems as well as on a proposed integrated culturing system, and how these systems modify these environmental factors. The responses to additions of feed pellets in fish farming are also indicated. The discussion is summarized in Table 1.

SEAWEED CULTIVATION

In seaweed cultivation, the efforts by man are usually limited to initial efforts of introducing seed plants to the cultivation area, and in some cases to provide a substrate. The main limiting factors for seaweed and algal production are solar energy and nutrients. Few wastes are produced by increasing production at the first trophic level. Increased nutrients in the environment will increase production, but this may lead to a shift in production from seaweeds to phytoplankton, since phytoplankton have a more efficient nutrient uptake which may lead to a decrease in light for the seaweeds (Kautsky 1990). Seaweed cultivation will reduce the amounts of dissolved nutrients in the water. During the entire cultivation period, including harvesting of the seaweeds, there is a removal of dissolved nutrients from the system.

MUSSEL LONG-LINE REARING

The cultivator sets out a substrate for the attachment and growth of mussels, after which the mussels need little attention until harvesting. The origin of nutrients in food is from the naturally-occurring plankton and particulate organic matter, which the mussels consume when filtering the water passing through the culture. Their filtering activity keeps the water clear. Some of the nutrients that are consumed are released as dissolved nutrients through the metabolism of the mussels. In contrast to a fish farm, the nutrients released by the mussels thus do not represent a net addition to the environment (Kautsky and Wallentinus 1980). Even if mussel excretion slightly increases the concentration of dissolved nutrients in the local water area, there is always a decrease in the total nutrient content in the water due to the removal of plankton through mussel filtration (Kautsky and Evans 1987, Folke and Kautsky 1989). Just as with seaweeds, the net effect of a mussel culture will always

TABLE 1

Site-specific inputs needed and outputs produced by seaweed, mussel, and fish cultures, as well as when these are integrated. The effects of solar energy, dissolved nutrients, phytoplankton, particulate organic matter (POM), work, and feed inputs on the different systems, and how these environmental factors are modified by the different types of cultivation are indicated. The last two columns show that the net effect of integrating the cultures will be a more efficient use of resources and less impact on the environment.

	Seaweeds		Mussels		Fish		Integrated Culture	
	Effect on cultivation	Addition or removal from environment	Effect on cultivation	Addition or removal from environment	Effect on cultivation	Addition or removal from environment	Effect on cultivation	Addition or removal from environment
Solar energy[1]	+	0	0	+	0	– –	+	0
Dissolved nutrients[2]	+	–	0	– (+)	0	+ +	+	0
Phytophankton[3]	– –	0	+ +	–	0	+ +	+	0
POM[4]	–	0	+ +	–	0	+ +	+	0
Work effort[5]	– –		– (+)		+ +		– +	
Feed input	– –		– –		+ +		– +	

Notes:

1. Solar energy has a positive effect on seaweed growth and on the integrated culture. Mussels improve the light quality in the water by removing particles, while fish cultivation decreases light penetration by adding to turbidity through production of POM and stimulation of algal blooms.
2. Nutrient inputs are positive for seaweeds and the integrated culture. They are added by fish but removed from the environment by seaweeds and at the harvest of mussels. Mussels may cause a temporary increase in dissolved nutrients in the water by excretion of digested plankton.
3. Phytoplankton competes with seaweeds for light and nutrients. They are used as food by mussels, which remove them from the environment. Excretion of nutrients from fish and mussels will increase phytoplankton in the environment.
4. Particulate organic matter is negative for seaweeds through shading, but constitutes food for mussels which clear the water from particles. Fragments of pellets and faeces in fish cultivation add POM to the environment.
5. Seaweed and mussel cultivation are more or less self-sustaining systems and only demand initial work. A fish cultivation is during the whole cultivation cycle dependent on inputs of work, i.e., feeding, cleaning, and medication. Integrated cultivation will also, but to a lesser degree, depend on regular input of work, i.e., feeding, cleaning, and maintaining the proportions of the different species to obtain optimal recirculation in the system.
6. Seaweed and mussel cultivation do not depend on external inputs of feed, as fish cultivation does, but instead utilize naturally occurring food and nutrients. In integrated cultivation, there will be an initially large dependence on external feed inputs to the fish, but when optimum production is reached, the recirculation of mussels as feed will decrease this input.

be a reduction of the pool of nutrients in the area, since substantial amounts are removed from the ecosystem at harvest.

FISH CAGE FARMING

While the human input to seaweed cultivation and mussel culturing is essentially limited to initial efforts in the provision of substrate, a fish cage farm, apart from needs for cages and seed fish, is heavily subsidized by inputs through continuous additions of feed and medicinal drugs (Folke and Kautsky 1989). Since the raw material for fabricating the feed pellets usually originates from other bodies of water, this results in a net addition to the environment of dissolved nutrients (Folke et al. 1994) and of organic matter in the form of feed fragments, particulate matter and fish faeces (Gowen and Bradbury 1987, Hall et al. 1990). Usually the environment around fish cages is characterized by very turbid water. If the wastes from the culture are not removed by currents or assimilated by natural ecosystem processes, water quality may deteriorate, leading to oxygen depletion, accumulation of toxic substances and promotion of diseases, which may even threaten the culturing activities themselves (Braaten et al. 1988, Troell et al. 1994).

INTEGRATED OPEN SEA AQUACULTURE

In general, the more a cultivation system resembles and fits with the processes and functions of ecosystems, the less one can expect undesirable environmental effects (Folke and Kautsky 1992). By introducing parts of food webs in the culture, as is the case in some parts of Asia, ecological complexity is increased (Yan and Yao 1989, Wang 1990, Zhao 1992). Such cultures will function more as natural ecosystems, which means that energy subsidies and costs can generally be reduced per unit of production (Mitsch and Jørgensen 1989). By creating feedbacks between the culturing and the ecosystem, such as recycling of waste products, environmental effects will be considerably reduced.

In Figure 1 we have suggested how different types of coastal sea cultivation could be combined and integrated in coastal sea aquaculture. Nutrients originating in land run-off, e.g., from agriculture, lead to increased production of both seaweeds and plankton in the sea. The plankton would stimulate the production of mussels in the area, which would at the same time reduce algal blooms and turbidity in the water. Fish cage farming in the area would further increase mussel production as mussels usually grow better adjacent to salmonid farms where they can utilize particulate feed fragments from pellets (Wallace 1980). The mussels would also tend to keep the water clear around the fish cages and reduce sedimentation.

Fragments of feed pellets added to the fish culture and fish waste would be filtered from the water by the mussels, which could then be harvested and partly recycled as fish feed. The higher the proportion of mussels used in feed, the lower would be the release of new nutrients, since less nutrients will be imported from other areas. The seaweed cultivation would remove part of the remaining dissolved nutrients.

In the case of integrated culturing, it is important to emphasize that the yield and culturing of each species must not be evaluated and compared with a monoculture, in isolation from the other species. It is the total ecological economic benefits

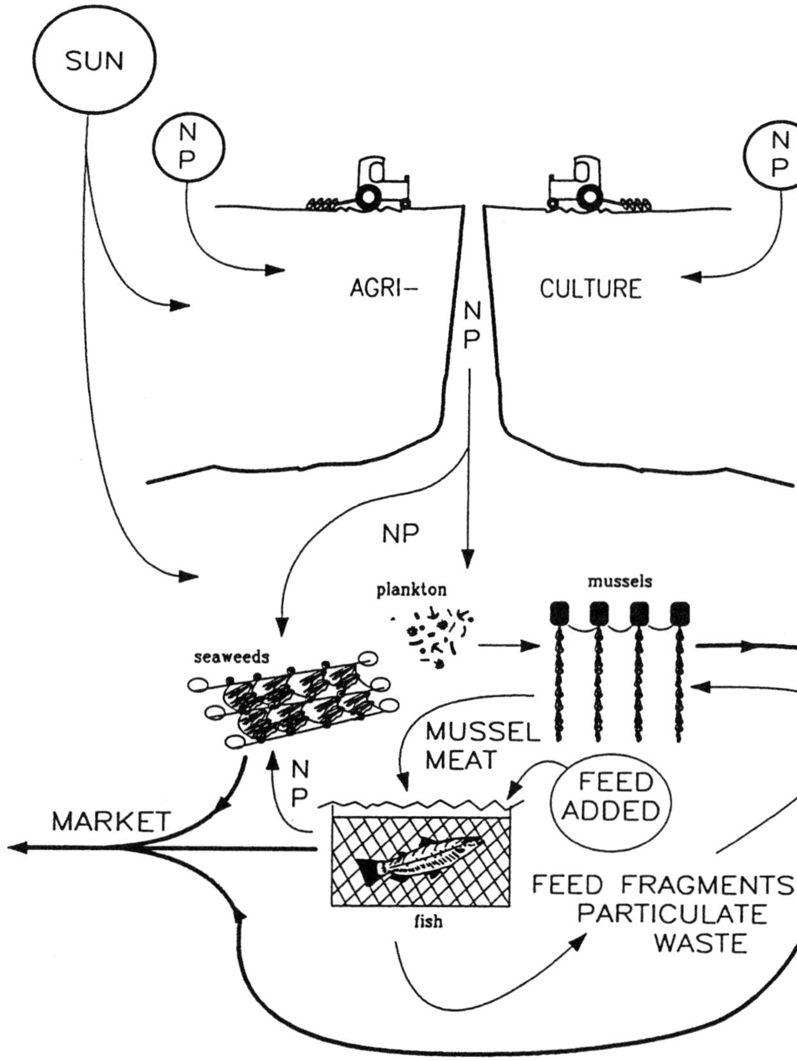

FIGURE 1. A tentative suggestion for a combined aquaculture system of seaweeds, mussels and salmon, where wastes from one type of culture are used as resources in the other. This would not only reduce waste problems and other environmental impacts, but it would also contribute to the improvement of environmental quality by transforming excess nutrients from domestic sources into biomass and marketable products. N = nitrogen, P = phosphorus

of the whole integrated system that needs to be evaluated, and whether integrated culturing would lead to an increased yield/unit resource input.

If such integrated aquaculture systems could be placed in coastal areas where there are excess nutrients from domestic sources, these nutrients could be utilized by the culture, and be transformed into useful biomass which is removed from the system at harvest. If this is possible, the culture would improve environmental quality instead of contributing to environmental degradation, as is the case with the presently applied farming technology (Kautsky and Folke 1989; Folke and Kautsky 1991, Folke et al. 1994).

Ecologically, there are no arguments against this, but of course practical feasibility tests need to be carried out. For a successful management of integrated coastal aquaculture, the ecology of culturing needs to be understood. This includes information on abiotic limitations and biological characteristics of the species. Critical problems will be to dimension the different types of cultivation to each other, and to phase them with annual dynamics, taking into account harvesting times, etc. Since it might be possible to dimension mussel and fish cultures to each other, seaweed should probably be cultivated to an extent which takes care of dissolved nutrients, and the size of the mussel culture should also be related to the amounts of plankton and suspended organic matter in the surrounding waters. Our plans are to empirically test the ideas of an integrated seaweed-mussel-fish culture in coastal waters of Chile.

CONCLUSIONS

As is well known, present production systems have mainly focused on output, that is the yield, with consequences of increased pollution of the surrounding environment, a fact that is very true for the rapidly expanding salmonid cage-farming industry. Integrated aquaculture of fish, mussels, and seaweed, based on ecological engineering, will more resemble natural ecosystems, which means that energy subsidies and costs, and environmental impacts, can be reduced per unit of production. Such efficiency-increasing development is very different from the throughput-increasing growth we have seen especially in the intensive salmon- and shrimp-farming industries during the last decade. It is of importance to try to develop a sustainable pattern of aquaculture production. In our opinion this is only possible by explicitly making use of ecosystem processes and functions, by integrating the culturing to these processes and functions, in other words, by applying the basic principlies of ecological engineering.

ACKNOWLEDGMENTS

This study was performed with partial support from SAREC (Swedish Agency For Research Cooperation with Developing Countries), and the Swedish Council for Forestry and Agricultural Research.

REFERENCES

Braaten, B., Poppe, T., Jacobsen, P. and K. Maroni 1988. "Risks from self-pollution in aquaculture: Evaluation and consequences." In E. Grimaldi and H. Rosenthal (eds.), *Efficiency in aquaculture production: Disease control.* Edizione del Sole. Seme S.p.A, Milano. 24 ORE. pp. 139-165.

Folke, C. 1988. "Energy economy of salmon aquaculture in the Baltic Sea." *Environmental Management* 12:525-537.

Folke, C. and N. Kautsky. 1989. "The role of ecosystems for a sustainable development of aquaculture." *Ambio* 18:234-243.

Folke, C. and N. Kautsky. 1991. "Ecological economic principles for aquaculture development." In. C.B. Cowey and C.Y. Cho (eds.) *Feeding fish in our water: Nutritional strategies in management of aquaculture waste.* Guelph, Ontario: University of Guelph Press. pp. 207-222.

Folke, C. and N. Kautsky. 1992. "Aquaculture with its environment: Prospects for sustainability." *Ocean and Coastal Management* 17:5-24.

Folke, C., N. Kautsky, and M. Troell. 1994. "The cost of eutrophication from salmon farming: Implications for policy." *Journal of Environmental Management.* 40:123-182.

Hall, P.O., L. Andersson, O. Holby, S. Kollberg, and M-O. Samuelsson. 1990. "Chemical fluxes and mass balances in a marine fish cage farm; I. Carbon." *Mar. Ecol. Prog. Ser.* 61:61-73.

Kautsky, L. 1990. Factors limiting seaweed production. In E.C. de Oliveira and N. Kautsky (eds.) *Cultivation of seaweeds in Latin America.* IFS workshop: São Paolo 1989. pp. 9-16.

Kautsky, N. and C. Folke. 1989. "Management of coastal areas for a sustainable development of aquaculture." *Biota* 5:1-11.

Kautsky, N. and C. Folke. 1990. "Environmental and ecological limitations for aquaculture." In R. Hirano and J. Hanyo (eds.), *Proceedings from the Second Asian Fisheries Forum, Tokyo April 17-22, 1989.* pp. 245-248. Asian Fisheries Society: Manila, Phillipines.

Kautsky, N. and I. Wallentinus. 1980. "Nutrient release from a Baltic *Mytilus*-red algal community and its role in benthic and pelagic productivity." *Ophelia Suppl.* 1:17-30

Kautsky N. and S. Evans 1987. "Role of biodeposition by *Mytilus edulis* in the circulation of matter and nutrients in a Baltic coastal ecosystem." *Mar. Ecol. Prog. Ser.* 38:201-212.

Li S. 1987. "Energy structure and efficiency of a typical chinese integrated fish farm." *Aquaculture* 65:105-118.

Liu C. 1987. Rice-azolla-fish system. Fujian Academy of Agricultural sciences. (FAAS) Fuzhov, Fujian, China. Project Report presented at Third Asian Fisheries Forum, 26-30 Oct. 1992. Singapore.

Ma S. and Yan J. 1989. "Ecological engineering for treatment and utilization of wastewater." In W. Mitsch and S.-E. Jørgensen (eds.), *Ecological engineering: An introduction to ecotechnology.* New York: Wiley Interscience.

Mitsch, W. and S.-E. Jørgensen (eds.) 1989. *Ecological engineering: An introduction to ecotechnology.* New York: Wiley Interscience.

Troell, M., N. Kautsky, and C. Folke 1994. "Integrated aquaculture for reduced environmental impacts and improved resource use." In K. Koop (ed.) *Ecology of marine aquaculture*, pp. 105-114. Stockholm: International Foundation for Science.

Yan J. and Yao H. 1989. "Integrated fish culture management in China." In W. Mitsch and S.-E. Jørgensen (eds.), *Ecological engineering: An introduction to ecotechnology.* New York: Wiley Interscience.

Wallace, J. 1980. "Growth rates of different populations of the edible mussel, *Mytilus edulis,* in north Norway." *Aquaculure* 19:303-311.

Wang J-K. and G. S. Jacob. 1990. Pond design and water management strategy for an integrated oyster and shrimp production system. Paper presented at Third Asian fisheries forum, 26-30 Oct. 1992, Singapore."

Zhao W. 1992. "A polyculture system of *Macrobracium rosenbergii* with silver and bighead carps." Paper presented at Third Asian Fisheries Forum, 26-30 Oct. 1992 Singapore.

Chapter 37

THE PERSONAL LABORATORY: THINKING TOOLS

Jorge de Sousa Pires

ABSTRACT

Around 1900, everybody who wanted to buy and drive a car would have had to know a lot about cars, car engines, tires — this knowledge was a necessity. Today, many car drivers hardly know where the motor is. The more advanced things are the less knowledge is necessary about their inner workings — this is true of the telephone, indeed, of any device in the home.

Personal computers have been around for only ten years but they are already very advanced. However, in people's minds, computers are still computers, i.e., devices that compute things, and which need be programmed. Nothing could be more wrong. Today's personal computers are so advanced that you do not need at all to know anything about how they work to use them successfully. Moreover, most computers today are used for word processing, not for computations!

Ecologists, to mention one group, patiently collect a lot of data on the Environment and this amount of data will grow — but the important issue is not to collect more and more data, but rather to analyze and distill this information into knowledge.

The paper will focus on how to use a computer to extract knowledge from the collected data through statistics, graphics, rotating plots, modeling, presentation . . .

BACKGROUND

For years, computers have been used as powerful calculation machines, dedicated to complex tasks such as solving differential equations or calculations of trajectories of missiles. The advent of the personal computer has brought an incredible and powerful resource to the desk of the common man. Unfortunately, "incredible" is the word here, as few people have realized or believe that they have this much power at their fingertips. Very few authors have engaged in a true, efficient way of dealing with computers and real knowledge or useful tools, but one such interesting author is Bork (1985). Bookshelves are filled with strange hints about learning about the World through LOGO, through Pascal or through other strange techniques which will never be adopted by an intelligent and time-efficient human being.

Still, many universities have not realized this, and there are actually very few places in the world where there is a course on "How to Solve Problems with a Personal Computer." It is taken for granted that a course on personal computers is the same as teaching students how this or that operating system works and that word

0-87371-990-5/97/$0.00+$.50

processing is what students need a personal computer for. Worse, most universities think that solving problems with a personal computer is the same as solving them with large computers and this invariably leads to more courses on FORTRAN, Pascal, or C.

An alternate but likewise unfortunate state of matters comprises the never-ending development of small programs to show this or to show that — mostly programs without any interest for the student to adopt as a tool for life and so the whole thing is forgotten as soon as that particular course has come to an end. These small programs are invariably constructed in more Pascal or in the "authoring language of the day" — some people even manage to get financed to develop a new authoring language! From a national point of view, this is a waste of time and money (and teacher capacity).

THE TASK OF A TEACHER

The problem with all of this is that teachers may come to easily forget their task. First, because they become entangled with the intricacies of programming for a long time, and, secondly, because much of this work is bound to remain with that particular teacher without any impact on teaching (most often not even used by colleagues at the same department) — a waste of time and energy. So, what is the task of a teacher? Clearly, it is not to embellish this or that operating system — a task for the software house which developed that particular system. *I maintain that the task of the teacher is to, at all times, pick the best tools available to teach his or her discipline*. Not the simple things of the discipline — what needs effort (and a computer) are the most difficult things of that discipline. However, when it comes to personal computers, the lack of maturity is obvious.

There is the issue of which manufacturer, which brand of computer. But is this an issue? Certainly not for a teacher, rather for marketing people. What you need is not primarily the computer that is accepted by the industry, rather the one with the lowest threshold for teaching and learning — your task is still to teach your discipline.

If teachers are to be the resource for society we like to say we are, then we must think for ourselves. Unfortunately, when it comes to personal computers, teachers have somehow "accepted" being led by the marketing of smart software houses.

The first example in history was that of spreadsheets — they were marketed towards economics, so most technical people discarded them, because "they were not good" tools for mathematics, physics, etc. They were discarded not because of actual knowledge, but because marketing said they were for economics. Much evidence exists today that this is not at all so — spreadsheets may be used for just about anything, be it to teach/learn about the "conjugaison" of French verbs or for advanced research in Semiconductor Physics.

The second proof of this line of reasoning regards graphical interfaces — strong marketing was directed against the Macintosh, the Atari, etc. The saying was that command languages were faster and more direct (clearly against all evidence provided by cognitive science since the 1960s). Now, in 1990, marketing again brings out Windows — and then people "accept" the new truth of the day. I sustain that this is very, very dangerous for the university as a whole, because the truth has been provided by the research of cognitive science for a long time now and that is what we should commit to!

Summarizing, I advocate and suggest that you: a) do not develop just another program, b) do not deal with small program surrogates, c) try to give students a tool for life and d) do not think "through" marketing departments.

SOLVING PROBLEMS WITH PERSONAL COMPUTERS

No Pascal, no programming, no authoring languages . . . how can one then use a personal computer? And yes, that is the question — actually I would say, it really is the issue. A personal computer is a very strong amplifier of the capabilities of the common wo/man and should be used for that purpose much more extensively. However, teachers at all stages in the education system have never been told (who has?) how to use the personal computer for improved teaching and students have never been told (or discovered) how to use the personal computer for improved learning.

So, at one end of the scale, advertisements are there bragging about those extreme megahertz, making people dwell a lot on irrelevant discussions; once the computer is brought home, it will eventually be used for . . . word processing (at 25 MHz!).

To trigger a discussion on these matters, I will put forward a set of examples as showcases.

Knowledge is a complicated issue, and its use is much dependent on the surrounding society. Knowing how to start a fire by rubbing sticks together is quite irrelevant for modern western society, but could be extremely useful in the Amazonian forest.

Modern society, with its word processors and copying machines and a multitude of authors and editors, produces thousands of printed lines every minute. Much of this will eventually reach the desk of the reader. The biggest challenge is to discard as much of this information noise as possible but still retain the important information, extract knowledge and keep up with progress.

Decision-making, for one thing, relies on knowledge (at least *careful* decision-making). But societies are becoming highly technical, increasingly complicated and very much interdependent. Ruling one country cannot be achieved in isolation — most economies are now so interwoven that it is really difficult for a limited number of persons to decide on what is best; ministers have been known to overlook things . . . However, the men are not to blame at all times: it is a complicated World, and the best we can do is to try to take into consideration as many factors as possible. Unfortunately, mankind has not been made for this. Our cognitive processes, however remarkable and extremely capable, cannot cope with so many things at the same time (try predicting 10 moves ahead the next time you play chess).

The personal computer brings some of the rescue. We can simulate decisions and see what the effects will be. We can use statistics, history (a company's income and expenses) and, together with a robust model, try to predict the near future or see whether there is an embodied oscillation that, although not very easy to detect, is present and might bring the system to complete failure (bankruptcy).

Here again we must be careful. A simple line regression should be used with great care for extrapolation. A line does not per se predict anything, it just averages history. So, it is very necessary to know your tools — a strong case for a university!

Computer systems have been totally unimportant for decision-making (has your personal computer ever helped you with any decision?) and I sustain that the main reasons for this fact are partly because they do not have enough data about your life and partly because they are extremely precise. Generally speaking, a computer needs an equation to solve any task. But decision-making makes do without equations, rather with a feeling for this and a feeling for that.

The facts are already there, the figures, the incomes and the expenses. What we really need now is a tolerant system that can cope with intuition, feelings and such *imprecise* knowledge.

EVOLUTION OF COMPUTERS TO THINKING TOOLS

We have had computers for *calculation*. Then, mankind realized that computer memory is good to store lots of data; so we created *databases* and *database managers*. Printers are good output devices, they are patient and can gladly print one hundred letters filled with more or less nonsense; so we created *word processors*. Nowadays, personal computers are used for word processing most of the time. Computers are not used mainly for calculation any more, they are there to generate information noise and (hopefully) knowledge.

I believe the next stage is using the personal computer as a *Thinking Tool*. Complex things, huge amounts of data, visualization tools in 2D, 3D, 4D . . . image processing, hypertext, multimedia — a conglomeration of techniques that blend together to enable the knowledge worker to sit and think in an efficient manner. Storing lots of data of (past) history may also solve another interesting problem: you could pass on much of your experience and knowledge to your children. Not in the biological database, the chromosomes, rather in an electronic database! (This is what AI — artificial intelligence — has been promising us for 20 years now.)

As we become increasingly dependent on high quality knowledge, we must be able to access that and at the touch of a finger.

SIMULATION

Simulation is what comes to mind as a first step of Thinking Tools. Simulation is quite a good tool for understanding, especially if the thinker "programs the model" from the very beginning. This fact was realized a long time ago and most of the pedagogical uses of programming languages reflect this wish: he who programs must know what to do.

However, programming languages are so inferior and so far from our cognitive processes, and the thinker becomes so entangled with the intricacies of the programming language (LOGO, Pascal, C . . .) that the problem at study quickly falls to the background. In any case, this is a very inefficient way of learning and has never been adopted by any student outside a course. His teacher most certainly does not learn new things this way either, so the use of a programming language must always be considered more or less a serious waste of time for learning.

General simulation packages, so called content-free software, provide an altogether new way of dealing with new knowledge. A spreadsheet may well be learned from an early age and used in almost any discipline — this should become second

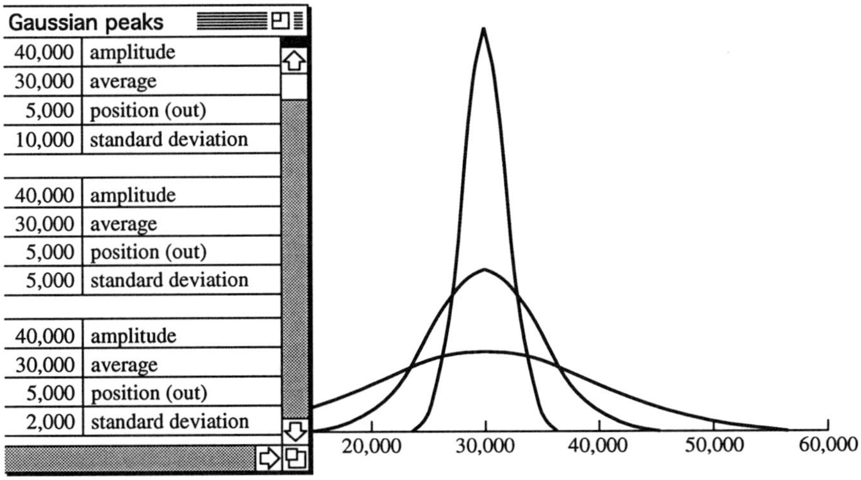

FIGURE 1. Gaussian peaks with different standard deviations. Your Personal Laboratory of Statistics — change values at left and peaks will change accordingly.

nature to the user. The same spreadsheet program may be used to compile and learn facts of history and in much the same way to deal with geography or facts about the environment (pH, temperature, time, places . . .). These applications do not necessarily involve so many calculations, but interesting computations can be made upon such data.

The same spreadsheet program may then be used in much the same way to learn statistics. Suppose the researcher (student) is to understand what a Gaussian Population is, a distribution given by the equation of the Gaussian distribution:

$$f(x) = \frac{e^{-(x-\mu)^2/2\sigma^2}}{\sigma\sqrt{2\pi}}$$

where μ is the average and σ its standard deviation. It is a simple matter to write such an equation and then immediately start playing with both μ and σ to understand what these quantities represent. Everything on-screen, without complications, without any programming.

WORKBENCH, EXCEL, STELLA AND MACSPIN . . .

A wise graphics environment makes difficult things simple. Collecting data with a personal computer becomes simple and efficient: you use symbols on-screen and have much more time to discuss what good measuring techniques are about: sensors, shielding, correlation, averaging . . .

In this category, the WorkBench program represents the state of the art in simplicity. WorkBench generates rows of numbers and can graph them as well (see Figure 2). Moreover, exporting the data to a general package, such as the Wingz spreadsheet, enables the researcher to obtain graphs at the click of a mouse:

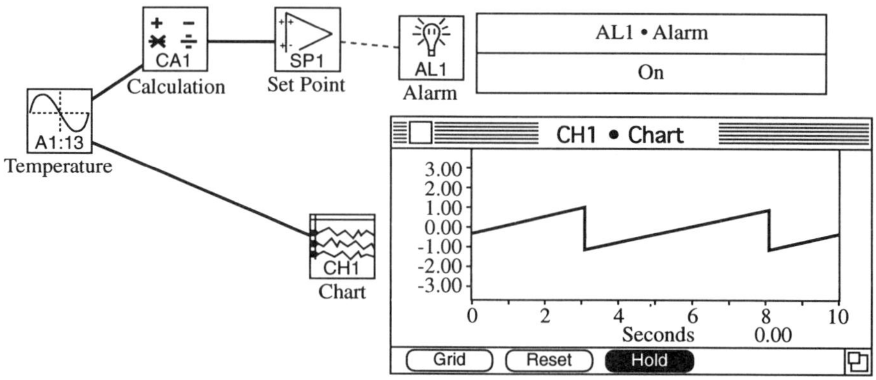

FIGURE 2. Using WorkBench for data collection. Data is on-screen within seconds — to be compared with hours for conventional methods. And no programming!

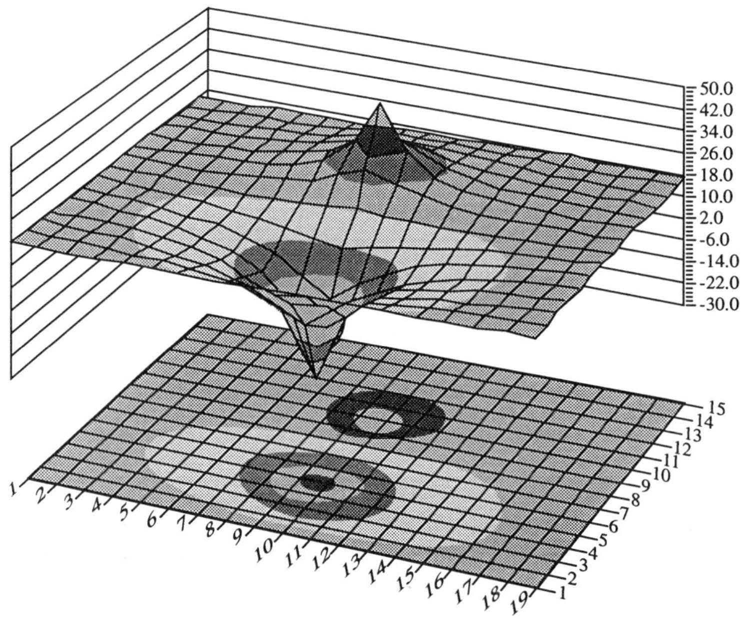

FIGURE 3. The temperature gradient of a body (a lake, etc.), modelled or measured in a spreadsheet (Wingz). It is in full color, in 3D and in a Spreadsheet!

But there are other general tools: the revolutionary Stella where you solve differential equations by *painting* them on-screen! This is a very important aspect of modern personal computing and I would like to elaborate a little further on the issue: it is not necessary at all for the user to know anything about differential equations — s/he is using knowledge about the environment, facts such as temperature and pH-values of the waters under consideration. I guess that if you give it a little thought, you might understand what Figure 4 is about even without any explanation. This is

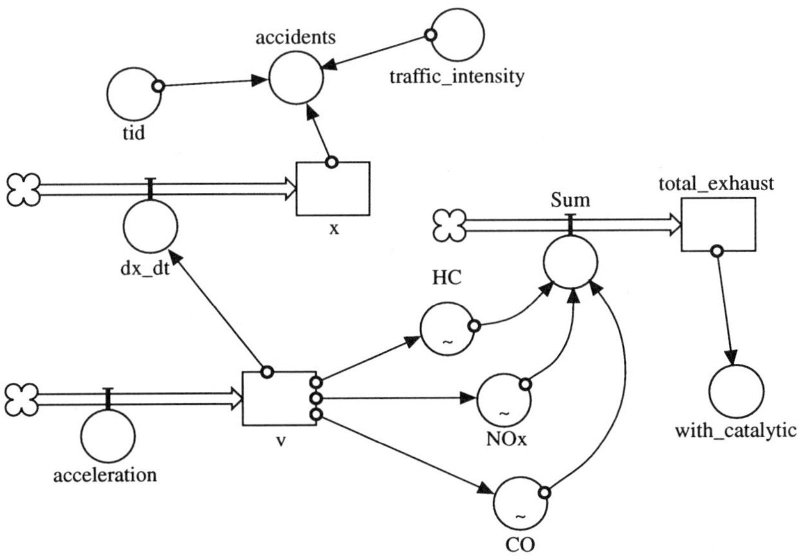

FIGURE 4. Using Stella to understand more. This is one of the first programs to deal with quite complicated issues (Systems Dynamics) with great ease.

what makes Stella so attractive to biologists, environmentalists and other non-mathematicians.

It is also possible to export the solutions to the MacSpin program where you rotate all together in four dimensions, make a trip in time, etc.

HYPERTEXT SYSTEMS

There is nothing visionary about this anymore: hypertext systems are here to stay. They convey information in very flexible and new ways — the reader does not have to engage on a linear description imposed by the author, rather he may jump around, browse through some of the topics and delve into some others — everything with a click of the mouse.

Moreover, he may add his views and make annotations in the material much in the same manner that he would do in a normal book. With some very large differences: he may add sound ("How do the birds in this habitat sound?"), video clips ("Look how beavers wander about!") and do simulations which would otherwise be totally impossible ("Let us raise the temperature of this lake and see what happens!")

Bill Atkinson's visionary program, HyperCard, and its cousin SuperCard, are examples of content-free software that may be used for a multitude of things. But there are clear advantages with HyperCard in an educational and research setting: HyperCard has been free for years and everybody may use it. So the students are bound to have it, you may distribute or sell your models (just as any other compendia) and they will elaborate and finish the models, use them with understanding — this is a very different situation to that of conventional methods! See Figure 5, next page.

My idea of electronic books relies really on these thoughts: you introduce the "book" and the readers read AND write into it — forever!

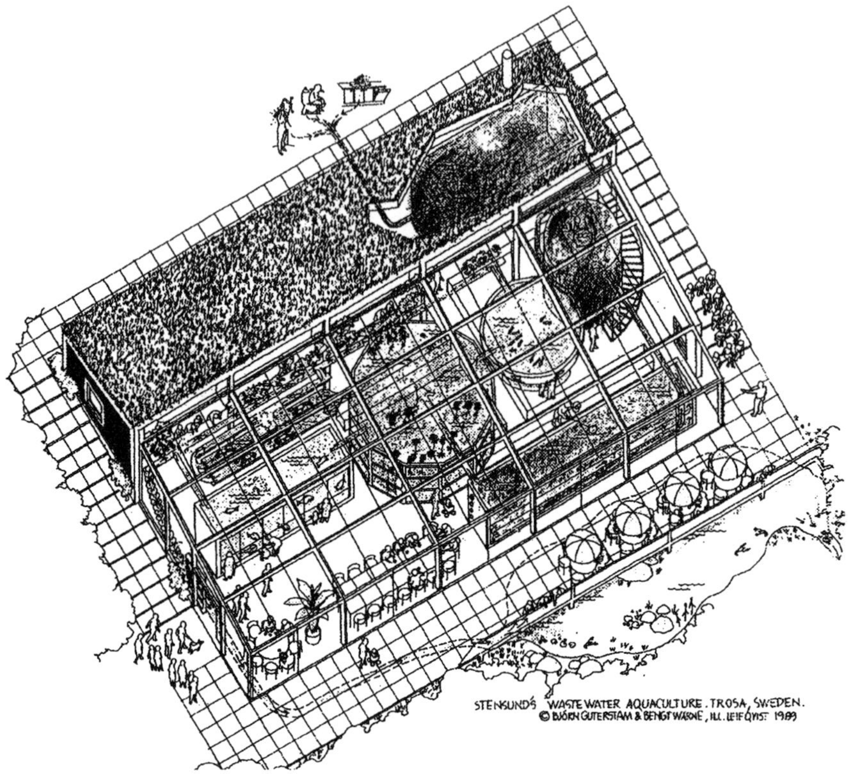

FIGURE 5A. Using HyperCard to convey information in a different manner. You may add sound and animation to the picture as well. Clicking with the mouse on an object brings more information:

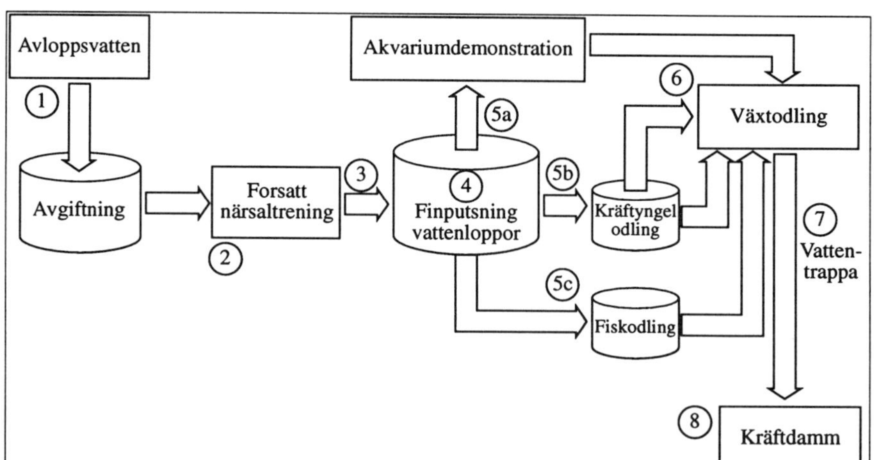

FIGURE 5B. Explanation on demand is one of the strongest points with hypertext. To see this figure you have to be curious and click on the picture above!

THE SAME PERSONAL COMPUTER IN ALL STAGES OF EDUCATION

On a nation-wide perspective, having a low-threshold machine from kindergarten to the university would mean having a system to which teachers (in all stages!!!) get acquainted with less cost of educating them to use the machine. This is especially important for the school system with its thousands of teachers that simply cannot attend courses over and over again. Sending them to learn about *the* operating system should be paid by the manufacturer of such an incomplete system, not by public resources!

There are simple computers that may be used in Kindergarten and extremely competent computers that are for use in industry and universities. Yet, the Apple Macintosh is the only professional computer that can be used throughout all stages of education. This represents a considerable money-saving alternative for a nation or a school system (less costs for training and re-training, greater satisfaction).

Please note that this is an almost unbelievable phenomenon: that the same professional operating system can be used throughout all stages of education and also in industry. Even in a rapidly changing world, this has been so since 1984.

The point I want to stress here is not that the Macintosh is the only computer that can do this (a strong point for Apple's salespeople). The issue here is that of national perspective: to find a smooth tool that delivers at all stages with a minimum of energy. A tool based on cognitive research, not on the marketing strategies of software or hardware firms. All such tools will mature and stay with us for a long time. The telephone is just another such tool: you learn it in kindergarten and the tool stays with you forever, you forget about it completely.

DEVELOPING COUNTRIES SHOULD AVOID UNNECESSARY STEPS AND TAKE NEW ONES

There is an issue which should be raised. The pedagogics outlined here (de Sousa Pires, 1989) also provide developing countries with a tremendous money-saving alternative. It is generally easy to say that all students of digital electronics should have access to a well-equipped laboratory with a multitude of integrated circuits, oscilloscopes, power supplies, analyzers, etc. It is easy to say, but not so easy to implement in reality. Developing countries? I would say that these issues apply just the same to all our universities in times of shrinking budgets.

MAKING THE MOST OF A PERSONAL COMPUTER

The personal laboratory also takes care of part of this problem. By "soldering" integrated circuits on-screen, students (and developers in the industry) can simulate much, much more than would be possible in the laboratory. Also, they can do much of this at home, at their own pace and when creativity is at the very top. Coming to the laboratory with a schematic diagram that really works implies that the logic of the whole circuit is correct (you have tested it!) and what is to be learned in the laboratory is how to make connections with real wires, where the power supply goes, and other such practical things. This is one of the major advantages for a student and a beautiful experience for a teacher. If you have never seen this, try it.

From an economics point of view, the same personal computer that an hour ago was used for digital electronics supplementing an oscilloscope, is now being used for a statistics course and, in the afternoon, will be used for word processing. This is money-saving and a good use of scarce resources.

THE ELECTRONICS HANDBOOK

A final note should be made on my own book (de Sousa Pires, 1989), not so much because it is my book, but because it is an attempt to bridge most of the gaps of modern education and research in a very holistic approach indeed (or so do I think). People trying to take care of the environment will start using computers and suddenly jump into words like byte, formatting, LCD displays. Researchers will try to fit regression lines to a set of data point — but how do you do it? What is regression? And then you want to monitor a set of parameters using a personal computer — and you will be confronted with many terms of the analog and the digital worlds of today. So, what the modern wo/man needs is easy access to knowledge of this kind: a little of mathematics, a little of computer science and a bit of statistics. I believe this is what is necessary to get started, then people will become more and more interested and eager to know more; actually, you must! Do not believe that you are a statistician just because you have bought a tremendous (software) package for statistics!

Also, to solve problems in the modern era of personal computers you need to know which programs to use. Herein lies one of the biggest problems of modern times, for it is difficult for anyone to cope with the thousands of programs which see the light of day each year.

The *Electronics Handbook* was written and designed in such a manner (almost like hypertext) in order to give a clue to most of these issues. It could be used in an introductory course on personal computers, to show which programs there are and how to use them creatively. It is to be used by every researcher who needs a quick reference source to the modern world of computing, mathematics, and statistics.

As a final remark, this book is being implemented on CD-ROM by Växjö University, Sweden (Lundberg et al., 1990), using HyperCard, in order that you may add more knowledge to it and slowly develop your own book. Have fun!

REFERENCES

Bork, Alfred. 1985. *Personal Computers for Education*, Harper & Row, Publishers, Inc.

de Sousa Pires, Jorge. 1989. *Electronics Handbook*, Studentlitteratur AB, Lund, Sweden.

Lundberg, Lennart and Rune Körnefors. 1990. Växjö University, P.O. Box 5053, S-350 05 VÄXJÖ, Sweden.

Figures 5A and 5B represent the Stensund wastewater treatment plant; courtesy of Dr. Björn Guterstam.

Chapter 38

THE ENVIRONMENTAL TECHNOLOGY CENTER

Karen H. Lerum

The Environmental Technology Center (ETC) is a visitor's center to be located in Sandane, Norway, at 61° N. lat., in a marine, west coast climate.

The Center will be a prototype for the demonstration of new technology incorporating various renewable energy systems: passive/active solar collection, a transparent hyperinsulating material, an integrated heat-collecting and distributing construction system, and a sewage and wastewater treatment system designed by Dr. John Todd (John Todd Research & Design). As well as being a technology demonstration and visitor's center, the ETC will also be a place for local and national education in environmental issues.

Pristine water as an end result of the wastewater and sewage treatment process means that the effluent can be used for fish hatcheries and grow-out ponds. There will also be a good opportunity to work with the reduction of nitrogen and phosphorus runoff from agriculture through a series of engineered marshes.

The Center will be bordered on three sides respectively by our client — the county's hydroelectric agency, Sogn og Fjordane Energiverk — a private camping facility, and a community-owned and operated swimming and fitness center. All of these groups will use different components of the Center, either by using its sewage treatment system, its energy systems, its swimming and fitness area, or its Visitor's Center buildings, which will accommodate 250 people, including lecturers, visitors, and school children.

This project has a great deal of meaning when one sees it in view of the Norwegian and international cooperation that will go into its design and execution. It is especially important with regard to its educational potential. As an exciting place to see how things work and to actually participate in scientific projects, in an aesthetic setting, we feel that this center can also be an educational prototype. Because this project combines an international group of architects, researchers, and educators, it can serve many different purposes and achieve various scientific and cultural goals. We see the center as a stepping-stone between the Scandinavian and world communities that will create a place and a way in which to introduce research of new technologies and to teach by means of seeing and doing.

0-87371-990-5/97/$0.00+$.50

Chapter 39

THE AWARENESS OF THE ENGINEER AS AN INTEGRAL PART OF ECOLOGICAL WASTEWATER TREATMENT

Alan Ismond

Ecological wastewater engineering is predicated on the assumption that the engineer has a working knowledge of nature and the environment. Whether the wastewater treatment system impacts the ecosystem positively or negatively is dependent on how well the engineer is capable of understanding the intricate workings of nature. As well, the engineer must be able to differentiate between the symptoms and the root cause of water pollution and degradation.

Prior to the concept of water treatment for the sustainability of nature, the philosophy was simply to dump effluents into nature. Man's awareness has expanded to comprehend that this is unacceptable if the quality of natural resources is to be maintained. The next step involved containment and destruction of pollutants. While this has been somewhat effective in preventing the pollution of the environment, it is an inefficient way to use resources. Once again man's awareness has expanded and pollutants are being extracted or bioconverted in order to restore the quality of water as well as recover useful byproducts. Thus far, the engineer has been concerned with abating the impact of wastewater. A new awareness is emerging where the effective treatment of effluents as well as the efficient use of primary resources is important. Engineers have traditionally treated the symptom without addressing the problem. The reduction and reuse of water as well as the recovery of waste byproducts is becoming important. Many modalities are being developed to try and accomplish these goals. To date, the physical or chemical characteristics of water have been used as a gauge of water quality and the degree of success of the treatment method. However, the engineer's awareness will once again have to expand in order to comprehend that water has other properties that must be considered as well.

A good analogy for exploring this concept is soil. In this century, farmers have generally considered soil to be an inert, dead medium to which chemicals are added in order to sustain plant life. The organic and in particular biodynamic farming movements have shown that soil is a living organism that must be nourished in order to sustain life (Koepf 1989). The same is true of water. This has been suggested by such people as Viktor Schauberger (Alexandersson 1990), Theodor Schwenk (Alexandersson 1990; Schwenk 1988), Rudolf Hauschka (1983), and Patrick Flanagan (1986).

Perhaps the greatest impediment at present is the attempt by wastewater engineers to use a mechanistic view of nature. This inevitably leads to the use of analytical tests for determining the quality of water. In the case of physical parameters, this is satisfactory, however, in the case of energetic properties, or vitality, this is inadequate. Some steps have been taken to bridge the gap between the use of analytical tests and the measurement of the more elusive properties of water. Theodore Schwenk has developed the drop test method where the quality of water as a whole can

0-87371-990-5/97/$0.00+$.50

be visualized. The interpretation of the results still require some intuitive understanding of nature.

Perhaps ecological wastewater engineer is at a major crossroads requiring a new paradigm. The complete comprehension of water requires a wholistic comprehension of nature. This may seem an awesome task if viewed as a process external to the engineer, where the engineer must analytically observe nature and correlate causes and effects. Since man is destroying and polluting the planet at an accelerated pace, a more effective and expedient approach will be required.

The new generation of engineers will learn that nature can be understood not by looking without but by looking within. The old adage "Know thyself" will prove to be the greatest tool that the engineer can rely on. Instead of using external instruments to measure fragmentary and partial properties of water, the engineer will become the diagnostic instrument as well as part of the solution. Man and nature are inextricably linked and the knowledge of one will lead to the knowledge of the other. The awareness required will be similar to the skills acquired by a master archer when shooting an arrow. He does not measure the wind speed, target distance, arrow weight, and bow tension with an array of instruments. Instead, he assimilates and integrates his surroundings and performs the task through a deep awareness of himself and the target. Intuitive and analytical awareness are balanced.

The history of industrialized man has gone from a purely physical view of nature to a more intellectual and emotional view. The effects of pollution and water misuse have been rationalized and addressed accordingly. The final step will require a wholistic or spiritual view of nature in order to fully comprehend the interactions in nature on all levels. The awareness of the engineer will be the critical factor in achieving this view of water, nature, and the world.

REFERENCES

Alexandersson, O. 1990. *Living Water: Viktor Schauberger and the Secrets of Natural Energy.* Bath, UK: Gateway Books.

Flanagan, P. and G.C. Flanagan 1986. *Elixir of the Ageless: You Are What You Drink.* Flagstaff, AZ: Vortex Press.

Hauschka, R. 1983. *The Nature of Substance.* London: Rudolf Steiner Press.

Koepf, H. H. 1989. *The Biodynamic Farm.* Hudson, NY: Anthroposophic Press. 1989.

Schwenk, T. 1988. *The Basis of Potentization Research.* NY: Mercury Press. Spring Valley.

Schwenk, T. and W. Schwenk. 1989. *Water — The Element of Life.* Hudson, NY: Anthroposophic Press.

Chapter 40

A LIVING SYMBOL OF PARTICIPATORY ECOLOGY: WASTEWATER, GROUNDWATER, AND AIR PURIFICATION SCHEMATICS FOR THE RENÉ DUBOS BIOSHELTER, PLANNED SOUTH TRANSEPT OF THE CATHEDRAL OF ST. JOHN THE DIVINE IN NEW YORK CITY

William C. Kinsinger, Paul S. Mankiewicz, and Julie A. Mankiewicz

ABSTRACT

The south transept of the world's largest gothic cathedral is being planned to incorporate ecologically engineered water and air filtration systems in an aesthetic, biologically diverse milieu. Situated on the highest point of Manhattan Island, and atop a natural spring, the traditional limestone edifice is to be enclosed within a south facing greenhouse containing falling water, climbing vines, and moss- and fern-covered ledges to demonstrate functional capacities for heating and air conditioning, and wastewater treatment and polishing.

By providing tests and demonstrations of ecologically engineered purification systems at the Cathedral, the second most visited tourist site in New York City, we hope to renew interest in those higher values which inspire the reach of science and urban design by providing an intuitive grasp of technologies based on an appreciative understanding of the functional, cooperative work daily effected by nature herself.

A PROBLEM TO SOLVE: THE ARCHITECTURE OF OPEN SYSTEMS

All structures modify and are modified by their surroundings. Every living structure modifies and conditions its own air and water quality by the integration of its "metabolism" with its surroundings. Yet the design of our buildings largely ignores the potential benefits of including living organisms as part of any systems for air and water conditioning.

Attempts have been made to build self-contained ecological modules. However, these efforts are only partially useful in ecological engineering or urban design, since all organisms and ecological systems are energetically and materially open. Materials flow in and out, but they are dramatically modified in the process. To date, little work has been done to connect interactively the large built structures of cities with the production of local environmental quality and biological diversity. Plans for the completion of the South Transept of the Cathedral of St. John the Divine have opened discussions on ecologically diverse large scale structures materially open to their surroundings.

0-87371-990-5/97/$0.00+$.50

While air and water are scrubbed by soil and plants, the materials out of which modern life is fabricated have excluded and compromised the capacities of ecological systems to fully affect water and atmospheric cycles of purification. Buildings have yet to exploit the efficiencies of these material cycles as purification systems in the urban landscape; Frederick Law Olmstead's "lungs of the city" are rudimentary organs at best.

PRINCIPLES OF BIOLOGICAL CLIMATE MODIFICATION: PARAMETERS OF ECOLOGICAL ENGINEERING

Most of the energy that passes through every plant, all built structures, and each of us, is heat, shortwave infrared radiation. Plants regulate their thermal loads, absorbing when cool, reradiating when hot. Also, understory leaves cool themselves by evaporation, as we do in sweating. Buildings require elaborate mechanical systems to achieve temperature regulation since their surfaces absorb large pulses of thermal radiation which are transferred to their interior by conduction.

Relationships between the temperature of a structure, ambient air temperature, solar radiation, and absorptance of radiation have very predictable outcomes. Because of the high absorptance of the materials of which cities are made, and because plants and water are largely excluded, temperature is greatly elevated on these surfaces.

These thermal relationships for both buildings and plants are well understood and may be expressed by these two thermodynamic equations (ASHREA 1977; Gates and Schmerl 1975):

Thermodynamics of buildings

$$t_\varepsilon = \tau o + \alpha/ho\ It — \varepsilon\ \Delta R/ho$$

Where:

t_ε = Sol- Air Temperature
τo = Outdoor temperature
α = absorptance of the surface
ho = heat transfer coefficient
It = solar radiation
ε = emittance of the surface
ΔR = difference between incident radiation incoming

Thermodynamics of plants

$$Qa = e\sigma T_l^4 + H\ (T_l - Ta) + LE + P$$

Where:

Qa = amount of radiation absorbed
e = emissivity of leaf (tendency to radiate heat)
σ = Boltzmann constant for blackbody radiation
T_l = leaf surface temperature
Ta = air temperature
H = convection coefficient
L = heat of evaporation, about 580 cal/gram water
E = amount of water evaporated by leaf
P = photosynthesis

Direct rays of the sun heat the ground, a building, or an organism depending on the albedo of their surfaces, their "color" to solar light and heat, which determines the reflectance or absorption of radiant energy. Direct sunlight heats a black surface to the maximum, white nearer the minimum. The graph on the left in Figure 1 expresses sol-air temperature, the effective rise in the ambient temperature caused by absorbtance of radiant energy.

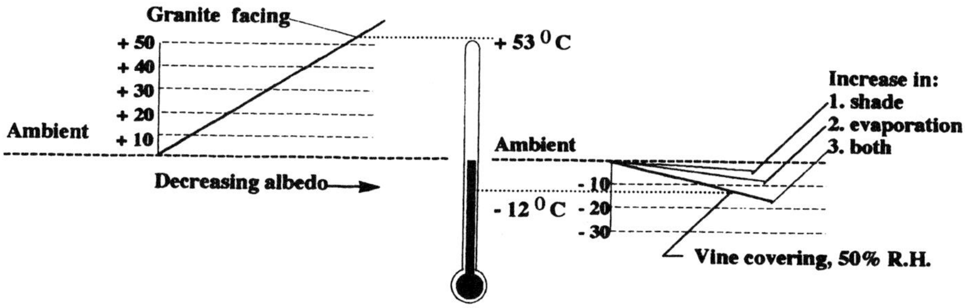

FIGURE 1. Left, the relationship between the albedo of a surface and temperature increase compared with ambient, in °C. A granite wall under certain conditions gives a temperature rise of 53°C. Right, the cooling effect of vines covering the same granite wall, due to shade, evaporation, and the combined effect.

The graph to the right in Figure 1 expresses the lowering of temperature below ambient caused independently by: (1) increased shading and (2) increased evaporation, a function of humidity. Plot 3 represents the combined effect.

Relating the two graphs, at the same ambient temperature, a bare granite face is 65°C warmer than a vine-covered surface.

SURFACE AREA AND BIOLOGICAL FILTRATION OF ROOTS AND LEAVES

Plants grow towards light, water, and minerals. In so doing, these organisms continually steepen the gradients that govern exchange, while structuring the flow of light, atmosphere, and water. In the process of growing, plants partition their environment into a staggering complexity of atmosphere, water and soil substrate.

Leaves partition the air such that they are capable of absorbing large amounts of materials present in parts per ten thousand. The sail of leaves over a single grass or tree can be four to eight times as great as the footprint, or crown perimeter. At the microscopic scale, the area of mesophyll cells within a leaf is ten to fifty times the leaf area.

Roots, although much less easy to measure, cover some 40 to 200 times the surface — footprint or crown perimeter — area. A single rye plant was measured to have some 565 km of root, 11,300 km counting root hairs.

A community of mosses appears to be a smooth green carpet growing on rock or logs. Magnified, ten to thirty or more leaves may be seen on each stem, greatly increasing the scale of the functional surface that filters water and captures light. Microscopically, leaves reveal rows of lamellae, stacks of cells which stand upright from the leaf surface, greatly multiplying surface area.

The immense amount of surface area of bryophytes, the mosses and their relatives, is apparent on microscopic inspection. Water that falls on these land plants or flows over their surfaces is pulled by surface tension into close contact with the plant cell wall. Pollutants and nutrients are adsorbed in the process. With roughly

5% of the earth's surface covered by such organisms, predictions based on their surface area would indicate that these diminutive plants contribute substantially to local, regional, and perhaps global air and water quality.

By trickling a small flow of water down a near vertical wall covered by mosses and providing the same air circulation common in office structures, a two part air filtration system is devised in which the water readily absorbs indoor air pollutants, acting as an accumulator, and the bryophytes, by virtue of their immense surface area, remove pollutants from the water. While considerable in their own right, the absorption surface of most indoor plants is much less than that of a wall of bryophytes.

The filtration capacity of these structures is largely unknown. We estimate that they are some 10,000 times the capacity of any technological system yet devised since they are structured to grow into the waste or pollutant stream, a capacity we have yet to devise in any known machine.

THE ECONOMICS OF INTEGRATED SYSTEMS

In natural systems there is no waste. The byproducts of one organism are the valued inputs of another. Nature moves with elegance to minimize the expenditure of time and energy, to lower the "cost" of meeting its requirements. Applied to the design of the south transept, the incorporation of vines and mosses, aside from providing a capacity for the filtration of air and water, also provides temperature regulation in the form of evaporative cooling — air conditioning. This can be directly compared to the equivalent amount of cooling capacity which must be purchased in the form of mechanical equipment, and then operated with purchased electric power.

As the schematic drawing depicts, layers of vines near the outer glazed skin and a layer of mosses and ferns on the inner wall of the south transept provide the cooling equivalent of an air conditioning system rated at over 300 tons operating continuously during the twelve hours of sunlight in summer. In a four month cooling season this offsets the requirement for over one million kilowatts of electrical energy. If this electricity were generated from petroleum fuels, 550 m^3 of fuel oil would otherwise be required at a cost of over U.S.$100,000.

CONCLUSION

In general, existing technologies of heating, cooling, and air purification must produce more pollution and waste than they ameliorate, following the general laws of systems. However, as argued here, by adding plants, microbes and water to this equation, local environmental quality can be improved, and pollution abated. The "wastes" of plants — oxygen, humus, leaf, fruit and seed — support the rest of the biosphere.

In the process of structuring themselves to capture sunlight, plants also necessarily regulate their surroundings, creating an envelope within which air temperature, movement and humidity are held within bounds. Many species on earth, including our own, have found the ameliorated environment under trees and vines or alongside mossy streams most appealing.

FIGURE 2 Southtransept, Cathedral of St. John the Divine.

This paper underscores the potential of the Réne Dubos Bioshelter to act as nature does, changing its circumstances as it changes itself, transforming wastewater and pollution into nutriment for the structural Gothic sensibilities of awe, wonder, and inspiration.

REFERENCES

Gates, D.M. and R. B. Schmerl, eds. 1975. *Perspectives in Biophysical Ecology.* New York: Springer-Verlag.

ASHREA Handbook, 1977 Fundamentals, American Society of Heating, Refrigerating and Air-Conditioning Engineers, Inc., New York.

*Case Studies**

* Editors' note: The case studies, as we noted in the Foreword, were an opportunity for the participants to go out together, look at real problems, and benefit from one another's knowledge and experience to propose ecological egineering solutions. The groups spent one day in the field with their hosts and, that same night, the conclusions of each group were presented to the conference.

As the papers in this section achieved more or less their final form ten hours after the participants first became aaquainted with the problem, they may not be as detailed or polished as the other contributions. The day itself, though, was widely reported by the participants to be one of the most stimulating that they had experienced at a conference. We think we speak for the participants in recommending this form of activity at future gatherings of ecological engineers.

SUBURBAN ECOVILLAGE FARSTA

Host: Per Olof Jägbeck, City of Stockholm
Chair: Susan Peterson, Ecological Engineering Associates, Falmouth, Massachusetts
Secretary: Robert Crosby, Biorealis Systems, Inc., Eagle River, Alaska

The City of Stockholm is considering an ecological village at this 2 ha-site south of the city. The site is currently a park. It is unfavorable for development. It is cold, damp, and low. It has been mined for topsoil down to a hard clay base; ground water varies from 70 cm below the surface to 35 cm above it; there are shallow pools and ditches throughout the site; the site receives run-off from the highway, the bike path, surrounding apartment complexes and school yard, and a large rock outcrop. It is not a good location for community flower or vegetable gardens. It is a wetland, with typical wetland grasses and shrubs. The area serves many functions typical of natural and constructed wetlands. It appears to drain a substantial area (approximately 10 ha), thus providing storm protection for adjacent properties; it provides habitat for birds, insects, and frogs; it is open space used by local residents in transit via the bike path and for walking and pet exercise; it cleans the runoff water passing through it.

The plan for the ecological village was to incorporate as many ecologically sound features as possible in 15 duplexes to be built by the owners. The assumption was that 120 people could eventually live on the site; the design should be for 150.

The general consensus of the group which reviewed the Farsta site was that it is physically and economically inappropriate for an ecological village. Besides the unfortunate physical characteristics, a major sewer line transverses the site, making hook-up less expensive than any water recycling or on-site treatment could possible be. While there was strong feeling that an ecovillage near Stockholm center would be an excellent educational addition, the applied ecologists in the group felt that use of this site would convey the idea that ecological systems were impractical and more expensive than traditional housing.

If the site is to be developed, the group felt that there would have to be major modifications in topography (with careful consideration of the area's storm abatement function), with some site drainage (either ponds or ditches), and the addition of soil. If the decision is to ignore the sewer line in favor of on-site systems, then the drainage ditches and ponds would serves as secondary and tertiary treatment for wastewater. The treated water could be used on-site for gardening, car washing, toilet flushing, etc.

Perhaps the strongest feelings in the group was that the houses should not be built in the low, damp areas, but rather upon the west side of the large rock outcrop, using multi-layered houses to reach into the light and away from the damp.

General assumptions:

1. The houses should be built using technology developed for very damp and cold areas, since no basements are possible on the wet soils.

0-87371-990-5/97/$0.00+$.50

2. That solid waste composting be incorporated to enhance soil quality for flower and vegetable gardens, but not to replace the soil mined from the area.
3. That water conservation devices be required in all households.
4. That there be a strong emphasis on reusing the nutrients carried in the wastewater.
5. That the houses be designed for maximum solar gain.
6. That the existing impermeable clay soils be bermed to provide needed pond or ditch structures.
7. That noise control devices adjacent to the highway would be needed if the site were to be a pleasant place to live.
8. That up to 60 cm of soil would be needed to restore the site, or the site ought to be landscaped as wetland, or a combination.

The alternatives for wastewater treatment, all of which assume use of nutrient-rich water on site for irrigation or ponds, include:

1. Separate piping for graywater, with treatment by a vertical sand filter or greenhouse, ditches and ponds; composting toilets.
2. Separate piping for graywater; toilet systems which separate feces and urine into storage tanks which are transported by contracting farmers to agricultural land.
3. Combined wastewater piping, with treatment by a vertical sand filter or greenhouse, ditches, and ponds.

Several members of the group felt it was important to provide a response for our hosts that fit in with the constraints we were given — that is, to help design ecologically sound systems which would serve duplexes built in rows, served by driveways, in the low ground of the southern half of the site. The suggestion was made that we could design self-sufficient systems for each duplex on 1050 m^2 with combined wastewater piping to a digester followed by ditches to a 240 m^2, 3-m deep pond with vegetables and flower gardens.

AN URBAN FRINGE AREA, UTTRINGE HAGE

Host: Per Carlsson, Trädgårdsstaden Arkitekter AB
Chair: Ann-Mari Jansson, Systems Ecology, Stockholm University
Secretary: Joan Wilder, Center for Restoration of Waters @ Ocean Arks International, Falmouth, Massachusetts.

The project area, Uttringe Hage/Karlskronaviken, is part of an old housing and weekend cottage area. Typical of many urban areas on the fringes of Stockholm, the area is being planned for redevelopment, through the Land Redevelopment Project. There are currently approximately 100 people living there, with expansion expected to accommodate up to 700 people. There is no sewage system for the area now, and the waste goes directly into Lake Uttran.

The site to be developed is 50 ha. It is surrounded on three sides by Lake Uttran, which is 10 km long. Much of the sea level area of the site is wetlands, and there is also a lot of hilly land from which beautiful views of the lake are seen. The area has rich vegetation. We saw duckweed in the wetlands, and the water was red with iron. We thought the iron might be used for the precipitation of phosphorus. Willow, aspen, and birch trees were scattered throughout the area.

In the past the lake was used for the communal dumping of sewage and it is polluted. There are fish kills in the summer months, and algal growth and toxic sediments in the lake floor prevent the growth of flora. The lake is still used for recreation but people are very concerned about their health and that of the pond.

RECOMMENDATIONS

The group made several recommendations for the implementation of plans for the treatment of sewage in the area. It was unanimously agreed that the water treatment solution should be environmentally sound, and the obstacles to this did not seem overwhelming.

The first recommendation was to build an aquaculture wastewater treatment facility such as the one at Stensund, or Ocean Arks in the U.S., designed to be more energy efficient than the Stensund prototype, perhaps by using less concrete. After leaving the facility, the effluent would be directed to run through an ecologically engineered marsh. The marsh would have an impervious clay layer at the bottom and would have high levels of iron and aluminum for the precipitation of phosphorus. Precautions should be made to prevent the shortcutting of water through the system. The marsh would have retaining pools where the effluent would remain for periods of time for longer treatment. For more detailed design specifications for this marsh, please contact Karlfriedrich (Charlie) Hofmann, Büro für biologische Planung und Analyse, Dorfstraße 4, D-72074 Tübingen, Germany.

The second recommendation is to build individual advanced septic systems for each building. There are several new models for these septic tanks, which should be researched. For more information on such designs, please contact Pavel Saga, Center of Ecological Information, Udernicka 1931, Praha 4 149 00 (Czechoslovakia),

0-87371-990-5/97/$0.00+$.50

or Gerald Navara, Institut für Pflanzenphysiologie, Universität Wien, Althanstr. 14, A-1091 Wien, Postfach 285 (Austria).

The effluent from the individual septic tanks would then be piped to flow through a marsh just like the one suggested in the first recommendation.

ECOVILLAGE NACKA

Host: Varis Bokalders, Architect, Royal Institute of Technology
Chairperson: John Todd, Center for the Restoration of Waters @ Ocean Arks International, Falmouth, Massachusetts, U.S.A.
Secretary: Karen Lerum, Via Prosjekt A/S, Sogndal, Norway.

The Ecovillage of Nacka, 15 kilometers from the center of Stockholm, will be a housing development of 20 two-story duplexes and have approximately 60 inhabitants. The goal of this development is to have as low an impact on the environment and to be as self-sufficient as possible. It will be studied by the Swedish National Board of Housing in order to establish guidelines for further urban developments. All of the processes used in this project are meant to be monitored and controlled locally by the Environment and Health Authority of Nacka Municipality.

The project will be built in such a way that there will be easy access to the sewage and wastewater treatment system, with special emphasis upon the following:

- Energy conservation and the use of renewable energy sources.
- Local food production (gardening and livestock).
- Separation of garbage for recycling and composting.
- Local drinking water supply.
- Local, ecological wastewater treatment.
- Use of building materials that do not induce allergic reactions.

Nacka Ecovillage will be located next to an already existing housing development. This housing development is connected to the conventional municipal sewage and wastewater treatment system. The Ecovillage residents do not want to be connected to the same municipal treatment system, except as a back-up for emergency situations within their own system, as they feel that it does not adequately address the requirements of the site and the surrounding environment. They feel that a "natural" treatment of household-generated graywater, such as an enclosed aquaculture treatment system, coupled with composting toilets and an engineered marsh, will more adequately satisfy the environmental needs by returning the ecologically treated, (vs. chemically treated), clean water to the existing ground water system and to the nearby lake. This lake has already been polluted by the existing sewage system, which allowed sewage ground water infiltration. As a result, Nacka regulatory officials will no longer issue permits for infiltration of wastewater from toilets.

The physical location of the Ecovillage site also makes its own demands. It is situated in a low-lying meadow, bordered on the northern side by a slightly elevated forest which is used for recreation, a road to the west, a low-rising hill to the east and the detached housing development to the south. The meadow soil has a very high clay content, as well as a high water table, with some saltwater intrusion. This is a coastal area of Sweden, with a very high level of annual precipitation. The tasks of the group working on this project were fairly straightforward:

0-87371-990-5/97/$0.00+$.50

1. Treat the household graywater by non-chemical means
2. Use the high ground water level in a constructive way (vs. just draining the soil)
3. Treat potential water-borne pathogens generated by households
4. Use the center of the property for productive crops and as a social space
5. Suggest a method for recycling clean wastewater
6. Suggest a method of making good use of the waste nutrients for production of food or plant crops.

The initial proposals made in our meeting with the committee and Swedish board members varied from the biorotor (proposed by the Berlin contingent), to the eco-filter, (Swiss solution), to the solar greenhouse, primary/forest fertilizer and the aquaculture/eco-filter plus aeration process (U.S., Swedish and Norwegian).

The treatment processes decided upon work as follows (Figure 1): Process A is the solids treatment system. This is achieved by a toilet which separates solids (fecal matter) from liquids (urine). The fecal matter is then composted, using worms to help in the decomposition process.

There are two kinds of liquids treatment systems, one for household graywater and the other for urine. In Process B, the graywater treatment system, water from showers, bathrooms and kitchen sinks is cleaned in the following steps: The liquid from these three sources (1 on diagram) is piped into a treatment area where it first goes through a septic tank, creating sludge (step 2). Then it goes through a biorotor, which is a drum filled with small plastic balls to provide surface area, and is given much aeration (step 3). A small part of the biorotor effluent is then sent through an ultra-violet treatment — which kills any remaining bacteria (step 4) — and stored in a tank (step 5) before being used in clothes washing (step 6). The wastewater from the washing machines is then recycled back (step 7) through the sludge trap and biorotor. The effluent from the biorotor that does not cycle through the clothes washing loop emerges (step 8) in the greenhouse, where it is combined with the urine from Process C (step 9).

Process C, the urine treatment process, has two basic stages. First, it is separated from solids by the toilet and collected in a holding tank (C1 in diagram). From this tank, three-fourths is taken out for crop fertilization by local farmers in the spring and summer months. The remaining one-third is then piped into the aquaculture system in the greenhouse (step C2).

There, combined with the cleaned graywater, the nutrients in the remaining urine are taken up by the zooplankton, fish, and water hyacinths (9 in diagram). Step 10 in the diagram shows the water from the aquaculture system being piped up to a series of plant beds, in which the water is then further cleaned and sent back (step 11 in diagram) into the aquaculture system. With step 12, the water is then sent by a tip-bucket (pulsator) in pulses through a series of watersteps, where it is further oxygenated. The pulses help spread it through the last process, a polishing marsh (step 13), which it goes through to emerge as pristine effluent. In step 14, the water is returned to the ground water. In step 15 it waters gardens. Ultimately, the cleaned water will find its way to the nearby polluted lake, where it will help to revitalize that body of water.

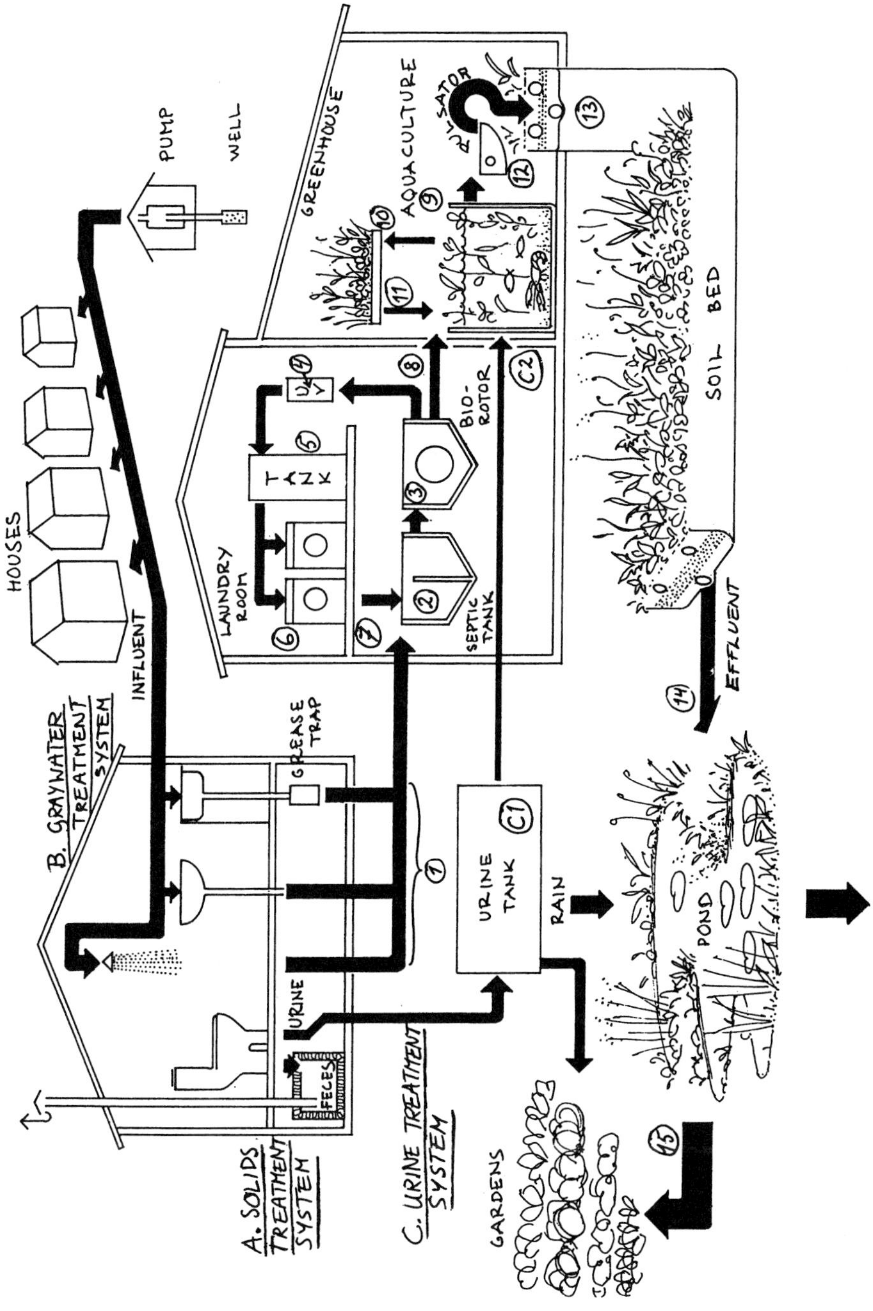

FIGURE 1. Wastewater flows proposed for Ecovillage Nacka.

PHARMACEUTICAL INDUSTRY WASTEWATER, KABI

Host: Anders Eriksson, Kabi Pharmacia
Chair: Karl Ehrlich, Aquaresearch, Ltd., North Hatley, Quebec, Canada
Secretary: Karen Schwalbe, The Center for the Restoration of Waters @ Ocean Arks International

Kabi manufactures five key biomedical products,which are all produced by fermentation, in the town of Strängnäs. This process consists of cell culture, separation, and drying. Most of the waste comes from the biosynthetic manufacturing; the part from the chemical processes is small. The raw materials involved are hydrolyzed protein, glucose, mineral salts, buffer salts, ammonium sulfate, and aluminum chloride.

Waste treatment is run in an activated sludge plant and currently operated on the principles of a chemostat, without a return activated sludge, similar to their production process.

The manufacturing plant is situated on the shore of Lake Mälaren, a eutrophic lake. Water from the lake is used as drinking water.

Expansion at Kabi has resulted in the construction of a new wastewater treatment plant (Table 1). There are two treatment plants nearby. One treats the wastewater from Strängnäs; the other treats wastes from another pharmaceutical company.

TABLE 1.
Pollutant Removal at Kabi Wastewater Treatment Plant

	Flow	BOD_7	Nitrogen	TP
Design	500 m^3	500 kg/d		
Current influent	50 m^3	225 kg/d	59 kg/d	1.6 kg/d
(% design)	(10%)	(45%)		
Goal		25 kg/d	50 kg/d	1 kg/d
Current effluent		24 kg/d	48 kg/d	0.6 kg/d
(% goal)		(96%)	(96%)	(60%)

Recommendations from the group included optimizing the current treatment process, improving public relations, and utilizing the nutrient resources.

Current data collected was not sufficient for biological process evaluation. We felt that different information was needed to monitor the bacterial processes. We also suggested more data collection at internal points around the aeration tank. The additional parameters suggested were orthophosphate (PO_4), soluble carbonaceous BOD and total BOD, COD, nitrate-nitrogen (NO_3-N), alkalinity, pH, O_2, mixed liquor suspended solids (MLSS), total organic carbon (TOC), dissolved organic carbon (DOC), and total heterotrophic cell count.

Ammonium sulfate, used in the extraction of the proteins in the fermentation process, contributes to most of the nitrogen loading in the waste stream. We sug-

0-87371-990-5/97/$0.00+$.50

gested Kabi look into the option of treating this waste separately from the main waste stream.

The waste produced is likely a monoculture of *E. coli*, because the fermentation process is sterile. We proposed bacterial augmentation to assure the presence of the appropriate microbial communities.

Kabi has the benefit of a nontoxic waste stream low in metals and toxins, with no occurrence of pathogens. It seems a logical step to recycle the nutrients through the production of food crops and/or sport fish. The public relation potential of such a project would be great. The production of fish, food, and decorative crops would be visual proof of output of good quality water.

There would be high visibility within the community of Strängnäs if Kabi were to use ecological engineering for polishing its existing effluent. Even more so, there is the possibility of treating the effluent from the other two treatment plants along with the effluent from Kabi by ecological engineering.

Our final suggestion was the immediate implementation of a pilot project using raw influent to assess the impact of ecological engineering on recovering the nutrients in the Kabi treatment process.

TULLGARN ROYAL CASTLE

Host: Olof Gillberg, National Board of Public Buildings
Chairperson: Karin Sundblad, Linköping University, Sweden
Secretary: Nancy Jack Todd, Ocean Arks International, U.S.A.

Tullgarn Castle is located on the Baltic Sea coast at Haggnäs Bay. The premises are maintained by the National Board of Public Buildings. The castle, which is still a tourist attraction, has not been in use since 1950. The surrounding park, designed along the lines of the English landscape park, is beautifully maintained. The treatment plant is intended to treat the waste from eight single family houses with about forty people and a tourist restaurant which is open only in the summer. It is situated in a low-lying meadow and consists of two stabilization ponds not far from the bay into which the water is discharged.

The primary pond is still operating as a stabilization pond. The second has been invaded and largely taken over by stands of macrophytes such as *Phragmites* and *Typha* as well as some alder (*Alnus* sp.). The initial problem we were asked to address was: should the weeds in the second pond be perceived as a problem and, if so, should they be removed? Our host, Olof Gillberg, had been told that, because the ponds are so far from the road, it would cost about one half million Swedish crowns (ca. U.S. $85,000) to remove the reeds mechanically.

It should be noted that it is claimed that at present the ponds are working well enough, with a removal efficiency as follows:

BOD	ca. 55%
P_{tot}	55%
N_{tot}	65%

As a member of the group pointed out, this plant in itself is not much of a problem. But seen in the context of the thousands of others like it along the coast, the problem is magnified and should be seen as a model for addressing comparable situations in the future.

We decided to approach the problem with three levels of recommendations. The first was that of minimal interference, meaning just enough maintenance to keep the ponds functional. Because the second pond has reached a stage sufficiently overgrown to be considered more a marsh than a stabilization pond we thought it advisable to restore it to its former function. We debated as to whether to dig out the pond or whether to harvest the rushes leaving the root systems intact. We decided to recommend that the rushes be harvested annually, in August at the time of greatest biomass, using a scythe. Said biomass, we decided, should be used and could either be made into a slow decomposing compost for use on the trees in the castle park or used as a mulch where needed. The second part of this minimal strategy would be to shorten the 300 m underground discharge pipe to the bay considerably, discharging the effluent into the existing natural marsh at surface level. Because of a

0-87371-990-5/97/$0.00+$.50

blocked pipe between the two ponds, another channel between them should be dug at surface level.

The above step we chose to consider as maintenance but only that and not a solution that was either ecological or imaginative. The second strategy was to view the wastewater and its products as a potential resource and to find a way to use it. The most obvious uses are a combination of irrigation and cultivation.

The drainage water could be used to fertilize such water-tolerant tree crops as alder, willow (*Salix* sp.), and birch (*Betula* sp.). Tree crops require 80 kg of fertilizer per hectare per year as opposed to 120 for wheat. Alder is already present and could be extended and planted in several adjacent areas. Willow, although trickier initially, is a more valuable crop. One member calculated the possible value over a 24-year span starting in 1991.

This could eventually yield 276 tons of dry matter per year, producing a crop of considerable economic value from a formerly non-utilized resource. It was pointed out that alder, as a crop, reduces nitrogen imbalance and that in Latvia it is claimed that the sale of 20 m^3 of alder will buy the harvester a chain saw.

The third and favorite recommendation of the group encourages the consideration of waste not only as a potential economic resource but as an educational resource — as information. Whereas the first, or the first and second approaches would be suitable for most comparable sites in the area, because the castle is an established tourist attraction, we suggest the treatment area be designed as an ecological education center in the form of a water park. We felt that the contrast between the formal but rather sterile and dated landscaping of the castle park would provide an interesting and instructive contrast for a prototype design that, at its ideal, would imitate the maximum integration of human and natural achieved by the Chinese designed ecosystems. As the area is also famous for bird migrations a further, if subliminal, educational message could be conveyed in the juxtaposition of formal, integrated, and wild ecosystems. Minimal costs, design, and maintenance criteria for such a water park might result in a small exhibition museum/garden made up of shore and marsh vegetation native to the area.

We think it would be more effective and interesting to create a park that would be beautiful and useful as well. We would design some of the pond effluent into meandering streams with planted banks that would incorporate falling water of flowforms. If, again using the Chinese model, as many niches could be filled, with a diversity of life forms (such as trees, shrubs, vegetables, flowers, fish, crayfish, frogs, and water birds), the taboo demanding that wastes be taken out of sight would be broken, as it is in the treatment plant at Stensund. Ideally a small greenhouse would eventually be designed to overwinter some of the less cold-tolerant species like water hyacinth (*Eichhornia crassipes*) and goldfish (*Carassius auratus*) that could add to the aesthetic quality of the park during the tourist season.

Knowing that nothing urgently needs to be done at this site allowed us the opportunity to envision the potential of what could be done that could anticipate and alleviate future problems. In brief, our recommendation is the evolution from semi-operational wastewater pond, to economic resource utilization through fertilization and cultivation of tree crops, to the Tullgarn Castle Water Park.

NYA NYKÖPING

Host: Jan Älfvåg, Nya Nyköping project
Chair: Vidar Lerum, Via Prosjekt, Sogndal, Norway
Secretary: Pat Ryan, Center for the Restoration of Waters @ Ocean Arks International, Falmouth, Massachusetts.

Participants: Thorkil Boisen, Denmark; Janis Zaloksnis, Latvia; Anders Lind, Sweden; Hamish Stewart, Denmark; Koit Jatsa, Estonia; Paul Mankiewicz, U.S.; Julie A. Mankiewicz, U.S.; William J. Mitsch, U.S.

Nya (new) Nyköping is a town which will be developed over a ten year period, beginning in 1993. It will be a substantial extension of the existing town of Nyköping, on the Baltic Coast. The projected population of the new town will be 15,000 people, roughly 5000 family units. The region includes several hectares of wetlands and pine forests, and much of the soil is soft clay extending for several meters above the granite bedrock. This poses a challenge for creating sound buildings over the long-term. A channel separates the northern and southern areas of the development. About a one hour train or car ride from Stockholm, the development will be an appealing one for people who wish to improve their quality of life and still maintain their economic base. The developers further hope to establish locally-based businesses, creating jobs in the area at a rate of about 300/year.

A group of 15 participants from the international conference Ecological Engineering for Wastewater Treatment spent a day as guests of the developers of this project, and offered several recommendations for enhancing the economic viability of the project while maintaining as much ecological diversity and integrity as possible. We worked within the parameters of accepting that the development could not be radically altered in its design, and so offered design modifications most of which, in our opinion, could be easily implemented.

The new town will depend on surface water, as is the case in most of Sweden. The current sewage treatment plant is located 5 km northeast of Nyköping. It is a conventional plant and is being used to maximum capacity at present. The plant removes about 90% of the phosphorus, but has little success in removing nitrogen. High nutrient loading of the Baltic is the result. Transporting the wastes of Nya Nyköping to the plant would be very expensive due to the need to build long pipelines, as well as the high cost of expanding the loading capacity of the present plant.

Our group recommends that several decentralized sewage treatment plants be built between the development clusters for the new town. Waste treatment systems of the solar-aquatics type would be low-cost, environmentally sound and aesthetically pleasing for the community, and could offer a source of economic byproducts, for example flowers, trees, and fish. The effluent could be released into the wetlands for a final polishing. We calculated that the wetlands could absorb an increase of water at the rate of 3.6 mm/day. Less than 100 ha of wetland area could serve as a final polishing stage in the treatment process for wastewater.

0-87371-990-5/97/$0.00+$.50

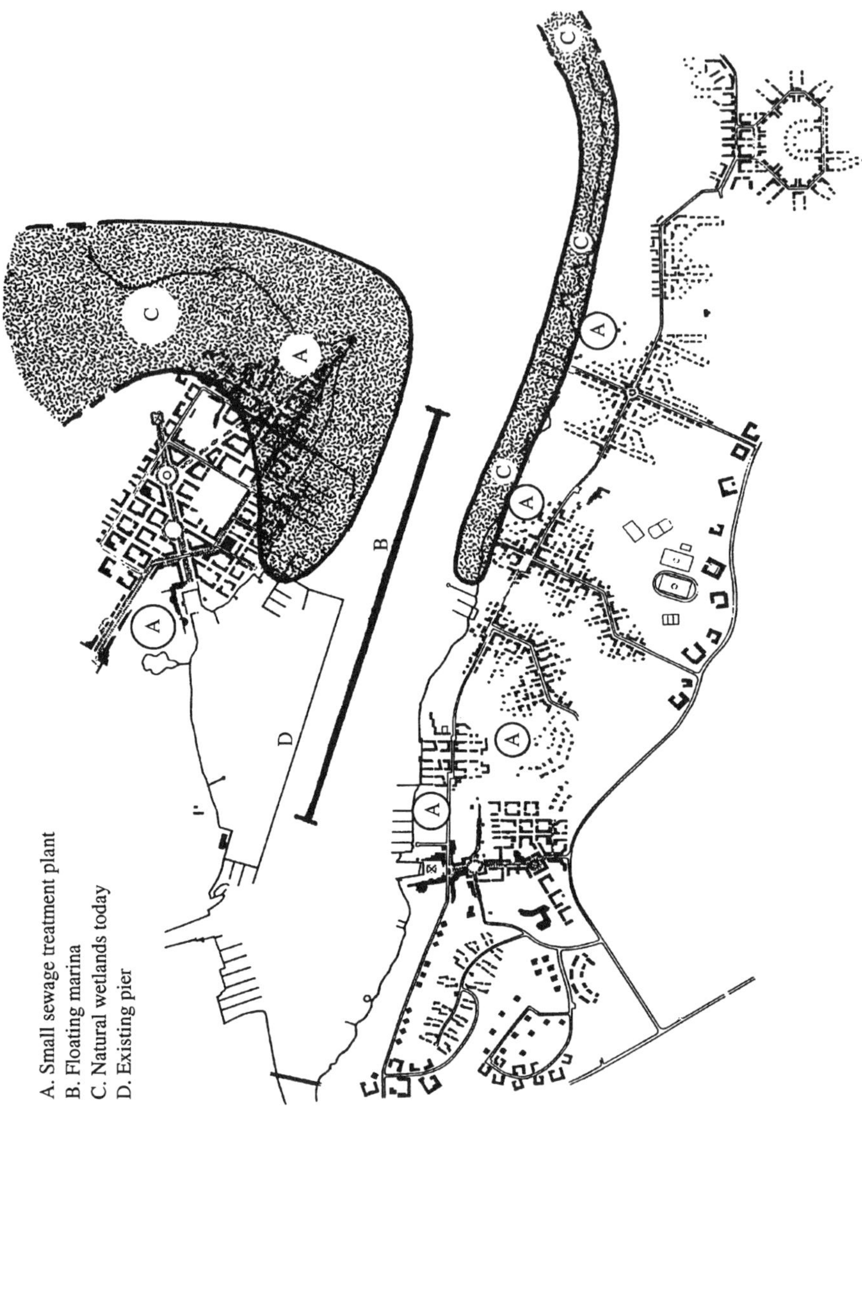

FIGURE 1. Nya Nyköping as planned, and with suggestions from the case study group.

We also suggested that rather than destroy much of the existing wetlands for the creation of a marina, that a floating marina be created, with access by boat or by a walkway. Destroying the wetland would increase the nutrient loading, and thus the eutrophication, of the channel. By our calculations, the wetlands currently absorb a distinct amount of nutrients, as well as provide a home for several bird and animal species which enhance the beauty of the region.

There is little space for residents to grow their own food in the current plans. We propose that rooftop greenhouse design be incorporated whenever possible. They would serve the dual purpose of growing fresh food year round, as well as help to retain radiant energy in the winter. For residents who build homes that are not fit for rooftop gardening, we recommend that some agricultural fields in the green belts be reserved for garden allotments (*koloniträdgårdar*).

As a further recommendation, which we feel could be a real plus in marketing the new community, we would propose that fewer roads be built than is currently planned. If a shuttle boat were used to transport residents from the southern and the northern side of the channel into the city center, there would be little need to own a car, as the express train to Stockholm would be readily available. Ecological design would allow residents to walk or bicycle easily to a local workplace, cultural center, or marketplace.

Another recommendation is to build a floating solar collector on the northern side of the floating pier in the channel. Water could be heated to a comfortable temperature for swimming in a protected lagoon. Water from the river would be channeled through the lagoon after passing through ecologically designed, engineered marshes, thus providing a clearer, cleaner, and warmer water for outdoor swimming in the summer.

The developer told us that one or two of the existing farms will be kept as demonstration farms or living museums. The group recommends that these farms be turned into eco-farms that will carry out functions essential to the new development of Nya Nyköping. On these farms we envision that digesters will transform old newspapers, organic waste from kitchens, and septage from low-flush toilets into a light-weight soil to be used in rooftop greenhouses. As a byproduct of this process, the eco-farms will also produce bio-gas to be used as fuel for the small buses.

GLOSSARY

Activated sludge process Sludge withdrawn from secondary treatment following the activated sludge process; it consists mostly of microbial biomass, with some inorganic solids. Return sludge is recycled to the head of the process to reinoculate it with the microbes; waste (excess) sludge is removed for further treatment.

Activated sludge A biological secondary wastewater treatment process in which a mixture of wastewater and activated sludge is agitated and aerated. The activated sludge is subsequently separated from the treated wastewater (mixed liquor) by sedimentation and wasted or returned to the process as needed.

Aerobic (syn. oxic) Free oxygen (O_2) is present (see Hofmann, this volume).

Algae Aquatic photosynthesizing organisms with cell nuclei, never differentiated into root, stem and leaves, contain chlorophyll *a* as the primary photosynthetic pigment, and no true vascular system. The algae range in size from single cells to plants over 100 m in length (kelps, *Macrocystis* sp.).

Anaerobic (1) Without oxygen. (2) No free oxygen (O_2) or nitrate (NO_3^-) present (see Hofmann, this volume).

Anoxic No free oxygen (O_2), but with nitrate (NO_3^-) present (see Hofmann, this volume).

Aquaculture The regulation and cultivation of water plants and animals for human use and consumption.

ATS Aquatic Treatment Systems (see *constructed wetlands 1, C.W. with free water surface*).

Autotrophic Making its own food from carbon dioxide (CO_2) by photosynthesis, as a green plant, or by chemosynthesis, as any of certain bacteria.

AWT Advanced wastewater treatment. Any physical, chemical, biological treatment process used to accomplish a degree of treatment larger than that achieved by secondary treatment. See also *tertiary treatment.*

Batch feeding Feeding an enclosed system with a fixed load, a batch, which remains in the system for a certain retention time, after which the system is emptied and reloaded, or simply reloaded, depending on the fate of the load during retention. Cf. *continuous flow* and *plug flow.*

Biocoenosis A term primarily used in Eastern Europe to refer to the living part of a biogeocoenosis, i.e., the living part of an ecosystem. (pl. biocoenoses)

Biogas Gas mixture, generally of 60% methane (CH_4) and 40% carbon dioxide (CO_2), produced by anaerobic microbial decomposition of organic material, i.e., fermentation.

Biotechnology (1) Genetic manipulation of species (Mitsch, this volume). (2) Manipulation of an organism's genetic composition, using techniques of molecular biology, to produce organisms which serve human purposes.

BOD Biochemical oxygen demand. The quantity of oxygen required to biochemically oxidize organic matter dissolved in water, measured under specified conditions. Indicates the load of organic matter in the water. The higher the BOD, the less dissolved oxygen is available for aquatic organisms.

COD Chemical oxygen demand. A quantitative measure of the amount of oxygen required for the chemical oxidation of carbonaceous (organic) material in wastewater, using inorganic dichromate or permanganate salts as oxidants in a two-hour test. Since the ratio of COD to BOD is usually constant for a given effluent, and COD is much faster to measure, COD is often the parameter monitored once the ratio is established.

Constructed wetland (C.W.) Engineered marsh system. Many different designs have been proposed and implemented. On the basis of their hydrology and emphasis on different purification processes, three main concepts can be distinguished:

> **Constructed wetland with free water surface**. FWS-systems are similar to stabilization ponds. Long detention times together with biochemical processes promoted by the plants will reduce pathogens, suspended solids, and BOD. Free floating or rooted plants prevent algal growth. This type of system is often described as Aquatic Treatment Systems (ATS).
>
> **Constructed wetland (C.W.) with subsurface flow**
> (1) *Horizontal flow systems.* In contrast to FWS-systems, the purification processes in these systems take place underground where water flows laterally through the medium (soil or an artificial substrate) and the root zone of the vegetation. The soil can be described as a mechanical biofilter where exudates from the plants, especially oxygen, improve purification. The MPI (Max Planck Institute)-System conceptualized by K. Seidel stresses the role of the plant (including nutrient uptake) and uses inert coarse material as the rooting medium. The MPI system is characteristically formed in trenches, to facilitate harvest. The RZM (Root Zone Method) evolved by Kickuth uses finer material in order to adsorb phosphor on soil particles. RZM systems are constructed in a rectangular shape with water movement across the short axis.
>
> Synonymous terms for C.W.s with subsurface horizontal flow are: *vegetated submerged filter, soil filter trenches, biological — macrophytic systems, hydrobotanical systems, marsh beds.*
>
> (2) *Vertical flow systems.* In contrast to (1), wastewater is spread out on the soil surface from where it percolates down through the medium. Most of the purification processes take place in the soil. The role of the vegetation is to preserve the infiltration capacity but biochemical processes such as nitrification and denitrification are also stimulated by the plants. This wetland system is evolved for high hydraulic loads and no clear-cut limits can be drawn between

this type of intensive land treatment and more extensive types of land treatment systems such as irrigation systems for crop production.

Synonymous terms for C.W.s with vertical flow systems are: *Rapid Infiltration Systems, Infiltration Wetlands.*

Continuous flow A method of feeding a reactor with substances on a continuous basis, not interrupted.

Cytophagas Chemo-organotrophic bacteria of the family Cytophagaceae, living off of organic compounds and characterized by gliding motility. Some are able to use complex polysaccharides.

Denitrification The anaerobic biological reduction of nitrate nitrogen (NO_3^-) to nitrogen gas (N_2).

Detritus Nonliving, particulate matter of biological origin, biological litter.

Diatoms Microalgae with 10 to 500 μm-long cells of the Diatomaphyceae (former Bacillariophyceae) taxonomical class, with silica-impregnated, bivalved shells, which are overlapping like the valves of a petri dish; freshwater, marine and soil habitats.

DO Dissolved oxygen (O_2), measured in ppm (mg/l) or in percent of saturation. The amount of oxygen that can be dissolved in water increases with decreasing temperature.

Ecological engineering (1) The management of nature (H.T. Odum). (2) The design of human society with its natural environment for the benefit of both (Mitsch and Jørgensen, in Mitsch, this volume). (3) A specially designed and operated system for production processes and the management of the natural environment for the benefit of both (See also Yan and Ma, this volume).

Ecosystem A biological community with its associated abiotic environment. See also *biogeocoenosis.*

Ecotechnology Applied ecological engineering or science.

Enterobacteria Bacteria capable of living anaerobically, with small, rod-shaped cells. The classically representative species is *Escherichia coli* (*E. coli*), found in very high numbers in human intestines and, therefore, feces, so it is used as an indicator of sewage pollution.

Environmental engineering The application of traditional environmental techniques to mitigate environmental problems.

FWS-systems See *constructed wetlands with free water surface.*

Halophyte A plant that grows naturally in soils containing large amounts of salt or alkali, e.g., *Salicornia* sp. and *Spartina* sp.

Helophyte Marsh plant with its roots in the soil or mud below the water level and its aerial shoots emerging above the water surface, e.g., the common reed (*Phragmites australis*).

Heterotrophic Unable to manufacture food from carbon dioxide (CO_2), and so obtaining carbon from organic compounds.

Hydraulic loading The amount of water or wastewater applied to a given treatment process, usually expressed as volume per unit time or volume per unit time per unit surface area.

Hydrophytes Plants that grow naturally in water or very water- saturated soils. Adaptations include finely divided submerged leaves, large floating leaves, aerenchyma (large, intercellular air spaces), and reduced root systems, e.g., water hyacinth (*Eichhornia crassipes*), *Elodea canadensis,* yellow water lily (*Nuphar lutea*).

Hydroponics Plant cultivation in a liquid culture solution rather than soil.

Imhoff tank A deep, two storied wastewater treatment tank, originally patented by Carl Imhoff. It consists of an upper continuous-flow sedimentation chamber and a lower sludge-digestion chamber. The floor of the upper chamber slopes steeply to trapped slots through which solids may slide into the lower chamber. The lower chamber receives no fresh wastewater directly, but is provided with gas vents and with means for drawing digested sludge from near the bottom.

Intensive indoor system Space-efficient enclosed system.

Lysimeter A structure containing a mass of soil and designed to permit measurements of water draining through the soil.

Mineralization The process by which elements combined in organic form in living or dead organisms are eventually reconverted into inorganic forms to be made available for a fresh cycle of plant growth. The mineralization of organic compounds occurs through metabolism in living animals and microorganisms.

Natural treatment systems Treatment systems in which specific ecological environments found in nature are utilized for the treatment of wastewater (Tchobanoglous, this volume).

Nitrification The biological or chemical oxidation of ammonia nitrogen (NH_3, NH_4^+) to nitrate nitrogen (NO_3^-).

Nutrients In the context of this book, essential mineral elements taken up and used by plants to produce organic molecules (e.g., by photosynthesis).

Organic matter Chemical substances containing carbon and hydrogen, often of animal or plant origin.

PAH See *polyaromatic hydrocarbons.*

PAR Photosynthetic Active Radiation, 400 to 700 nm, about 45% of the solar radiation. The wavelength spectrum at which chloroplasts absorb light.

Pathogens Disease-producing organisms.

p.e. See *person equivalent.*

Person equivalent A quantity of wastewater equivalent to that produced by one person of a household. Also called population equivalent.

Photosynthesis The conversion of carbon dioxide to organic compounds in the presence of light.

Plankton Small, usually microscopic, plants (phytoplankton) and animals (zooplankton) in aquatic systems. In contrast to nekton (e.g., fish), plankton drift with water movements.

Plug Flow (1) Flow in which fluid particles are discharged from a tank or pipe in the same order in which they entered it. The particles retain their discrete identities and remain in the tank for a time equal to the theoretical detention time. (2) One-way flow without any mixing.

Polishing A general term for those tertiary wastewater treatment or AWT processes that are applied as a last treatment step, e.g., lagoons, aquaculture ponds, or fields for cleaning, reuse of nutrients, and/or the water.

Polyaromatic hydrocarbons Oil- and water-soluble chemicals, commonly found in wastewater, which are by their nature volatile, e.g., benzo(a)pyrene, phenanthrene. Usually toxic to organisms and capable of affecting the entire food chain.

Polyculture An aquaculture system using a diverse composition of plant and animal species.

Primary production The production of biomass by green plants or autotrophs generally.

Primary treatment Wastewater treatment involving the settling of solids by passing it through a tank slowly enough that they settle out of suspension by gravity.

Q_{10} The ratio of the velocity of a reaction at a given temperature to that of the same reaction at a temperature 10°C lower. This ratio is approximately 2.0 for most biological reactions.

Redox potential Substances vary in their tendencies to give up electrons and become oxidized. This tendency is measured electrically in reference to a standard substance, hydrogen. Redox potentials are useful in monitoring the extension of the oxidized parts of sediments or soils.

Rhizosphere Narrow zone of soil subject to the influence of living plant roots, as manifested by leakage or exudation of substances that effect microbial activity.

RZM Root zone method; see *constructed wetlands*.

Salt marsh A marsh periodically flooded by salt water. Common plants are the marsh grasses of the genus *Spartina*.

Secondary treatment A biological process in an aerobic reactor where organic matter in solution is oxidized or incorporated into living cells, usually bacteria.

Septage The sludge produced in individual on-site wastewater treatment systems such as septic tanks and cesspools.

Sewage Household and industrial wastewater that contains human waste.

Sludge (1) The accumulated solids separated from liquids, such as water or wastewater, during processing. (2) Organic deposits on bottoms of streams or other bodies of water. (3) The removed material resulting from chemical treatment, coagulation, flocculation, sedimentation flotation, and/or biological treatment of water or wastewater.

Submersed plant A plant whose roots are in the bottom of water bodies and with most of its vegetative growth under water, e.g., species of the genuses *Elodea, Isoëtes, Myriophyllum, Potamogeton.* The genus *Ceratophyllum* is free-living in the water without roots.

Sulfide A compound of sulfur which the reduced ion S^{2-} forms with other elements or radicals, e.g., hydrogen sulfide — H_2S, and metal sulfides.

Suspended solids (SS) (1) Insoluble solids that either float on the surface of, or are suspended in water, wastewater, or other liquids. (2) Solid organic or inorganic particles (colloidal, dispersed, coagulated, flocculated) physically held in suspension by agitation.

Tertiary treatment The treatment of wastewater beyond the secondary biological stage. The term normally implies the removal of nutrients, such as phosphor and nitrogen, and of a high percentage of suspended solids. Advanced wastewater treatment (AWT) is replacing this term.

Testate amoeba Protozoan belonging to the taxonomic class Rhizopodea, with shells, called tests. In order to construct the shells the amoebas glue together sand grains and inorganic detritus (e.g., diatom shells), depending on what material is available. Distribution in fresh and marine waters, and especially common in soil.

Thermodynamics Physical laws describing the behavior of energy.

Thomson weir Instrument for measuring water flow rates. The flow rate can be determined from the height of the water as it flows through a V-shaped notch before a cascade.

Trickling filter Very coarse filter used to provide secondary treatment of wastewater. A film of aerobic microorganisms on the filter media metabolizes the organic material in the wastewater trickling downwards to underdrains; biofilm that sloughs off is subsequently removed by sedimentation.

Virus The smallest (10 to 300 nm in diameter) life form, capable of producing infections and diseases.

VOCs Volatile organic compounds, e.g., xylene, toluene.

Wetland Land in which the water table is at or above the ground level long enough each year to maintain saturated conditions and the growth of corresponding vegetation.

INDEX

D

E

F

G

H

R

S

T

U

V

W

Z